BEHAVIOUR AND PHYSIOLOGY OF FISH

This is Volume 24 in the

FISH PHYSIOLOGY series
Edited by Anthony P. Farrell and Colin J. Brauner
Honorary Editor: William S. Hoar and David J. Randall

A complete list of books in this series appears at the end of the volume

BEHAVIOUR AND PHYSIOLOGY OF FISH

Edited by

KATHERINE A. SLOMAN
School of Biological Sciences
University of Plymouth
Plymouth, Devon, United Kingdom

ROD W. WILSON
School of Biosciences
Hatherly Laboratories
University of Exeter
Exeter, United Kingdom

SIGAL BALSHINE
Department of Psychology, Neuroscience, and Behaviour
McMaster University
Hamilton, Ontario, Canada

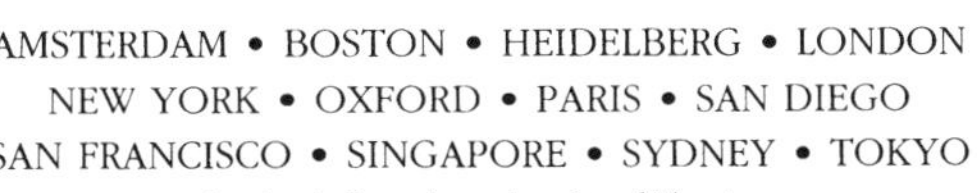

Front Cover Photograph: The photo shows the mating position of the hamlet fish (Hypoplectrus unicolor, Serranidae) as they release gametes. The fish makes a unique mating sound while dispersing gametes (Lobel, 2002). These fish are simultaneous hermaphrodites, maintain long-term pair bonds, mate at sunset at least several times, and switch sex roles in between each mating event. Photo by P. S. Lobel, of fish located on a reef offshore of the Discovery Bay Marine Laboratory, Jamaica, West Indies.

Elsevier Academic Press
525 B Street, Suite 1900, San Diego, California 92101-4495, USA
84 Theobald's Road, London WC1X 8RR, UK

This book is printed on acid-free paper. 

For all information on all Elsevier Academic Press publications visit our Web site at www.books.elsevier.com

ISBN-13: 978-0-12-350448-7
ISBN-10: 0-12-350448-1

Printed and bound in the United Kingdom

Transferred to Digital Printing, 2010

CONTENTS

3. The Physiology of Antipredator Behaviour: What You do with What You've Got

Mark Abrahams

4. Effects of Parasites on Fish Behaviour: Interactions with Host Physiology

Iain Barber and Hazel A. Wright

5. Social Interactions

Jörgen I. Johnsson, Svante Winberg, and Katherine A. Sloman

6. Circadian Rhythms in Fish

Irina V. Zhdanova and Stéphan G. Reebs

9. Reproductive Pheromones
Norm Stacey and Peter Sorensen

10. Anthropogenic Impacts Upon Behaviour and Physiology
Katherine A. Sloman and Rod W. Wilson

CONTRIBUTORS

The numbers in parentheses indicate the pages on which the authors' contributions begin.

MARK ABRAHAMS *(79), Department of Zoology, University of Manitoba, Winnipeg, Manitoba, Canada*

SIGAL BALSHINE *(xiii) Department of Psychology, Neuroscience, and Behaviour, McMaster University, Hamilton, Ontario, Canada*

IAIN BARBER *(109), Institute of Biological Sciences, University of Wales Aberystwyth, Aberystwyth, Wales, United Kingdom*

VICTORIA A. BRAITHWAITE *(1), Institute of Evolutionary Biology, School of Biological Sciences, Ashworth Laboratories, University of Edinburgh, Edinburgh, Scotland, United Kingdom*

STEVEN J. COOKE *(239), Centre for Applied Conservation Research, Department of Forest Sciences, University of British Columbia, Vancouver, British Columbia, Canada*

A. P. (TONY) FARRELL *(239), Faculty of Agricultural Sciences and Department of Zoology, University of British Columbia, Vancouver, British Columbia, Canada*

MICHAEL C. HEALEY *(239), Institute for Resources Environment and Sustainability, University of British Columbia, Vancouver, British Columbia, Canada*

SCOTT G. HINCH *(239), Department of Forest Sciences, Faculty of Forestry, University of British Columbia, 3022 Forest Sciences Centre, Vancouver, British Columbia, Canada*

JÖRGEN I. JOHNSSON *(151), Department of Zoology, Göteborg University, Göteborg, Sweden*

PHILLIP S. LOBEL *(39), Department of Biology, Boston University and Marine Biological Laboratory, Woods Hole, Massachusetts*

RUI F. OLIVEIRA *(297), Unidade de Investigação em Eco-Etologia, Instituto Superior de Psicologia Aplicada, Lisboa, Portugal*

STÉPHAN G. REEBS *(197), Département de Biologie, Université de Moncton, Moncton, Canada*

GIL G. ROSENTHAL *(39), Department of Biology, Texas A & M University, College Station, Texas*

KATHERINE A. SLOMAN (151, 413), *School of Biological Sciences, University of Plymouth, Plymouth, Devon, United Kingdom*

PETER SORENSEN *(359), Department of Fisheries, Wildlife and Conservation Biology, University of Minnesota, St. Paul, Minnesota*

NORM STACEY *(359), Department of Biological Sciences, University of Alberta, Edmonton, Alberta, Canada*

ROD W. WILSON *(413), School of Biosciences, Hatherly Laboratories, University of Exeter, Exeter, United Kingdom*

SVANTE WINBERG *(151), Evolutionary Biology Centre, Department of Comparative Physiology, Uppsala University, Uppsala, Sweden*

HAZEL A. WRIGHT *(109), Institute of Biological Sciences, University of Wales, Aberystwyth, Aberystwyth, Wales, United Kingdom*

IRINA V. ZHDANOVA *(197), Department of Anatomy and Neurobiology, Boston University School of Medicine, Boston, Massachusetts*

PREFACE

Volume 6 of the Fish Physiology series was the first to focus specifically on fish behaviour in relation to physiology. Almost thirty five years later, we are dedicating another volume of this internationally recognised series to the interrelations between behaviour and physiology in fish. Within the intervening period, several volumes had Chapters that considered fish behaviour; however, it is only in recent years that the integrative approach to fish behaviour and physiology has dramatically increased. The present volume (24) brings together these disciplines in a comprehensive review of the available literature with an additional introductory overview. The progression of Chapters focuses on different aspects in the life history of a fish, each written by scientists who are bridging the gap between behaviour and physiology in their own specialised discipline. In addition to contributing to our current knowledge on both fish behaviour and physiology, we hope that this volume will excite the future use of multidisciplinary approaches to understand the interplay between behaviour and physiology in fish.

The present Fish Physiology volume is a result of considerable effort and enthusiasm by the Chapter authors for which we are truly grateful. Without their dedication this book would not have materialised. Twenty referees, including Grant Brown, Jonathan Evans, Michael Fine, Richard Handy, Andrew Hendry, Felicity Huntingford, Masayuki Iigo, Rosemary Knapp, Jens Krause, Robin Liley, Edward Little, Carin Magnhagen, Anne Magurran, Justin Marshall, Steve McCormick, Tom Pottinger, Robert Poulin, Javier Sánchez-Vázquez, Alex Scott, and one referee who would like to remain anonymous, provided constructive guidance and have played an integral part in this project. We would also like to thank the series editors, Tony Farrell and Colin Brauner, for their never-ending support and Andrew Richford at Elsevier for publishing advice and encouragement. Our editorial association was initiated through McMaster University, Ontario where the idea for this volume was born. Our final thanks go to Chris Wood for recognising the potential for this volume.

Katherine A. Sloman
Rod W. Wilson
Sigal Balshine

BEHAVIOUR AND PHYSIOLOGY OF FISH: AN INTRODUCTION

Niko Tinbergen, a pioneer of the study of animal behaviour, argued that to fully understand behaviour, research on the function of behaviour must be married with understanding the development, causation, and evolution of behaviour. Although these sentiments were presented several decades ago (see Tinbergen, 1963), interest in causation of behaviour had largely waned. Instead, most behavioural scientists have targeted questions about how behavioural mechanisms underlie an animal's relationship with its environment. Physiologists too have generally turned their back on studies examining the "regulation of behaviour" or the "mechanisms of behaviour," possibly frustrated with an apparent lack of ubiquitous principals. Today that gap between the two disciplines of behaviour and physiology is finally beginning to be filled. A renaissance of research has recently emerged, providing a much needed link between behavioural studies and physiology.

The integration of physiological laboratory studies and empirical behavioural field observations has been partly led by technical advances related to the neuroscience and molecular biology revolutions. Molecular genetic techniques such as real time PCR, gene sequencing, cellular physiology, proteomics, paternity analyses, and various new biochemical techniques to study hormone action, have literally transformed the fields of behavioural ecology and physiology. Only recently has attention been given to sequencing the whole genomes of fish species (zebrafish and two pufferfish species, medaka, salmon, and tilapia); these projects will undoubtedly shed light on the molecular mechanisms underlying many behavioural and physiological functions. The growth in neuroscience has also generated a watershed of techniques including EEG, EOG, ECG, intra and extracellular electrophysiological recordings, electromyogram biotelemetry, neuroimaging of neurochemical circuits, and gene expression levels in the brain. These techniques and tools have profoundly enabled the interplay between pure physiology and pure behavioural studies.

Fishes are particularly useful organisms to utilise in bridging the gap between behaviour and physiology. Fish comprise the most speciose vertebrate order with over 25,000 species and an unrivaled diversity in life history patterns, breeding systems, sensory systems, as well as environmental requirements. Hence, fish provide an almost endless test-bed for either single species studies or comparative analyses of links between behaviour and physiology. Four fish groups may be particularly useful: zebrafish, sticklebacks, cichlids, and salmonids. As a consequence of their small body size and relatively large clutches, zebrafish have emerged as the major model system to study vertebrate development and gene function. Sticklebacks and cichlids have long been the major model organisms used by evolutionary and behavioural biologists. These two fish groups have undergone recent and extreme evolutionary radiations generating a large number of distinct populations (sticklebacks) and species (cichlids) accompanied by fascinating morphological, physiological, and behavioural changes. Salmonids are another model group because of their interesting life history, ecology, and distribution, as well as the economic importance of both wild and aquaculture salmonids.

The new era of interdisciplinary research and the recent proliferation of studies linking behaviour and physiology spawned the idea for this volume on Behaviour and Physiology for the Fish Physiology series. We wanted to document the active bridging of these two often separated areas of fish biology. Our aims were to provide a comprehensive review of the available literature, highlight the need for further links between behavioural and physiological studies, and stimulate new ideas by suggesting many possible avenues for future research.

We selected 10 topics mainly organised along a lifespan/life history theme, and attempted to highlight the most active areas where fish behaviour and physiology are closely related. Each Chapter was commissioned by research experts in that specialised subdiscipline and comprises a significant contribution to our scientific knowledge on both behaviour and physiology. The first two Chapters as well as the last Chapter of this book do not fit neatly into a life-span framework. However, the effects of cognition, communication, and pollution on physiological and behavioural processes play integral roles throughout the life-span of fish, and we felt that including such chapters was imperative.

In Chapter 1, Victoria Braithwaite outlines current knowledge on cognitive processes in fishes with particular emphasis on learning and memory capacities. Although only a few studies have been published on the neural basis of cognition, these imply a strong homology between fish brains and the brains of other vertebrates, suggesting that fishes will be useful for more general studies of the evolution of cognitive processes such as memory and learning. In Chapter 2, Gil Rosenthal and Phillip Lobel review the rapidly growing field of fish communication, including visual, chemical, acoustic, and mechanosensory signaling. They show how our understanding of communication in all modalities has been bolstered by the use of playback techniques. They also stress that fish neurobiology is in its infancy and it would be wise to pay special attention to the role of early experience and ontogeny in general when considering communication abilities in fishes.

Spending time and energy avoiding predators and parasites is a necessity in fishes and other organisms. In Chapter 3, Mark Abrahams eloquently reviews the various options available to fish when avoiding predators and describes why the "grow big fast" rule is not the only one that makes sense. Using hormonal and genetic manipulation of fish that elevate and sustain growth hormone levels, salmon have been shown to be much more willing to risk exposure to a predator in order to gain access to additional food (Du *et al.*, 1992; Devlin *et al.*, 1994). Unlike predators, parasites more often than not have only sub-lethal physiological effects on hosts but these can still impact behaviour and population regulation. In Chapter 4, Iain Barber and Hazel Wright provide in-depth coverage of how fish parasites influence the behaviour and physiology of their hosts. For example, sticklebacks infected with a particular parasite (*Schistocephalus solidus*) increase food intake rates, apparently to compensate for the nutritional demands of the parasites (Ranta, 1995).

In Chapter 5, Jörgen Johnsson, Svante Winberg, and Katherine Sloman summarise recent research on the complex interrelationships between social behaviour and physiology in fish. They discuss the many interacting physiological factors that determine social status such as energetic status, metabolic rate, growth hormone and steroid levels. They show the short and long term physiological consequences of social status (neuroendocrinological stress response) and how these might be related to fitness (via growth rates and reproduction). They also explore how both biotic and abiotic environmental variation influence social relations, and how genetics may influence these interactions and generate stress-coping phenotypes with distinct physiological and behavioural characteristics.

Virtually all organisms show daily (circadian) or annual rhythms in their behaviour and metabolism. In Chapter 6, Irina Zhdanova and Stéphan Reebs describe our knowledge of fish circadian rhythms. An elaborate network of photosensitive, coexisting central and peripheral circadian oscillators has been identified in fishes, and a large number of circadian clock genes have recently been discovered in zebrafish. In fish, melatonin (secreted by the pineal gland) has long been considered the principal circadian hormone, and in most fish it is secreted only at night, as secretion is suppressed by bright light. They also discuss how other possible environmental factors (temperature, water chemistry, food availability, social interactions, and even predation risk) may act as synchronisers of the daily cycles. Migration for many fish species occurs on an annual rather than a daily cycle. In Chapter 7, a diverse team of scientists

(Scott Hinch, Steven Cooke, Michael Healey, and Tony Farrell) review current knowledge on the mechanisms underlying migration in fishes and use sockeye salmon as a model system. For example, in salmon the decision to leave the ocean seems to be closely related with the initiation of gonadal maturation (Ueda *et al.*, 2000); gonadotropin-releasing hormone (GnRH) triggering gonad growth and the onset of gonadal growth is associated with the shift from foraging to homing migration (Ueda and Yamauchi, 1995). Although a small number of fish species migrate (2.5%), the number of individuals that migrate is vast. Many of these species are important in fisheries, and migratory species seem to be at twice the risk of extinction as non-migratory species (Riede, 2004). Hence, understanding migration is crucial to successful management and conservation of many fish species.

Reproduction is the topic of the next two Chapters. In Chapter 8, Rui Oliveira reviews neuroendrocrine mechanisms of discrete within-sex reproductive variance, often called alternative reproductive tactics (ARTs). He finds broad support for the notion that like sex differentiation (male vs female), sex steroids provide the proximate control of ARTs. Brantley *et al.* (1993) found that conventional males had significantly higher levels of circulating 11-ketotestosterone (or KT, the most potent androgen in fish) compared to parasitic males. Oliveira updates the analysis, finding an even stronger association between ARTs and KT. However, differences in KT levels may in fact be a consequence, not a cause, of ARTs, as the various male morphs will have vastly different social experiences. Oliveira reviews how conventional and parasitic males differ in terms of steroid binding globulins, steroidogenic enzymes, steroid receptors, and in terms of various other neurochemical systems. In Chapter 9, Norm Stacey and Peter Sorensen review the evidence for reproductive pheromones in fishes. Pheromones are basically chemical odours or signals derived from hormones and other essential metabolites that coordinate or influence conspecific behaviour and/or physiology. Three chemical classes of reproductive pheromones have been identified in fishes; bile acids, sex steroids, and F prostaglandins. Using a number of model fish groups such as salmonids and carp, the authors show how sex steroid and prostaglandin pheromones, fairly ubiquitous among fishes, are both sensitive and specific in their behavioural and physiological effects on conspecifics.

The impact of human activities on aquatic ecosystems has unfortunately increased and intensified in recent years. In Chapter 10, Katherine Sloman and Rod Wilson explore the anthropogenic impacts upon fish behaviour and physiology; in particular they cover the effects of pollution (chemical contaminants) on fish cognition, sensory processes, predator-prey interactions, social behaviour, and reproduction of fishes. They examine the influence of inorganic chemicals (including metals) and organic contaminants on behaviour *via* physiological pathways. They show that there is likely to be differential sensitivity to toxicants depending on the route of exposure and social status. For example, subordinate trout have higher uptake rates of waterborne copper and silver than dominant trout. However, if a toxicant is present in the diet, then a dominant fish is likely to accumulate toxicants faster as dominants may eat larger quantities of the available contaminated food. They end their Chapter with a plea for further interlinking of laboratory and field-based approaches in ecotoxicological studies that address not simply the effects of a single chemical but the cocktail of chemicals present in the aquatic environment.

Traditionally, behaviour and physiology have been considered two separate fields of enquiry with the majority of available literature focusing primarily on one or the other. The Chapters of this book underscore the fact that behaviour and physiology are inextricably linked. They advocate the need for interdisciplinary approaches to fully understand behaviour and physiology and to unravel many of the unanswered questions. New technologies and a growing recognition of the impact of physiology on life history has led behavioural ecologists to re-embrace Tinbergen's message, and the embryonic development of a more mechanistic study of behaviour has certainly begun in earnest (Ricklefs and Wikelski, 2002; Costa and Sinervo, 2004). Understanding the pattern of behaviour is no longer sufficient; there is an increased and growing interest in understanding the physiological mechanisms underlying particular behaviours. Physiologists too have increased their emphasis on field studies and recognise the need to assess the fitness implications of different physiological functions linked to behaviour. In conclusion, we hope that this book will contribute to our knowledge on both behaviour and physiology, but will more importantly highlight the increasing need for additional multidisciplinary research in this area.

Sigal Balshine
Katherine A. Sloman
Rod W. Wilson

REFERENCES

Brantley, R. K., Wingfield, J. C., and Bass, A. H. (1993). Sex steroid levels in *Porichthys notatus*, a fish with alternative reproductive tactics, and a review of the hormonal bases for male dimorphism among teleost fishes. *Horm. Behav.* **27,** 332–347.

Costa, D. P., and Sinervo, B. (2004). Field physiology: Physiological insights from animals in nature. *Ann. Rev. Physiol.* **66,** 209–238.

Devlin, R. H., Yesaki, T. Y., Biagi, C. A., Donaldson, E. M., Swanson, P., and Chan, W-K. (1994). Extraordinary salmon growth. *Nature* **371,** 209–210.

Du, S. J., Gong, Z., Fletcher, G. L., Shears, M. A., King, M. J., Idler, D. R., and Hew, C-L. (1992). Growth enhancement in transgenic Atlantic salmon by the use of an "all fish" chimeric growth hormone gene construct. *Biotechnology* **10,** 176–181.

Ranta, E. (1995). *Schistocephalus* infestation improves prey-size selection by three-spined sticklebacks, *Gasterosteus aculeatus*. *J. Fish Biol.* **46,** 156–158.

Ricklefs, R. E., and Wikelski, M. (2002). The physiology-life history nexus. *Trends Ecol. Evol.* **17,** 462–468.

Riede, K. (2004). Global register of migratory species—From global to regional scales. German Federal Agency for Nature Conservation, Project 808 05 081. Final Report.

Tinbergen, N. (1963). On aims and methods of Ethology. *Zeitschrift fur Tierpsychologie* **20,** 410–433.

Ueda, H., and Yamauchi, K. (1995). Biochemistry of fish migration. *In* "Environmental and Ecological Biochemistry" (P. W. Hochachka and T. P. Mommsen, Eds.), pp. 265–279. Elsevier, Amsterdam.

Ueda, H., Urano, A., Zohar, Y., and Yamauchi, K. (2000). Hormonal control of homing migration in salmonid fishes. "Proceedings of the 6th International Symposium on the Reproductive Physiology of Fish," pp. 95–98. Bergen.

1

COGNITIVE ABILITY IN FISH

VICTORIA A. BRAITHWAITE

1. BACKGROUND

Animal cognition has seen a tremendous number of advances in the last two decades. New techniques and technologies have revolutionised the way in which data is gathered; for example, it is now possible to impair specific regions of the brain to determine the mechanisms that underlie certain behaviours (Broglio *et al.*, 2003; Shiflett *et al.*, 2003) or to track animals very precisely as they navigate around their natural environment (Armstrong *et al.*, 1996; Biro *et al.*, 2002). As a consequence, we now have a much better understanding of cognitive processes and factors that affect cognition. Despite the many studies of cognitive ability in birds and mammals, until recently there were relatively few studies investigating cognition in fish (reviewed in Odling-Smee and Braithwaite, 2003a). This imbalance, however, is beginning to be redressed and a new interest in this field appears to be fuelled by the growing evidence that fish are capable of quite complex cognitive capacities. For example, fish are able to combine detailed spatial

Behaviour and Physiology of Fish: Volume 24
FISH PHYSIOLOGY

DOI: 10.1016/S1546-5098(05)24001-3

relationships to form a mental map (e.g., Rodríguez *et al.*, 1994; Vargas *et al.*, 2004) and several species have the capacity for complex, flexible learning and memory (reviewed in Braithwaite, 1998; Odling-Smee and Braithwaite, 2003a). Those interested in the evolution of cognitive ability are now paying more attention to fish and this interest has begun to determine a number of similarities between fish and other vertebrate cognitive processes (Shettleworth, 1998; Laland *et al.*, 2003). Concerns for fish welfare are also generating an interest in their cognitive capacities. As we search for answers to questions such as whether fish have the mental capacity for suffering, we turn to studies of fish cognition for answers (Braithwaite and Huntingford, 2004; Chandroo *et al.*, 2004; Huntingford *et al.*, in press).

Far from being stereotyped and invariant, the behavioural repertoires and learning and memory abilities observed in fish suggest that this taxonomic group is remarkably flexible (Laland *et al.*, 2003). Also, similarly to other vertebrate groups, variation in learning and memory ability arises between species and, in some cases, between different populations within a species (Odling-Smee and Braithwaite, 2003b). For example, shoaling species have been shown to be capable of recognising and remembering individual conspecifics, and they can ascribe competitive abilities to these individuals (Metcalfe and Thomson, 1995; Utne-Palme and Hart, 2000; Griffiths, 2003; Chapter 5, Section 2 of this volume). Furthermore, learning plays a key role in antipredator behaviours because it allows fish to adjust their responses to potential predators (Kelley and Magurran, 2003). Fish also show variation in the types of orientation information that they learn and remember (Girvan and Braithwaite, 1998; Odling-Smee and Braithwaite, 2003b).

Given the diversity of fish species, it seems surprising that it has taken until now for us to appreciate just how cognitively competent many fish are. Part of the problem is that it has taken many years to shrug-off the image that fish are slow-moving with a short-term memory and only capable of basic behaviours. Previously, it was recognised that a number of fish species could habituate and learn associations between certain stimuli and a response (MacPhail, 1982; Laming and Ebbeson, 1984; Rooney and Laming, 1987). However, these forms of learning are generally considered to be basic and earlier research found that fish could continue to perform these stimulus-response learning tasks even in the absence of their forebrain (Figure 1.1). This led to the conclusion that learning processes in fish were simple and did not involve complex neuronal processing (Overmeir and Hollis, 1983; Overmeir and Papini, 1986). Our understanding of how fish control their behaviour and make decisions began to change in the late 1980s. Lesion experiments, in which the forebrain was destroyed, revealed that fish can still feed and learn simple associations in the absence of this structure (Rooney and Laming, 1988; Laming and McKinney, 1990; Salas *et al.*,

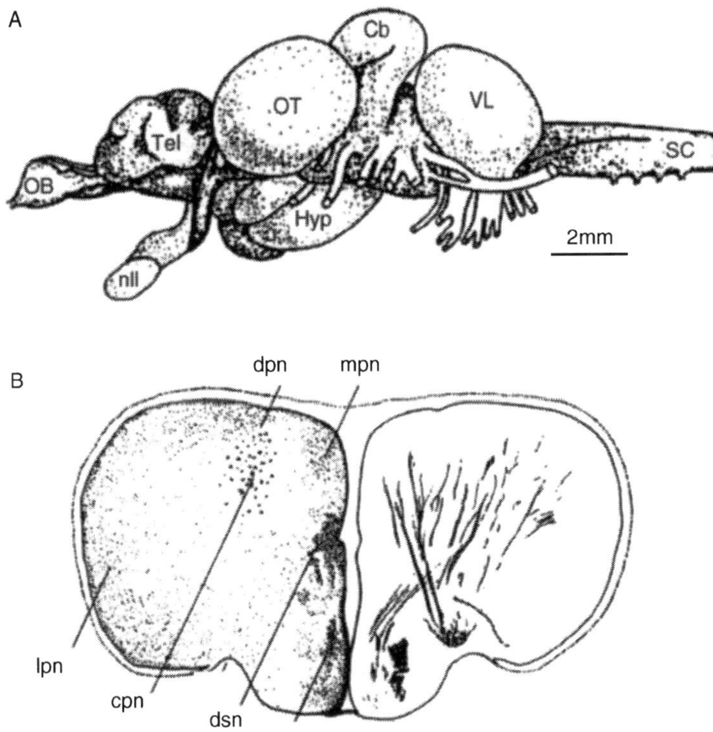

Fig. 1.1. Schematic diagrams of the brain of a general teleost fish. (A) Lateral view of the goldfish brain showing the key large-scale structures, Hindbrain (Cb, cerebellum; VL, vagal lobe; SC, spinal cord), midbrain (Hyp, hypothalamic lobe; OT, optic tectum), and forebrain (OB, olfactory bulb; nII, optic nerve; Tel, telencephalon). (Adapted from Broglio *et al.*, 2003.) (B) Transverse section through the telencephalon of a typical teleost (cpn, central pallial nucleus; dpn, dorsal pallial nucleus; lpn, lateral pallial nucleus; mpn, medial pallial nucleus; dsn, dorsal subpallial nucleus; vsn, ventral subpallial nucleus). (Adapted from Nieuwenhuys, 1967.)

1996a,b). However, some researchers began to realise that certain forms of learning were affected in the absence of the forebrain; for example, a fish with lesions to this region shows impaired aversion learning (Overmeir and Papini, 1986).

The reductionist approach of ablating the whole forebrain provided a useful route with which to begin exploring the neural bases of various learning and memory capacities. However, as techniques and methods within experimental neuroscience have become more refined, large-scale lesion studies have become somewhat redundant. The need for fine-scale lesioning to refine our understanding of how the teleost brain functions became clear when a number of studies began to demonstrate that fish were not reliant on stimulus–response learning alone. Over the last two decades, there has been increasing evidence of a surprising capacity for complex learning and behaviour in intact, nonlesioned teleosts. For instance, teleosts are now known to associate events with times and places (Reebs, 1996, 1999; Chapter 6, Section 4 of this volume) and they can learn to avoid sites that are perceived as dangerous (Huntingford and Wright, 1989; Kelley and Magurran, 2003). With these cognitive capacities identified, a renewed interest in fish neurobiology has led some researchers to use fine-scale lesioning to determine which regions of the forebrain are necessary for complex behaviour (Broglio *et al.*, 2003).

Although there are studies of learning and memory in both elasmobranch and teleost fish, this Chapter will primarily concentrate on the latter group. MacPhail (1982) and Kotrschal and associates (1998) reviewed a number of studies on elasmobranch fishes and there have been some recent studies in this field (New, 2001). However, the majority of advances in the last two decades have focused on our understanding of teleost cognitive processes and, owing to the breadth of this work, this Chapter will concentrate on cognition within the teleost fishes.

This review begins with an introduction of the concept of cognition, and then discusses the techniques used to study cognition in other vertebrates to illustrate how one can use similar approaches to investigate fish cognition. This is followed by a description of the general anatomy of the teleost brain indicating where certain types of information are processed. Recent observations indicate that like other vertebrates, several species of fish show lateralisation in the brain; that is, certain types of stimuli and information are preferentially processed on one side of the brain or the other. A number of examples of this recent lateralisation work are reviewed. The following sections then give a range of examples that illustrate the learning capacities shown by different teleosts. As there is considerable breadth to this work, the examples are divided into four sections: (a) learning in relation to foraging, (b) recognition of conspecifics, (c) social learning, and (d) spatial learning. This Chapter concludes with a review of recent work that demonstrates how small-scale lesion experiments have led to some impressive steps forward in our understanding of how the fish brain controls spatial cognition and behaviour.

2. COGNITION

Cognition is the process by which an animal internalises information about an experience through its sensory systems (perception), how this information is learned, and how it remembers the experience through recall (memory). Thus, it is not a single process but rather consists of three interacting aspects (perception, learning, and memory). There have been many ways in which cognition has been studied (Pearce, 1997; Shettleworth, 1998). Perception, learning, and memory can be studied as independent processes but it is also possible to work on their product: mental representations. Many regard mental representations of events and places to underpin animal cognition (Gallistel, 1990). For example, several researchers propose that many animals form mental maps, an internal representation of the topographical environment (Healy, 1998). The animal uses this mental map to make decisions about the route it should take to reach a particular destination. Representations, therefore, provide the animal with a general framework for making decisions, and they are flexible in that new information can be added or updated. As these representations coordinate remembered information from multiple events, they are generally considered to require complex neural processing and so typically have been linked to animals with well-developed cognitive abilities.

As the examples in this paper will highlight, fish can perform a wide range of cognitive tasks that enable them to make complex decisions. This capacity for decision-making generates flexible behaviour that, in many cases, enables the fish to fine-tune and adapt their behaviour to their local environment. In this way, fish can hone the efficiency with which they perform routine day-to-day tasks such as foraging, spatial behaviour, and antipredator responses.

There are a range of techniques that experimental psychologists have devised to quantify cognitive abilities. One favoured approach is to train an animal to perform some form of task. The animal can then be given a series of tests where parts of the task are manipulated to see whether this affects performance. In this way, learning and memory ability or the capacity to form a mental representation can be measured (Wasserman, 1984). Experimental psychologists typically use rats and pigeons as their model species. More recently, however, evolutionary biologists have recognised the value of techniques commonly used by experimental psychologists (e.g., classical or Pavlovian conditioning, and instrumental conditioning such as Skinner boxes) and have adapted these procedures to test a range of species both in the laboratory and in the field (Shettleworth, 1998). These methods can provide data on what an animal has learned and for how long it remembers it. Furthermore, such techniques are versatile; for example, even fish can be trained to associate a light with the delivery of a particular stimulus (Portavella *et al.*, 2004).

At many points, this Chapter will rely on data based on spatial learning and memory. *Spatial ability* refers to the way in which an animal learns and encodes information about its home range or local environment. It enables a territorial animal to define its boundaries or allows an animal to quickly locate shelter when threatened. Owing to the importance of spatial ability in the daily behavioural repertoire of many animals, spatial learning and memory provide an ideal method for assaying cognitive ability. This approach is also favoured because it is relatively straightforward to train an animal to move through a maze, and the maze can be varied in design from simple (e.g., a T maze) to more complex (e.g., a 16-arm radial maze) to allow a range of learning and memory abilities to be tested. Furthermore, it has been possible with birds and mammals to determine which regions of the brain are involved in spatial cognition; more recently, this has also been possible in fish (Healy, 1998; Broglio *et al.*, 2003). Thus, there are considerable benefits to studying spatial behaviour and this has led to an impressive body of literature (Healy, 1998).

In the last decade, a number of other specialised forms of fish learning have also been studied; one particularly fruitful area is that of social learning (Brown and Laland, 2003). Here work has addressed how fish learn from each other. Much of this work relies on fish being able to recognise individuals. Individual recognition has been shown in a number of fish species. For example, both sticklebacks and guppies join each other to form pairs or small groups to approach a predator and inspect it, probably to determine how much of a threat it represents (Milinski *et al.*, 1990a,b; Dugatkin and Aifieri, 1991a,b). The choice of partner for this dangerous task is critical, and there is good evidence that fish learn to recognise partners that cooperate and individuals that do not (Milinski *et al.*, 1990b). Therefore, this Chapter will also review recent observations of learned individual recognition and its consequences for our understanding of social and competitive interactions among fish. However, before considering some of the many examples of learning behaviour, this Chapter focuses on the teleost brain, paying attention to the various structures and emphasising how the general plan is similar to other vertebrates. In particular, it highlights the forebrain, the region that is proposed to confer complex neuronal processing.

3. THE TELEOST BRAIN

Fish brains follow the same basic plan as other vertebrate brains in that they have a hind brain that contains the cerebellum, a midbrain (or mesencephalon), and a forebrain (or telencephalon). General teleost brain anatomy can be seen in some more detail in the schematic diagrams shown in Figure 1.1. The diversity of brain specialisation within the teleosts is impressive, but much of this specialisation has been within regions of the brain that process sensory information. Many of the basic brain structures are present, and perhaps some of the most exciting recent observations have been those showing clear homologies of specific areas and their roles in fish and other vertebrates (see Broglio *et al.*, 2003 for a review). Given the large number of teleost species (it has been estimated that there are over 30,000 species; Kullander, 1999) and their radiation into almost every type of aquatic habitat, fish provide an excellent opportunity for investigating how different environments have shaped the evolution of the brain and, in particular, the evolution of diverse sensory systems (Kotrschal *et al.*, 1998).

3.1. General Introduction to Fish Brain Anatomy

Fish brains are housed within a braincase that, curiously, in many species is disproportional in size to the brain itself, and the space that is left is filled with a fatty tissue (Kotrschal *et al.*, 1998). Teleost brains share many of the same classical subdivisions seen in most vertebrates; a brainstem, the hindbrain, the mesencephalon (midbrain), a cerebellum, a pair of optic lobes, the paired cerebral hemispheres of the telencephalon (forebrain), and the associated olfactory bulbs (several of these structures can be identified in Figure 1.1A).

Somatosensory information does not ascend through the spinal cord, but instead it reaches the brain primarily through specialised cranial nerves, in particular the tirgeminus (V), facialis (VII), vagus (X) and three lateral line nerves. Within the brainstem, the main centres for the somatosensory systems (other than vision and olfaction) are organised into a series of horizontal columns. In one of these dorsally positioned columns, auditory, gustatory, and lateral line sensory information is processed, and an extra column is found in species of fish that process electrosensory information (Kotrschal *et al.*, 1998). Specialisations of particular sensory systems can sometimes be identified within this region because of the increased size of parts of the dorsal column.

The brainstem and the tegmentum of the mesencephalon and diencephalon (the region of the midbrain containing the thalamus and the hypothalamus) are continuous with each other. The integration of various sensory inputs occurs in the paired inferior lobes of the hypothalamus and these affect fish behaviour through the secretion of hormones. The cerebellum varies enormously in size across different species of teleost. It serves a range of functions (Finger, 1983), and is particularly prominent in electrosensitive fish such as the elephantnose fish, *Gnathonemus* (Maler *et al.*, 1991); the central acoustic area is found in the ventral surface of the cerebellum. The paired hemispheres of the optic tectum arise from the roof of the mesencephalon. The two telencephalon hemispheres of the forebrain arise from the rostral section of the neural tube and, similarly, the olfactory bulbs arise from the rostral tip. There are large projection neurons connecting the telencephalon and the diencephalon via the olfactory tracts.

3.2. Teleost Forebrain Structure

During development, the forebrains of teleosts arise through eversion rather than evagination as with all other nonactinopterygian vertebrates. The eversion occurs when, during development, the dorsal parts of the neural tube start to thicken while the ventral regions remain relatively unchanged. As the thickening progresses, the dorsal area begins to curve and envelop the ventral area. This sequence of development results in a forebrain that has a thin, membranous roof and no internal ventricles (MacPhail, 1982). Once developed, different regions of the forebrain can be described. For example, there are subdivisions between the pallial areas towards the roof of the forebrain, and subpallial areas underneath (Figure 1.1B). There are three clearly distinct subpallial regions: a ventral nucleus, a dorsal nucleus, and a lateral nucleus (Nieuwehuys, 1967). The pallial region can be further subdivided into central and peripheral regions; here the distinction between these areas is attributed to the central region having large scattered neurons (Figure 1.1B). The peripheral area is composed of three regions: the medial, dorsal, and lateral pallial nuclei (MacPhail, 1982). This latter structure, the lateral pallium, will be further discussed later in the Chapter because it has attracted the attention of researchers working on teleost spatial learning and memory.

The forebrain receives sensory information from a variety of sources, visual and olfactory cues, lateral line information, picked up by mechanoreceptors and electroreceptors (Finger, 1980). There is remarkable similarity between lateral line organisation in fish and auditory systems in terrestrial vertebrates (see also Chapter 4, Section 2.3). Both rely on mechanosensory receptors that utilise hair cells to pick up information. However, both types of reception are represented independently in many species of fish, for example, catfish (*Pylodictus olivaris*) (Knudsen, 1977).

The teleost telencephalon plays a key role, which is to process and coordinate sensory and motor information. However, as MacPhail (1982) discusses, there would appear to be sufficient space for a number of other higher-level processes to occur within this structure. Given the increasing numbers of observations that fish are capable of relatively complex cognitive processing, it is the telencephalon that has become a focal area for fish cognition studies.

Fish forebrains have also proved to be useful structures for comparative studies in which the relationship between fish brains and their environment are studied. Such comparisons have indicated that a number of forebrain adaptations are ecologically driven. For example, in cichlids, variation in telencephalon or forebrain size appears to relate closely to the challenges of spatial environmental complexity (Kotrschal *et al.*, 1998). Van Staaden *et al.* (1994) and Huber *et al.* (1997) examined the brains of 189 species of cichlids from the East African Lakes and Madagascar, and found that species living in complex habitats created by shallow rock and vegetation had comparatively larger telencephalons than those species living in midwater.

4. PERCEPTION AND BRAIN LATERALISATION

In addition to divisions within the brain that give rise to specialised areas, there are also subtle differences between the two hemispheres. Differential use of each hemisphere of the brain is known as *lateralisation*. There are several studies showing lateralisation in the visual sensory system; for example, some animals use their left eye, and hence the right hemisphere of their brain, when observing social stimuli but their right eye and the left hemisphere when encountering novel objects. Such preferential eye use is found in birds, reptiles, amphibians, and fishes (Rogers, 1989; Deckel, 1995; McKenzie *et al.*, 1998; Sovrano *et al.*, 1999; Dadda *et al.*, 2003; see Rogers and Andrew, 2002 for a review). Current theories suggest that lateralisation in the visual system initially arose in response to the development of laterally placed eyes with little binocular overlap in the visual field and complete decussation (crossing-over of nerve fibres) at the optic chiasma. It is also proposed that it provides an enhanced ability to perform two tasks at the same time, such as foraging while watching out for predators (Rogers *et al.*, 2004).

Visual lateralised behaviour in fish has been studied in a variety of ways but these different approaches have revealed consistent results (Facchin *et al.*, 1999). A common technique used to investigate lateralisation is the "detour test." Here, subjects swim down a channel to approach an open field area; the fish can then swim, or detour, to the left or the right in order to view a scene which is partially obscured by an obstacle. When the scene consists of a shoal of the opposite sex or a predator, 90% of the individuals tested show a preference to turn to the left so that they view the scene with their right eye. However, when fish are faced with a shoal of individuals of the same sex or if the detour barrier causes them to lose sight of the stimulus goal, they turn to the right (Bisazza *et al.*, 1997). Motivation and current state also affects the fish; for example, lateralisation is enhanced in females viewing a shoal of males if they have been deprived of male contact for 2 months (Bisazza *et al.*, 1998). Similarly, the motivation to perform predator inspection behaviour in mosquito fish (*Gambusia holbrooki*) is affected by the side on which the companion fish is seen. Mosquito fish will move closer to the predator if the companion fish is on their left side (Bisazza *et al.*, 1999).

An alternative technique to assay lateralisation is to monitor the angle at which fish view objects. Miklosi *et al.* (1997) report that zebrafish (*Brachydanio rerio*) use their right eye to inspect novel scenes but swap to the left eye on subsequent viewing. When mosquito fish are placed in a round arena with a predator in the middle they tend to swim in a clockwise direction, enabling them to fixate on the predator with the right eye (Bisazza *et al.*, 1997). Also, when a mirror is placed such that their own image is on their left side, mosquito fish are more likely to inspect predators than when the mirror is placed on their right (Bisazza *et al.*, 1999). Other work has also shown female poeciliids tend to fixate on conspecifics using their left eye (Sovrano *et al.*, 1999). Taken together, these results indicate that the preferential use of either eye is not necessarily fixed but stimuli resulting in an emotive response generally cause fish to view them with the right eye, whereas other scenes are generally observed using the left eye (Bisazza *et al.*, 1998; Facchin *et al.*, 1999).

Heuts (1999) suggested that interspecific variation in lateralised escape responses in fishes could be affected by the predation risk experienced by the fish. A recent investigation has confirmed this among populations of the Panamanian bishop (*Brachyrhaphis episcopi*). Fish collected from regions with high predation pressure used their right eye to view a live predator restrained behind a clear Perspex barrier. The predator is presumably perceived as a potential threat, which invokes an emotive response. Fish sampled from low predation sites showed no significant preference for either eye when exposed to the same predator (Brown *et al.*, 2004).

In shoaling species, individuals need to simultaneously monitor both predators and shoal mates. For populations that have had very little contact with predators, such responses are unlikely to arise. If lateralisation initially evolved to cope with different information coming from each eye, leading to left and right hemisphere specialisations and a reduction of the problems associated with divided attention (Griffiths *et al.*, 2004), then exposure to different levels of predation pressure may affect the degree of lateralised specialisation. This interaction between the environment and lateralised cognitive processing would appear to present a number of directions for future research.

Lateralised behaviour in fish is not solely restricted to the visual sense: blind Mexican cave fish (*Astyanax fasciatus*) have been shown to exhibit lateralisation in their mechanosensory sense. Eye formation in blind Mexican cave fish arrests during development; consequently, these fish have no access to visual cues. These fish are dependent on their lateral line to provide them with information on the topography of their perpetually dark environment. As they swim around their environment, the fish detect water displacement patterns with their lateral line organ (LLO). When a fish moves forwards it displaces water, and the lateral line detects small differences in water flow patterns as the displaced water is reflected off objects within the environment. The fish move slowly around their environment as they explore a new area, but then increase their swimming velocity once they have learned the position of objects (Teyke, 1985, 1989).

An experiment investigating lateralised behaviour in this species also revealed a right-side bias to novelty (Burt de Perera and Braithwaite, 2005). These fish show a preference for using the right side of the LLO when passing a novel landmark, whether they swim in a clockwise or anticlockwise direction. However, this right-side preference wanes once the fish become familiar with the landmark. Similarly to vision, the lateralised use of the right LLO can be related to use of the left hemisphere of the brain, because the neurones associated with the lateral line course bilaterally from the hindbrain nucleus to the midbrain with contralateral predominance (McCormick, 1989). Relatively little is known about laterality of nonvisual sensory modalities, but given the independence of visual and lateral line sensory systems, this suggests that lateralisation is a deep-rooted phenomenon.

5. LEARNING IN FISHES

Like other animals that live in the complex and variable natural world, fish rely on their ability to change and adapt their behaviour through learning and memory. It is now widely recognised that fish can learn and remember different types of information (Huntingford, 2003). For instance, they can reliably locate places to forage, react quickly and appropriately when threatened by a predator, and identify potential partners for reproduction. Similarly, they can also assess competitors, and can communicate and recognise dominant or subordinate status, which helps prevent conflicts escalating into potentially harmful fights.

Fish can learn about their environment through a series of trial and error processes; alternatively, they can learn to adapt their behaviour by observing the behavioural responses of others within that environment (*social learning*). To survive from one day to the next, fish rely on their ability to generate decisions that produce appropriate behavioural responses. How finely tuned these decisions are will, in effect, determine the cognitive capacity of that individual (Laland *et al.*, 2003).

Advances have been made in a number of areas relating to fish learning. Interestingly, relatively less attention has been given to memory processes, yet these two are inextricably linked; you cannot have memory without an initial learning phase, and there is little value to learning if the information will not be recalled through memory at some later event. This section will review what we now know about fish learning in a variety of contexts. For example, how do fish learn to forage or to recognise certain characteristics associated with specific individuals, such as attractiveness (Witte and Nottemeier, 2002) or competitive ability (Hoare and Krause, 2003). Although Chapter 3 reviews antipredator behaviour, this Chapter includes a few additional examples to emphasise the role that cognition plays in terms of learning how to recognise, react, and avoid the unwanted attention of predators.

5.1. The Role of Learning and Memory in Foraging

Fish have been shown to discover foraging patches by individual exploration or by observing the behaviour of other foragers (Pitcher and Magurran, 1983; Pitcher and House, 1987). A key concept associated with foraging theory hinges on the ability of the animal to assess food patch profitability and to adjust their behaviour accordingly. Foraging theory has used Charnov's (1976) Marginal Value Theorem to make predictions about how long an animal will stay at one food patch before moving to another. This predicts that food patches, regardless of their original profitability, should be exploited to a certain critical level, at which the rate of energy gain is equal to the average rate of gain across all feeding patches in the environment (Charnov, 1976). Thus, to make the correct decision of "Should I stay?" or "Should I leave for an alternative patch?" the animal needs to retain a memory of the profitability of previously encountered patches (Warburton, 2003). Such a behavioural strategy allows the animal to forage at an optimal rate within its current environment. There is some evidence that fish do learn and remember this type of information; for example, bluegill sunfish (*Lepomis macrochirus*) use prior experience of patch profitability to influence how long they spend in a particular food patch (Wildhaber *et al.*, 1994).

An ability to learn also plays a role in how quickly individuals hone their handling skills for different types of prey. When a fish's ability at attacking, manipulating, and ingesting prey is quantified over a series of repeated trials, the fish's performance generates a typical learning curve. The fish are initially slow to consume novel prey, but over a few successive trials their ability to locate and ingest the prey improves until it reaches a point where it cannot consume the prey any faster (Colgan *et al.*, 1986; Mills *et al.*, 1987; Croy and Hughes, 1991; Hughes and Croy, 1993). Croy and Hughes (1991) demonstrated that fifteen-spined sticklebacks (*Spinachia spinachia*) needed between five and eight trials to learn how to handle a new type of prey. Similar numbers of trials were also required by bluegill sunfish as they learned how to consume *Daphnia* (Werner *et al.*, 1981). Thus, this type of learning occurs over fairly short periods of time.

Fish can track changes in food availability in different ways and may rely on a range of learning and memory systems to help them achieve this (Warburton, 2003). For example, they can learn and remember cues or landmarks associated with particular food patches (Braithwaite *et al.*, 1996; Odling-Smee and Braithwaite, 2003b) or they can learn to use the foraging behaviour of other fish around them (Brown and Laland, 2003; Griffiths, 2003). The duration of the memories associated with these different variables should be tuned to the environment in which a fish finds itself. In a frequently changing environment, the value of a memory may be relatively low because, within a relatively short space of time, the memorised information may no longer be relevant. Thus, in a highly heterogeneous environment, memory durations would be expected to be short and a fish would be expected to continually update its memory through learning. In contrast, in a stable undisturbed environment, memories would be predicted to be longer-term. An example illustrating that these predictions are met can be seen in the way that sticklebacks forget the handling skills needed to capture and consume different types of prey (Mackney and Hughes, 1995). In a comparison of the memory characteristics of three different populations of sticklebacks, Mackney and Hughes (1995) showed that three-spined sticklebacks (*Gasterosteus aculeatus*) sampled from a pond remembered specific prey handling skills for 25 days; however, the same species, but an anadromous population from an estuary, was found to have a decreased memory duration of only 10 days. Similarly, a short 8-day memory was observed in a population of marine fifteen-spined sticklebacks (*Spinachia spinachia*). The differences in memory duration appear to reflect the changeable nature of the environment that these different populations experience, from the most stable (the pond population) to very variable (the anadromous and marine populations). Interestingly, all three groups learned how to handle the prey at similar rates; it was only the duration of memory for these skills that differed.

How fish learn to handle live prey and then switch from one prey type to another has become an area of interest for fisheries managers, who annually release many millions of fish in attempts to restock threatened or dwindling natural populations. Worryingly, the vast majority of these fish perish shortly after release (Olla *et al.*, 1998; Brown and Laland, 2001). A proportion of this mortality occurs because some fish are incapable of effectively switching from feeding on their reliable and safe pellet food to less predictable, moving live prey. After release, hatchery fish have been found to consume less food, eat fewer types of prey, and experience poor growth compared to their wild counterparts (Sosiak *et al.*, 1979; Ersbak and Haase, 1983; Bachman, 1984). There are even reports of fish consuming small stones owing to their similarity in appearance to commercially produced food pellets, and in some cases this behaviour can persist even 6 weeks after alternative live prey are available (Ellis *et al.*, 2002). This applied problem raises an interesting dilemma: not only do the hatchery fish have to *learn* how to feed on live prey, but they also need to *forget* about feeding on pellets (Ellis *et al.*, 2002).

Various attempts have been made to try and improve the pellet-to-live prey dietary transition (Olla *et al.*, 1998). Many of these rely on exposing hatchery fish to live prey prior to release. A problem here is the labor intensiveness and cost of this remedial training. An alternative approach may be to manipulate the hatchery environment in a way that promotes learning. It is already well established in captive reared birds and mammals that environmental enrichment promotes cognitive capacity and behavioural flexibility (Hunter *et al.*, 2002; Kempermann *et al.*, 2002; Bredy *et al.*, 2003; Rabin, 2003). Recent evidence suggests that environmental enrichment may have similar effects in hatchery reared fish (Braithwaite and Salvanes, 2005; Brown *et al.*, 2003). For example, juvenile cod (*Gadus morhua*) reared with experience of variable food and spatial cues are significantly faster at investigating and consuming live prey than cod from the same brood stock reared in standard, constant, unchanging hatchery conditions (Braithwaite and Salvanes, 2005). These differences arise even though both groups have been fed on a standard diet of food pellets. Thus, introducing some variation into the standard rearing environment generates fish that are more efficient at the pellet-to-live food transition.

5.2. Learning to Recognise Conspecifics

Being able to recognise other individuals, and to associate certain behaviours with those individuals, could be beneficial for a number of reasons. There are a variety of ways in which conspecifics may be recognised. For example, there is good evidence that some species of fish can recognise and preferentially associate with specific individuals that are familiar (Krause *et al.*, 2000). However, it is also possible that fish learn to discriminate between fish using factors other than individual recognition. For example, fish can discriminate between general competitive skills. European minnows (*Phoxinus phoxinus*) have been found to prefer to associate with individuals that are poor competitors (Metcalfe and Thomson, 1995). To do this, the minnows must be able to discriminate between good or poor competitors. It is not necessarily individual competitive abilities that are recognised and remembered, but rather the overall competitive ability associated with that group or school. This type of discrimination would not require as much complex cognitive processing as recognising specific individuals and associating particular behaviour patterns to those fish. Presumably, the advantage for the minnows to discriminate between good and poor competitors is that a fish feeding on a small limited food patch has a better chance of securing more food if it is surrounded by poor competitors.

Where specific individuals are recognised, the extent of the recognition can vary from a preference for familiar individuals of the same species (Krause *et al.*, 2000) to familiar individuals from mixed species schools (Ward *et al.*, 2003). Recognition of particular individuals can be important when repeated interactions occur. For example, sticklebacks prefer to inspect a predator with a partner that they are familiar with and perceive to be cooperative (Milinski *et al.*, 1990a,b). Similarly, Utne Palm and Hart (2000) found that the levels of aggression between sticklebacks that jointly forage on the same resource varies depending on how much the sticklebacks have interacted. As the fish become more familiar with each other, they are less aggressive and more cooperative in a foraging task (Figure 1.2). The effect of familiarity increased over a 4-week period: after 2 weeks the fish showed intermediate levels of aggression, but after 4 weeks the fish showed almost no aggression. Similarly, the breakdown of the individual recognition also showed a time-lag response with the sticklebacks showing some recognition after 2 weeks, but apparently no recognition after 4 weeks.

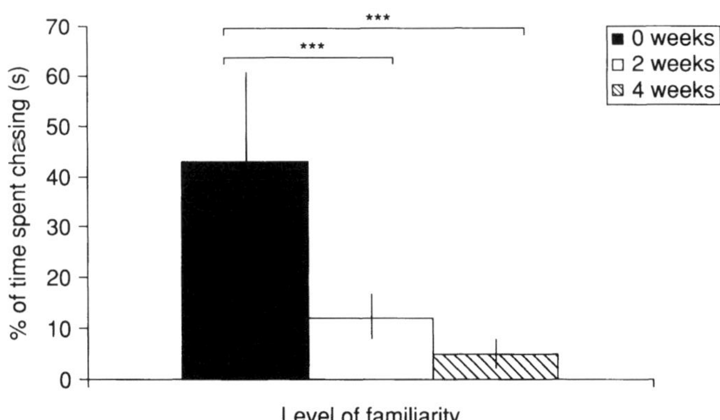

Fig. 1.2. Levels of aggression in pairs of sticklebacks sharing a food resource. Aggression was greatest in pairs that were unfamiliar, and least in pairs that had become familiar over a 4-week period. Bars and asterisks indicate significant differences between the different groups. ***$P < 0.01$). (Adapted from Utne-Palm and Hart, 2000.)

In her review, Griffiths (2003) distinguishes between two further forms of learned recognition. The examples of stickleback predator inspection behaviour represent one of these: condition-dependent recognition. Here, individuals are recognised based on prior experience with specific types of cue (for example, size, colour, competitive ability, or ability to cooperate during inspection behaviour). The alternative form of learned recognition is condition-independent. In this case, the learned recognition takes a relatively long time to develop, days rather than hours. For example, Griffiths and Magurran (1997) demonstrated that small groups of guppies took 12 days to learn to recognise each other. By actively staying within a school of familiar individuals, fish gain a number of advantages; schools of familiar fathead minnows (*Pimephales promelas*) are more cohesive and cooperative than schools made up of unfamiliar minnows (Chivers *et al.*, 1995). Thus, schools of familiar individuals are likely to be more efficient and effective at foraging and predator avoidance. Griffiths (2003) points out that there is reliable evidence that a number of fish species have the capacity to discriminate between familiar and unfamiliar conspecifics; for example, bluegill sunfish (Brown and Colgan, 1986), three-spined sticklebacks (Van Havre and Fitzgerald, 1988), and guppies (Magurran *et al.*, 1994). Yet, surprisingly little attention has been given to the role of familiarity in schooling behaviour despite the recognised capacity for discrimination between familiar and unfamiliar fish. Griffiths (2003) also points out that almost all the studies of familiarity in fish schools are based on laboratory observations. Work that has tried to address this issue in the field suggests that there may be little tendency for familiar individuals to associate with one another (Helfman, 1984; Hilborn, 1991; Klimley and Holloway, 1999). However, long-term associations may exist between subsets of members of a school (Barber and Ruxton, 2000). The discrepancy between field and laboratory derived results would seem worthy of further study; it seems curious that fish should have the capacity to perceive another fish as familiar without that information playing some type of role. The advancement of a number of marking techniques for fish, such as elastomer tags or PIT tags for larger fish, may ease future work investigating schooling behaviour in the field.

An ability to learn and remember the fighting ability of neighbours or rivals may also confer a number of advantages to fish. Johnsson and Åkerman (1998) used a series of pairwise interactions between rainbow trout (*Oncorhynchus mykiss*) to test the prediction that fish that had observed an aggressive interaction between two individuals should learn and remember the different fighting abilities of these fish. They hypothesised that this information could be used to settle a conflict faster if the observer was subsequently paired with a fish it had previously watched. Their results showed that when a fish lost, the interaction was both shorter and less aggressive if it was paired against a "known" opponent compared with when it was naïve to its opponents fighting ability. However, observers that won interactions had a faster increase in their aggression levels compared to naïve fish. These results suggest that fish that have previously observed their opponent are capable of making a faster decision when they have *a priori* knowledge of their opponent's fighting ability. They may also reflect a slight artifact of the experimental design; the trout were size-matched and so differences in fighting ability were relatively small, which led to more challenges between individuals than would be expected if the fish varied in size. These results demonstrate that trout can observe, learn, and remember the fighting ability of fish they interact with (see also Chapter 5, Section 2.1).

Oliveira and colleagues (1998) have similarly demonstrated that male Siamese fighting fish (*Betta splendens*) can watch and assess aggressive interactions between neighbouring fish. Information on the relative fighting ability of neighbours is then used in decisions about future agonistic interactions between the neighbours and the observers (Figure 1.3). Further research revealed a subtle twist to this finding: threat displays given by the males were found to vary depending on who is observing. In the presence of a female observer, males were found to reduce the overall aggression within their display, for example by decreasing the number of bites (Doutrelant *et al.*, 2001). Males that were aware of the presence of a female also included

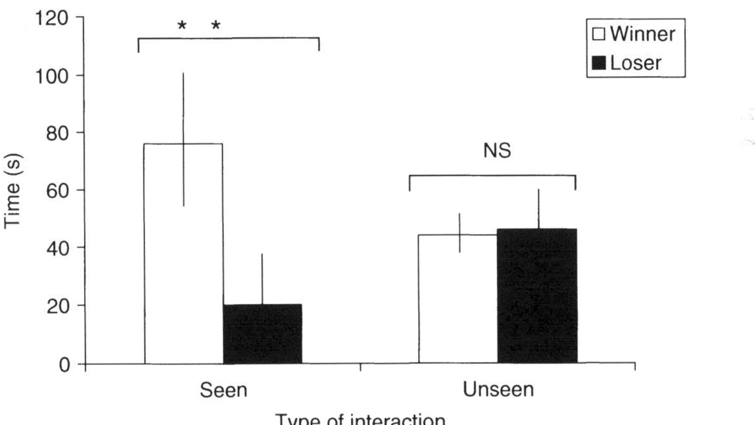

Fig. 1.3. Mean latency to display in male Siamese fighting fish when they had either "eavesdropped" their opponents fighting prowess by observing the opponent fight before (seen) or when they were naïve to their partners fighting ability (unseen). For each type of interaction, the male was paired with a former winner (white bars) and a former loser (black bars). **Significant difference with $P < 0.01$. NS, nonsignificant difference of $P > 0.05$. (Adapted from Oliveira *et al.*, 1998.)

more sexual components in their display to the rival male. It is not clear whether the inclusion of some sexual display simply reflects an arousal response generated in reaction to observing a female, or whether higher-level decision making is also included. It may be possible in future experiments to manipulate the attractiveness of female observers in an effort to tease apart reflex reactions to display over cognitively driven decisions.

5.3. Social Learning

Social learning is a specialised form of learning where an animal acquires a new behaviour by watching or interacting with other animals (Heyes, 1994). It is generally assumed that animals that can perform social learning will be at an advantage because individuals can quickly become locally adapted without incurring the costs that are often associated with trial-and-error learning (Giraldeau *et al.*, 1994). Thus, rather than an individual taking the time to learn that a particular patch is the most profitable or that an area should be avoided because it is frequented by predators, it can obtain vital local knowledge by observing others. At one time, social learning was believed to be so cognitively demanding that it was only to be found in "intelligent" species (Brown and Laland, 2003). As such, fish were largely overlooked; however, there are now a number of empirical field and laboratory studies demonstrating that fish can exhibit competent social learning (for a review, see Brown and Laland, 2003).

Social learning in fish has been investigated for a range of behaviours from mate choice to foraging (Brown and Laland, 2003). There are numerous examples of studies investigating social learning in antipredator responses (e.g., Kelley and Magurran, 2003). As highlighted in Sections 4 and 5.2. above, antipredator behaviour clearly has a learned component to it. Learning how best to react in the presence of a predator is an excellent example where trial-and-error learning is extremely expensive: make a wrong decision and it may be your last. Thus we could predict that there will be strong selection for animals to socially learn about antipredator responses, and as some reviews illustrate, fish can socially learn about both predator presence and identity (Suboski and Templeton, 1989; Brown and Laland, 2001).

The schooling tendency of various fish species may help mediate rapid antipredator responses. Here, the three sensory systems that are likely to play key roles in rapid responses are vision, the lateral line system, and olfaction (Webb, 1980; Brown, 2003). Observing that a close neighbour is responding to seeing a predator, detecting recently released alarm substances, or picking up information on a sudden change in the movement of nearby fish could quickly allow individual fish to adjust their own behaviour

(Treherne and Foster, 1981; Brown and Warburton, 1999; Chapter 3 of this volume). As a school of fish reacts to the presence of a predator on repeated occasions, individuals within the school can learn from those around them how to respond in a coordinated fashion (Brown and Warburton, 1999). Ultimately, this can lead to extraordinary levels of precision swimming that are observed when a school of fish suddenly forms a tight, rapidly moving ball in the presence of a predator. Such antipredator responses appear to confuse the predator and make it difficult for it to track an individual fish (Treherne and Foster, 1981). Just how complex this information transfer needs to be has been the focus of study (Couzin *et al.*, 2005). Using models to investigate how information spreads through a group, it has been proposed that shoaling information could, in some situations, effectively spread without explicit signaling or complex forms of communication (Couzin *et al.*, 2005).

Social learning is important for several species of fish that are known to possess chemical alarm cues. When these alarm cues are detected they can promote overt antipredator responses in conspecifics (Grant, 2003; Chapter 3 of this volume). These cues provide reliable information regarding local predation risk. Many of the fish capable of producing alarm substances do not have an innate recognition of predators, but instead they need to learn this information. To do this, they need to pair alarm cues with the visual and/or chemical cues associated with the predator (Suboski, 1990; Smith, 1999). Once learned (and they may learn this association after one exposure only), the fish are then primed to react to alarm substances released into the environment by conspecifics (Grant, 2003).

The way in which some fish learn to move efficiently around an environment to find specific resources is now known to be facilitated by social learning (Brown and Laland, 2003). For example, Warner (1988, 1990) demonstrated the mating grounds that bluehead wrasse (*Thalassoma bifasciatum*) migrate to remain constant across generations. To investigate whether each new generation socially learned the migration and the location of the mating sites, Warner removed a complete population and replaced these with another group of fish. These transplanted individuals established new mating grounds, and it was these newly established grounds that were subsequently used by the following generations. Such site fidelity for mating grounds across generations can only be reinforced through social learning.

Helfman and Schultz (1984) have observed similar evidence for social learning in another species of reef fish, the French grunt (*Haemulon flavolineatum*). French grunts make daily migrations to feeding areas, and after an initial observation that small groups of grunts appear to be joined on these daily migrations by juveniles, they investigated whether the juveniles socially learn routes to the feeding areas. By moving juveniles between groups and observing their subsequent swim paths when older group members were removed, Helfman and Schultz (1984) demonstrated that these young fish learn their migration paths by following older group members. Both the French grunt and the bluehead wrasse examples highlight how the day-to-day spatial movements of some reef fish lend themselves well to field studies of social learning.

Laland and Williams (1997, 1998) have brought studies of spatial social learning into the laboratory. They found that "naïve" guppies (*Poecilia reticulata*) were able to find their way to a foraging area by following "demonstrators" that had previously been trained to swim a particular route. After a period of overlap between the knowledgeable demonstrators and the naïve observers, the demonstrators were removed. Despite the missing tutors, the formerly naïve observers were able to swim the correct route themselves. This illustrates that the observer guppies had socially learned the route to the food area. Being able to assay this type of social learning within the controlled confines of a laboratory tank has provided an excellent method for addressing a variety of related questions (reviewed in Brown and Laland, 2003).

The role of social learning during foraging behaviour is one example that has benefited from detailed laboratory-based observations. Here, the experimenter can control a number of variables to determine which factors play a role. Such experiments have revealed that the size of the social group and the type of foraging environment are important. For example, Day and colleagues (2001) were able to show that when a food patch is visually isolated from the shoal, smaller groups of fish discover it more quickly than larger groups. However, when the food patch is visible through a transparent barrier, larger groups of fish locate the food faster than smaller groups. This curious result is proposed to reflect the motivation to shoal (Day *et al.*, 2001). Fish that join large shoals are reluctant to leave them as larger shoals provide more protection. However, fish that are members of small shoals are less compelled to stay together, which provides opportunities for individuals to move away from the shoal and to find food over wider areas.

A refinement of these types of approach has investigated how different individuals use foraging information obtained by watching others. Coolen and colleagues (2003) demonstrated that sympatrically occurring species of stickleback learn about the relative profitability of feeding patches in different ways. These differences were attributed to variations in response to predators in their microhabitat. Nine-spined sticklebacks (*Pungitius pungitius*), which have very little body plating to protect them from predators, use information provided by others to assess food patch quality. The nine-spined sticklebacks can learn about patch profitability by watching the behaviour of conspecifics and, interestingly, by also observing heterospecific three-spined sticklebacks.

Three-spined sticklebacks, on the other hand, were able to locate food patches by observing other fish (both conspecific and heterospecific) feeding, but they were unable to detect information on the relative profitability of a patch. The well-developed armor plating of the three-spined sticklebacks presumably places these fish at a lower risk of predation compared to the less well-defended nine-spined sticklebacks. These differences in predator defense may force the nine-spined stickleback to be more cautious, and individuals that can detect not just the location of a food patch but also its relative profitability before leaving shelter to forage will be at an advantage (Coolen *et al.*, 2003).

In an extension of these initial observations, van Bergen and colleagues (2004) continued to compare both species of stickleback. They showed that nine-spined sticklebacks adjust their foraging decision-making to preferentially use reliable information. In particular, they discovered that fish were able to discriminate between reliable and unreliable information. The sticklebacks were shown to prefer more recent, more reliable observations of conspecifics foraging at a particular location, even when this meant ignoring information that they had previously obtained when they had sampled the patch for themselves.

These results are very timely because, as with several other examples used in this Chapter, they indicate that some species of fish have the capacity to gauge the reliability of information. Perceiving differences in information is fairly straightforward, but understanding that one piece of information is more reliable than another requires a higher level of information processing. The approaches that Coolen *et al.* (2003) and van Bergen *et al.* (2004) have used are very versatile, and they are particularly elegant because they allow studies of both individual and social learning to be carried out simultaneously. Despite the fact that many animals are likely to rely on both forms of learning, most studies investigate these two forms of learning in isolation. The experimental approach of Coolen *et al.* also provides an excellent technique for assessing memory durations; for example, how long will a fish remember the relative profitabilities of two or more foraging patches? This is a technique that was successfully used by Milinski (1994), but little has been done with it since.

5.4. Spatial Learning

Many aspects of day-to-day existence rely on an animal's ability to move between different biologically important areas. Fish are no exception to this and many have the ability to move efficiently through an environment using reference points to guide their movements. This can be achieved in a number of ways. This section will look in detail at the sorts of cues that fish use to help guide their movements around their home range.

Resources are frequently distributed in a nonrandom way within an environment. An ability to move from one point to another taking the shortest possible route, or alternatively a route that will least likely expose a fish to a predator, will be favoured. In some instances this is achieved by the animal making a series of body-centered, or egocentric, movements or turns (Rodríguez *et al.*, 1994; Odling-Smee and Braithwaite, 2003b). In other cases, the animal will orient itself by keeping track of its own movement relative to one or more external reference points, and sometimes these may be used to form a map (Rodríguez *et al.*, 1994). When moving over much longer distances, for example during seasonal migrations, fish may use a compass rather than a map to allow them to maintain a particular course or bearing (see Section 5.4.2. and Chapter 7, Section 12 of this volume).

One of the most frequently cited examples of impressive spatial ability in fish is the escape response first described in rock-pool gobies by Aronson (1951, 1971). When threatened, the gobies jump from their current tide pool into a neighbouring one. The accuracy with which they are able to locate adjacent pools is remarkable. Aronson was able to demonstrate through a series of experiments that the gobies must be learning the local topography during periods of high tide when the fish could swim around and explore the intertidal area. If the gobies were prevented from exploring the topography at high tide, they were unable to accurately locate pools to jump into when they were later startled. These results suggest that the gobies can learn relatively complex spatial relationships and then use their spatial memory to recall this information to allow them to escape. It is interesting that these experiments are quoted so frequently because, although it is clear the fish must be using spatial learning and memory to behave this way, we still do not know what types of spatial cues are being used. It seems remarkable that the initial experiments were performed more than a half-century ago, and yet we are still unsure of the mechanisms that allow the fish to escape so effectively. What are the spatial cues that the fish learn at high tide? Do the fish form a mental map of the tidal pools and do they rely on landmarks to guide their chosen escape direction?

Virtually all the initial studies investigating fish homing behaviour were performed in the field using capture-mark-recapture techniques to estimate the area over which fish ranged (Green, 1971; Carlson and Haight, 1972; Hallacher, 1984). Experiments in which marked fish were displaced different distances showed that a number of fish species had a strong homing response, even after relatively large displacements. For example, yellowtail rockfish (*Sebastes flavidus*) were able to find their way back to their point of capture after being displaced more than 20 km (Carlson and Haight, 1972). Similarly, it was observed that the memory for home could persist over long periods of time. This was demonstrated by Green (1971) when he showed that intertidal sculpin (*Oligocottus maculosus*) could return to the site they were initially caught despite a 6-month period in captivity. More recently, more precise, small-scale spatial behaviours have been documented; planktivorous reef fish have been shown to adopt sophisticated search patterns while they forage on patches of prey. By following marked individuals, Noda and colleagues (1994) observed that individual fish repeatedly visited three specific foraging sites. While at a site, the fish swam slowly and followed a stereotypic pattern before swimming swiftly on to the next site. The authors suggest that the fish are using spatial memory to direct their foraging behaviour efficiently within each site and to avoid revisiting recently depleted areas. Their observations certainly seem to implicate spatial memory, but alternative explanations could also explain some of their observations. For example, the use of olfactory cues associated with their prey could indicate when a particular feeding area is food-rich and worth visiting, but falling numbers of prey and a decrease in odor cues could signal when the fish should leave for a new foraging site.

5.4.1. Learning about Landmarks

There are a number of examples that highlight how landmarks can be used by different species of fish (for review see Braithwaite, 1998). Field observations have demonstrated the use of visual landmarks to guide fish along a specific route on coral reefs (Reese, 1989). Many fish perform daily migrations along the reef to reach spawning sites, and these daily migrations typically follow a fixed route. If a head of coral from the reef is removed, several species of butterfly fish (family Chaetodontidae) will stop and begin a pattern of searching behaviour, apparently looking for the missing landmark (Reese, 1989). This is consistent with the fish using a navigational strategy known as piloting where the animal has a list of landmarks that will direct it to its goal. It moves from one landmark to the next working its way down the list. Brown surgeon fish (*Acanthurus nigrofuscus*) appear to use a similar technique as they too can be deflected from their swim path when landmarks are displaced (Mazeroll and Montgomery, 1998). Field experiments such as these are logistically difficult, and it is perhaps not surprising that the majority of work addressing landmark use by fish comes from controlled laboratory experiments.

A number of species of fish have been trained to use landmarks to direct their behaviour towards some goal point. In the simplest cases, the fish are trained to use a visual cue as a beacon (i.e., a single landmark that identifies the location of a specific goal). Warburton (1990) demonstrated that goldfish can learn to use such a beacon to locate a hidden food item. Fish can also swim from one beacon to the next; for example, three-spined sticklebacks were observed to follow small plant landmarks to find their way through a series of doors in a maze (Girvan and Braithwaite, 1998). Juvenile Atlantic salmon (*Salmo salar*) can also learn to follow moveable food patches labeled with unique visual landmarks to indicate the position of a food reward (Braithwaite *et al.*, 1996).

Given the diversity in fish sensory systems it should be no surprise that there are numerous examples of landmark learning in fish where the cues are not visual but instead the fish make use of other sensory modalities (see Chapter 2 for more details of sensory systems). Some species of electric fish such as the African mormyrid (*Gnathonemus pertersii*) can learn to locate a single hole in a wall that allows the fish to move between two halves of a tank. The fish use electrolocation to guide their movements and to map the position of this hole (Cain *et al.*, 1994). Mexican blind cavefish use their lateral line organ to help them navigate by detecting disturbances in their self-induced flow field (see Section 4), and these fish can use their lateral line organ to discriminate between differently shaped openings in a wall placed in the middle of their tank (von Campenhausen *et al.*, 1981). Even fish that are more usually dependent on vision can learn to use direction of flow to direct their movements; three-spined sticklebacks, for example, will use flow direction in an artificial stream to learn the position of a hidden food patch. Here, reversing the direction of flow within the stream causes the fish to swim to the opposite, previously nonrewarded end of the stream (Braithwaite and Girvan, 2003).

Although limited, there are also compelling examples that fish will learn nonvisual landmarks in the field. For example, anadromous forms of salmon are known to learn the specific odours of their natal stream as they migrate towards the marine environment (see Chapter 7). This phase is marked by a period of elevated production of thyroxine, and it has been suggested that the thyroxine is a key factor promoting olfactory learning (Dittman and Quinn, 1996). The fish learn the signature odours of their river system at several points on their outward migration, and these river-specific smells then form a sequence, or list, of learned odor-landmarks. When the fish are ready to begin their return migration, the salmon make their way to the coast and then reverse the remembered sequence of odours to allow them to follow the odour landmarks to reach their natal stream (Quinn, 1985; Hansen *et al.*, 1987; Heard, 1996). One possibility is that thyroxine surges trigger transient neural changes in the peripheral olfactory system, permitting windows of sensitivity, which prime the salmon for olfactory learning. Animals typically go through a phase of learning when they encounter new environments; it is possible then that thyroxine may be in some way preparing the fish for a period of learning. This possibility could be tested by artificially increasing thyroxine levels while quantifying learning capacity of the salmon. Further evidence that there may be a link between thyroxine and learning is that structural reorganisation in certain brain circuits has been shown to occur during the downstream migration when levels of thyroxine are high (Ebbesson *et al.*, 2003; Chapter 7 of this volume).

Very recently, auditory cues have been proposed to play a role in reef fish orientation (Tolimieri *et al.*, 2000; Simpson *et al.*, 2004). Some species of reef fish produce demersal eggs that hatch into free-swimming pelagic larvae and develop in the open sea for a short period of time before returning to the benthos. In some cases, juveniles are apparently able to home and settle on their natal reef (Jones *et al.*, 1999; Swearer *et al.*, 1999). Underwater sound can cover long distances with little attenuation and it is highly directional. Given these properties, sound could be used as an orientation cue. Reefs can generate a certain amount of biological sound; for example, nocturnal activities of snapping shrimps, urchins, and fish have been recorded and can be so loud that they have been described as an "evening chorus" (Tolimieri *et al.*, 2000). Using light traps that either emit a reef noise through loud speakers or are silent, two independent studies have shown that benthic reef fish are more attracted to the traps emitting noise (Tolimeiri *et al.*, 2000; Simpson *et al.*, 2004). In an extension of this work, Simpson and colleagues (2005) have found that different reef fish species respond to different sound spectra. However, it remains to be determined whether it is the noise alone that allows the fish to hone so precisely to their natal reef or whether other cues play a role. Furthermore, there are several aspects of these observations that remain unresolved. Are the fish using the whole spectrum of noise or are some parts more important than others? How do the fish learn the signature noise of their home reef? These questions are currently being investigated by Simpson and colleagues in a series of field experiments running on the Great Barrier Reef in Australia.

5.4.2. Navigating by a Compass

Fish are likely to rely on more than one source of spatial information when learning about locations or routes. Fish, like other vertebrates, are apt to give more attention to the most reliable or abundant sources of information (Healy, 1998). The use of more than one cue provides the fish with back-up information. In this way, if the spatial environment becomes altered in some way, the fish can switch to using alternative spatial cues (Reese, 1998; Vargas *et al.*, 2004). Compasses (such as the position of the sun or magnetic cues) provide a relatively stable, unchanging source of spatial information, which can be used on its own or in combination with landmarks or a map (Goodyear, 1973).

A few studies have demonstrated that fish can learn to use compass information. For example, sun-compass responses have been shown in fish trained to locate shelter in one of several circularly arranged refuges in outdoor arenas under clear skies (Hasler *et al.*, 1958; Schwassmann and Braemer, 1961; Hasler, 1971). Goodyear (1973) also demonstrated that mosquitofish (*Gambusia affinis*) have a sun-compass response (Goodyear and Bennet, 1979). When moved to unfamiliar surroundings, the mosquitofish use a sun compass to set them on a heading that is at right angles to the shore from which they were captured. This movement towards the shallow water close to the shore is thought to help the fish avoid other piscine predators (Goodyear and Ferguson, 1969).

There are also alternative forms of compass information (e.g., polarised light and magnetic cues). Experiments using polarising filters have demonstrated that juvenile rainbow trout (*Oncorhynchus mykiss*) can learn to orient using polarised light cues (e.g., Hawryshyn *et al.*, 1990); however, this ability appears to be lost in older fish. Natural polarised light patterns are at their most intense at sunrise and sunset, with the most obvious band of polarised light, the E-vector, being directly overhead at these times of day. Given that several species of reef fish perform their daily migrations first thing in the morning and at dusk, it would be an effective form of compass for these fish to use, but to date it has not been tested (Quinn and Ogden, 1984; Mazeroll and Montgomery, 1998).

The ability for animals to detect and use the Earth's magnetic field as a compass was proposed over a century ago (see Alerstam, 2003). Just how this information is detected and where it is processed, however, has been a hard problem to resolve. A significant contribution to this field came from work by Walker and colleagues (1997) who used the rainbow trout as their model organism. For the first time, they were able to demonstrate behavioural and electrophysiological responses to magnetic fields, and they identified an area in the fish's snout where candidate magnetoreceptor cells were located. Their results showed that the trout detect the magnetic field using magnetite, a biogenically produced iron oxide crystal. Tiny microscopic magnetite crystals were located near the basal lamina of the olfactory epithelium. Using a fluorescent lipophilic dye to trace the nerve pathways, the ros V nerve was proposed to be the most likely route through which information from the magnetoreceptor cells was passed to the brain (Walker *et al.*, 1997). These results were impressive not just in their detail but also because they allowed us to finally determine how some animals are able to detect magnetic fields. A magnetic compass is likely to underlie the ability of salmonids to migrate substantial distances out at sea while maintaining a constant bearing (see also Chapter 7, Section 9.3).

6. THE LATERAL TELENCEPHALIC PALLIUM AND SPATIAL LEARNING

For the last two decades, avian and mammalian biologists have been interested in the role of a small region of the brain called the hippocampus. This area that is in the midbrain of mammals, but on the roof of the brain in birds, has been clearly linked with learning and memory processes. In particular, it is associated with spatial learning and memory. Like birds and mammals, fish also have a capacity for spatial cognition, and this knowledge has prompted the search for a hippocampus homologue within the teleost brain. Such an area has now been identified within the forebrain: the lateral pallium (Vargas *et al.*, 2000; Rodríguez *et al.*, 2002a). It is well-established that, in birds and mammals, when the hippocampus is damaged or lesioned, spatial learning is impaired. Work by a team in Seville, Spain, has now shown that lesioning the lateral pallium in fish causes them to lose their ability to find their way to a previously learned goal location (Rodríguez *et al.*, 2002b). These striking similarities between teleost brains and those of other vertebrates suggests that there may be much more conservation in ancestral patterns of organisation in the brain than originally proposed.

It was observations made by Vargas and colleagues that indicated the lateral pallium may play a significant role in spatial learning. They discovered an increase in protein synthesis in neurones in the lateral pallium after goldfish had been trained in a spatial task (Vargas *et al.*, 2000). Subsequent work testing goldfish with a number of different pallial lesions gave further support to the idea that the lateral pallium is involved in spatial learning. For example, Rodríguez *et al.* (2002b) showed that goldfish with lesions to the lateral pallium, or ablation of the telencephalon (where the lateral pallium is located), were unable to navigate around a previously learned radial maze. Fish that were similarly trained but were subjected to lesions of the medial or dorsal pallium were not spatially impaired (Figure 1.1).

In the teleost forebrain, the region next to the lateral pallium, the medial pallium is considered to be homologous to the pallial amygdala found in mammals that is associated with emotion and certain types of learning. When the teleost medial pallium is lesioned, the fish sometimes display disrupted or disorganised aggressive sexual and reproductive behaviour (Bruin, 1983). Others, who have stimulated this area rather than lesioning it, have shown that medial pallium stimulation generates arousal, defensive behaviour, and escape responses (Savage, 1971; Demski and Beaver, 2001; Quick and Laming, 1988). Given these results, and the knowledge that the mammalian amygdala is involved in generating emotion, Broglio and colleagues (2003) speculated that the fish medial pallium may have an

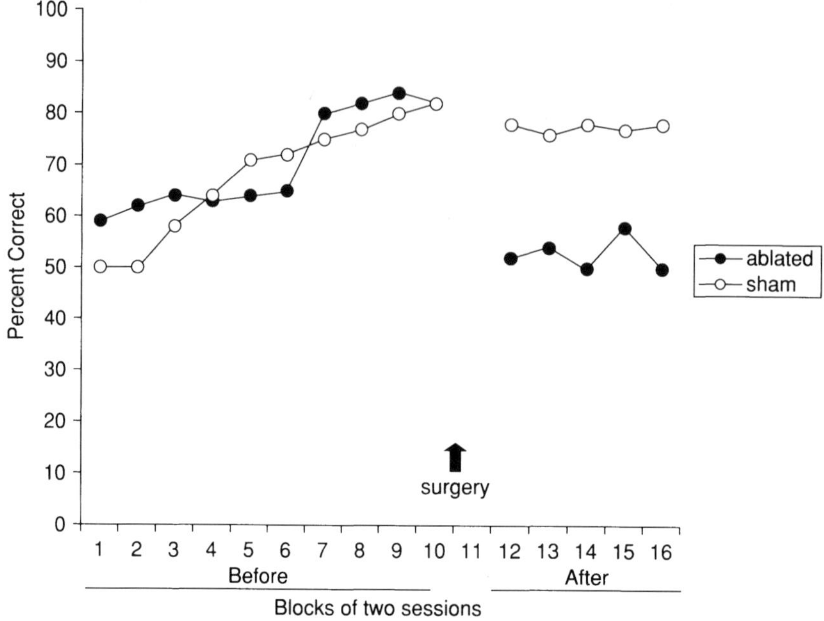

Fig. 1.4. Goldfish performance in a maze before and after surgery. Ablated group had their forebrain (telencephalon) aspirated. The sham group was handled in the same way, but no nervous tissue was damaged. Before surgery, during the acquisition or learning phase, there were no differences between the two groups (Mann-Whitney $U > 11$; $P > 0.11$). After surgery, however, the performance of the ablated group significantly deteriorated (Wilcoxon $Z = 2.07$; $P < 0.05$) and their postsurgery performance decayed to levels close to chance (56.7 ± 4.2%). (Adapted from Salas *et al.*, 1996b.)

important role in spatial memory that is linked to motivational effects such as learning to avoid a place frequented by a predator, or learning the location of a territory held by an attractive mate. In support of this, lesions to the medial pallium, but not the lateral pallium, impaired the associative learning capacity of goldfish that were trained to avoid a location that was paired with an electric shock (Portavella *et al.*, 2002). Here the medial pallium lesioned fish took longer to learn the task and forgot it more quickly than fish with sham or lateral pallium lesions.

These different examples again highlight the importance of studies of spatial learning and memory. Using spatial ability (i.e., accuracy at completing a maze) as the behavioural assay, the effects of different fine-scale lesions have identified which regions of the fish forebrain are necessary for processing different types of spatial information. Comparative approaches to investigating the role of the hippocampus have proved particularly fruitful in studies comparing different species of birds and mammals (Shettleworth, 1998). Given the diversity and specialisations seen in so many fish species, a comparative approach to fish cognition and neuroanatomy may also provide new insights into the development and plasticity of learning mechanisms.

7. CONCLUSION

There are many instances where perception, learning, and memory are key processes that enable fish to adapt to their local environment. Fish are a remarkably diverse group of organisms that exhibit an enormous range of specialisations in their sensory systems, as well as in their learning and memory capacities. This Chapter attempted to indicate different ways in which learning and memory processes allow fish to adapt to their local environment so that they can survive, mature, and reproduce. Recent work investigating the neural bases of cognition has demonstrated that there are striking similarities between fish brains and those of other vertebrates. Identifying homologous brain structures that share similar processing properties strongly suggests a degree of conservation between fish brains and other vertebrates. Studies of fish cognition may therefore provide a useful tool with which to further investigate the evolution of vertebrate learning and memory systems.

ACKNOWLEDGEMENTS

I would like to thank Nichola Brydges, Culum Brown, Sue Healy, Kate Henderson, Kevin Laland, Anne Gro Vea Salvanes, and two anonymous referees for stimulating discussions and comments on earlier drafts. I also thank The Leverhulme Trust, NERC, and BBSRC for their continued support of the research that my group and I do.

REFERENCES

Alerstam, T. (2003). Animal behaviour—The lobster navigators. *Nature* **421,** 27–28.

Armstrong, J. D., Braithwaite, V. A., and Ryecroft, P. (1996). A flat-bed passive integrated transponder antenna array for monitoring behaviour of Atlantic salmon parr and other fish. *J. Fish Biol.* **48,** 539–541.

Aronson, L. R. (1951). Orientation and jumping behaviour in the Gobiid fish Bathygobius soporator. *Am. Mus. Novit.* **1486,** 1–22.

Aronson, L. R. (1971). Further studies on orientation and jumping behaviour in the Gobiid fish Bathygobius soporator. *Ann. N. Y. Acad. Sci.* **188,** 378–392.

Bachman, R. A. (1984). Foraging behaviour of free-ranging wild and hatchery brown trout in a stream. *Trans. Am. Fish. Soc.* **113,** 1–32.

Barber, I., and Ruxton, G. D. (2000). The importance of stable schooling: Do familiar sticklebacks stick together? *Proc. R. Soc. Lond. B* **267,** 151–155.

Biro, D., Guilford, T., Dell'Omo, G., and Lipp, H. P. (2002). How the viewing of a familiar landscape prior to release allows pigeons to home faster: Evidence from GPS tracking. *J. Exp. Biol.* **205,** 3833–3844.

Bisazza, A., De Santi, A., and Vallortigara, G. (1999). Laterality and cooperation: Mosquitofish move closer to a predator when the companion is on the left side. *Anim. Behav.* **57,** 1145–1149.

Bisazza, A., Facchin, L., Pignatti, R., and Vallortigara, G. (1998). Lateralisation of detour behaviour in poeciliid fish: The effect of species, gender and sexual motivation. *Behav. Brain Res.* **91,** 157–164.

Bisazza, A., Pignatti, R., and Vallortigara, G. (1997). Laterality in detour behaviour: Interspecific variation in poeciliid fish. *Anim. Behav.* **54,** 1273–1281.

Braithwaite, V. A. (1998). Spatial memory, landmark use and orientation in fish. *In* "Spatial Representation in Animals" (Healy, S. D., Ed.), pp. 86–102. Oxford University Press, Oxford.

Braithwaite, V. A., Armstrong, J. D., McAdam, H. M., and Huntingford, F. A. (1996). Can juvenile Atlantic salmon use multiple cue systems in spatial learning? *Anim. Behav.* **51,** 1409–1415.

Braithwaite, V. A., and Girvan, J. R. (2003). Use of waterflow to provide spatial information in a small-scale orientation task. *J. Fish Biol.* **63,** S74–S83.

Braithwaite, V. A., and Huntingford, F. A. (2004). Fish and welfare: Can fish perceive pain and suffering? *Anim. Welf.* **13,** S87–S92.

Braithwaite, V. A., and Salvanes, A. G. V. (2005). Environmental variability in the early rearing environment generates behaviourally flexible cod: Implications for rehabilitating wild populations. *Proc. R. Soc. Lond. B.* **272,** 1107–1113.

Bredy, T. W., Humpartzoomian, R. A., Cain, D. P., and Meaney, M. J. (2003). Partial reversal of the effect of maternal care on cognitive function through environmental enrichment. *Neurosci.* **118,** 571–576.

Broglio, C., Rodríguez, F., and Salas, C. (2003). Spatial cognition and its neural basis in teleost fishes. *Fish Fish* **4,** 247–255.

Brown, C., and Warburton, K. (1999). Social mechanisms enhance escape responses in the rainbow fish, Melanotaenia duboulayi. *Env. Biol. Fish* **56,** 455–459.

Brown, C., and Laland, K. N. (2001). Social learning and life skills training for hatchery reared fish. *J. Fish Biol.* **59,** 471–493.

Brown, C., and Laland, K. N. (2003). Social learning in fishes: A review. *Fish* **4,** 280–288.

Brown, C., Davidson, T., and Laland, K. (2003). Environmental enrichment and prior experience improve foraging behaviour in hatchery-reared Atlantic salmon. *J. Fish Biol.* **63** (Suppl. 1), 187–196.

Brown, C., Gardner, C., and Braithwaite, V. A. (2004). Population variation in lateralised eye use in the poeciliid Brachyraphis episcopi. *Proc. R. Soc. Lond. B.* **271,** S455–S457.

Brown, G. E. (2003). Learning about danger: Chemical alarm cues and local risk assessment in prey fishes. *Fish Fish* **4,** 227–234.

Brown, J. A., and Colgan, P. W. (1986). Individual and species recognition in centrarchid fishes: Evidence and hypotheses. *Behav. Ecol. Sociobiol.* **19,** 373–379.

Bruin, J. P. C., and de (1983). Neural correlates of motivated behaviour in fish. *In* "Advances in Vertebrate Neuroethology" (Ewert, J. P., Capranica, R. R., and Ingle, D. J., Eds.), pp. 969–995. Plenum Press, New York.

Burt de Perera, T., and Braithwaite, V. A. (2005). Laterality in a non-visual sensory modality – the lateral line of fish. *Curr. Biol.* **15,** 241–242.

Cain, P., Gerin, W., and Moller, P. (1994). Short range navigation in the weakly electric fish Gnathonemus petersii L. *Ethol.* **96,** 33–45.

Campenhausen, C. V., Reiss, I., and Weissert, R. (1981). Detection of stationary objects by the blind cave fish Anoptichthy jordani. *J. Comp. Physiol. A* **143,** 369–374.

Carlson, H. R., and Haight, R. E. (1972). Evidence for a home site and homing of adult yellowtail rockfish, Sebastes flavidus. *J. Fish. Res. Bd. Can.* **29,** 1011–1014.

Chandroo, K. P., Duncan, I. J. H., and Moccia, R. D. (2004). Can fish suffer?: Perspectives on sentience, pain, fear and stress. *Appl. Anim. Behav. Sci.* **86,** 225–250.

Charnov, E. L. (1976). Optimal foraging theory: The marginal value theorem. *Theoret. Pop. Biol.* **9,** 129–136.

Chivers, D. P., Brown, G. E., and Smith, R. J. F. (1995). Familiarity and shoal cohesion in fathead minnows (Pimephales promelas): Implications for antipredator behaviour. *Can. J. Zool.* **73,** 955–960.

Colgan, P. W., Brown, J. A., and Osratti, S. D. (1986). Role of diet and experience in the development of feeding behaviour in large-mouth bass (Micropterus salmoides). *J. Fish Biol.* **28,** 161–170.

Coolen, I., van Bergen, Y., Day, R. L., and Laland, K. N. (2003). Species differences in adaptive use of public information in sticklebacks. *Proc. R. Soc. Lond. B* **270,** 2413–2419.

Croy, M. I., and Hughes, R. N. (1991). The role of learning and memory in the feeding behaviour of the fifteen spined staickleback (Spinachia spinachia L.). *Anim. Behav.* **41,** 149–160.

Couzin, I. D., Krause, J., Franks, N. R., and Levin, S. A. (2005). Effective leadership and decision-making in animal groups on the move. *Nature* **433,** 513–516.

Dadda, M., Sovrano, V. A., and Bisazza, A. (2003). Temporal pattern of social aggregation in tadpoles and its influence on the measurements of lateralised responses to social stimuli. *Physiol. Behav.* **78,** 337–341.

Day, R., MacDonals, T., Brown, C., Laland, K. N., and Reader, S. M. (2001). Interactions between shoal size and conformity in guppy social foraging. *Anim. Behav.* **62,** 917–925.

Deckel, A. W. (1995). Lateralisation of aggressive responses in Anolis. *J. Exp. Zool.* **272,** 194–200.

Demski, L. S., and Beaver, J. A. (2001). Brain and cognitive function in teleost fishes. *In* "Brain, Evolution and Cognition" (Roth, G., and Wulliman, M. F., Eds.), pp. 297–332. Wiley, New York.

Dittman, A. H., and Quinn, T. P. (1996). Homing in Pacific salmon: Mechanisms and ecological basis. *J. Exp. Biol.* **199,** 83–91.

Doutrelant, C., McGregor, P. K., and Oliveira, R. F. (2001). The effect of an audience on intrasexual communication in male Siamese fighting fish, Betta splendens. *Behav. Ecol.* **12,** 283–286.

Dugatkin, L. A., and Aifieri, M. (1991a). TIT FOR TAT in guppies (Poecilia reticulata): The relative nature of cooperation and defection during predator inspection. *Evol. Ecol.* **5,** 300–309.

Dugatkin, L. A., and Aifieri, M. (1991b). Guppies and the TIT FOR TAT strategy: Preference based on past interaction. *Behav. Ecol. Sociobiol.* **28,** 243–246.

Ebbesson, L. O. E., Ekstrom, P., Ebbesson, S. O. E., Stefansson, S. O., and Holmqvist, B. (2003). Neural circuits and their structural and chemical reorganisation in the light-brain-pituitary axis during parr-smolt tranformation. *Aquaculture* **222,** 59–70.

Ellis, T., Hughes, R. N., and Howell, B. R. (2002). Artificial dietary regime may impair subsequent foraging behaviour of hatchery-reared turbot released into natural environment. *J. Fish Biol.* **61,** 252–264.

Ersbak, K., and Haase, B. L. (1983). Nutritional deprivation after stocking as a possible mechanism leading to mortality in stream-stocked brook trout. *N. Am. J. Fish. Manag.* **3,** 142–151.

Facchin, L., Bisazza, A., and Vallortigara, G. (1999). What causes lateralisation of detour behaviour in fish? Evidence for asymmetries in eye use. *Behav. Brain Res.* **103,** 229–234.

Finger, T. E. (1980). Nonolfactory sensory pathway to the telencephalon in a teleost fish. *Science* **210,** 671–673.

Finger, T. E. (1983). Organisation of the teleost cerebellum. *In* "Fish Neurobiology" (Davis, R. E., and Northcutt, R. G., Eds.), pp. 261–284. The University of Michigan Press, Ann Arbor.

Gallistel, C. R. (1990). "The organisation of learning." M.I.T., Cambridge, MA.

Giraldeau, L. A., Caraco, T., and Valone, T. J. (1994). Social foraging: Individual learning and cultural transmission of innovations. *Behav. Ecol.* **5,** 35–43.

Girvan, J. R., and Braithwaite, V. A. (1998). Population differences in learning and memory in three-spined sticklebacks. *Proc. R. Soc. Lond. B* **265,** 913–918.

Goodyear, C. P. (1973). Learned orientation in the predator avoidance behaviour of mosquitofish, Gambusia affinis. *Behaviour* **45,** 191–223.

Goodyear, C. P., and Bennett, D. H. (1979). Sun compass orientation of immature bluegill. *Trans. Am. Fish. Soc.* **108,** 555–559.

Goodyear, C. P., and Ferguson, D. E. (1969). Sun compass orientation in the mosquitofish, Gambusia affinis. *Anim. Behav.* **17,** 636–640.

Grant, G. E. (2003). Learning about danger: Chemical alarm cues and local risk assessment in prey fishes. *Fish Fish* **4,** 227–234.

Green, J. M. (1971). High tide movements and homing behaviour of the tidepool sculpin Oligocottus maculosus. *J. Fish. Res. Bd. Canada* **28,** 383–389.

Griffiths, S. W. (2003). Learned recognition of conspecifics by fishes. *Fish Fish* **4,** 256–268.

Griffiths, S. W., and Magurran, A. E. (1997). Familiarity in schooling fish: How long does it take to acquire? *Anim. Behav.* **53,** 945–949.

Griffiths, S. W., Brockmark, S., Hojesjo, J., and Johnsson, J. I. (2004). Coping with divided attention: The advantage of familiarity. *Proc. Royal Soc. London B.* **271,** 695–699.

Hallacher, L. E. (1984). Relocation of original territories by displaced black-and-yellow rockfish, Sebastes Chrysomelas, from Carmel Bay, California. *Calif. Fish Game* **70,** 158–162.

Hansen, L. P., Døving, K. B., and Jonsson, B. (1987). Migration of farmed adult Atlantic salmon with and without olfactory sense, released on the Norwegian coast. *J. Fish Biol.* **30,** 713–721.

Hasler, A. D. (1971). Orientation and migration. *In* "Fish Physiology" (Hoar, W. S., and Randall, D. J., Eds.), pp. 429–510. Academic Press, London.

Hasler, A. D., Horrall, R. M., Wisby, W. J., and Braemer, W. (1958). Sun orientation and homing in fishes. *Limnol. Oceanogr.* **111,** 353–361.

Hawryshyn, C. W., Arnold, M. G., Bowering, E., and Cole, R. L. (1990). Spatial orientation of rainbow trout to plane-polarised light: The ontogeny of E-vector discrimination and spectral sensitivity characteristics. *J. Comp. Physiol. A.* **166,** 565–574.

Healy, S. D. (1998). "Spatial Representation in Animals." Oxford University Press, Oxford.

Heard, W. R. (1996). Sequential imprinting in chinook salmon: Is it essential for homing fidelity? *Bull. Nat. Res. Inst. Aqua.* **2**(Suppl.), 59–64.

Helfman, G. S. (1984). School fidelity in fishes: The yellow perch pattern. *Anim. Behav.* **32,** 663–672.

Helfman, G. S., and Schultz, E. T. (1984). Social transmission of behavioural traditions in a coral reef fish. *Anim. Behav.* **32,** 379–384.

Heuts, B. A. (1999). Lateralisation of trunk muscle volume, and lateralisation of swimming turns of fish responding to external stimuli. *Behav. Proc.* **47,** 113–124.

Heyes, C. M. (1994). Social learning in animals: Categories and mechanisms. *Biol. Rev.* **69,** 207–231.

Hoare, D. J., and Krause, J. (2003). Social organisation of fish shoals. *Fish Fish.* **4,** 269–279.

Huber, R., Van Staaden, M. J., Kaufman, L. S., and Liem, K. F. (1997). Microhabitat use, trophic patterns and the evolution of brain structure in African cichlids. *Brain Behav. Evol.* **50,** 167–182.

Hughes, R. N., and Croy, M. I. (1993). An experimental analysis of frequency-dependent predation (switching) in the 15-spined stickleback, Spinachia spinachia. *J. Anim. Ecol.* **62,** 341–352.

Hunter, S. A., Bay, M. S., Martin, M. L., and Hatfield, J. S. (2002). Behavioural effects of environmental enrichment on harbour seals (Phoca vitulina concolor) and grey seals (Halichoerus grypus). *Zoo Biol.* **21,** 375–387.

Huntingford, F. A. (2003). Learning in Fishes. *Fish Fish.* **4,** 197–198.

Huntingford, F. A., and Wright, P. J. (1989). How sticklebacks learn to avoid dangerous feeding patches. *Behav. Proc.* **19,** 181–189.

Huntingford, F. A., Adams, C., Braithwaite, V. A., Kadri, S., Pottinger, T., Sandoe, P., and Turnbull, J. (in press). Fish welfare: A broad overview of current research. *J. Fish Biol.*

Johnsson, J. I., and Åkerman, A. (1998). Watch and learn: Preview of the fighting ability of opponents alters contest behaviour in rainbow trout. *Anim. Behav.* **56,** 771–776.

Jones, G. P., Milicich, M. J., and Lunow, C. (1999). Self-recruitment in a coral reef fish population. *Nature* **402,** 802–804.

Kelley, J. L., and Magurran, A. E. (2003). Learned predator recognition and antipredator responses in fishes. *Fish Fish.* **4,** 216–226.

Kempermann, G., Gast, D., and Gage, F. H. (2002). Neuroplasticity in old age: Sustained five-fold induction of hippocampal neurogenesis by long term environmental enrichment. *Ann. Neurol.* **52,** 135–143.

Klimley, A. P., and Holloway, C. F. (1999). School fidelity and homing synchronicity of yellowfin tuna (Thunnus albacares). *Mar. Biol.* **133,** 307–317.

Knudsen, E. I. (1977). Distinct auditory and lateral line nuclei in the midbrain of catfishes. *J. Comp. Neurol.* **173,** 417–432.

Kotrschal, K., Van Staaden, M. J., and Huber, R. (1998). Fish brains: Evolution and environmental relationships. *Rev. Fish. Biol. Fish.* **8,** 373–408.

Krause, J., Butlin, R. K., Peuhkuri, N., and Pritchard, V. L. (2000). The social organisation of fish shoals: A test of the predictive power of laboratory experiments for the field. *Biol. Rev.* **75,** 477–501.

Kullander, K. O. (1999). Fish species – how and why. *Rev. Fish Biol. Fish.* **9,** 325–352.

Laland, K. N., Brown, C., and Krause, J. (2003). Learning in fishes: From three-second memory to culture. *Fish Fish.* **4,** 199–202.

Laland, K. N., and Williams, K. (1997). Shoaling generates social learning of foraging information in guppies. *Anim. Behav.* **53,** 1161–1169.

Laland, K. N., and Williams, K. (1998). Social transmission of maladaptive information in the guppy. *Behav. Ecol.* **9,** 493–499.

Laming, P. R., and Ebbesson, S. O. E. (1984). Arousal and fright responses and their habituation in the slippery dick, Halichoeres bivittatus. *Experientia* **40,** 767–769.

Laming, P. R., and McKinney, S. J. (1990). Habituation in goldfish (Crassius auratus) is impaired by increased interstimulus interval, interval variability, and telencephalic ablation. *Behav. Neurosci.* **194,** 869–875.

Mackney, P. A., and Hughes, R. N. (1995). Foraging behaviour and memory window in sticklebacks. *Behaviour* **132,** 1241–1253.

Macphail, E. M. (1982). "Brain and Intelligence in Vertebrates." Clarendon Press, Oxford.

Magurran, A. E., Seghers, B. H., Shaw, P. W., and Carvalho, G. R. (1994). Schooling preferences for familiar fish in the guppy Poecilia reticulata. *J. Fish Biol.* **45,** 401–406.

Maler, L., Sas, E., Johnston, S., and Ellis, W. (1991). An atlas of the brain of the electric fish Apteronotus leptorhynchus. *J. Chem. Neuroanat.* **4,** 1–38.

Mazeroll, A. I., and Montgomery, W. L. (1998). Daily migrations of a coral reef fish in the Red Sea (Gulf of Aqaba, Israel): Initiation and orientation. *Copeia* **4,** 893–905.

McCormick, C. A. (1989). Central lateral line pathways in bony fish. *In* "The Lateral Line: Neurobiology and Evolution" (Coombs, S., Görner, P., and Münz, H., Eds.), pp. 341–363. Springer Verlag, New York.

McKenzie, R., Andrew, R. J., and Jones, R. B. (1998). Lateralisation in chicks and hens: New evidence for control of response by the right eye system. *Neuropsych.* **36,** 51–58.

Metcalfe, N. B., and Thomson, B. C. (1995). Fish recognise and prefer to shaol with poor competitors. *Proc. R. Soc. Lond. B* **259,** 207–210.

Miklosi, A., Andrew, R. J., and Savage, H. (1997). Behavioural lateralisation of the tetrapod type in the zebrafish (Brachydanio rerio). *Phys. Behav.* **63,** 127–135.

Milinski, M. (1994). Long-term memory for food patches and implications for ideal free distributions in sticklebacks. *Ecology* **75,** 1150–1156.

Milinski, M., Pfluger, D., Kulling, D., and Kettler, R. (1990a). Do sticklebacks cooperate repeatedly in reciprocal pairs? *Behav. Ecol. Sociobiol.* **27,** 17–21.

Milinski, M., Kulling, D., and Kettler, R. (1990b). Tit for tat: Sticklebacks 'trusting' a co-operating partner. *Behav. Ecol.* **1,** 7–12.

Mills, E. L., Widzowski, D. V., and Jones, S. R. (1987). Food conditioning and prey selection by young yellow perch (Perca flavsecens). *Can. J. Fish. Aquat. Sci.* **44,** 549–555.

New, J. G. (2001). Comparative neurobiology of the elasmobranch cerebellum: Theme and variations on a sensorimotor interface. *Environ. Biol. Fish.* **60,** 93–108.

Nieuwenhuys, R. (1967). Comparative anatomy of olfactory centres and tracts. *In* "Sensory Mechanisms" (Zotterman, Y., Ed.), pp. 1–64. Elsevier, Amsterdam.

Noda, M., Gushima, K., and Kakuda, S. (1994). Local prey search based on spatial memory and expectation in the planktivorous reef fish, Chromis chrysurus (Pomacentridae). *Anim. Behav.* **47,** 1413–1422.

Odling-Smee, L., and Braithwaite, V. A. (2003a). The role of learning in fish orientation. *Fish Fish.* **4,** 235–246.

Odling-Smee, L., and Braithwaite, V. A. (2003b). The influence of habitat stability on landmark use during spatial learning in the threespine stickleback. *Anim. Behav.* **65,** 701–707.

Oliveira, R. F., McGregor, P. K., and Latruffe, C. (1998). Know thine enemy: Fighting fish gather information from observing conspecific interactions. *Proc. R. Soc. Lond. B* **265,** 1045–1049.

Olla, B. L., Davis, M. W., and Ryer, C. H. (1998). Understanding how the hatchery environment represses or promotes the development of behavioural survival skills. *Bull. Mar. Sci.* **62,** 531–550.

Overmeir, J. B., and Hollis, K. (1983). The teleostan telencephalon and learning. *In* "Fish Neurobiology, Volume 2 Higher Brain Functions" (Davis, R., and Northcutt, G., Eds.), pp. 265–284. University of Michigan Press, Ann Arbor.

Overmeir, J. B., and Papini, M. R. (1986). Actors modulating the effects of teleost telencephalon ablation on retention, relearning and extinction of instrumental avoidance behaviour. *Behav. Neurosci.* **100,** 190–199.

Pearce, J. M. (1997). "Animal Learning and Cognition." Psychology Press, Sussex.

Pitcher, T. J., and House, A. C. (1987). Foraging rules for group feeders: Area copying depends upon food density in shoaling goldfish. *Ethology* **76,** 161–167.

Pitcher, T. J., and Magurran, A. E. (1983). Shoal size, patch profitability and information exchange in foraging goldfish. *Anim. Behav.* **31,** 546–555.

Portavella, M., Vargas, J. P., Torres, B., and Salas, C. (2002). The effects of telencephalic pallial lesions on spatial, temporal, and emotional learning in goldfish. *Brain Res. Bull.* **57,** 397–399.

Portavella, M., Torres, B., and Salas, C. (2004). Avoidance response in goldfish: Emotional and temporal involvement of medial and lateral telencephalic pallium. *J. Neurosci.* **24,** 2335–2342.

Quick, I. A., and Laming, P. R. (1988). Cardiac, ventilatory and behavioural arousal responses evoked by electrical brain stimulation in the goldfish (Carassius auratus). *Physiol. Behav.* **43,** 715–727.

Quinn, T. P. (1985). Salmon homing: Is the puzzle complete? *Envir. Biol. Fish.* **12,** 315–317.

Quinn, T. P., and Ogden, J. C. (1984). Field experience of compass orientation in migrating juvenile Grunts (Haemulidae). *J. Exp. Mar. Biol. Ecol.* **81,** 181–192.

Rabin, L. A. (2003). Maintaining behavioural diversity in captivity for conservation: Natural behaviour management. *Anim. Welf.* **12,** 85–94.

Reebs, S. G. (1996). Time-place learning in golden shiners (Pisces: Cyprinidae). *Behav. Proc.* **36,** 253–262.

Reebs, S. G. (1999). Time place learning based on food but not on predation risk in a fish, the inanga (Galaxias maculatus). *Ethology* **105,** 361–371.

Reese, E. S. (1989). Orientation behaviour of butterflyfishes (family Chaetodontidae) on coral reefs: Spatial learning of route specific landmarks and cognitive maps. *Enviro. Biol. Fish.* **25,** 79–86.

Rodríguez, F., Durán, E., Vargas, J. P., Torres, B., and Salas, C. (1994). Performance of goldfish trained in allocentric and egocentric maze procedures suggest the presence of a cognitive mapping system in fishes. *Anim. Learn. Behav.* **22,** 409–420.

Rodríguez, F., Dúran, E., Vargas, J. P., Broglio, C., Gómez, Y., and Salas, C. (2002a). Spatial memory and hippocampal pallium through vertebrate evolution: Insights from reptiles and teleost fish. *Brain Res. Bull.* **57,** 499–503.

Rodríguez, F., López, J. C., Vargas, J. P., Gómez, Y., Broglio, C., and Salas, C. (2002b). Conservation of spatial memory function in the pallial forebrain of amniotes and ray-finned fishes. *J. Neurosci.* **22,** 2894–2903.

Rogers, L. J. (1989). Laterality in animals. 3, 5–25. *Int. J. Comp. Psychol.* **3,** 5–25.

Rogers, L. J., and Andrew, R. J. (2002). "Comparative Vertebrate Lateralisation." Cambridge University Press, Cambridge, UK.

Rogers, L. J., Zucca, P., and Vallortigara, G. (2004). Advantages of having a lateralised brain. *Proc. R. Soc. Lond. B* **271,** S420–S422.

Rooney, D. J., and Laming, P. R. (1987). Teleost telencephalic involvement with habituation of arousal responses. *Behav. Brain Res.* **26,** 223–231.

Rooney, D. J., and Laming, P. R. (1988). Effects of telencephalic ablation on habituation of arousal responses within and between daily training sessions in goldfish. *Behav. Neural Biol.* **49,** 83–96.

Salas, C., Broglio, C., Rodríguez, F., López, J. C., Portavella, M., and Torres, B. (1996a). Telencephalic ablation in goldfish impairs performance in a spatial constancy problem but not in a cued one. *Behav. Brain Res.* **79,** 193–200.

Salas, C., Rodríguez, F., Vargas, J. P., Durán, E., and Torres, B. (1996b). Spatial learning and memory deficits after telencephalic ablation in goldfish trained in place and turn maze procedures. *Behav. Neurosci.* **110,** 965–980.

Savage, G. E. (1971). Behavioural effects of electrical stimulation of the telencephalon of goldfish, Carassius auratus. *Anim. Behav.* **19,** 661–668.

Schwassmann, H. O., and Braemer, W. (1961). The effect of experimentally changed photoperiod on the sun orientation rhythm of fish. *Physiol. Zool.* **34,** 273–326.

Shettleworth, S. J. (1998). "Cognition, Evolution, and Behaviour." Oxford University Press, Oxford.

Shiflett, M. W., Smulders, T. V., Benedict, L., and De Voogd, T. J. (2003). Reversible inactivation of the hippocampal formation in food-storing black-capped chickadees. *Hippocampus* **13,** 437–444.

Simpson, S. D., Meekan, M. G., McCauley, R. D., and Jeffs, A. (2004). Attraction of settlement-stage coral reef fishes to reef noise. *Mar. Ecol. Prog. Ser.* **276,** 263–268.

Simpson, S. D., Jeffs, A., Meekan, M. G., Montgomery, J., and McCauley, R. (2005). Homeward sound. *Science* **308,** 221.

Smith, R. J. F. (1999). What good is smelly stuff in the skin? Cross-function and cross taxa effects in fish 'alarm substances'. *In* "Advances in Cemical Signals in Vertebrates" (Johnstone, R. E., Müller-Schwarze, D., and Soresen, P. W., Eds.), pp. 475–488. Kluwer Academic, New York.

Sovrano, V. A., Rainoldi, C., Bisazza, A., and Vallortigara, G. (1999). Roots of brain specialisations: Preferential left-eye use during mirror-image inspection in six species of teleost fish. *Behav. Brain Res.* **106,** 175–180.

Suboski, M. D. (1990). Releaser-induced recognition learning. *Psychol. Rev.* **97,** 271–284.

Suboski, M. D., and Templeton, J. I. (1989). Life skills training for hatchery fish: Social learning and survival. *Fish. Res.* **7,** 343–352.

Swearer, S. E., Caselle, J. E., Lea, D. W., and Warner, R. R. (1999). Larval retention and recruitment in an island population of coral reef fish. *Nature* **402,** 799–802.

Teyke, T. (1985). Collision with and avoidance of obstacles by blind cave fish Anoptichthy jordani. *J. Comp. Physiol. A* **157,** 837–843.

Teyke, T. (1989). Learning and remembering the environment in blind cave fish Anoptichthy jordani. *J. Comp. Physiol. A* **164,** 655–662.

Tolimieri, N., Jeffs, A., and Montgomery, J. C. (2000). Ambient sound as a cue for navigation by the pelagic larvae of reef fishes. *Mar. Ecol. Prog. Ser.* **207,** 219–224.

Treherne, J. E., and Foster, W. A. (1981). Group transmission of predator avoidance-behaviour in a marine insect – the 'Trafalgar Effect.' *Anim. Behav.* **29,** 911–917.

Utne-Palm, A. C., and Hart, P. J. B. (2000). The effects of familiarity on competitive interactions in threespined sticklebacks. *Oikos* **91,** 225–232.

van Bergen, Y., Coolen, I., and Laland, K. N. (2004). Nine-spined sticklebacks exploit the most reliable source when public and private information conflict. *Proc. R. Soc. Lond. B* **271,** 957–962.

Van Staaden, M. J., Huber, R., Kaufman, L. S., and Karel, F. L. (1994). Brain evolution in cichlids of the African Great Lakes: Brain and body size, general patterns and evolutionary trends. *Zoology* **98,** 165–178.

Van Havre, N., and Fitz Gerald, G. J. (1988). Shoaling and kin recognitionin the three-spine stickleback (Gasterosteus aculeatus L.). *Biol. Behav.* **13,** 190–201.

Vargas, J. P., López, J. C., Salas, C., and Thinus-Blanc, C. (2004). Encoding of geometric and featural spatial information by goldfish (Carassius auratus). *J. Comp. Psychol.* **118,** 206–216.

Vargas, J. P., Rodríguez, F., López, J. C., Arias, J. L., and Salas, C. (2000). Spatial memory and dorsolateral telencephalic pallium of goldfish (Carassius auratus) Ph.D., thesis, University of Sevilla, Seville, Spain.

Walker, M. M., Diebel, C. E., Haugh, C. V., Pankhurst, P. M., Montgomery, J. C., and R.G.C. (1997). Structure and function of the vertebrate magnetic sense. *Nature* **390,** 371–376.

Warburton, K. (1990). The use of local landmarks by foraging goldfish. *Anim. Behav.* **40,** 500–505.

Warburton, K. (2003). Learning of foraging skills by fish. *Fish Fish.* **4,** 203–216.

Ward, A. J. W., Axford, S., and Krause, J. (2003). Cross-species familiarity in shoaling fishes. *Proc. R. Soc. Lond. B* **270,** 1157–1161.

Warner, R. R. (1988). Traditionality of mating-site preferences in coral reef fish. *Nature* **335,** 719–721.

Warner, R. R. (1990). Male versus female influences on mating-site determination in a coral-reef fish. *Anim. Behav.* **39,** 540–548.

Wasserman, E. A. (1984). Animal intelligence: Understanding the minds of animals through their behavioural 'ambassadors.' *In* "Animal Cognition" (Roitblat, H. L., Bever, T. G., and Terrace, H. S., Eds.), pp. 45–60. Lawrence Erlbaum Associates, Hillsdale, NJ.

Webb, P. W. (1980). Does schooling reduce fast-start response latencies in teleosts? *Comp. Biochem. Physiol. A* **65,** 231–234.

Werner, E. E., Mittelbach, G. G., and Hall, D. J. (1981). The role of foraging profitability and experience in habitat use by the bluegill sunfish. *Ecology* **62,** 116–125.

Wildhaber, M. L., Green, R., and Crowder, L. B. (1994). Bluegills continuously update patch giving-up times based on foraging experience. *Anim. Behav.* **47,** 501–503.

Witte, K., and Nottemeier, B. (2002). The role of information in mate-choice copying in female sailfin mollies (Poecilia latipinna). *Behav. Ecol. Sociobiol.* **52,** 194–202.

2

COMMUNICATION

GIL G. ROSENTHAL
PHILLIP S. LOBEL

1. INTRODUCTION

Communication is defined as an interaction between a *signaler*, who produces a sensory stimulus or signal, and a *receiver*, who perceives the signal and makes a consequent behavioural decision. Communication is fundamentally a series of steps between one animal's brain and another: from the production of behaviours and strategies on the signaling end to perception and behavioural response on the receiving end.

Fish have made a central contribution to our understanding of communication. Communication underwater generally presents an instructive

Behaviour and Physiology of Fish: Volume 24
FISH PHYSIOLOGY

DOI: 10.1016/S1546-5098(05)24002-5

comparison to communication in terrestrial habitats; further, fish occupy an unparalleled range of sensory environments, from the nearly lightless sound fixing and ranging (SOFAR) channel of the deep ocean to the bright, noisy shallows of a coral reef. Although several other taxa are able to perceive electrical fields, it is only fish that use electrocommunication. Finally, several model systems have proven amenable to genetic (*Poecilia, Xiphophorus,* Rift Lake cichlids) or neurobiological (*Opsanus, Astatotilapia,* and the weakly electric fish) analysis of communication systems.

In many ways, however, communication research has lagged in fish relative to other animals. The technical difficulties of studying fish in nature have dichotomised the field into observational work by snorkelers and scuba divers, and experimental work on fish in aquaria—often from selectively bred aquarium lines. Without more rigorous studies of wild fish, we may be overlooking much of the sophistication of fish communication systems. In addition, it is only recently that work on learning and individual differences (e.g. Wilson *et al.*, 1993; Lachlan *et al.*, 1998; Engeszer *et al.*, 2004) has refuted the dogma that all fish behaviour is stereotyped and "hard-wired" (reviewed in Laland *et al.*, 2003). New techniques and a new appreciation of the cognitive world of fish (see Chapter 1) should prove fruitful in bringing to bear the comparative and manipulative power of fish systems on nearly all aspects of communication.

Here we focus on discussing how communication signals are produced and the constraints they face in the range of sensory environments in which they are used. This chapter is by no means intended as an exhaustive survey of studies on fish communication. The large number of observational reports by aquarists and divers are given particularly scant treatment; these do, however, highlight how many interesting systems are candidates for future study.

A particular challenge in animal communication is distinguishing between a *signal* (a stimulus that has evolved in an explicitly communicative context) and a *cue* (a stimulus whose communicative function is incidental). Signals are easy to identify if they involve behaviours performed only in the presence of an appropriate receiver; many acoustic calls and motor displays fall into this category. Other stimuli, like sex hormones released during courtship, are more difficult to class. The situation is further complicated by the fact that some communication signals may have evolved in order to enhance transmission of a cue to a receiver. For example, close-range courtship motor displays may waft olfactory cues to the receiver. Once a signal has been produced, it must be transmitted to the receiver, who then must detect it against background, identify it, and respond. An effective signal must therefore be both detectable in the environment in which it is produced and distinct from signals that would elicit a different response from the receiver.

2. VISUAL COMMUNICATION

A disproportionate amount of work on fish communication has concentrated on vision. This is due to two principal factors: the striking diversity of form, colour, and pattern displayed by coral reef fish and many tropical freshwater fish; and, as Douglas and Partridge (1997) observed, the fact that fish occupy "almost every conceivable visual environment," from lightless caves and ocean canyons to blackwater streams and sunny reefs. Paradoxically, the fact that water is a poor medium for visual signals has made visual communication in fish a fruitful area of study, because this has imposed strong selection on the evolution of visual systems and signals.

Research on visual communication is both aided and hampered by the fact that humans are highly visual creatures; underwater, moreover, the other modalities that fish use are largely inaccessible to us without special instrumentation. The fallacy of "seeing is believing" has often resulted in a lack of rigor compared to studies of other modalities. With the exception of a growing body of quantitative studies on the colour component of communication, most work on visual communication in fish (and in general) does not take a quantitative approach to characterising the colour, spatial, or temporal properties of stimuli, backgrounds, or ambient light conditions, whether characterising signals themselves or evaluating receiver response (Rosenthal and Ryan, 2000). Nevertheless, the fact that visual stimuli can be readily characterised and experimentally manipulated through dummies, models, video, and computer animation (Rowland, 1999; Rosenthal, 1999, 2000), has given us substantial insight into this mode of communication in fish.

Although surprisingly few studies have explicitly isolated visual components of communication signals, fish have been shown to use visual communication extensively in the context of mate choice (Andersson, 1994; Rosenthal, 1999; Cummings *et al.*, 2003; Kingston *et al.*, 2003), agonistic displays (McKinnon and McPhail, 1996), as well as recognising conspecifics (Katzir, 1981), shoalmates (Turnell *et al.*, 2003), similar shoalmates (Engeszer *et al.*, 2004), and individuals (Balshine-Earn and Lotem, 1998). The characid *Hemigrammus erythrozonus* has a visual antipredator display (see Section 7). Deep-sea anglerfish use bioluminescent lures to snare prey (Hastings and Morin, 1991) and the monacanthid *Paraluteres prionurus* uses shape and colour patterns to mimic a toxic pufferfish (Caley and Schluter, 2003). We discuss the production mechanisms of both bioluminescent and reflected-light signals, and then turn to the transmission and perception of visual signals in the aquatic environment.

2.1. Production: Bioluminescence

The vast majority of visual signals, including those produced by most fish, involve modifying the distribution of light generated by the sun. Many marine fish, however, produce bioluminescent signals themselves, either endogenously or, more commonly, by concentrating bacterial symbionts inside specialised light organs (Hastings and Morin, 1991). Bioluminescence is particularly useful at night and in the deep sea, where exogenous light is scarce.

Bioluminescence has evolved independently about 13 times in teleosts, exclusively in marine taxa. Hastings and Morin (1991) estimated that 189 teleostean genera, or 4.9% of the total, contained luminescent species including the bristlemouths (genus *Cyclothone*), one of the most abundant vertebrates on Earth. Bioluminescence serves a variety of functions, including prey capture and camouflage (see below) and can play a role in shoaling and courtship.

Bioluminescence involves an enzymatic reaction whereby an oxygenase, luciferase, oxidises a luciferin, producing oxyluciferin and emitting light (Hastings and Morin, 1991). The system therefore requires a fresh supply of luciferin, which can be acquired through the diet (Mallefet and Shimomura, 1995) or synthesised endogenously. The resulting light typically has a peak wavelength of 500 nm (Hastings and Morin, 1991), close to the peak transmission of clear seawater and the peak spectral sensitivities of many fish visual systems.

Most bioluminescent fish, including all the known bioluminescent species in Lophiiformes, Gadiformes, and Beryciformes, rely on bacterial symbionts for light production. Many of these bacteria are facultative symbionts and are also found at high concentrations in the plankton (Nealson and Hastings, 1979). The design of light organs in fish with bacterial symbionts is constrained by their function as bacterial chemostats and as regulators of the bacteria's bioluminescent activity (Nealson, 1977). The bacterial light organ generally consists of a highly vascularised gland which supplies oxygen and nutrients to large numbers of bioluminescent bacteria, coupled with mechanisms that selectively occlude the light. Tightly packed tubules or highly convoluted epithelia typically provide surface area for bacteria (Herring and Morin, 1978). In other taxa, such as the midshipmen (Batrachoidae: *Porichthys* spp.), lanternfish, and some apogonids, the luciferin–luciferase system is produced endogenously (Hastings and Morin, 1991; Gon, 1996), with bioluminescence under neural control (LaRiviere and Anctil, 1984).

Fish have evolved a variety of structures to control the spectral, spatial, and temporal properties of emitted light. Both bacterial light organs and photophores can contain reflectors, filters, lenses, and light guides (Hastings and Morin, 1991). Species with bacterial light organs use "shutters" controlled by specialised musculature to create flashes of light. Lampeyes (Anomalopidae) can use both musculature derived from the levator maxilla to draw a black pigmented curtain over the surface of the organ, or rotate the entire organ into a suborbital pocket. The fish can cycle these movements up to 100 times per minute, creating a "blinking" effect which is used both for prey detection and intraspecific signalling (Hastings and Morin, 1991). In the trachichthyid genus *Paratrachichthys,* bacterial organs are located around the anus, opening onto the rectum. The light is distributed anteriorly and posteriorly by translucent, collimating muscles "piping" light to various areas of the body in a manner analogous to fiber optics (Herring and Morin, 1978).

Three deep-sea genera of loosejaws (Malacosteidae) couple a luciferin luciferase system, emitting typical blue-green light, to red fluorescent protein (Campbell and Herring, 1987) and a series of filters to produce far-red light (maximum emission >700 nm) from suborbital photophores (Widder *et al.*, 1984). This light is invisible to most other deep sea fish, providing them with a private channel with which to locate prey and signal to conspecifics. Loosejaws detect these wavelengths not using a long-wavelength sensitive opsin, but with a novel, long-wavelength photosensitiser, derived from dietary chlorophyll (Douglas *et al.*, 1999).

The light organs of ponyfish *Leiognathus splendens* (Leiognathidae) enclose bacteria within tubules in a glandular epithelium of an esophageal diverticulum. The light organ is silvered dorsally and half-silvered ventrally, so that it acts as a parabolic mirror to direct light onto the translucent musculature, where it is diffused over the ventral surface (Hastings and Morin, 1991). In some leiognathid species, multiple "shutters" on the wall of the light organ are used to mechanically control the duration and intensity of light emission. A large shoal of these fish was observed to emit a high-frequency, synchronised light display, collectively creating a highly conspicuous signal (Woodland *et al.*, 2002). The congener *L. elongatus* appears to be sexually-dimorphic, with females lacking a light organ. In natural shoals, luminous individuals appear to be socially dominant over nonluminous ones (Sasaki *et al.*, 2003).

Although bioluminescence affords the signaler the luxury of controlling the temporal characteristics of the signal, it has several limitations. Most bioluminescent signals are spatially simple and monochromatic; more important, they are fairly ineffective in brightly lit environments where vision plays a dominant role. Most visual signals, therefore, involve forms, patterns, colours, and movement that reflect available light.

2.2. Production: Reflected-Light Signals

Between the aquarium hobby, underwater tourism, and animated movies, the bright colour patterns of many fish account for a nonnegligible fraction of the world's disposable income. To our visually biased human

Umwelt, or perceptual universe, the remarkable colour patterns of coral reef fish are one of the most striking examples of flamboyance in nature. It is not only their conspicuousness that catches our eye, but also the profusion of colours and patterns across thousands of species. We are just beginning to understand the proximate and ultimate forces underlying this diversity of patterns.

Reflected-light signals are characterised by spectral reflectance as a function of space and time. Reflectance is defined as the light reflected back from a target illuminated by a uniform white light source, as a function of wavelength. Reflectance can change as a function of the angle of illumination, spatial position, and time. Many fish express complex colour patterns; over the course of a courtship or aggressive display, they vary the colour and intensity of these patterns and engage in changes in position, orientation, and posture. The amount of information that can be conveyed in a display is thus considerable.

The colour sensitivity of fish ranges from around 340 nm (the near ultraviolet) to 750 nm (the near infrared), with considerable variation in visual sensitivity among species. Fish produce colours across this spectrum (Marshall, 2000; Losey *et al.*, 2003). Spectral variation in signals is produced either by pigments, which absorb portions of the light visible to receivers, or by morphological structures that selectively scatter light as a function of wavelength. Pigments are typically used to generate long-wavelength colours (yellows, oranges, and reds), whereas morphological structures are used for shorter-wavelength colours (blues, indigos, violets, and ultraviolets).

2.2.1. Pigment-Based Colours

The chemical nature of a pigment determines the wavelengths of light it can absorb. Many pigments in fish are organic compounds containing long chains of conjugated double bonds, such as carotenoids and pterins. In general, small molecules absorb only high-energy (short) wavelengths, whereas larger molecules additionally absorb longer wavelengths of light. Carotenoids and pterins typically absorb shorter wavelengths, thereby reflecting yellows, oranges, and reds. Carotenoids bound to proteins can absorb middle wavelengths while reflecting both red and blue to produce purple. Melanins are large proteins that absorb most visible and ultraviolet light to produce black. Pigments are housed in chromatophores, within small packets called chromatosomes (Bagnara and Hadley, 1973).

Fish cannot synthesise carotenoids; they must be obtained from dietary sources. In freshwater systems, notably three-spine sticklebacks (*Gasterosteus aculeatus*) and guppies (*Poecilia reticulata*), males vary in carotenoid expression (Houde, 1997; Boughman, 2001). Female sticklebacks prefer redder males (Boughman, 2001). The intensity of red colour is determined by the concentration of astaxanthin carotenoids in the skin, whereas the chroma (colour purity) of red is determined by the ratio of astaxanthins to tunaxanthin (Wedekind *et al.*, 1998). Variation in astaxanthin deposition also appears to underlie the variation in the intensity and quality of orange colouration in guppies (Hudon *et al.*, 2003). The area of long-wavelength patches in guppies is highly heritable, whereas the intensity of colour patches depends on carotenoid uptake from the diet and on parasite load (Houde and Torio, 1992; Houde, 1997) and is correlated with male condition (Nicoletto, 1991, 1993).

Intriguingly, although long-wavelength colours in freshwater fish often appear to be limited by the availability of dietary carotenoids, this does not appear to be the case in marine fish. First, there is far less intraspecific variation between individuals in colour intensity in reef fish than has been reported for the aforementioned freshwater systems (G.G. Rosenthal, unpublished data). Second, many deep-water and nocturnal marine fish are bright red in colour, whereas fish in analogous freshwater niches are usually pale and unpigmented. Perhaps dietary pigments are more reliably available in marine environments.

2.2.2. Structural Colours

Short-wavelength (green to ultraviolet) colours are often produced structurally, using small (<300 nm) particles embedded into a transparent layer backed by melanin pigment. Short wavelengths are scattered (Rayleigh scattering) and reflected back to the receiver, whereas longer wavelengths are absorbed by the melanin. Adding filters to the transparent layer can further restrict the range of wavelengths being reflected (Bagnara and Hadley, 1973).

Interference colouration is likely to account for the shimmering iridescent colours found in many fish, notably poeciliids, cyprinodonts, and cichlids. This process involves a thin layer of transparent material with a high index of refraction coating the skin. Light hitting the surface at an angle is partially reflected to produce a primary reflection; much of the remaining light is refracted, reflected by the boundary at the bottom of the transparent material, then refracted again to produce a secondary reflection. The wavelengths for which these primary and secondary reflections are in phase depend on the viewing angle, the width of the transparent layer, and its refractive index. The apparent colour can thus change substantially with viewing angle (Denton, 1970). Ultraviolet (UV) colouration in birds is produced by coupling multilayer interference with a scattering mechanism (Finger *et al.*, 1992). Many fish produce signals that either reflect primarily in the UV (Figure 2.1) or show secondary ultraviolet peaks (Losey *et al.*, 1999), and it would be worthwhile to determine how these are produced.

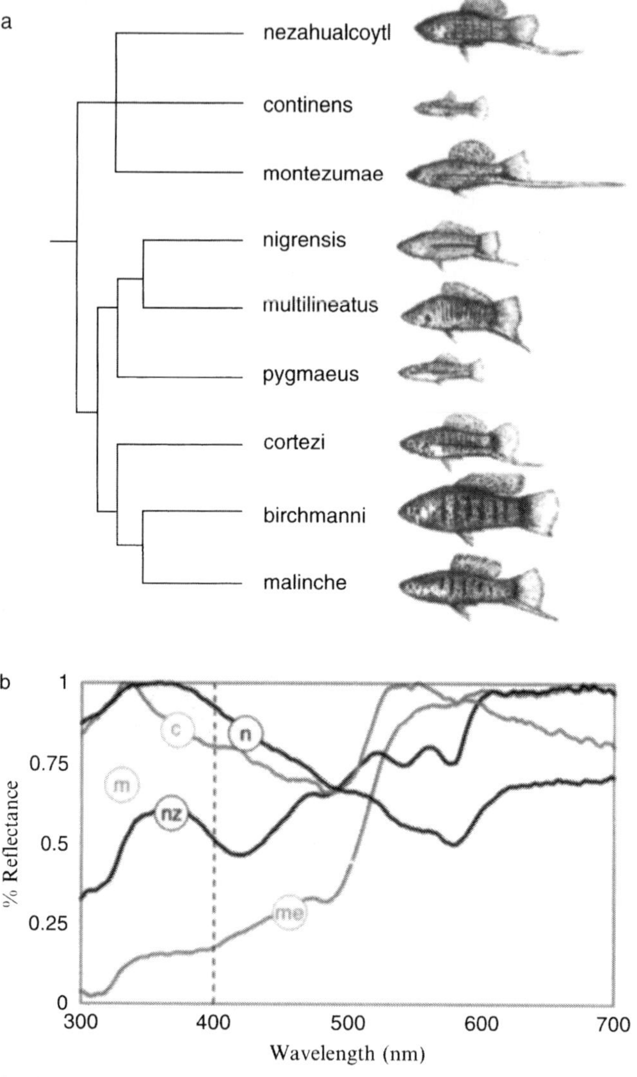

Fig. 2.1. Visual signal diversity in northern swordtails (Poeciliidae: genus *Xiphophorus*). (a) Diversity of body size and ornamentation in males. The sword ornament shows considerable interspecific variation and has been secondarily lost three times (Ryan and Rosenthal, 2001). (b) Normalised reflectance spectra of swords in male *X. nigrensis* (n), *X. multilineatus* (m), *X. cortezi* (c), *X. nezahualcoytl* (nz), and *X. malinche* (me). The proportion of reflectance in the ultraviolet (UV) is positively correlated with the density of UV-insensitive predators. Modified from Cummings *et al.* (2003).

White colouration is often produced by guanine, which forms microcrystalline platelets that reflect all wavelengths. If guanine platelets are packed in a dense array, they produce the specular (mirrorlike) reflectance characteristic of silvery scales. These dense arrays can be concentrated in iridosomes, which intensify pigment colours by reflecting wavelengths transmitted by chromatosomes (Bagnara and Hadley, 1973).

The spectral curves produced by fish colours can be characterised as simple (reflecting strongly in one region of the spectrum) or complex (with separate reflectance peaks in distinct portions of the spectrum) (Marshall, 2000). Simple colours are either step-shaped with high reflectance of all visible wavelengths above or below a cutoff point, or peak-shaped, reflecting within a particular range of wavelengths. Long-wavelength colours tend to be step-shaped; short-wavelength colours are peak-shaped (Marshall *et al.*, 2003). Complex spectral curves contain multiple peaks and/or steps. The purple expressed by labroids, particularly tropical wrasses and parrotfish, is one of the most complex animal colours, with distinct peaks around 400 and 580 nm and a step around 730 nm. Marshall (2000) suggested that this colour might be produced by combining several interference colours together.

The colours produced by fish are not randomly distributed across the spectrum. In a broad survey of Hawaiian reef fish, Marshall *et al.* (2003) found that peak-shaped colours were clustered around 400 and 500 nm, a region containing few step-shaped colours, and that the distribution of colours was clustered into seven to eight distinct peaks.

Many fish can change colour patterns over a range of timescales. Barlow (1963) identified nine different patterns in male *Badis badis* (Badidae) expressed according to social context and dominance status. Simultaneously hermaphroditic *Serranus subligarius* (Serranidae) alternate from a banded black-and-white male pattern to a solid black female pattern during a sex-changing courtship interaction (Demski and Dulka, 1986). Males in several species of swordtails (Poeciliidae, genus *Xiphophorus*) express conspicuous vertical bars on the flank, which they intensify over the course of courtship and agonistic interactions (Morris *et al.*, 1995). Colour change can operate over longer time scales as well. Male *Astatotilapia* ("*Haplochromis*") *burtoni,* for example, express a drab, femalelike pattern when subdominant and reproductively inactive, and a colourful, high-contrast, sexually-dimorphic pattern when dominant and reproductive (Fernald, 1990). These shorter-term, reversible colour changes are regulated via neural control of chromatophore cells, by aggregating or dispersing pigment granules within the cells (Bagnara and Hadley, 1973).

Short-term temporal changes in reflected visual signals can also be effected by raising or lowering the fins, as in many poeciliids, or by postural changes that change the appearance of interference colours. Many signal interactions—courtship displays, aggressive displays, and cleaning solicitation displays—involve characteristic motor patterns that may also be providing information in nonvisual modalities (see Section 7). In some cases, fish have evolved specialised motor control systems dedicated to communication. Male Siamese fighting fish *Betta splendens* (Osphronemidae) perform a dramatic frontal display to females and to other males, which involves a pronounced flaring of the gill covers. Both the musculoskeletal apparatus (Ma, 1995a) and motor control neurons (Ma, 1995b) appear to have been modified for this display. Female green swordtails, *Xiphophorus helleri*, attend specifically to the spatiotemporal visual characteristics of the male courtship display (Rosenthal *et al.*, 1996). Sticklebacks *Gasterosteus aculeatus* attend to variation in courtship tempo (Rowland, 1995). Despite the importance of temporal patterns in fish communication, we are not aware of any studies comparable to Fleishman's (1986) quantitative treatment of the dewlap displays of anoline lizards, in which Fourier analysis was used to quantify temporal conspicuousness relative to a naturally moving background.

Ontogenetic colour changes are also quite common. Many sex-changing reef fish (see also Chapter 8) have strikingly different colour patterns according to sex and social status. In the bluehead wrasse *Thalassoma bifasciatum* (Labridae), both females and initial-phase males have yellow and black horizontal stripes; terminal males, which are derived from either females or initial-phase males, have completely different colours and spatial patterns: a blue head, black-and-white vertical bars, and a green posterior (Warner, 2001). Many pomacentrids and pomacanthids also have conspicuous juvenile colour patterns strikingly different from the adult phase.

The mechanisms underlying such ontogenetic changes, as well as the development of spatial patterns in general, are poorly understood. Turing (1952) proposed a hypothetical molecular mechanism, the reaction-diffusion system, whereby spot-and-stripe patterns arise as a result of instabilities in the diffusion of multiple morphogenetic chemicals in the skin during early development. Differences in boundary and initial conditions, and in the number and diffusion properties of each morphogen, can interact to produce an array of stripe-and-spot patterns.

Kondo and Asai (1995) and Painter *et al.* (1999) showed that a reaction-diffusion system was consistent with the development of the body pattern of juvenile angelfish (*Pomacanthus* spp.), in which the body pattern is not fixed on the skin over ontogeny; rather, the size of the black melanophore colouration between yellow stripes is maintained constant, by continuous rearrangement of spatial patterns. Shoji *et al.* (2003) further suggested that a reaction-diffusion system could act in concert with small differences in diffusion anisotropy of morphogens (caused by structures with directional conformation such as scales) to induce directionality in stripe patterns. A molecular mechanism conforming to the reaction-diffusion model has yet to be identified, although Asai *et al.* (1999) argued that the zebrafish *leopard* gene may be acting as a component of a reaction-diffusion system.

Parichy (2003) provides a concise review of the developmental genetics of stripe formation in *Danio* (Cyprinidae). In the zebrafish *D. rerio*, blue stripes consist of a mixture of xanthophores and melanophores; white stripes are a mixture of xanthophores and iridophores. Although there is a fairly detailed picture of the mechanisms whereby pigment cells are recruited and aggregated, the forces underlying the spatial arrangement of stripes remain unclear. The lateral line system might be involved in recruiting stripes, as it is in salamander larvae; the zebrafish mutant *puma* exhibits defects in both lateral line and stripe development (Parichy, 2003).

Regular horizontal or vertical stripes are only a tiny subset of the spatial patterns produced by fish. Many species, like the aptly named Picasso triggerfish (Balistidae: *Rhinecanthus rectangulus*) are characterised by complex, aperiodic patterns that are highly conserved between individuals. The planktonic larval stage and long generation time of most reef fish make them intractable as models in developmental biology. The West African killifish of the subfamily Aplocheilinae (*Aphyosemion, Fundulopanchax*, and allied genera), in contrast, are easy to rear in the laboratory, express highly diverse, repeatable colour patterns (Amiet, 1987), and could prove an interesting model for mechanisms of pigment pattern development.

2.3. Visual Signals, the Visual Environment, and Visual Perception

For reflected-light signals, production and propagation entirely depend on ambient light levels and properties of the physical environment. Underwater, light quality and quantity varies with depth, water chemistry, and biotic factors like plankton density (Nybakken, 1977). Visually-oriented fish are thus an ideal system for studying environmental effects on communication (Lythgoe, 1988). If a fish cannot make its own light, it must work with the available light environment to produce communication signals. Strong associations between transmission characteristics of the environment, visual sensitivity, and body colour have been demonstrated in a variety of fish systems (Endler, 1991, 1992; Endler and Houde, 1995; Boughman, 2001; Cummings and Partridge, 2001). In haplochromine cichlids, eutrophication of clear lake water masks the apparent colour differences between two recently diverged species, promoting hybridisation (Seehausen *et al.*, 1997).

Microhabitat differences can drive sensory systems and thereby speciation. The adaptation of fish visual systems to divergent visual environments was the empirical cornerstone of Endler's (1992) "sensory drive" hypothesis, whereby tuning to an environment would drive receiver perception, signal production, and preference for a signal. Boughman (2001) showed that female sensitivity to red in sticklebacks correlated with the abundance of red light in the environment and with female preference for red versus black males, producing divergent selection for red and black males in divergent environments.

Perhaps more than other modalities, visual systems are fairly broadly tuned, with substantial overlap in the sensitivities of most animal visual systems. Most animals are sensitive to wavelengths between 400 and 700 nm and to the spatial frequencies of objects in their environment (Rosenthal and Ryan, 2000). In some cases, this has allowed signalers to exploit fairly permissive visual biases on the part of receivers. In swordtails, genus *Xiphophorus* females exhibit an ancestral preference for a conspicuous caudal ornament, the sword (Basolo, 1990, 1995). The preference appears to be a special case of a more general preference for large apparent size (Rosenthal and Evans, 1998). The signal is, however, exploited by predators (the Mexican tetra *Astyanax mexicanus*, Charicidae) with the same broad visual biases (Rosenthal *et al.*, 2001). In highly predated populations, female swordtails reduce their preference for the visible-light signals reflected by the sword (Rosenthal *et al.*, 2002).

The visual spectrum is not universally shared among receivers, however. The ultraviolet (UV; 320 to 400 nm) is the one area where visual sensitivity varies substantially across fish taxa, providing the opportunity for a degree of signal privacy. Siebeck and Marshall (2001), for example, found that only about 30% of reef-dwelling fish were sensitive in the ultraviolet. Swordtails are sensitive to UV wavelengths, whereas *A. mexicanus* are not. Male swordtails in highly predated populations exhibit bright UV stripes, which females attend to in mate choice, but which are relatively inconspicuous to predators (Figure 2.1; Cummings *et al.*, 2003). Similarly, Endler (1991) showed that male guppies court at times of day and in microhabitats that minimised visual conspicuousness to invertebrate predators while retaining conspicuousness to females.

3. ELECTRICAL COMMUNICATION

Electrocommunication and electrical signaling are reviewed in detail in excellent, short reviews by Hopkins (1999a,b) and Zakon (2003) and also mentioned in Chapter 3, Section 3.1. Therefore, only abbreviated treatment is given to this topic here. Electroreception is an ancestral trait in vertebrates, shared by lampreys, elasmobranchs, and sarcopterygians, and appears to have evolved secondarily in several fish and tetrapod lineages

(Bullock and Heiligenberg, 1986). Generating electric signals, however, is much more restricted and is unique to fish. Weakly electric fish, which include many marine skates, gymnotiforms, mormyriforms (Bennett, 1970), and stargazers (Uranoscopidae; Pickens and McFarland, 1964), produce low-voltage electric organ discharges (EODs) for communication and electrolocation. Strongly electric fish include the electric eel *Electrophorus electricus*, a gymnotid; electric catfish (Malapteruridae); and torpedo rays (Torpedinidae). These fish produce high-voltage EODs for prey capture, with some fish possessing a second, low-voltage organ for communication (Bennett, 1970). The vast majority of work on electrocommunication has focused on the gymnotiforms and mormyriforms, which live in murky water and use EODs (and often, acoustic signals) for courtship, species and individual recognition, and agonistic interactions (Zakon, 2003). Electrocommunication is arguably the best-understood sensory modality in fish, and has served as a canonical system in neuroethology. Electrical signals are "simple, mathematically describable, and already coded in the currency of the nervous system" (Zakon, 2003) and the circuits involved in electroreception are relatively straightforward. Furthermore, the diversity of gymnotiforms and mormyriforms has permitted a host of comparative studies.

3.1. Electrical Signal Production

Generation of electrical signals is a straightforward extension of the ability of excitable cells to generate action potentials. Electrocytes are modified muscle cells (one gymnotiform family, the Apteronotidae, has modified nerve cells) stacked in insulated columns (Bennett, 1970). The differentiation of electrocytes from muscle tissue is dependent on neural input (Unguez and Zakon, 2002).

EOD generation is described in Zakon (2003). The EOD is driven by cells in the pacemaker nucleus, a midline medullary nucleus; the output axons of the pacemaker nucleus synapse onto a pool of spinal motoneurons, the electromotoneurons. The frequency of the EOD is determined by the firing frequency of the pacemaker neurons. There are two broad classes of EODs: *pulses*, found in most electrocommunicative fish, are very short duration (1 to 3 ms) multiphasic EODs; and *waves*, sinusoidal signals lasting 3 to 5 ms, generated by apteronotid and sternopygid gymnotiforms (Hopkins, 1999a). Several wave-type species produce "chirps," rapid, transient increases in discharge frequency that are likely used in aggressive interactions (Dunlap *et al.*, 1998). These are typically associated with male–male encounters, but Tallarovic and Zakon (2002) have found that females produce short chirps in agonistic interactions with other females and emit long chirps in the presence of males, perhaps to advertise status or reproductive condition.

An excellent review by Hopkins (1999a) reviews the evolution, ontogeny, and design features of EODs and their underlying mechanisms.

Like sexually-dimorphic signals in other modalities, the expression of EODs can be modulated by sex steroids and corticosteroids (reviewed in Zakon, 1996, 2003), principally by varying the kinetics of electrocyte sodium and potassium channels and by acting on the pacemaker nucleus. These effects can vary substantially among species. Dunlap *et al.* (1998; Zakon and Dunlap, 1999) studied two wave-producing apteronotids differing in both the direction of the sexual dimorphism of the EOD and in the propensity to emit aggressive "chirp" signals. Brown ghost (*Apteronotus leptorhynchus*) males produce higher frequency EODs than females, and males chirp at a higher rate than females. Black ghost (*A. albifrons*) males discharge lower-frequency EODs than females, and both sexes chirp at equal rates. Accordingly, androgens raise EOD frequency in brown ghosts, but lower it in black ghosts; androgens induce chirping in female brown ghosts but not in black ghosts (Dunlap *et al.*, 1998; Zakon and Dunlap, 1999).

Sexual dimorphism can be associated with constraints on signal expression. The duration of the EOD of the South American pulse fish *Brachypomus pinnicaudatus* is twice as long in males as it is in females, at a cost of a corresponding drop in amplitude. Males in several species compensate for the amplitude loss by growing longer tails; as with the swordtails discussed in Section 2.2 (Rosenthal *et al.*, 2001), longer-tailed males are targets for predation (Hopkins *et al.*, 1990). Females prefer to mate with males producing longer-duration EODs (Hopkins *et al.*, 1990).

The various mechanisms modulating EOD polarity, frequency, duration, and structure have permitted a spectacular array of species-specific EODs in murky water to parallel the visual diversity found on coral reefs (reviewed in Hopkins, 1999a). Northeastern Gabon, for example, boasts a species flock of 21 mormyrids, each with a species-typical EOD (Figure 2.2; Hopkins, 1986). Morphologically similar species may be readily distinguished on the basis of their EOD (Hopkins, 1999a). In some species, complex multiphasic signals (like the triphasic signal of the mormyrid *Pollimyrus adspersus*) permit reliable recognition of individuals (Paintner and Kramer, 2003).

3.2. Transmission of Electrical Signals

Like visual signals, electrical signals are transmitted from signaler to receiver as a nonpropagating electric field, decreasing as the inverse cube of the distance from the emitter (Hopkins and Westby, 1986). This effectively limits the active space of a 10 to 20 cm long weakly electric fish to an ellipsoidal radius of about 1 m (Hopkins, 1999b). Signaling is also constrained by the conductivity of the medium. Hopkins (1999b) showed that species in the gymnotid genus

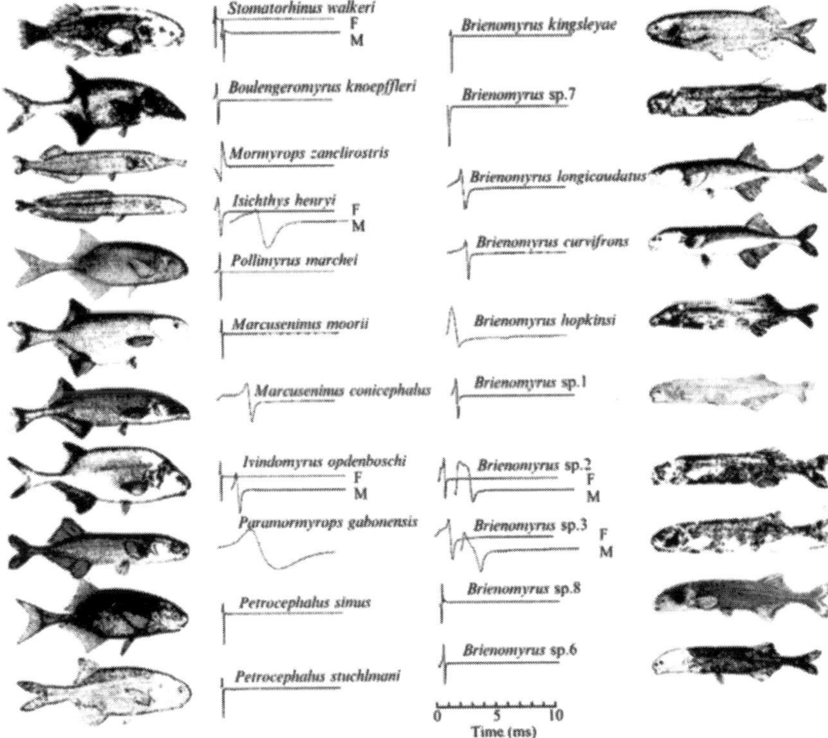

Fig. 2.2. Diversity of electric organ discharges (EODs) in mormyrid fish of northeastern Gabon. The fish are arranged by genus. Each EOD is plotted on the same time base with head-positivity upwards. The EODs vary according to three different dimensions: overall waveform, polarity, and duration. Waveforms can be monophasic, biphasic, triphasic, or tetraphasic and can show a variety of inflection points and plateaus. Note sexual dimorphism in some male (M) versus female (F) EODs. (Reprinted from Hopkins, C. D. Design features for electric communication. *J. Exp. Biol.* **202**, 1217–1228, with permission from the Company of Biologists Ltd.)

Brachypopomus are distributed according to water conductivity. Species in low-conductivity (high-resistance) water have three columns of electrocytes on each side of their caudal filament, whereas species in higher-conductivity have four or five. Hopkins (1999b) showed that this functions to match the impedance of the electric organ to local conductivity, and argued that differences in habitat conductivity might form a barrier to gene flow. The lack of electrocommunication in saltwater fish may be due to constraints on the amount of energy that can be devoted to electrocytes in a very high-conductivity environment. The energetic costs of producing EODs may account for the diel variation seen in EOD discharges, with substantial increases at night (when these fish are active) and in social contexts (Hopkins, 1999b).

Although active space is limited by the fact that electrical signals do not act as propagated waves, this does prevent signals from being degraded by reflection, refraction, and reverberation. The temporal fine structure encoded in species-typical signals can thus be reliably detected by receivers (Hopkins, 1999b).

3.3. Perception of Electrical Signals

Ampullary electroreceptors are broadly shared by electrosensitive fish and amphibians, and are primarily used for navigation and localising prey. They are sensitive to extremely low amplitudes (0.01 microvolts in marine species) of low-frequency electrical fields (8 to 30 Hz; Stoddard, 1999). Tuberous electroreceptors, in contrast, have evolved independently in the gymnotiforms and mormyriforms. These are at least two orders of magnitude less sensitive than ampullary organs, but respond to the higher frequencies characteristic of EODs (50 to 3,000 Hz; Stoddard, 1999). Mormyriforms have two types of tuberous electroreceptors: mormyromasts, which are amplitude coders used in electrolocation; and knollenorgans, which lock on to the temporal pattern of EOD discharge and are solely devoted to communication (Hopkins, 1986). Just as the complexity of the inner ear appears to have influenced speciation rates in frogs (Ryan, 1986), the presence of an organ dedicated to communication may facilitate divergence of communication systems in these fish. In many species, tuberous receptor sensitivities and EODs are narrowly tuned to one another (Heiligenberg, 1977).

Fish in three gymnotiform families have modified the ancestral monophasic wave signal, adding a second-wave phase that shifts the emitted spectrum above the peak sensitivity of the major electroreceptive predator, the electric eel *Electrophorus electricus*, which lacks a tuberous electroreceptive system (Stoddard, 1999). Sexually mature males in some species further amplify and extend this second phase. Predation thus appears to have, somewhat paradoxically, selected for greater complexity and diversification of signals (Stoddard, 1999). As do swordtail fish with ultraviolet swords (Section 1; Cummings *et al.*, 2003), these fish appear to have a degree of signal privacy from predators.

The evolution and diversification of electrocommunication shows interesting parallels with other modalities. In addition to the relative ease involved in characterising and playing back electrical signals, work on electrocommunication has been helped by the fact that the sense is utterly alien to us. We have no choice but to take a fish's-eye view of the production and perception of these signals, and we lack the vocabulary to describe them using anything but the rigorous, universal lexicon of physics. We would do well to sometimes "blind" ourselves and take this approach to other signaling modalities.

4. MECHANICAL COMMUNICATION

The lateral displays, tail beats, and "quivering" characteristic of many agonistic and courtship signals produce low-frequency mechanosensory cues (Nelissen, 1991; Bleckmann, 1993; Sargent *et al.*, 1998) detectable by the lateral-line system of the receiver (Bleckmann, 1993). The water displaced by these close-range motor displays is certain to produce mechanosensory stimulation.

Mechanosensory communication in fish has received scant attention, but there is compelling evidence to suggest that mechanosensory cues may play a critical role in courtship behaviour. An elegant pair of studies by Satou *et al.* (1994a,b) showed that lateral-line stimulation by courting female Hime salmon (*Oncorhynchus nerka*) was necessary to elicit a spawning response in males. Males spawned with an artificial model female only if it included both vibrational and visual cues (Satou *et al.*, 1994a). Further, males failed to spawn with the model if the function of their lateral line hair cells was blocked with cobalt (Satou *et al.*, 1994b).

Sargent *et al.* (1998) reported preliminary findings suggesting that the frequency of particle acceleration generated by the male courtship display of the green swordtail, *Xiphophorus helleri*, fell within the peak range of lateral line sensitivity (20 to 50 Hz). Female *X. helleri* attend to specific visual components of the courtship display (Rosenthal *et al.*, 1996) and lateral-line stimulation may also elicit female preference.

The ubiquity of close-range aggressive and courtship motor patterns in fish indicate that lateral-line signals are likely to play an important role in fish communication, particularly at times or in habitats where the visual component of these displays is difficult to detect. High-speed particle velocimetry (Kanter and Coombs, 2003) makes it feasible to measure mechanosensory cues. Recent studies on frogs (Narins, 2001) and hemipteran insects (Rodríguez *et al.*, 2004) have successfully employed synthetic playback of mechanical stimuli to identify behaviourally salient features. It is anticipated that these approaches can be applied fruitfully to fish communication.

5. CHEMICAL COMMUNICATION

Olfaction is highly developed in many fish, and the importance of chemical information to social interactions among fish has long been recognised. An overview of the fish olfactory system is given in Chapter 9, Section 3. Fish use olfactory cues to recognise conspecifics (Chapter 5, Section 5; McKinnon and Liley, 1987; Crapon de Caprona and Ryan, 1990; McLennan and Ryan, 1997, 1999), potential mates (MacGinitie, 1939; McLennan, 2003), offspring

(Künzer, 1964; Myrberg, 1966), and individuals (Todd *et al.*, 1967; Göz, 1941, cited in Bardach and Todd, 1970). Water-soluble sex hormones released by an individual can produce dramatic changes in the behaviour and physiology of receivers (Chapter 9; Liley and Stacey, 1983; Dulka *et al.*, 1987; Stacey *et al.*, 2003). It is interesting to note that male half-naked hatchetfish *Argyropelecus hemigymnus* (Sternoptychtidae) have the biggest nose, relative to body size, of any vertebrate. Jumper and Baird (1991) argued that this permitted these bathypelagic fish to localise mates despite their extremely low densities.

In general, very little is known about either signaling behaviour or signal structure in fish olfactory communication. A notable exception is chemical alarm cues, or *Schreckstoff*, which are discussed in Chapter 3. In nearly every case, it is difficult to ascertain whether chemical stimuli produced by fish are pheromones—chemical signals that have evolved in the context of communication among individuals—or simply the incidental byproducts of physiological processes. This difficulty arises both because of the nature of chemical signaling in water, and because of the structure of chemical stimuli themselves (see also Chapter 9, Section 2).

Many terrestrial animals expel pheromones from exocrine glands, and perform characteristic behaviours associated with the deposition or release of scent (Bradbury and Vehrencamp, 1998). Exocrine glands, and particularly specialised scent glands, are not widely reported in fish, nor are behaviours unambiguously associated with pheromone release. In addition to the club cells releasing alarm pheromones, some noteworthy exceptions suggest that these may in fact be more common than suspected in fish. In the internally-fertilising glandulocaudine characins, males have epidermal glandular tissue on both sides of the caudal peduncle, covered by caudal fin scales modified to form an apparent "pheromone pump" (Figure 2.3; Weitzman and Fink, 1985; Bushmann and Burns, 1994). Another group of characins, the tribe Compsurini of the subfamily Cheirodontinae, also has internal fertilisation and specialised caudal glands (Malabarba, 1998). Male peacock blennies, *Salaria pavo*, have an anal gland that releases pheromones attractive to females (Barata *et al.*, 2002).

Increased attention to chemical communication in fish will surely yield more examples of specialised chemical-release structures. Fish can also release chemical cues through the gills or through the urine. Rainbow trout (*Oncorhyncus mykiss*, Salmonidae) release urine in periodic bursts (Curtis and Wood, 1991) and urine release is affected by social status (Sloman *et al.*, 2004). Fish might thus be able to control the release of urineborne pheromones.

Chemical stimuli used by aquatic animals are typically water-soluble substances released directly into the water column (Atema, 1988), rather than deposited on the substrate as in terrestrial systems (Bradbury and

a

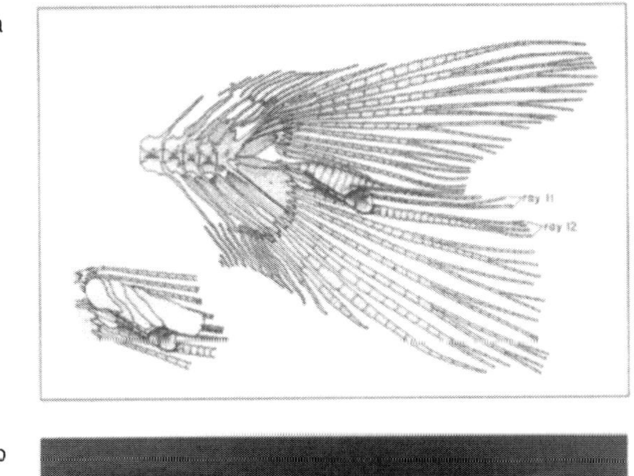

b

Fig. 2.3. Pheromone pump in the glandulocaudine characin *Mimagoniates lateralis*. (a) Caudal skeleton showing modified caudal fin rays 11 and 12. Modified caudal squamation is shown at left. (b) Caudal fin and glandular region of live adult male. Arrow indicates the pheromone pump. (Adapted from Weitzman, S. H., Menezes, N., and Weitzman, M. J. (1988). Phylogenetic biogeography of the Glandulocaudini (Teleostei: Characiformes, Characidae) with comments on the distributions of other freshwater fish in eastern and southeastern Brazil. *In* "Neotropical Distribution Patterns: Proceedings of a 1987 Workshop" (Heyer, W. R., and Vanzolini, P. E., Eds), pp. 379–427. Academia Brasilerra de Ciencias, Rio de Janeiro, with permission from the Smithsonian Institution, NMNH, Division of Fish.)

Vehrencamp, 1998). Diffusion transmission, widely used among small terrestrial animals, is 10,000 times slower in water than in air. This has the effect of dramatically reducing the maximum size of the active space within which a receiver can detect a signal (Bradbury and Vehrencamp, 1998). Fish must therefore rely on currents for signal transmission, using either currents present in the environment or generating currents through gill ventilation, finning, body movements, or urine release (cf. Atema, 1988). This produces "flavored eddies," which contain both mechanosensory and chemical information (Atema, 1996). The perception and localisation of olfactory cues thus depends on both the concentration and chemical structure of the signals, and the hydrodynamics of these eddies.

Our understanding of olfactory communication is frustrated by the fact that olfactory signals in fish remain largely a "black box" in terms of chemical structure, with a few notable exceptions including hypoxanthine-3 (N)-oxide as a component of alarm pheromone. Many chemical cues also function as endogenous signals (Carr, 1988). Sex hormones—steroids and prostaglandins—are important chemical cues in fish (Chapter 9; Stacey *et al.*, 2003). In some cases, released steroids appear to have a direct physiological effect, priming receivers for reproduction. Male goldfish exposed to female sex pheromone increase sperm production (Stacey *et al.*, 2003), and the pheromone released by male blue gouramis *Trichogaster trichopterus* induces female ovarian maturation (Degani and Schreibman, 1993).

Sex hormones function endogenously in signalers; their function as pheromones is therefore constrained by their function as hormones. Signalers may have some control as to the timing, context, and quantity of release of hormone cocktails. Yet the evolution of the chemical structure of these compounds, their ratios, and their concentrations is likely to be determined in large part by their physiological effects on the female reproductive system. It remains to be determined to what extent waterborne hormones are acting as genuine signals, rather than as cues incidentally released by reproductive individuals (Sorensen and Scott, 1994).

A series of studies by McLennan and Ryan (1997, 1999) showed that female response to male olfactory cues was a function of phylogenetic distance. Female swordtails *Xiphophorus nigrensis* responded to cues produced by the closely related *X. cortezi*, but not to the more distantly related *X. montezumae*. Females were therefore not only responding to species-specific cues, but responding differently to cues produced by males of more and less distantly related species. Moreover, response was asymmetric: *X. montezumae* females responded to *X. nigrensis* cues, but not vice versa (McLennan and Ryan, 1999).

Chemical cues appear to be both necessary and sufficient for species recognition in some *Xiphophorus* species: Crapon de Caprona and Ryan (1990) showed that female *X. pygmaeus* preferred male *X. nigrensis* based on chemical cues, but preferred conspecifics when either olfactory cues alone or both types of cues were provided. Male olfactory cues and female olfactory responses appear to be coevolving, with the male odorant evolutionarily labile and complex enough to encode species identity.

Chemical information can encode individual identity. Conditioning experiments on a single *Phoxinus* minnow (Göz, 1941) and on yellow bullhead catfish *Ictalurus natalis* (Todd *et al.*, 1967) showed individual recognition of conspecifics based on chemical cues (reviewed in Bardach and Todd, 1970). Fathead minnows *Pimephales promelas* use chemical cues to discriminate their shoalmates from unfamiliar individuals (Brown and Smith, 1994).

Reusch *et al.* (2001) showed that female three-spine sticklebacks (*Gasterosteus aculeatus*) use olfactory cues to select males with the highest diversity of major histocompatibility complex (MHC) alleles. Salmonids also appear to use olfactory cues to discriminate by MHC haplotype (Olsen *et al.*, 1998) and relatedness (Brown and Brown, 1995). The nature of the chemicals involved, and their dependence on other processes within the signaling individuals, remains to be determined in all these systems.

Except for the sex hormones described above, very little is known about constraints on olfactory signal production in fish. Female response to male olfactory cues in fathead minnows *Pimephales promelas* was dependent on male reproductive condition (Cole and Smith, 1992), and release of the sex hormones discussed above depends on reproductive state. Research on other taxa and in other modalities suggests that the production of chemical cues may also depend directly or indirectly on diet, social status, parasite load, and predation risk.

Neither the ontogeny of olfactory communication nor the properties of the signaling environment have received much attention in fish. Work on olfactory homing in salmon (Chapter 7; Scholz *et al.*, 1976; Nevitt *et al.*, 1994) suggests that individuals might imprint on chemical cues produced by conspecifics or relatives (see also Chapter 1, Section 5.4.1). Mirza and Chivers (2003) showed that *Pimephales promelas* can learn heterospecific alarm cues when exposed to those cues in the diet (i.e., via the feces and the urine) of a major predator, the northern pike *Esox lucius* (Chapter 3, Section 3.4).

In addition to current regimes, ambient water chemistry might constrain the production and perception of olfactory signals. High concentrations of humic acid, a ubiquitous constituent of lakes and streams, inhibit olfactory response to 17,20 beta-dihydroxy-4-pregnen-3-one (17,20 β-P) in goldfish, possibly via the humic acid chelating the steroid molecule (Hubbard *et al.*, 2002). Evidence suggests that female recognition of male conspecifics in the swordtail *Xiphophorus birchmanni* is masked by high concentrations of humic acid. The effect persists for several days when females are retested on male pheromones produced in clear water, suggesting that humic acid is acting on olfaction rather than on the chemical signal (H. S. Fisher and G. G. Rosenthal, unpublished data).

Brown *et al.* (2002) showed that acidifying freshwater pH to 6.0 from the natural value of 8.0 produced a loss of alarm response in two cyprinids, finescale dace (*Phoxinus neogaeus*) and *Pimephales promelas*. They failed to respond after acidified stimuli were buffered to pH 7.5, suggesting that the pheromone had undergone a nonreversible covalent change. Smith and

Lawrence (1998), however, found that decreasing pH from 7.0 to 5.0 did not affect response to alarm substance in *P. promelas.* Brown *et al.* (2002) argued that Smith and Lawrence's (1988) failure to find an effect might reflect increased sensitivity to alarm pheromone in their study population.

Just as fish occupy a host of visual environments, they occupy a broad range of physicochemical environments. Both the biochemical "soup" they live in (substances used by other organisms to communicate, metabolic byproducts) and their abiotic environment (dissolved gases and minerals, salinity, pH, temperature) are sure to affect the efficacy of chemical signals and their ability to discriminate between them. Knowledge of the chemistry of signals would shed light on the selective pressures acting on chemical communication in fish.

In summary, there is substantial knowledge surrounding the perception, production, and function of chemical stimuli in fish. Much of what is known from a functional standpoint is from natural systems where the underlying chemistry is unknown, and much of what is known from a mechanistic standpoint is from domestic strains where little can be inferred about communication from an ecological or evolutionary point of view. The fathead minnow *Pimephales promelas*, has an alarm pheromone of known structure and has been the subject of numerous studies on both behavioural ecology and behavioural mechanisms of alarm communication. Studies of chemical signaling would benefit from more systems providing both a clear picture of how olfactory communication evolves, and integrates mechanistic information about signal production and perception into a comparative and behavioural-ecological context. The glandulocaudine characins, a speciose, somewhat aquarium-friendly group with specialised scent glands (Weitzman and Fink, 1985) may provide a satisfying entry into the integrative study of chemical communication in fish.

6. ACOUSTIC COMMUNICATION

It has long been suggested that some fish might use sound to communicate (Darwin, 1874). The sonic behaviours of many fish, however, have often gone unnoticed (Lobel, 2001a), even in systems where other aspects of communication have been extensively studied such as haplochromine cichlids (Lobel, 1998; Ripley and Lobel, 2004) and *Hypoplectrus* [Serranidae] (Lobel, 1992). This has mainly been due to the technical difficulties and expense associated with making quality acoustic recordings underwater. Recent technological innovations, however, have invigorated interest in fish acoustic communication (e.g., Lobel, 2001b; Rountree *et al.*, 2002). For reviews of fish bioacoustics and communication, see Fine *et al.* (1977a), Myrberg (1980, 1981), and Hawkins and Myrberg (1983).

6.1. Sound Production

Fish can produce sounds using three distinctive mechanisms: fast muscles beating on the swimbladder, stridulating bones, and hydrodynamic sounds (Tavolga, 1971). In general, sounds are either long single pulses, a series of rapid short pulses, or tones (Tavolga, 1971). Among other taxa, sound production has been well-studied within two labroid families, the marine damselfish (Pomacentridae) and the freshwater cichlids. Sound production in cichlids and pomacentrids is widespread throughout the families, with eight of the recognised 14 pomacentrid genera and eight of the 20 cichlid subfamilies reported to produce sound (reviewed in Lobel 2001a; Rice and Lobel, 2004). Although the contexts of sounds and the general characteristics of sounds produced are similar in cichlids and pomacentrids, there are several important acoustic differences (Rice and Lobel, 2004). Pomacentrids have a call composed of a series of pulses wherein successive pulses increase in amplitude rapidly from the first few pulses within a call (Lobel and Mann, 1995). Cichlids also produce a call consisting of a series of pulses, but exhibit amplitude modulation between those pulses within a call (Lobel, 1998, 2001a). The sounds from both families show a linear correlation between numbers of pulses per call and call duration, suggesting a fixed mechanism that is responsible for sound production (Rice and Lobel, 2004). The structures of a typical pulsed sound pattern are shown in Figure 2.4.

Other fish produce sound patterns with varying structure. Hydrodynamic sounds are broadband noise patterns. Stridulatory sounds have much wider frequency ranges and often more irregular pulse patterns than basic swimbladder sounds. Swimbladder sounds can be expressed as a series of pulse, tones, or broadband patterns (Fish and Mowbray, 1970; Fine *et al.*, 1977a; Lobel, 2002).

Among teleosts with simple swimbladders, larger fish generally produce lower frequency sounds than smaller fish (e.g., Fine *et al.*, 1977a; Rowland, 1978; Myrberg *et al.*, 1993). This arises from the fact that larger swimbladders resonate at lower frequencies than smaller ones (Clay and Medwin, 1977; Urick, 1983). In contrast, some fish, such as toadfish (batrachoids), do not show a relationship between body size and fundamental frequency (Fine, 1978; Waybright *et al.*, 1990). In these systems, sounds may represent a forced response determined by contraction and relaxation of the sonic muscle (Fine *et al.*, 2001; Connaughton *et al.*, 2002).

The typical sound produced by most fish is a series of rapid low frequency (typically 200 to 600 Hz) pulses (Figure 2.4). Myrberg *et al.* (1965) noted that the duration of an aggressive sound was positively correlated with the intensity of the behavioural interaction. A simple comparison of the pulse rate in the courtship calls of two sympatric species *Copadichromis*

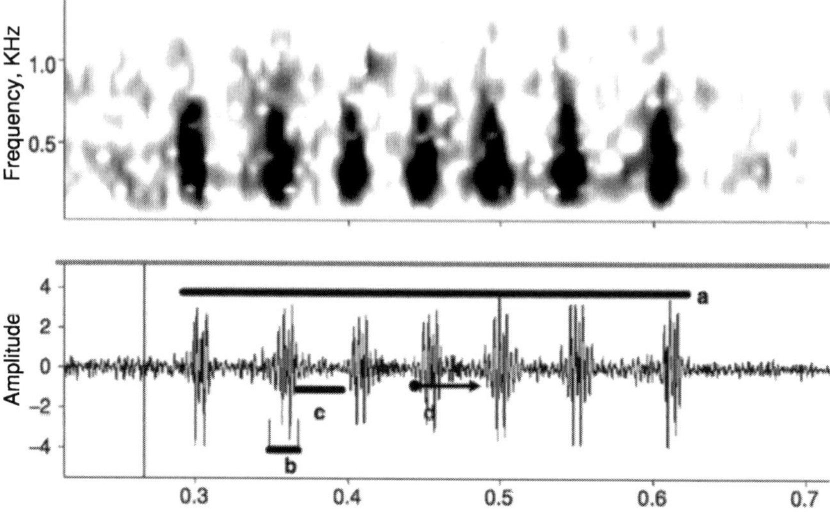

Fig. 2.4. Basic structure and definitions of a typical pulsed sound pattern produced by many species. This example is from a damselfish (*Dascyllus albisella*, Pomacentridae). Top panel is a sonogram showing individual pulses plotted as frequency (Hz) versus time with sound intensity indicated by darker shading. Bottom panel is an oscillogram showing sound amplitude versus time. (a) Overall call duration. (b) Individual pulse width. (c) Interpulse interval. (d) Pulse period.

conophorus and *Tramitochromis intermedius* in Lake Malawi revealed statistically significant differences in pulse rate (Lobel, 1998, 2001a). These data support the hypothesis that cichlid sounds are temporally structured in a way that contains information that could be used in the mate selection process. This type of acoustic coding is common in insects and amphibians and has been proposed for fish as well (Winn, 1964; Myrberg *et al.*, 1978; Spanier, 1979; Crawford, 1991). Of course, fish do produce other sound patterns that are not just simple pulse series. For example, tonal type sounds are produced by batrachoids, triglids, mormyrids, and ostracids; a broadband long duration sound is produced by the small serranid, *Hypoplectrus* spp.

6.2. Propagation and Detection of Sound

To our knowledge, there have only been two studies to date on the propagation of natural fish sounds in their natural environment (Horch and Salmon, 1973; Mann and Lobel, 1997) and one study that used playback of sounds to determine attenuation and possible range (Fine and Lenhardt, 1983). From both field and laboratory measurements, the active space of fish sounds ranged from 0.5 m for gobies (Lugli and Fine, 2003; Lugli *et al.*, 2003), about 3 m for toadfish (Fine and Lenhardt, 1983), and 7 to 12 meters for the damselfish and squirrelfish (Horch and Salmon, 1973; Mann and Lobel, 1997).

For recent reviews of fish hearing, see the works by Popper (2003) and Popper *et al.* (2003). The common range of fish hearing is roughly between 20 and 1000 Hz, although some hearing-specialist species can hear in ranges up to 10 KHz. One herring species can hear at 180 KHz (Hawkins, 1981; Mann *et al.*, 1997; Popper and Fay, 1998). The coupling of fish auditory perception with sound production, as well as the effects of the underwater signaling environment on acoustic communication, are among the most important unknowns in animal communication.

6.3. Acoustic Signal Function

The best known examples of fish acoustic signals involve sounds produced during the courtship behaviour of a few territorial species that lay demersal eggs in nests (e.g., Tavolga, 1958a,b,c; Myrberg *et al.*, 1978; Lobel, 2001a), as well as the courtship calls of some economically important species such as cod (Hawkins and Rasmussen, 1978; Hawkins and Amorim, 2000) and sciaenids (e.g., Connaughton and Taylor, 1996; Luczkovich *et al.*, 1999). A few studies have suggested that sounds might encode species identity, with measurable differences among sympatric species (Ladich *et al.*, 1992; Lobel and Mann, 1995; Lobel, 1998). Discrimination of conspecific and heterospecific sounds has been demonstrated in only one species, the pomacentrid *Stegastes partitus* (e.g., Myrberg *et al.*, 1986, 1993).

Sound patterns associated with the actual mating act (i.e., gamete release) have been reported for only a few species to date (Lobel, 1992, 1996; Lobel and Mann, 1995; Mann and Lobel, 1998; Hawkins and Amorim, 2000). This type of sound may play a key role in synchronisation of gamete release in order to maximise fertilisation (Lobel, 1992, 2002). Some examples of fish mating sounds are shown in Figure 2.5. The discovery of specific mating sounds has important application to the development of applied fisheries technology for mapping fish spawning sites. Specifically, it is now possible to deploy underwater recorders that can document the temporal and spatial patterns of sound production for those species that produce courtship and/or mating specific sounds (Lobel, 2001b). This unobtrusive sampling of reproductive status and behaviour suggests an important avenue for studies of fish in their natural environment. The potential for using species-specific sounds applied as a fisheries tool for locating and monitoring populations in the field is not new (e.g., Fish, 1954; Cummings *et al.*, 1964; Breder, 1968). However, it is only recently that computer and acoustic technology makes this type of quantitative monitoring feasible (Lobel, 2001b).

In many of the fish families that contain sound-producing species, it is usually only the males that vocalise. Frequently in such cases, there is a significant sexual dimorphism in the physiology and morphology of the

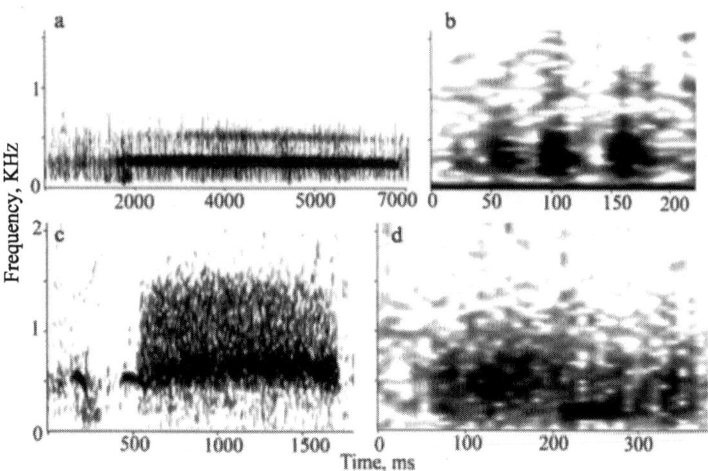

Fig. 2.5. Spawning sounds produced synchronously with gamete dispersal. Frequency scale is the same in all graphs, but time scale differs in each. (a) A tone type sound produced by *Ostracion meleagris* (duration, 6,213 ms; dominant frequency [DF], 258 Hz) (b) A pulse type sound by *Dascyllus albisella* (three pulses; 130 ms, duration; DF, 328 Hz). (c) A broadband sound by *Hypoplectrus nigricans* (duration, 1,581 ms; DF, 656 Hz). (d) A hydrodynamic sound by schools of fast swimming *Scarus iserti* (duration, 329 ms; DF (two peaks), 492 and 211 Hz). Size range of these fish is about 10 to 20 cm standard length (SL). (Adapted from Lobel, P. S. (2002). Diversity of fish spawning sounds and the application of passive acoustic monitoring. *Bioacoustics* **12**, 286–289.)

sound-producing mechanism as well (Knapp *et al.*, 1999; Bass *et al.*, 2000; Rice and Lobel, 2004). In many species of pomacentrids and cichlids, only males produce sound in courtship and agonistic contexts. Females in these groups have not been observed to produce sound (Amorim *et al.*, 2003; Rice and Lobel, 2004; Ripley and Lobel, 2004; and references therein). However, there is limited evidence to date that some cichlid species may include sound-producing females (Schwarz, 1974, 1980). In other fish, females may be capable of making sounds but do so much less often than the males (for example, in batrachoids, ophidiids, carapids, triglids, and some sciaenids). The sexual dimorphism in the neural and anatomical mechanisms responsible for sound production has been best described in batrachoid fish (Fine *et al.*, 1984, 1990; Bass and McKibben, 2003).

As noted above, many fish produce sounds associated with courtship. How does variation in the acoustic structure of fish sounds relate to mating success? Call rates, for example, are correlated with male success in diverse taxa including insects, anurans, birds, and mammals (Andersson, 1994).

Relatively little is known about the energetics of call production. In pomacentrids, vocalisations are often accompanied by a fast motor display, the signal jump, and are associated with the presence of fat reserves (see Section 7). It is possible, however, that when an acoustic call is not accompanied by vigorous swimming, it may not be very energetically costly at all. The calling rates in toadfish (Batrachoididae) have also been very well studied. In contrast to damselfish, male toadfish merely sit on the bottom in one spot and produce their courtship calls (e.g., Winn, 1967, 1972; Fish, 1972; Fine *et al.*, 1977b; Thorson and Fine, 2002). In an energetic study of calling rate costs in the oyster toadfish, *Opsanus tau*, Amorim *et al.* (2002) determined that sound production did not increase oxygen consumption significantly. They concluded that calling per se was not an expensive activity on a whole fish basis. Future work should strive to decouple the energetic cost of calling per se from that of the swimming that accompanies sound production.

6.4. Background Noise and Acoustic Communication

No treatment of acoustic communication would be complete without some mention of background noises. Recent studies of fish bioacoustics have increased awareness about the importance of the underwater acoustic environment and, once more, raise old questions regarding potential adverse impacts of noise pollution to fish and other aquatic animals (e.g,, Chapter 10, Section 3.3; Myrberg, 1978, 1980; Scholik and Yan, 2002a,b). In addition to its impacts on natural populations, acoustic noise may complicate interpretations of many aquarium studies of behaviour. Auditory sensitivity (Popper and Clark, 1976) and even viability (Banner and Hyatt, 1973) are compromised when fish are exposed to typical aquarium levels of noise; care must be taken to provide experimental subjects with the appropriate acoustic environment (e.g., Akamatsu *et al.*, 2002; Okumura *et al.*, 2002).

Some fish do reside in noisy habitats and have adapted their acoustic behaviour accordingly. Lugli *et al.* (2003) showed that two freshwater gobies, *Padogobius martensii* and *Gobius nigricans*, produce courtship sounds with maximal sound energy and have peak auditory sensitivity in the frequency window between the ambient turbulence and bubble noises in their environment.

7. MULTIMODAL COMMUNICATION

Partitioning communication by modality is analogous to partitioning visual signals into spatial, spectral, and temporal components. It serves as a useful heuristic tool, yet it also builds walls within what are, ultimately, integrated systems from a behavioural and cognitive point of view. Nearly all communication is constrained to be multimodal because generating information in one modality almost inevitably produces cues perceivable by other modalities.

Motor displays, in particular, produce a spatiotemporal (and sometimes chromatic) visual cue; they waft chemicals released via the gills and the urine, and they produce mechanosensory and often electrical cues. In the haplochromine cichlid *Astatotilapia burtoni* (Fernald and Hirata, 1975), the magnitude and velocity of the male agonistic display lowers water pressure enough to produce sounds by cavitation, the formation of a vapor pocket in the surrounding water. Generation of electrical signals is also often accompanied by auditory, visual, and mechanical cues (Moller, 2002). The mechanosensory contribution of response to motor displays deserves more attention (Sargent *et al.*, 1998).

Olfactory signaling, as discussed in Section 5, is inextricably tied to mechanoreception. Banded kokopu, *Galaxias fasciatus* (Galaxiidae), use the lateral line system (rather than bilateral comparison of olfactory stimuli) to track an odor source (Baker *et al.*, 2002), and this mechanism should apply to communication cues as well.

There have been few experimental studies that explicitly address multimodal signaling in fish. Crapon de Caprona and Ryan (1990) presented female pygmy swordtails *Xiphophorus pygmaeus* with both visual and olfactory cues from male conspecifics, which lack courtship displays and from courting male *X. nigrensis*, a closely related, allopatric species. When presented only with chemical cues, females preferred conspecifics; but with only visual cues, females preferred heterospecific males. Females failed to show a preference when presented with both cues. Hankison and Morris (2003) performed a similar experiment, but they tested *X. pygmaeus* on conspecifics and a sympatric heterospecific *X. cortezi*. Again, females preferred conspecific olfactory cues and heterospecific visual cues; however, when both cues were presented together, they chose conspecific males. In swordtails, females appear to select visual traits somewhat independent of species, and use olfactory traits to assess species identity.

McLennan (2003) came to a rather different conclusion in her study of chemical and visual communication in three-spine sticklebacks (*Gasterosteus aculeatus*). By shifting the timing of presentation of olfactory and visual cues, she found that chemical cues alerted receivers to the presence of the visual cues provided by male courtship. The olfactory cue was acting as a long distance message independent of line-of-sight, allowing a female to detect a male's presence before seeing him.

Chemical and visual cues are similarly combined in the antipredator system of the characid *Hemigrammus erythrozonus*. Rather than fleeing or freezing when a chemical alarm cue is presented, these fish perform a conspicuous motor display of "fin flicking," involving very rapid changes of the caudal, dorsal, and pectoral fins with no change in position. The display may function both as a visual alarm signal to prey and as a deterrent to predators (Brown *et al.*, 1999). This strategy is reminiscent of antipredator stotting behaviour in African ungulates (Caro *et al.*, 2004).

Communication may also exhibit tradeoffs among modalities. Nelissen (1978), studying Lake Tanganyikan cichlids, found that the number of sonic displays produced by a species was negatively correlated with the number of colour patterns that species could assume (Nelissen, 1978). In another labroid fish, the pomacentrid *Stegastes partitus*, acoustic and visual/mechanosensory signals appear to be closely integrated. Courtship calling rates are correlated with the male courtship display known as the "signal jump." This is a display in which a male swims vigorously, first rising in the water column and then swimming rapidly downward while producing a pulsed sound (Fishelson, 1964; Myrberg, 1972; Spanier, 1979; Lobel and Mann, 1995). In one damselfish species, *Stegastes partitus*, mate choice was associated with signal jump repetition rate (Knapp and Warner, 1991). A significant portion of a damselfish's daily call production occurs in low light or no visibility conditions (e.g., the "predawn chorus;" Mann and Lobel, 1995). Thus, it is likely that acoustic and visual cues provide some degree of redundancy for this mate-choice signal.

The little work on multimodal integration suggests that integration across sensory modalities plays a fundamental role in fish communication. The diversity within each communication channel reinforces the notion that not only are signals and receivers evolving in response to environmental and social forces within modalities, they are also evolving sophisticated interplays among modalities depending on phylogenetic history, the sensory environment, and the communicative context.

8. COMMUNICATION IN FISH: THE NEXT STEPS

We now have the techniques and the conceptual approaches to paint a detailed quantitative picture of signals and the signaling environment in all modalities. Genome projects in a variety of fish systems, coupled with the wealth of inter- and intraspecific genetic variation in signalers and receivers, should shed light on the molecular mechanisms underlying signal production and perception. Playbacks are now feasible in all modalities including, with some effort, presentation of synthetic chemical signals.

With the exception of a few electric fish systems and toadfish, the neurobiology of communication in fish is poorly understood, particularly with respect to visual and chemical signals. We know next to nothing about the ontogeny of communication in fish. What role do early experience and imprinting play in the development of communication systems? How do fish with planktonic larval stages develop a full repertoire of communicative behaviours?

Large gaps remain in our understanding of the mechanisms of chemical communication, particularly with respect to mate choice and species recognition; of the role of mechanosensory information in communication; of how spatiotemporal components of visual signals evolve in the signaling environment; and of how acoustic signals are produced and perceived in the wild. We have presented a fraction of the astonishing diversity of fish communication systems, with emphasis on a few taxa that we consider to be particularly promising candidates for future study. Fish inhabit almost every conceivable communicative environment, and are faced with almost every conceivable circumstance for communication. The range of solutions they have evolved to these challenges makes it so that any interesting question in animal communication is likely to find a fruitful model in fish.

ACKNOWLEDGEMENTS

We thank H. Fisher, G. Gerlach, E. Neeley, P. Sorensen, K. Warkentin, J. Webb, and B. Wong for comments and animated discussion. K. Sloman, S. Balshine, M. Fine, and an anonymous reviewer provided many critical and helpful suggestions. C. O'Keefe and M. Fayfman provided invaluable assistance with the preparation of the manuscript.

REFERENCES

Akamatsu, T., Okumura, T., Novarini, N., and Yan, H. Y. (2002). Empirical refinements applicable to the recording of fish sounds in small tanks. *J. Acoust. Soc. Am.* **112,** 3073–3082.

Amiet, J.-L. (1987). "Faune du Cameroun/Fauna of Cameroon Volume 2: Le genre *Aphyosemion* Myers (Pisces, Teleostei, Cyprinodontiformes)." Sciences Nat Venette Compiègne, France.

Amorim, M. C. P., McCracken, M. L., and Fine, M. L. (2002). Metabolic costs of sound production in the oyster toadfish, *Opsanus tau. Can. J. Zool.* **80,** 830–838.

Amorim, M. C. P., Fonseca, P. J., and Almada, V. C. (2003). Sound production during courtship and spawning of *Oreochromis mossambicus*: Male-female and male-male interactions. *J. Fish Biol.* **62,** 658–672.

Andersson, M. (1994). "Sexual Selection." Princeton University Press, Princeton, NJ.

Asai, R., Taguchi, E., Kume, Y., Saito, M., and Kondo, S. (1999). Zebrafish Leopard gene as a component of the putative reaction-diffusion system. *Mechan. Dev.* **89,** 87–92.

Atema, J. (1988). Distribution of chemical stimuli. *In* "Sensory Biology of Aquatic Animals" (Atema, J., Fay, R. R., Popper, A. N., and Tavolga, W. N., Eds.), pp. 29–56. Springer-Verlag, New York.

Atema, J. (1996). Eddy chemotaxis and odor landscapes: Exploration of nature with animal sensors. *Biol. Bull.* **191,** 129–138.

Bagnara, J. T., and Hadley, M. E. (1973). "Chromatophores and Colour Change." Prentice-Hall, Englewood Cliffs, NJ.

Baker, C. F., Montgomery, J. C., and Dennis, T. E. (2002). The sensory basis of olfactory serach behaviour in banded kokopu (*Galaxias fasciatus*). *J. Comp. Physiol. A* **188,** 553–560.

Balshine-Earn, S., and Lotem, A. (1998). Individual recognition in a cooperatively breeding cichlid: Evidence from video playback experiments. *Behaviour* **135,** 369–386.

Banner, A., and Hyatt, M. (1973). Effects of noise on eggs and larvae of two estuarine fish. *Trans. Am. Fish. Soc.* **108,** 134–136.

Barata, E. N., Nogueira, R., Serrano, R., Gomes, L., and Canario, A. V. M. (2002). A pheromonal role for the anal gland of male peacock blennies (*Salaria pavo*) in female attraction. *J. Comp. Biochem. Physiol. A.* **132,** S63–S70.

Bardach, J. E., and Todd, J. H. (1970). Chemical communication in fish. *In* "Advances in Chemoreception Volume I: Communication by Chemical Signals" (Johnston, J. W. J., Moulton, D. G., and Turk, A., Eds.), pp. 205–240. Appleton-Century-Crofts, New York.

Barlow, G. W. (1963). Ethology of the Asian teleost *Badis badis*. II. Motivation and signal value of the colour patterns. *Anim. Behav.* **11,** 97–105.

Basolo, A. L. (1990). Female preference predates the evolution of the sword in swordtail fish. *Science* **250,** 808–810.

Basolo, A. L. (1995). Phylogenetic evidence for the role of a preexisting bias in sexual selection. *Proc. R. Soc. Lond. B* **259,** 307–311.

Bass, A. H., and McKibben, J. R. (2003). Neural mechanisms and behaviours for acoustic communication in teleost fish. *Prog. Neurobiol.* **69,** 1–26.

Bass, A. H., Bodnar, D. A., and Marchaterre, M. A. (2000). Midbrain acoustic circuitry in a vocalising fish. *J. Comp. Neurol.* **419,** 505–531.

Bennett, M. V. L. (1970). Comparative physiology: Electric organs. *Ann. Rev. Physiol.* **32,** 471–528.

Bleckmann, H. (1993). Role of the lateral line in fish behaviour. *In* "The Behaviour of Teleost Fish" (Pitcher, T. J., Ed.), pp. 201–246. Chapman Hall, London.

Boughman, J. W. (2001). Divergent sexual selection enhances reproductive isolation in sticklebacks. *Nature* **411,** 944–947.

Bradbury, J. W., and Vehrencamp, S. L. (1998). "Principles of Animal Communication." Sinauer, Sunderland.

Breder, C. M., Jr. (1968). Seasonal and diurnal occurrences of fish sounds in a small Florida bay. *Bull. Am. Mus. Nat. Hist.* **138,** 327–378.

Brown, G. E., Adrian, J. C., Jr., Lewis, M. G., and Tower, J. M. (2002). The effects of reduced pH on chemical alarm signaling in ostariophysan fish. *Can. J. Fish. Aquat. Sci.* **59,** 1331–1338.

Brown, G. E., and Brown, J. A. (1995). Kin discrimination in salmonids. *Rev. Fish Biol. Fish.* **6,** 201–220.

Brown, G. E., and Smith, J. F. (1994). Fathead minnows use chemical cues to discriminate natural shoalmates from unfamiliar conspecifics. *J. Chem. Ecol.* **20,** 3051–3061.

Brown, G. E., Godin, J.-G. J., and Pedersen, J. (1999). Fin-flicking behaviour: A visual anti-predator alarm signal in a characin fish, *Hemigrammus erythrozonus*. *Anim. Behav.* **58,** 469–475.

Bullock, T. H., and Heiligenberg, W. (1986). "Electroreception." John Wiley and Sons, New York.

Bushmann, P. J., and Burns, J. R. (1994). Social control of male sexual maturation in the swordtail characin, *Corynopoma riisei*. *J. Fish Biol.* **44,** 263–272.

Caley, M. J., and Schluter, D. (2003). Predators favour mimicry in a tropical reef fish. *Proc. R. Soc. Lond. B* **270,** 667–672.

Campbell, A. K., and Herring, P. J. (1987). A novel red fluorescent protein from the deep sea luminous fish *Malacosteus niger*. *Comp. Biochem. Physiol. B* **86,** 411–417.

Caro, T. M., Graham, C. M., Stoner, C. J., and Vargas, J. K. (2004). Adaptive significance of antipredator behaviour in artiodactyls. *Anim. Behav.* **67,** 205–228.

Carr, W. E. S. (1988). The molecular nature of chemical stimuli in the aquatic environment. *In* "Sensory Biology of Aquatic Animals" (Atema, J., Fay, R. R., Popper, A. N., and Tavolga, W. N., Eds.), pp. 3–27. Springer-Verlag, New York.

Clay, C. S., and Medwin, H. (1977). "Acoustical oceanography: Principles and applications." John Wiley and Sons, New York.

Connaughton, M. A., and Taylor, M. H. (1996). Drumming, courtship and spawning behaviour in captive weakfish, *Cynoscion regalis*. *Copeia* **1996,** 195–199.

Connaughton, M. A., Fine, M. L., and Taylor, M. H. (2002). Weakfish sonic muscle: Influence of size, temperature and season. *J. Exp. Biol.* **205,** 2183–2188.

Crapon de Caprona, M.-D., and Ryan, M. J. (1990). Conspecific mate recognition in swordtails, *Xiphophorus nigrensis* and *X. pygmaeus*: Olfactory and visual cues. *Anim. Behav.* **39,** 290–296.

Cole, K. S., and Smith, R. J. F. (1992). Attraction of female fathead minnows, Pimephales *promelas*, to chemical stimuli from breeding males. *J. Chem. Ecol.* **18,** 1269–1284.

Crawford, J. D. (1991). Sex recognition by electric cues in a sound-producing mormyrid fish, *Pollimyrus isidori*. *Brain Behav. Evol.* **38,** 20–38.

Cummings, M. E., and Partridge, J. C. (2001). Visual pigments and optical habitats of surfperch (Embiotocidae) in the California kelp forest. *J. Comp. Physiol. A* **187,** 875–889.

Cummings, M. E., Rosenthal, G. G., and Ryan, M. J. (2003). A private ultraviolet channel in visual communication. *Proc. Biol. Sci.* **270,** 897–904.

Cummings, W. C., Brahy, B. D., and Herrnkind, W. F. (1964). The occurrence of underwater sounds of biological origin off the west coast of Bimini, Bahamas. *In* "Marine Bio-Acoustics" (Tavolga, W. N., Ed.), pp. 27–43. Pergamon Press, New York.

Curtis, B. J., and Wood, C. M. (1991). The function of the urinary bladder *in vivo* in the freshwater rainbow trout. *J. Exp. Biol.* **155,** 567–583.

Darwin, C. (1874). "The Descent of Man" 2nd ed. Caldwell Publications, New York.

Degani, G., and Schreibman, M. P. (1993). Pheromone of male blue gourami and its effect on vitellogenesis, steroidogenesis and gonadotropin cells in pituitary of the female. *J. Fish Biol.* **43,** 475–485.

Demski, L. S., and Dulka, J. G. (1986). Thalamic stimulation evokes sex colour changes and gamete release in a vertebrate hermaphrodite. *Experimentia* **42,** 1285–1287.

Denton, E. J. (1970). On the organisation of reflecting surfaces in some marine animals. *Philos. Trans. R. Soc. London B* **258,** 285–313.

Douglas, R. H., and Partridge, J. C. (1997). On the visual pigments of deep-sea fish. *J. Fish Biol.* **50,** 68–85.

Douglas, R. H., Partridge, J. C., Dulai, K. S., Hunt, D. M., Mullineaux, C.W, and Hynninen, P. H. (1999). Enhanced retinal longwave sensitivity using a chlorophyll-derived photosensitiser in *Malacosteus niger*, a deep-sea dragon fish with far red bioluminescence. *Vision Res.* **39,** 2817–2832.

Dulka, J. G., Stacey, N. E., Sorensen, P. W., and van der Kraak, G. J. (1987). A steroid sex pheromone synchronises male-female spawning readiness in goldfish. *Nature* **325,** 251–253.

Dunlap, K. D., Thomas, P., and Zakon, H. H. (1998). Diversity of sexual dimorphism in electrocommunication signals and its androgen regulation in a genus of electric fish, *Apteronotus*. *J. Comp. Physiol. A* **183,** 77–86.

Endler, J. A. (1991). Variation in the appearance of guppy colour patterns to guppies and their predators under different visual conditions. *Vis. Res.* **31,** 587–608.

Endler, J. A. (1992). Signals, signal conditions, and the direction of evolution. *Am. Nat.* **139,** S125–S153.

Endler, J. A., and Houde, A. E. (1995). Geographic variation in female preferences for male traits in *Poecilia reticulata*. *Evolution* **49,** 456–458.

Engeszer, R. E., Ryan, M. J., and Parichy, D. M. (2004). Learned social preference in Zebrafish. *Curr. Biol.* **14,** 881–884.

Fernald, R. D. (1990). *Haplochromis burtoni*: A case study. *In* "The Visual System of Fish" (Douglas, R. H., and Djamgoz, M. B. A., Eds.), pp. 443–463. Chapman and Hall, London.

Fernald, R. D., and Hirata, N. R. (1975). Non intentional sound production in a cichlid fish (*Haplochromis burtoni*, Gunther). *Experientia* **31,** 299–300.

Fine, M. L. (1978). Seasonal and geographic variation of the mating call of the oyster toadfish. *Opsanus tau*. *Oecologia* **36,** 45–57.

Fine, M. L., and Lenhardt, M. L. (1983). Shallow-water preparation of the toadfish mating call. *Comp. Biochem. Physiol. A* **76,** 225–231.

Fine, M. L., Winn, H. E., and Olla, B. (1977a). Communication in fish. *In* "How Animals Communicate" (Sebeok, T., Ed.), pp. 472–518. Indiana University Press, Bloomington, IN.

Fine, M. L., Winn, H. E., Joest, L., and Perkins, P. J. (1977b). Temporal aspects of calling behaviour in the oyster toadfish, *Opsanus tau*. *Fish. Bull.* **75,** 871–874.

Fine, M. L., Economos, D., Radtke, R., and McClung, J. R. (1984). Ontogeny and sexual dimorphism of the sonic motor nucleus in the oyster toadfish. *J. Comp. Neurol.* **225,** 105–110.

Fine, M. L., Burns, N. M., and Harris, T. M. (1990). Ontogeny and sexual dimorphism of the sonic muscle in the oyster toadfish. *Can. J. Zool.* **68,** 1374–1381.

Fine, M. L., Malloy, K. L., King, C. B., Mitchell, S. L., and Cameron, T. M. (2001). Movement and sound generation by the toadfish swimbladder. *J. Comp. Physiol. A* **187,** 371–379.

Finger, E., Burkhardt, D., and Dyck, J. (1992). Avian plumage colours: Origin of UV reflection in a black parrot. *Naturwiss.* **79,** 187–188.

Fish, M. P. (1954). The character and significance of sound production among fish of the western North Atlantic. *Bulletin Bingham. Oceanographic Coll.* **14,** 1–109.

Fish, J. F. (1972). The effect of sound playback on the toadfish. *In* "Behaviour of marine animals" (Winn, H. E., and Olla, B., Eds.), Vol. 2, pp. 386–434. Plenum Press, New York.

Fish, M. P., and Mowbray, W. H. (1970). "Sounds of the western north Atlantic fish" Johns Hopkins Press, Baltimore, MD.

Fishelson, L. (1964). Observation on the biology and behaviour of Red Sea coral fish. *Contributions to the Knowledge of the Red Sea* **30,** 11–26.

Fleishman, L. J. (1986). Motion detection in the presence and absence of background motion in an *Anolis* lizard. *J. Comp. Physiol.* **159,** 711–720.

Gon, O. (1996). Revision of the cardinalfish subgenus *Jaydia* (Perciformes, Apogonidae, *Apogon*). *Trans. Roy. Soc. South Africa* **51,** 147–194.

Göz, H. (1941). Ueber den Art und Individualgeruch bei Fischen. *Z. Vergl. Physiol.* **29,** 1–45.

Hankison, S. J., and Morris, M. R. (2003). Avoiding a compromise between sexual selection and species recognition: Female swordtail fish assess multiple species-specific cues. *Behav. Ecol.* **14,** 282–287.

Hastings, J. W., and Morin, J. G. (1991). Bioluminescence. *In* "Offprints from Neural and Integrative Animal Physiology" (Prosser, C. L., Ed.), pp. 131–170. Wiley-Liss, New York.

Hawkins, A. D. (1981). The hearing abilities of fish. *In* "Hearing and Sound Communication in Fish" (Tavologa, W. N., Popper, A. N., and Fay, R. R., Eds.), pp. 109–139. Springer-Verlag, New York.

Hawkins, A. D., and Amorim, M. C. P. (2000). Spawning sounds of the male haddock, *Melanogrammus aegelfinus*. *Env. Biol. Fish.* **59,** 29–41.

Hawkins, A. D., and Myrberg, A. A., Jr. (1983). Hearing and sound communication underwater. *In* "Bioacoustics, A Comparative Approach" (Lewis, B., Ed.), pp. 347–405. Academic Press, London.

Hawkins, A. D., and Rasmussen, K. J. (1978). The calls of gadoid fish. *J. Marine Biol. Assoc. UK* **58,** 891–911.

Heiligenberg, W. (1977). Principles of electrolocation and jamming avoidance in electric fish: A neuroethological approach. *In* "Studies in Brain Function" (Braitenberg, V., Ed.), pp. 1–85. Springer-Verlag, New York.

Herring, P. J., and Morin, G. (1978). Bioluminescence in fish. *In* "Bioluminescence in Action" (Herring, P. J., Ed.), pp. 273–329. Academic Press, New York.

Hopkins, C. D. (1986). Behaviour of mormyridae. *In* "Electroreception" (Bullock, T. H., and Heilingenberg, W., Eds.), pp. 527–576. John Wiley and Sons, New York.

Hopkins, C. D. (1999a). Design features for electric communication. *J. Exp. Biol.* **202,** 1217–1228.

Hopkins, C. D. (1999b). Signal evolution in electric communication. *In* "The Design of Animal Communication" (Hauser, M. D., and Konishi, M., Eds.), pp. 461–491. MIT Press, Cambridge, MA.

Hopkins, C. D., and Westby, G. W. M. (1986). Time domain processing of electric organ discharge waveforms by pulse-type electric fish. *Brain Behav. Evol.* **29,** 77–104.

Hopkins, C. D., Comfort, N. C., Bastian, J., and Bass, A. H. (1990). A functional analysis of sexual dimorphism in an electric fish, *Hypopomus pinnicaudatus*, order Gymnotiformes. *Brain Behav. Evol.* **35,** 350–367.

Horch, K., and Salmon, M. (1973). Adaptations to the acoustic environment by the squirrelfish *Myripristis violaceus* and *M. pralinius*. *Mar. Behav. Physiol.* **2,** 121–139.

Houde, A. E. (1997). *In* "Sex, Colour, and Mate Choice in Guppies" (Krebs, J. R., and Clutton-Brock, T., Eds.). Princeton University Press, Princeton, NJ.

Houde, A. E., and Torio, A. J. (1992). Effect of parasitic infection on male colour pattern and female choice in guppies. *Behav. Ecol.* **3,** 346–351.

Hubbard, P. C., Barata, E. N., and Canario, A. V. M. (2002). Possible disruption of pheromonal communication by humic acid in the goldfish, *Carassius auratus*. *Aquat. Toxicol.* **60,** 169–183.

Hudon, J., Grether, G. F., and Millie, D. F. (2003). Marginal differentiation between the sexual and general carotenoid pigmentation of guppies (*Poecilia reticulata*) and a possible visual explanation. *Physiol. Biochem. Zool.* **76,** 776–790.

Jumper, G. Y., and Baird, R. C. (1991). Location by olfaction—a model and application to the mating problem in the deep-sea hatchetfish *Argyropelecus hemigymnus*. *Am. Nat.* **138,** 1431–1458.

Kanter, M., and Coombs, S. (2003). Rheotaxis and prey detection in uniform currents by Lake Michigan mottled sculpin (*Cottus bairdi*). *J. Exp. Biol.* **6,** 59–60.

Katzir, G. (1981). Visual aspects of species recognition in the damselfish *Dascyllus aruanus* L. (Pisces, Pomacentridae). *Anim. Behav.* **29,** 842–849.

Kingston, J., Rosenthal, G. G., and Ryan, M. J. (2003). The role of sexual selection in maintaining a colour polymorphism in the pygmy swordtail *Xiphophorus pygmaeus*. *Anim. Behav.* **65,** 735–743.

Knapp, R. A., and Warner, R. R. (1991). Male parental care and female choice of the bicolour damselfish, *Stegastes partitus*: Bigger is not always better. *Anim. Behav.* **41,** 747–756.

Knapp, R., Marchaterre, M. A., and Bass, A. H. (1999). Early development of the motor and premotor circuitry of a sexually dimorphic vocal pathway in a teleost fish. *J. Neurobiol.* **38,** 475–490.

Kondo, S., and Asai, R. (1995). A reaction-diffusion wave on the skin of the marine angelfish *Pomacanthus*. *Nature* **376,** 765–768.

Künzer, P. (1964). Weitere Versuche zur Auslösung der Nachfolgereaktion bei Jungfischen von *Nannacara anomala* (Cichlidae). *Naturwissenschaften* **51,** 419–420.

Lachlan, R. F., Crooks, L., and Laland, K. N. (1998). Who follows whom? Shoaling preferences and social learning of foraging information in guppies. *Anim. Behav.* **56,** 181–190.

Ladich, F., Bischoff, C., Schleinzer, G., and Fuchs, A. (1992). Intra- and interspecific differences in agonistic vocalisations in croaking gouramis (Genus: *Trichopsis*, Anabantoidei, Teleostei). *Bioacoustics* **4,** 131–141.

Laland, K. N., Brown, C., and Krause, J. (2003). Learning in fish: From three-second memory to culture. *Fish Fish* **4,** 199–202.

LaRiviere, L., and Anctil, M. (1984). Uptake and release of super(3)H-serotonin in photophores of the midshipman fish, *Porichthys notatus*. *Comp. Biochem. Physiol.* **78,** 231–239.

Liley, N. R., and Stacey, N. E. (1983). Hormones, pheromones, and reproductive behaviour in fish. *In* "Fish Physiology, Vol. IX: Reproduction" (Hoar, W. S., Randall, D. J., and Donaldson, E. M., Eds.), pp. 1–63. Academic Press, New York.

Lobel, P. S. (1992). Sounds produced by spawning fish. *Env. Biol. Fish* **33,** 351–358.

Lobel, P. S. (1996). Spawning sound of the trunkfish, *Ostracion meleagris* (Ostraciidae). *Biol. Bull.* **191,** 308–309.

Lobel, P. S. (1998). Possible species specific courtship sounds by two sympatric cichlid fish in Lake Malawi, Africa. *Env. Biol. Fish* **52,** 443–452.

Lobel, P. S. (2001a). Acoustic behaviour of cichlid fish. *J. Aquaricult. Aquat. Sci.* **9,** 167–186.

Lobel, P. S. (2001b). Fish bioacoustics and behaviour: Passive acoustic detection and the application of a closed-circuit rebreather for field study. *Mar. Technol. Soc. J.* **35,** 19–28.

Lobel, P. S. (2002). Diversity of fish spawning sounds and the application of passive acoustic monitoring. *Bioacoustics* **12,** 286–289.

Lobel, P. S., and Mann, D. A. (1995). Spawning sounds of the damselfish, *Dascyllus albisella* (Pomacentridae), and relationship to male size. *Bioacoustics* **6,** 187–198.

Losey, G. S., Cronin, T. W., Goldsmith, T. H., Hyde, D., Marshall, N. J., and McFarland, W. N. (1999). The UV visual world of fish: a review. *J. Fish Biol.* **54,** 921–943.

Losey, G. S., McFarland, W. N., Loew, E. R., Zamzow, J. P., Nelson, P. A., and Marshall, N. J. (2003). Visual biology of Hawaiian coral reef fish. I. Ocular transmission and visual pigments. *Copeia* **2003,** 433–454.

Luczkovich, J. J., Sprague, M. W., Johnson, S. E., and Pullinger, R. C. (1999). Delimiting spawning areas of weakfish *Cynoscion regalis* (family Sciaenidae) in Pamlico Sound, North Carolina using passive hydroacoustic surveys. *Int. J. Anim. Sound Record* **10,** 143–160.

Lugli, M., and Fine, M. L. (2003). Acoustic communication in two freshwater gobies: Ambient noise and short-range propagation in shallow streams. *J. Acoustl. Soc. Am.* **114,** 512–521.

Lugli, M., Yan, H. Y., and Fine, M. L. (2003). Acoustic communication in two freshwater gobies: The relationship between ambient noise, hearing thresholds and sound spectrum. *J. Comp. Physiol. A* **189,** 309–320.

Lythgoe, J. N. (1988). Light and vision in the aquatic environment. *In* "Sensory Biology of Aquatic Animals" (Atema, J., Fay, R. R., Popper, A. N., and Tavolga, W. N., Eds.), pp. 75–82. Springer-Verlag, New York.

Ma, P. M. (1995a). On the agonistic display of the Siamese fighting fish. I. The frontal display apparatus. *Brain Behav. Evol.* **45,** 301–313.

Ma, P. M. (1995b). On the agonistic display of the Siamese fighting fish. II. The distribution, number and morphology of opercular display motoneurons. *Brain Behav. Evol.* **45,** 314–326.

MacGinitie, G. E. (1939). The natural history of the blind goby, *Typhlogobius californiensis*. *Amer. Mid. Naturalist* **21,** 489–505.

Malabarba, L. R. (1998). Monophyly of the Cheirodontinae, characters and major clades (Ostariophysi: Characidae). *In* "Phylogeny and Classification of Neotropical Fish" (Malabarba, L. R., Reis, R. E., Vari, R. P., Lucena, Z. M. S., and Lucena, C. A. S., Eds.), pp. 193–233. Porto Alegre, Edipucrs.

Mallefet, J., and Shimomura, O. (1995). Presence of coelenterazine in mesopelagic fish from the Strait of Messina. *Mar. Biol.* **124,** 381–385.

Mann, D. A., and Lobel, P. S. (1995). Passive acoustic detection of sounds produced by the damselfish, *Dascyllus albisella* (Pomacentridae). *Bioacoustics* **6,** 199–213.

Mann, D. A., and Lobel, P. S. (1997). Propagation of damselfish (Pomacentridae) courtship sounds. *J. Acoust. Soc. Am.* **101,** 3783–3791.

Mann, D. A., and Lobel, P. S. (1998). Acoustic behaviour of the damselfish *Dascyllus albisella*: behavioural and geographic variation. *Env. Biol. Fish* **51,** 421–428.

Mann, D. A., Lu, Z., and Popper, A. N. (1997). Ultrasound detection by a teleost fish. *Nature* **389,** 341.

Marshall, N. J. (2000). The visual ecology of reef fish colours. *In* "Signaling and Signal Design in Animal Communication" (Espmark, Y., Amundsen, T., and Rosenqvist, G., Eds.), pp. 83–120. Tapir Academic Press, Trondheim.

Marshall, N. J., Jennings, K., McFarland, W. N., Loew, E. R., and Losey, G. S. (2003). Visual biology of Hawaiian coral reef fish. II. Colours of Hawaiian coral reef fish. *Copeia* **2003,** 455–466.

McKinnon, J. S., and Liley, N. R. (1987). Asymmetric species specificity in responses to female sexual pheromone by males of two species of *Trichogaster* (Pisces: Belontiidae). *Can. J. Zool.* **65,** 1129–1134.

McKinnon, J. S., and McPhail, J. D. (1996). Male aggression and colour in divergent populations of the threespine stickleback: Experiments with animations. *Can. J. Zool.* **74,** 1727–1733.

McLennan, D. A. (2003). The importance of olfactory signals in the gasterosteid mating system: Sticklebacks go multimodal. *Biol. J. Linnean Soc.* **80,** 555–572.

McLennan, D. A., and Ryan, M. J. (1997). Responses to conspecific and heterospecific olfactory cues in the swordtail *Xiphophorus cortezi*. *Anim. Behav.* **54,** 1077–1088.

McLennan, D. A., and Ryan, M. J. (1999). Interspecific recognition and discrimination based upon olfactory cues in swordtails. *Evolution* **53,** 880–888.

Mirza, R. S., and Chivers, D. P. (2003). Fathead minnows learn to recognise heterospecific alarm cues they detect in the diet of a known predator. *Behaviour* **140,** 1359–1369.

Moller, P. (2002). Multimodal sensory integration in weakly electric fish: A behavioural account. *J. Physiol.-Paris* **96,** 547–556.

Morris, M. R., Mussel, M., and Ryan, M. J. (1995). Vertical bars on male *Xiphophorus multilineatus*: A signal that deters rival males and attracts females. *Behav. Ecol.* **6,** 274–279.

Myrberg, A. A., Jr. (1966). Parental recognition of young in cichlid fish. *Anim. Behav.* **14,** 565–571.

Myrberg, A. A., Jr. (1972). Ethology of the bicolour damselfish, *Eupomacentrus partitus*, (Pisces; Pomacentridae). A comparative analysis of laboratory and field behaviour. *Anim. Behav. Mono.* **5,** 199–283.

Myrberg, A. A., Jr. (1978). Underwater sound: Its effects on the behaviour of sharks. *In* "Sensory Biology of Elasmobranch fish" (Mathewson, R., and Hodgson, T., Eds.), pp. 391–417. US Government Printing Office, Washington, DC.

Myrberg, A. A., Jr. (1980). Sensory mediation of social recognition process in fish. *In* "Fish Behaviour and its Use in the Capture and Culture of Fish" (Bardach, J. E., Magnuson, J. J., May, R. C., and Reinhart, J. M., Eds.), pp. 146–178. International center for Living Aquatic Resources Management, Manila, Philippines.

Myrberg, A. A., Jr. (1981). Sound communication and interception of fish. *In* "Hearing and Sound Communication in Fish" (Tavologa, W. N., Popper, A. N., and Fay, R. R., Eds.), pp. 109–138. Springer-Verlag, New York.

Myrberg, A. A., Jr., Kramer, E., and Heinecke, P. (1965). Sound production by cichlid fish. *Science* **149,** 555–558.

Myrberg, A. A., Jr., Spanier, E., and Ha, S. J. (1978). Temporal patterning in acoustical communication. *In* "Contrasts in Behaviour" (Reese, E. S., and Lighter, F. J., Eds.), pp. 137–179. Wiley-Interscience Publishing, New York.

Myrberg, A. A., Jr., Mohler, M., and Catala, J. D. (1986). Sound production by males of a coral reef fish (*Pomacentrus partitus*). *Anim. Behav.* **33,** 411–416.

Myrberg, A. A., Jr., Ha, S. J., and Shamblott, M. J. (1993). The sounds of bicolour damselfish (*Pomacentrus partitus*): Predictors of body size and a spectral basis for individual recognition and assessment. *J. Acoust. Soc. Am.* **94,** 3067–3070.

Narins, P. M. (2001). Vibration communication in vertebrates. *In* "Ecology of Sensing" (Barth, F., and Schmidt, A., Eds.), pp. 127–148. Springer-Verlag, Berlin.

Nealson, K. H. (1977). Autoinduction of bacterial luciferase: Occurrence, mechanism and significance. *Arch. Microbiol.* **112,** 73–79.

Nealson, K. H., and Hastings, J. W. (1979). Ecology and control of bacterial bioluminescence. *Microbial Rev.* **43,** 496–518.

Nelissen, M. H. J. (1978). Sound production by some Tanganyikan cichlid fish and a hypothesis for the evolution of their communication mechanisms. *Behaviour* **64,** 137–147.

Nelissen, M. H. J. (1991). Communication. *In* "Cichlid Fish: Behaviour, Ecology and Evolution" (Keenleyside, M. H. A., Ed.), pp. 225–240. Chapman and Hall, London.

Nevitt, G. A., Dittman, A. H., Quinn, T. P., and Moody, W. J. (1994). Evidence for a peripheral olfactory memory in imprinted salmon. *PNAS* **91,** 4288–4292.

Nicoletto, P. F. (1991). The relationship between male ornamentation and swimming performance in the guppy, *Poecilia reticulata. Behav. Ecol. Sociobiol.* **28,** 365–370.

Nicoletto, P. F. (1993). Female sexual response to condition-dependent ornaments in the guppy, *Poecilia reticulata. Anim. Behav.* **46,** 441–450.

Nybakken, J. W. (1977). "Marine Biology: An Ecological Approach." Addison Wesley Longman, Menlow Park.

Okumura, T., Akamatsu, T., and Yan, H. Y. (2002). Analyses of small tank acoustics: Empirical and theoretical approaches. *Bioacoustics* **12,** 330–332.

Olsen, K. H., Grahn, M., Lohm, J., and Langefors, A. (1998). MHC and kin discrimination in juvenile Artic charr, *Salvelinus alpinus* (L.). *Anim. Behav.* **56,** 319–327.

Paintner, S., and Kramer, B. (2003). Electrosensory basis for individual recognition in a weakly electric, mormyrid fish, *Pollimyrus adspersus* (Gunther, 1866). *Behav. Ecol. Sociobiol.* **55,** 197–208.

Painter, K. J., Maini, P. K., and Othmer, H. G. (1999). Stripe formation in juvenile *Pomacanthus* explained by a generalised turing mechanism with chemotaxis. *PNAS* **96,** 5549–5554.

Parichy, D. M. (2003). Pigment patterns: Fish in stripes and spots. *Current Biology* **13,** R947–R950.

Pickens, P. E., and McFarland, W. N. (1964). Electric discharge and associated behaviour in the stargazer. *Anim. Behav.* **12,** 362–367.

Popper, A. N. (2003). Effects of anthropogenic sound on fishes. *Fisheries* **28,** 24–31.

Popper, A. N., and Clark, N. L. (1976). The auditory system of the goldfish (*Carassius auratus*): Effects of intense acoustic stimulation. *Comp. Biochem. Physiol. A* **53,** 11–18.

Popper, A. N., and Fay, R. R. (1998). The auditory periphery in fish. *In* "Comparative Hearing: Fish and Amphibians" (Fay, R. R., and Popper, A. N., Eds.), pp. 43–100. Springer-Verlag, New York.

Popper, A. N., Fay, R. R., Platt, C., and Sand, O. (2003). Sound detection mechanisms and capabilities of teleost fish. *In* "Sensory Processing in Aquatic Environments" (Collin, S. P., and Marshall, N. J., Eds.), pp. 3–38. Springer-Verlag, New York.

Reusch, T. B. H., Haberli, M. A., Aeschlimann, P. B., and Milinski, M. (2001). Female sticklebacks count alleles in a strategy of sexual selection explaining MHC polymorphism. *Nature* **414,** 300–302.

Rice, A. A., and Lobel, P. S. (2004). The pharyngeal jaw apparatus of the Cichlidae and Pomacentridae: Function in feeding and sound production. *Rev. Fish Biol. Fish.* **13,** 433–444.

Ripley, J. L., and Lobel, P. S. (2004). Correlation of acoustic and visual communication in the Lake Malawi cichlid *Tramitichromis intermedius*. *Env. Biol. Fish.* **71,** 389–394.

Rodríguez, R. L., Sullivan, L. E., and Cocroft, R. B. (2004). Vibrational communication and reproductive isolation in the *Enchenopa binotata* species complex of treehoppers (Hemiptera: Membracidae). *Evolution* **58,** 571–578.

Rosenthal, G. G. (1999). Using video playbacks to study sexual selection. *Env. Biol. Fish* **56,** 307–316.

Rosenthal, G. G. (2000). Design considerations and techniques for constructing video stimuli. *Acta Ethol.* **3,** 49–54.

Rosenthal, G. G., and Evans, C. S. (1998). Female preference for swords in *Xiphophorus helleri* reflects a bias for large apparent size. *PNAS* **95,** 4431–4436.

Rosenthal, G. G., and Ryan, M. J. (2000). Visual and acoustic communication in nonhuman animals: A comparison. *J. Biosci.* **25,** 285–290.

Rosenthal, G. G., Evans, C. S., and Miller, W. L. (1996). Female preference for a dynamic trait in the green swordtail, *Xiphophorus helleri*. *Anim. Behav.* **51,** 811–820.

Rosenthal, G. G., Flores Martinez, T. Y., García de León, F. J., and Ryan, M. J. (2001). Shared preferences by predators and females for male ornaments in swordtails. *Am. Nat.* **158,** 146–154.

Rosenthal, G. G., Wagner, W. E. J., and Ryan, M. J. (2002). Secondary loss of preference for swords in the pygmy swordtail *Xiphophorus nigrensis* (Pisces: Poeciliidae). *Anim. Behav.* **63,** 37–45.

Rountree, R., Goudey, C., and Hawkins, T. (2002)."Conference Proceedings: 8–10 April 2002 An International Workshop on the Applications of Passive Acoustics in Fisheries." Massachusetts Institute of Technology, Cambridge, MA.

Rowland, W. J. (1978). Sound production and associated behaviour in the jewel fish, *Hemichromis bimaculatus*. *Behaviour* **64,** 125–136.

Rowland, W. J. (1995). Do female stickleback care about male courtship vigour? Manipulation of display tempo using video playback. *Behaviour* **132,** 951–961.

Rowland, W. J. (1999). Studying visual cues in fish behaviour: A review of ethological techniques. *Env. Biol. Fish.* **56,** 285–305.

Ryan, M. J. (1986). Neuroanatomy influences speciation rates among anurans. *PNAS* **83,** 1379–1382.

Ryan, M. J., and Rosenthal, G. G. (2001). Variation and selection in swordtails. *In* "Model Systems in Behavioural Ecology" (Dugatkin, L. A., Ed.), pp. 138–148. Princeton University Press, Princeton, NJ.

Sargent, R. C., Rush, V. N., Wisenden, B. D., and Yan, H. Y. (1998). Courtship and mate choice in fish: Integrating behavioural and sensory ecology. *Am. Zool.* **38,** 82–96.

Sasaki, A., Ikejima, K., Aoki, S., Azuma, N., Kashimura, N., and Wada, M. (2003). Field evidence for bioluminescent signaling in the pony fish, *Leiognathus elongates*. *Env. Biol. Fish.* **66,** 307–311.

Satou, M., Takeuchi, H. A., Takei, K., Hasegawa, T., Matsushima, T., and Okumoto, N. (1994a). Characterisation of vibrational and visual signals which elicit spawning behaviour in the male hime salmon (landlocked red salmon). *J. Comp. Physiol. A* **174,** 527–537.

Satou, M., Takeuchi, H. A., Nishii, J., Tanabe, M., Kitamura, S., Okumoto, N., and Iwata, M. (1994b). Behavioural and electrophysiological evidences that the lateral line is involved in the inter-sexual vibrational communication of the Hime salmon (landlocked red salmon, *Oncorhynchus nerka*). *J. Comp. Physiol. A* **174,** 539–549.

Scholik, A. R., and Yan, H. Y. (2002a). The effects of noise on the auditory sensitivity of the bluegill sunfish, *Lepomis macrochirus*. *Comp. Biochem. Physiol. A* **133,** 43–52.

Scholik, A. R., and Yan, H. Y. (2002b). Effects of noise on auditory sensitivity of fish. *Bioacoustics* **12,** 186–188.

Scholz, A. T., Horrall, R., Cooper, J. C., and Hasler, A. D. (1976). Imprinting to chemical cues: The basis for home stream selection in salmon. *Science* **192,** 1247–1249.

Schwarz, A. L. (1974). The inhibition of aggressive behaviour by sound in a cichlid fish, *Cichlasoma centrarchus*. *Z. Tierpsychol.* **35,** 508–517.

Schwarz, A. L. (1980). Sound production and associated behaviour in a cichlid fish, *Cichlasoma centrarchus*, II. Breeding pairs. *Env. Biol. Fish.* **5,** 335–342.

Seehausen, O., Van Alphen, J. J. M., and Witte, F. (1997). Cichlid fish diversity threatened by eutrophication that curbs sexual selection. *Science* **277,** 1808–1811.

Shoji, H., Mochizuki, A., Iwasa, Y., Hirata, M., Watanabe, T., Hioki, S., and Kondo, S. (2003). Origin of directionality in the fish stripe pattern. *Develop. Dynam.* **226,** 627–633.

Siebeck, U. E., and Marshall, N. J. (2001). Ocular media transmission of coral reef fish – can coral reef fish see ultraviolet light? *Vis. Res.* **41,** 133–149.

Sloman, K. A., Scott, G. R., McDonald, D. G., and Wood, C. M. (2004). Diminished social status affects ionoregulation at the gills and kidney in rainbow trout (*Oncorhynchus mykiss*). *Can. J. Fish. Aquat. Sci.* **61,** 618–626.

Smith, R. J. F., and Lawrence, B. J. (1988). Effects of acute exposure to acidified water on the behavioural response of fathead minnows, *Pimephales promelas*, to alarm substance (Schreckstoff). *Environ. Toxicol. Chem.* **7,** 329–335.

Sorensen, P. W., and Scott, A. P. (1994). The evolution of hormonal sex pheromones in teleost fish: Poor correlation between the pattern of steroid release by goldfish and olfactory sensitivity suggests that these cues evolved as a result of chemical spying rather than signal specialisation. *Acta Physiol. Scand.* **152,** 191–205.

Spanier, E. (1979). Aspects of species recognition by sound in four species of damselfish, genus *Eupomacentrus* (Pisces: Pomacentridae). *Zeitshrift für Tierpsychologie* **51,** 301–316.

Stacey, N., Chojnacki, A., Narayanan, A., Cole, C., and Murphy, C. (2003). Hormonally derived sex pheromones in fish: Exogenous cues and signals from gonad to brain. *Can. J. Physiol. Pharmacol.* **81,** 329–341.

Stoddard, P. K. (1999). Predation enhances complexity in the evolution of electric fish signals. *Nature* **400,** 254–256.

Tallarovic, S. K., and Zakon, H. H. (2002). Electrocommunication signals in female brown ghost electric knifefish, Apteronotus leptorhynchus. *J. Comp. Physiol. A* **188,** 649–657.

Tavolga, W. N. (1958a). The significance of underwater sounds produced by males of the gobiid fish, *Bathygobius soporator*. *Physiol. Zool.* **31,** 259–271.

Tavolga, W. N. (1958b). Underwater sounds produced by males of the blenniid fish, *Chasmodes bosquianus*. *Ecology* **39,** 759–760.

Tavolga, W. N. (1958c). Underwater sounds produced by two species of toadfish *Opsanus tau* and *Opsanus beta*. *Bull. Mar. Sci. Gulf and Caribbean* **8,** 278–284.

Tavolga, W. N. (1971). Sound production and detection. *In* "Fish Physiology" (Hoar, W. S., and Randall, D. J., Eds.), pp. 135–205. Academic Press, New York.

Thorson, R. F., and Fine, M. L. (2002). Crepuscular changes in emission rate and parameters of the boatwhistle advertisement call of the Gulf toadfish, *Opsanus beta*. *Env. Biol. Fish.* **63,** 321–331.

Todd, J. H., Atema, J., and Bardach, J. E. (1967). Chemical communication in the social behaviour of a fish, the Yellow Bullhead, *Ictalurus natalis*. *Science* **158,** 672–673.

Turing, A. (1952). The chemical basis of morphogenesis. *Phil. Trans. R. Soc. Lond. B* **237,** 37–72.

Turnell, E. R., Mann, K. D., Rosenthal, G. G., and Gerlach, G. (2003). Mate choice in zebrafish (*Danio rerio*) analyzed with video-stimulus techniques. *Biol. Bull.* **5,** 225–226.

Unguez, G. A., and Zakon, H. H. (2002). Skeletal muscle transformation into electric organ in *S. macrurus* depends on innervation. *J. Neurobiol.* **53,** 391–402.

Urick, R. J. (1983). "Principles of underwater sound." McGraw-Hill, New York.

Warner, R. R. (2001). Synthesis: Environment, mating systems, and life history allocations in the bluehead wrasse. *In* "Model Systems in Behavioural Ecology: Integrating Conceptual, Theoretical, and Empirical Approaches" (Dugatkin, L. A., Ed.), pp. 227–244. Princeton University Press, Princeton, NJ.

Waybright, T. D., Kollenkirchen, U., and Fine, M. L. (1990). Effect of size and sex on grunt production in the oyster toadfish. *Soc. Neurosci. Abstr.* **16,** 578.

Wedekind, C., Meyer, P., Frischknecht, M., Niggli, U. A., and Pfander, H. (1998). Different carotenoids and potential information content of red colouration of male three-spined stickleback. *J. Chem. Ecol.* **24,** 787–801.

Weitzman, S. H., and Fink, S. V. (1985). Xenurobryconin phylogeny and putative pheromone pumps in glandulocaudine fish (Teleostei: Characidae). *Smithsonian Contributions to Zoology* **i–iii,** 1–121.

Weitzman, S. H., Menezes, N., and Weitzman, M. J. (1988). Phylogenetic biogeography of the Glandulocaudini (Teleostei: Characiformes, Characidae) with comments on the distributions of other freshwater fish in eastern and southeastern Brazil. *In* "Neotropical Distribution Patterns: Proceedings of a 1987 Workshop" (Heyer, W. R., and Vanzolini, P. E., Eds.), pp. 379–427. Academia Brasilerra de Ciencias, Rio de Janeiro.

Widder, E. A., Latz, M. F., Herring, P. J., and Case, J. F. (1984). Far-red bioluminescence from two deep-sea fish. *Science* **225,** 512–514.

Wilson, D. S., Coleman, K., Clark, A. B., and Biederman, L. (1993). The shy-bold continuum: An ecological study of a psychological trait. *J. Comp. Psychol.* **107,** 250–260.

Winn, H. E. (1964). The biological significance of fish sounds. *In* "Marine Bio-Acoustics" (Tavolga, W. N., Ed.), pp. 213–231. MacMillan, New York.

Winn, H. E. (1967). Vocal facilitation and the biological significance of toadfish sounds. *In* "Marine Bio-acoustics" (Tavolga, W. N., Ed.), pp. 283–304. Pergamon Press, New York.

Winn, H. E. (1972). Acoustic discrimination by the toadfish with comments on signal systems. *In* "Behaviour of Marine Animals: Current Perspectives in Research" (Winn, H. E., and Olla, B. L., Eds.), Vol. 2, pp. 361–385. Plenum Press, New York.

Woodland, D. J., Cabanban, A. S., Taylor, V. M., and Taylor, R. J. (2002). A synchronised rhythmic flashing light display by schooling *Leiognathus splendens* (Leiognathidae: Perciformes). *Mar. Freshwater Res.* **53,** 159–162.

Zakon, H. H. (1996). Hormonal modulation of communication signals in electric fish. *Dev. Neurosci.* **18,** 115–123.

Zakon, H. H. (2003). Insight into the mechanisms of neuronal processing from electric fish. *Curr. Op. Neurobiol.* **13,** 744–750.

Zakon, H. H., and Dunlap, K. D. (1999). Sex steroids and communication signals in electric fish: A tale of two species. *Brain Behav. Evol.* **54,** 61–69.

3

THE PHYSIOLOGY OF ANTIPREDATOR BEHAVIOUR: WHAT YOU DO WITH WHAT YOU'VE GOT

MARK ABRAHAMS

1. INTRODUCTION

If you are a small fish, it is not an unreasonable assumption that you may potentially be eaten by any predator in whose mouth you fit (Figure 3.1). On the flip side, potential food for a small fish is usually restricted to that which fits in your mouth. By staying small you are limiting yourself to a restrictive subset of potential food items. With this very simplistic view of life as a small fish, it is possible to generate two rules for success:

1. Get big.
2. Do it fast.

If small fish conform to these rules, then I predict that being small is only a brief phase of a fish's life, and that ultimately all small fish, regardless of

Behaviour and Physiology of Fish: Volume 24
FISH PHYSIOLOGY

DOI: 10.1016/S1546-5098(05)24003-7

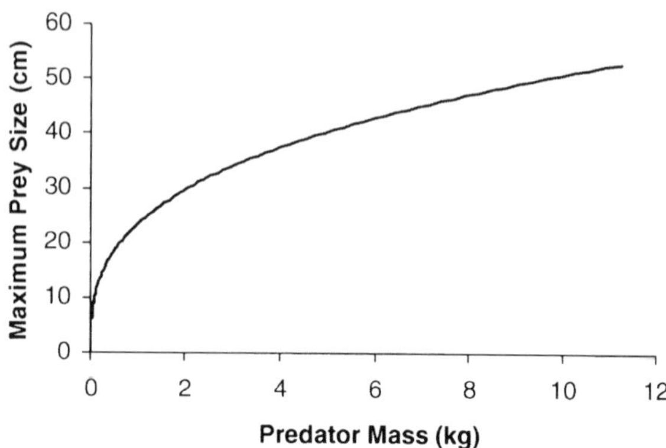

Fig. 3.1. The relation between the body mass of a brown trout predator (kg) and the maximum body length of Arctic charr prey (cm) it can consume. (Adapted from Nilsson, P. A., and Brönmark, C. (2000). Prey vulnerability to a gape-size limited predator: behavioural and morphological impacts on northern pike piscivory. *Oikos* **88**, 539–546, with permission from Blackwell Publishing.)

species, will seek to become as large as possible. You don't need to know a great deal about the diversity of different fish species to know that this prediction is wrong. Indeed, from an evolutionary perspective there is a trend for more derived species to be smaller than their ancestral form within North American fishes (Knouft and Page, 2003). A great many species never become large. And this is not due to some form of environmental constraint. Bringing these fish into captivity will not change the final outcome. These fish just remain small.

From the perspective of predator–prey interactions, it seems that a great many species just don't get it. They start small, grow at a relatively slow rate, and never achieve a very large final size. How come?

One potential explanation is that for many species, predation is just not a big deal. As a consequence, there may be little opportunity for selective predation by predators to impose such a phenotype on these species. A more pressing evolutionary force may be available habitat. Size of species is affected to a certain extent by the type of habitat, with smaller species tending to be found in smaller streams (Page and Burr, 1991). The vast majority of streams within North America are small (Leopold *et al.*, 1964), meaning that there is far more habitat available for small species (Knouft and Page, 2003). Another possibility is that these species really do want to become large, but are blocked by some sort of phylogenetic constraint. Or being small may not actually be that bad when dealing with predators. This Chapter will examine the latter argument.

In considering the role of physiology on antipredator behaviour, this Chapter considers the mechanisms as they relate to the three components that affect the probability of death from predation: the probability of being killed by a predator once detected, the probability of being detected by a predator, and the time spent vulnerable to predation (Lima and Dill, 1990).

2. THE PROBABILITY OF DEATH ONCE DETECTED BY A PREDATOR

2.1. The Benefits of Being Small

A popular tenet of many martial arts is to use the large size of your opponent to your own advantage. When you are a small fish being pursued by a much larger predator, there is a benefit associated with Newton's conservation of angular momentum. In simple terms, an object containing large mass (the predator) requires more energy to change direction than does an object containing less mass (the prey). This concept has been formalised in a mathematical theorem known as the Homicidal Chauffeur (Isaacs, 1965). This game posits the idea that you are trapped in a parking lot with a driver who is determined to kill you by running you down with their car and asks what are you going to do to provide yourself with the longest life expectancy (neither leaving the parking lot, phoning for assistance, nor acquiring some sort of military device are options).

The solution to the Homicidal Chauffeur is to veer perpendicular to the direction of motion of the car just before it gets close enough to strike you. The speed and mass of the car determine its minimum turning radius and so survival is determined by remaining within that turning radius. For a small fish trying to escape a predator, a single move of this sort may provide the opportunity for escape. This may come in the form of moving outside the detection range of the predator, finding a less vulnerable position within a group, or escaping into either a weed bed or rocky crevice.

Is there evidence that the Homicidal Chauffeur game is indeed a reasonable description of predator–prey interactions in fish? To adequately test this hypothesis, it is important to have good information about the angle at which the predator attacks the prey and the subsequent direction of escape by the prey, with the prediction that fleeing prey should delay their response to the last possible moment, and when they do they should flee in a direction perpendicular to that of the approaching predator. Although such research has not been addressed specifically with respect to this model, work has been done to investigate how the angle of escape is affected by the angle of approach by a predator. Webb and Skadsen (1980) found that minnows were attacked by pike (*Esox lucius*) with a mean angle of 82 degrees. No data were presented on their "escape angle." Domenici and Blake (1993) found that angelfish (*Pterophyllum eimekei*) have a preferential angular zone of escape between 130 and 180 degrees from a stimulus (sound emitted by two hydrophones). They argue such a response maximises the distance between the prey and an approaching predator and reflects subsequent changes of direction associated with the initial escape response (See also Domenici, 2002). Although the physics of predator–prey interactions are relatively simple, the underlying biology generates some complications.

In a recent paper, Odell *et al.* (2003) show that guppy (*Poecilia reticulata*) populations subject to a range of predator pressures exhibit corresponding changes in the physiology that underlies their ability to accelerate rapidly. They used second- and third-generation individuals that were derived from high and low predation regions within Trinidad. From these individuals, they obtained information on aerobic and burst swimming performance, morphological parameters (including an estimate of the swimming muscle mass), and enzymes associated with aerobic activity (citrate synthase), anaerobic activity (lactate dyhydrogenase), and speed of muscle fiber contraction (myofibrillar ATPase). Their data demonstrated that the predation regime significantly increased the swimming muscle mass and lactate dyhydrogenase activity, indicating the greater requirement for burst swimming performance in the presence of predators.

2.2. Avoidance at Close Range—The Mauthner System

When a predator is approaching rapidly and is very close, movement perpendicular to its direction may be a small fish's only hope for survival. Virtually every species of fish contains a reflex response that makes a rapid, appropriate response possible. This has been demonstrated by Eaton and Emberley (1991) in which a negative relation exists for the angle of initial orientation for the prey fish and the angle of the escape turn. All this is made possible by the Mauthner system (Eaton *et al.*, 1997; Zottoli and Faber, 2000). The Mauthner system consists of two very large myelinated nerve fibers located on either side of the fish's brainstem (Figure 3.2). These cells are connected to hair cells that are capable of detecting pressure waves on one side of the body, and then stimulating contraction of the muscles on the opposite side (Eaton *et al.*, 1997). In addition, once one cell has been stimulated, it acts to prevent the contralateral cell from responding. The net effect is to initiate a rapid response that should move the animal away from an approaching predator (Canfield and Rose, 1996). The response is also extremely rapid, with Mauthner cells firing within 3 to 4 ms of detecting the stimulus (Canfield and Rose, 1996). Contrast this with the human withdrawal response (i.e., the reflex that moves

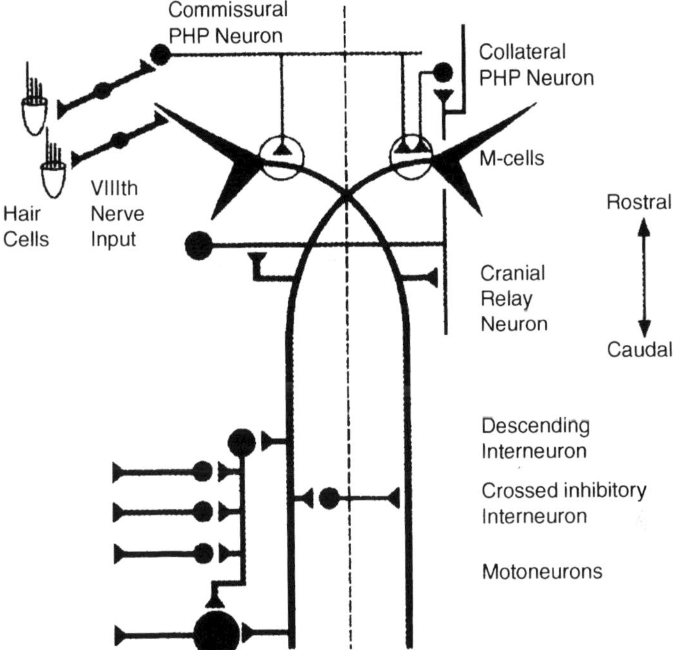

Fig. 3.2. Schematic illustration of Mauthner cells and their input and output neurons. Note that sensory neurons activate motoneurons on the side opposite stimulation and that the activated Mauthner cell inhibits the cell on the other side of the body. (Reprinted from Zottoli, S. J., and Faber, D. S. (2000). The Mauthner Cell: What has it taught us? *Neuroscientist* **6**, 25–37, with permission from Sage Publications.)

your hand away from a hot surface), which takes 19 to 24 ms (Ganong, 2001). However, as seen below, the distance at which this reaction is likely to be initiated will probably be when the predator is very close to the prey and hence this reaction must be very fast. Some species are able to extend the range of detection of pressure waves by having sensory cells detect pressure changes within the swimbladder (Schellart, 1992). In essence, the Mauthner system represents the last chance that a prey has of escaping a predator.

3. MAXIMISING THE PROBABILITY OF DETECTING A PREDATOR—THE DETECTION GAME

Ideally, prey would rather not find themselves in a position where they are relying upon a very rapid reflex response to escape a predator that is now very close. Indeed, whoever detects who first may well determine the outcome of a predator–prey interaction. It is important to note that should the prey detect the predator first, an immediate escape response is not always expected. Rather, the prevailing view among behavioural ecologists is that prey should make an adaptive response that weighs the benefits of remaining in its current location versus the costs of potentially being killed by a predator (Ydenberg and Dill, 1986). If the prey does detect the predator first, it may employ one of two options. The first is to communicate this information directly to the predator to allow both the prey and the predator to avoid a needless waste of energy (Hasson, 1991; Martin and López, 2001). This, of course, presents a conundrum because the prey must provide honest information to the predator that it has indeed been detected, such as precise information regarding its location. Although there are examples of such behaviour in fish (Sweatman, 1984), this does not seem to be the most prevalent system. The alternative is that once detected, small fish make an economic decision as to how close to allow the predator to approach prior to initiating their escape.

Most animals have no option but to incur some level of predation risk associated with most activities. As a consequence, it is reasonable to assume that both predator and prey are adapted to mutually detect and avoid each other. However, the mechanism of adaptation is constrained by variation in relative size. By definition, prey will be small and therefore present a smaller target to their predator. This may allow them to avoid detection until they get relatively close to a predator. In contrast, predators will be larger and therefore should have larger sensory apparatus that should increase the range at which prey are detected. From this perspective, predator–prey interactions are an information war. The weapon of choice will be dependent upon the dominant sensory modality.

Fish are capable of detecting their predators and prey using at least four different sensory modalities: electrical, sound (including pressure), chemical, and visual. This section will briefly discuss each sensory system with respect to predator–prey interactions, and then conclude with a discussion as to how fish may employ information from multiple sensory sources to make decisions associated with predator detection. For more detail on the sensory apparatus of fish, see Chapter 2.

3.1. Electrical

The ability of fishes to detect electrical information from within their environment was first described by Kalmijn (1966). To date, the ability to detect such information is limited to sharks, skates, and rays, as well as some catfishes, mormyrids, and Gymnotiformes. Two distinct mechanisms are known for using electrical information, passive and active detection. Passive detection requires that the fish detect the electrical field that naturally exists around all organisms. These fields are generated by a variety of processes that include the underlying electrophysiology and hence the voltage differential that exists between organisms and their environment, streaming potentials generated as a consequence of water moving along an animal's skin, and the induced voltages that are generated as an organism moves through the earth's magnetic field (Kalmijn, 1988). Because such electrical fields decline in intensity with the cube of distance, the effective distance of such information is limited. Kajiura and Holland (2002) demonstrated that juvenile scalloped hammerhead sharks (*Sphyrna lewini*) ranging between 53 and 80 cm in total length have a detection range of up to 30 cm on either side of their head, or approximately one-half body length.

Employing a self-generated electrical field, electric fishes in South America (Gymnotiformes) and Africa (Mormyridae) can use an active electrical detection system. These groups are not closely related but are convergent in a number of characters, including their sensory abilities. This system detects other animals as they pass through and distort this electrical field. As with the passive system, the active form of detection also has a very short range. Nelson and MacIver (1999) measured maximum detection ranges for 14 to 18 cm long black ghost knifefish (*Apteronotus albifrons*) at 2 cm when foraging for the small invertebrate prey *Daphnia magna*. Given the very short detection range of both active and passive electroreception, it is unlikely they provide much benefit to potential prey in the form of early detection of predators.

3.2. Sound and Pressure

The physical properties of water are considerably different from air, meaning that pressure waves associated with either sound or the movement of an object have a relatively low attenuation rate, and hence a greater range than in air. Sound also moves four times faster in water than in air (Dusenbery, 1992), making it potentially a very important source of long-range information that may provide early warning of an approaching predator. Yet the qualities that may make this an important source of information may also be its downfall. Because of low attenuation rates, sound and pressure within water will contain a large amount of background noise that can potentially mask important information (Blaxter, 1988).

With respect to sound and pressure, a distinction is made between near- and far-field behaviour. Near-field behaviour is the pressure wave generated by an object moving through an incompressible medium such as water and is analogous to the bow wave generated by a moving boat. Generally, near-field waves have a lower frequency and tend to attenuate at shorter distances than do sound (far-field waves). Fish are capable of detecting low frequency (i.e., 10 Hz) sound and pressure through the lateral line and inner ear sense organs at distances up to 100 m (Kalmijn, 1988). The lateral-line system also contains organs that are all oriented in different directions so that their combined response allows the orientation to the pressure wave to be determined (Coombs *et al.*, 1988). If an effective system exists for discriminating important information from background noise, the lateral line could provide effective early warning of an approaching predator, particularly for pelagic fish species.

The ability to detect sound and pressure is significantly enhanced by the presence of a swimbladder (Schellart, 1992). Due to differences in density, the swimbladder will vibrate more readily to environmental sound than other parts of the body, and this vibration can be transmitted to the ears by three specialised bones (the Weberian ossicles) in ostariophysans, anterior projections of the swimbladder in herrings, squirrelfish, and sciaenids, and connection via an auxiliary air bubble attached to the saccule in mormyrids (Fritzsch, 1999). Indeed, the swimbladder can be so important for hearing that some have speculated that the shape of the swimbladder is an adaptation that permits fish to detect the direction of sound (Barimo and Fine, 1998).

3.3. Vision

Within a predator–prey context, vision presents a very interesting paradox. As individuals become larger, so do their eyes. Larger eyes are better at gathering information from the environment. In a predator–prey context, this means a greater ability to detect potential predators and prey at greater distances. The paradox is that the larger eye, and hence larger size of an individual, then makes it visually more apparent to others that may be potential prey or predators.

The ability of the eye to provide visual information is quantified by visual acuity, which is the ability of the eye to resolve two points as separate objects. This is quantified by the following relation:

$$\frac{1}{\phi} = \frac{f}{s}$$

where Φ is the angular separation of the individual receptors with its inverse being proportional to visual acuity, f is the focal distance within the eye, and s is the separation distance between retinal elements (Goldsmith, 1990). Because the focal distance within the eye is the distance between the lens and the retina, increasing this distance (and hence the size of the eye) will increase visual acuity. Reducing the separation distance (s) between visual receptors is constrained by a minimum size limit of 1 to 2 μm: receptors smaller than this size cannot function effectively as a light waveguide (Kirschfield, 1976;

Goldsmith, 1990). Because there appear to be limited options to manipulating s, it may be assumed that the only option to increase visual acuity ($1/\Phi$) is to increase f by increasing the size of the eye (Kiltie, 2000).

From the predator's perspective, the issue then is whether visual acuity, and hence the distance of detection, increases at a rate greater than the distance at which their increased size makes them apparent to their prey. To the author's knowledge, no such information currently exists and the reason may well be that a number of other factors will become important. The ability to detect an object will not solely be determined by visual acuity, but will also be affected by the characteristics of the target itself, specifically how it contrasts against the visual background. Most species are well designed to avoid detection by reducing their visual contrast, although there are exceptions. In particular, these would include species that are sexually dimorphic, with one species possessing colors that simultaneously advertise their presence to both potential mates and predators (Pocklington and Dill, 1995). Other physical characteristics of the environment, such as ambient light levels and the clarity of the water, will modify visual ability. Finally, the sensory abilities of both predator and prey may not be limited to the abilities of individuals alone. Fish in shoals are capable of using the senses of all individuals present, providing pooled information that increases their ability to detect any external stimuli (Godin *et al.*, 1988).

3.4. Chemical

To chemically detect a predator, it is important that there be a chemical reliably associated with a predator both in space and time. Although this seems obvious, it does impose some important conditions on the chemical. In particular, for fish that occupy relatively small bodies of water, the chemical must either rapidly degrade or diffuse to levels which are not detectable. This is critical because otherwise the correlation between the presence of the chemical and the likelihood of encountering a predator will diminish. With a low correlation, detection of the chemical will provide little or no useful information regarding the presence of predators and the risk associated with remaining within a particular habitat.

There is abundant literature that describes the chemical detection of predators (reviewed in Kats and Dill, 1998). The best known chemical for detecting predators or recent predation is alarm substance (AS) or Schreckstoffe (Smith, 1992). This chemical is possessed by fish within the superorder Osatariophysi and is generally agreed to be hypoxanthine 3-N oxide (Pfeiffer *et al.*, 1985). The specific components that make this chemical biologically active have now been determined (Brown *et al.*, 2000, 2003). AS resides within the club cells of the epidermis of these fish and can only be released when the fish has suffered physical damage that ruptures these cells (i.e., when it has been injured by an attacking predator). Fish that are capable of detecting AS often exhibit a fright reaction that can include a range of specific antipredator behaviours such as evasive swimming maneuvers, reduced motion, and movement to different habitats. In the field, AS will cause individuals to avoid the area for up to 12 hours after release (Mathis and Smith, 1992; Chivers and Smith, 1994). Other individuals will move into this location after only 3 hours, demonstrating that detection of the chemical cause fish to remember a location as being dangerous long after the chemical is no longer present in their environment (see also Chapter 1).

Fish need not only rely upon injured conspecifics to release this chemical. Some predators such as pike (*Esox lucius*) will release this chemical after they have consumed an Ostariophysid (Brown *et al.*, 1995a,b). The value of this chemical is not restricted only to species that produce this pheromone. Recent experiments have demonstrated that species that do not produce AS are capable of detecting and responding to this chemical, including some salmonids (Brown and Smith, 1998) and sticklebacks (Brown and Godin, 1997).

Crucian carp (*Carassius carassius*) are also able to chemically detect their predators, although their response to this information is not restricted only to antipredator behaviour. Petterson *et al.* (2000) have demonstrated that when carp detect the odor of a predator that has consumed prey containing AS, they respond by modifying their pattern of growth. In the presence of these cues, individuals from populations that contain predators will alter their pattern of growth so that they become deeper bodied. The importance of this tactic is discussed further in Section 4.3.

AS is not the only chemical known to generate antipredator behaviour. Some salmonids are also known to respond to the presence of the amino acid L-serine that is commonly found on the skin of mammals (Alderdice *et al.*, 1954; Idler *et al.*, 1956). When this chemical is detected, these fish will cease moving and feeding, and it is believed that such a response will then prevent their response to many types of gear associated with sport fishing. Such a response indicates that animals are capable of detecting and using any information that is reliable in signaling the presence of danger.

3.5. Sensory Compensation

Information within an ecosystem is now recognised as a valuable commodity, and one that organisms are prepared to pay for (Koops and Abrahams, 2003). This payment is usually in the form of sensory apparatus, such as their sensory neurons. Maintenance of these structures is expensive, and there is now evidence that varying environmental conditions can alter investment in different sensory systems. Huber and Rylander (1992) examined eye size and the number of optic nerve fibers for six different species of minnows in the genera *Notropis* and *Cyprinella* in Texas and Oklahoma. They found that individuals that occupied turbid water had reduced eye size and as few as half the number of optic nerve fibers as those that occupied clear water. African cichlids also exhibit considerable variation in the extent to which their brains are developed for vision, olfaction, and mechanosensation (van Staaden *et al.*, 1995). In a review of the brain morphology of 189 species from Lakes Malawi, Tanganyika and Victoria, van Staaden *et al.* (1995) found that variation in light penetration within these lakes was the most important factor accounting for variation in visual and olfactory structures. Similarly, Brandstätter and Kotrschal (1990) found a negative correlation of olfactory and visual abilities in cyprinid brains. Kotrschal *et al.* (1998) have also found that cichlid species occupying locations with limited light penetration (e.g., turbid water or very deep locations) tended to invest more heavily in chemosenses and less so in visual detection. Schellart (1992) has reviewed the sensory capabilities of 63 marine and freshwater species and found that poor performance by one sense is usually compensated by enhanced abilities of other senses.

From this perspective, the evolutionary function of AS has been debated. Does it have as its primary function the role of communication (meaning that this chemical is a pheromone)? If so, have individuals evolved the ability to *produce* this chemical? Or does this chemical serve some other primary function, but individuals have evolved the ability to *detect* this chemical because its presence in the environment should be highly correlated with the presence of a predator?

If this chemical has evolved as a pheromone, then we must answer the evolutionary puzzle as to how such a system would evolve, because the only mechanism by which this pheromone can be released into the water is by an injury to the skin that ruptures the club cells. If this injury occurs from predation, the sender is likely in the jaws of a predator and therefore unlikely to benefit from this chemical. Two potential mechanisms have been proposed to provide a benefit to the sender. The first assumes that such individuals are likely in close contact with group members, and hence will receive a benefit either by increasing the survival of their kin or through reciprocal altruism that increases their own chance of survival (Smith, 1986). Another mechanism is that other predators are likely attracted by this chemical. These predators will then fight over the individual releasing this chemical, providing some finite chance of escape (Chivers *et al.*, 1996).

Such explanations have recently become suspect as fish apparently become less responsive to these chemical as the conditions of their environment approach natural conditions. Magurran *et al.* (1996) and Irving and

Magurran (1997) demonstrated that the response of European minnows to AS diminished within laboratory experiments as they added increasing structure to their environment. When these fish were placed in their natural environment, they exhibited no response to this chemical, leading them to conclude that fish may detect AS, but their response to this chemical depends upon context (but see Smith, 1997).

If AS is a signal, then individuals have the ability to detect it to obtain more information about the presence of predators. If this is the case, then there may be conditions under which individuals should react to this information and others where they will not. In contrast, if AS is a pheromone, individuals should react to its presence under all conditions. One simple explanation as to how such a system may operate is a sensory compensation model (Hartman and Abrahams, 2000). This model argues that individuals will react to some threshold concentration of AS, but the threshold concentration will be adjusted downwards (i.e., they will become more likely to react to AS) as fish perceive their environment to become more dangerous, and upwards as fish gain access to additional sources of information, such as vision (Figure 3.3). In laboratory experiments, Hartman and Abrahams (2000) manipulated risk of predation through the presence or absence of physical cover, or by altering the hunger levels of fish (hungry fish act as if

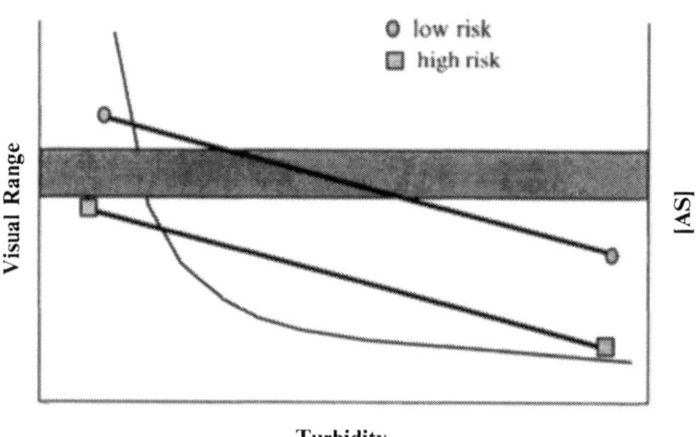

Fig. 3.3. The sensory compensation model. The grey bar indicates the concentration of alarm substance (AS) released. As indicated by the two parallel lines, the concentration of AS necessary to generate an antipredator response decreases as the level of predation risk increases. This threshold concentration also decreases in response to diminishing visual information due to increased turbidity. When faced with a high risk of predation, minnows should respond to a given concentration of AS in both clear and turbid water. With low risk, minnows should only respond to AS in turbid water. (Reprinted from Hartman, E. J., and Abrahams, M. V. (2000). Sensory compensation and the detection of predators: The interaction between chemical and visual information. *Proc. Roy. Soc. B* **267,** 571–575, with permission from the Royal Society.)

they perceive their environment as being less dangerous). The value of visual information was manipulated by running the experiments in either clear or turbid water. Consistent with the model predictions, fish reacted to the introduction of AS under all high-risk conditions and when access to visual information was limited. However, increased access to visual information and reduced risk of predation consistently eliminated their response to AS. These data are consistent with the interpretation by Magurran *et al.* (1996) that AS is a cue, not a pheromone, and that fish combine information from multiple senses in assessing the level of predation risk associated with their habitat. This model further assumes that most freshwater species with well-developed eyes preferentially use visual information to assess the risk of predation because it will provide more detailed information about predator location, intent, and size (Chivers *et al.*, 2001).

Similar results have been observed by Smith and Belk (2001), in which they found that western mosquitofish (*Gambusia affinis*) rely primarily upon visual cues when engaging in various dangerous behaviours such as predator inspection, but use chemical and visual information additively for general avoidance behaviour. However, the relative importance of visual and chemical information in detecting predators is still being debated (see Wisenden and Thiel, 2002; Brown and Magnavacca, 2003; Wisenden *et al.*, 2004).

4. TIME SPENT VULNERABLE TO PREDATION

4.1. Temperature and the Abiotic Environment

Most fish are ectothermic, and hence their physiology is greatly influenced by their thermal environment. However, the response of most ectothermic fishes is not a simple consequence of how the ambient environment influences the rate of biochemical reactions but rather has an associated optimum (Huey and Kingsolver, 1989). From the prey's perspective, they should be seeking such an environment, but not necessarily if it happens to be the same environment that is also optimal for their ectothermic predators.

Within temperate ecosystems, not all fish have the same optimal temperature. Freshwater fishes of North America have been categorised into three general thermal preferences (cold-, cool-, and warm-water guilds), with these preferences approximating the different stratified and climatic thermal regimes (Magnuson and DeStasio, 1996). Curiously, not all physiological functions appear to have the same thermal optima, and as a consequence it has been argued that some fish have a daily vertical migration that allows them to take advantage of different thermal regions to maximise energetic intake (Wurtsbaugh and Neverman, 1988). This migration involves fish moving from daytime, deep, cold regions to nighttime warm, shallow regions where they actively feed. It should be noted that a variety of different hypotheses have been posed to explain such shifts, including taking advantage of different light conditions that correlate with this vertical migration (see reviews by Lampert, 1989; Neilson and Perry, 1990). However, such a process does raise the intriguing possibility that use of shallow waters during the night or crespuscular periods may challenge visually-oriented predators.

It is known that the European minnow (*Phoxinus phoxinus*) and some salmonids become nocturnal at low water temperatures (Greenwood and Metcalfe, 1998). Although the reason for such a phenomenon is unknown, Greenwood and Metcalfe (1998) do speculate that many of these fishes' predators are diurnal endotherms. The low water temperature would therefore make them less able to escape attack and hence they become nocturnal to avoid exposure to such predators. Although this seems to be a reasonable explanation, there is no information that directly links temperature to predator–prey interactions in fish, but its impact is known for anurans. Anderson *et al.* (2001) found that increasing temperature was both a blessing and a burden for tadpoles. With increasing temperature, the tadpoles' growth rate increased. However, this growth had the associated cost of an increased mortality rate from predators. The underlying mechanism responsible for this result is not clear, although if movement rates are affected by temperature, this can generate such a result. Increased movement will increase encounter rates with both predators and prey (Werner and Anholt, 1993).

It is not unreasonable to assume a similar system operates for prey fish. As temperature increases, the energetic demands for fish will increase (Elliott, 1976). It is also known that increasing temperature will increase rates of movement for brown trout (*Salmo trutta*; Alanärä *et al.*, 2001). If both predator and prey are influenced in the same positive fashion by increasing temperature, then provided the temperature is neither becoming physiologically lethal nor significantly reducing dissolved oxygen levels (see below), the rate of predator–prey interactions should increase. However, changing temperature also generates an interesting caveat. Alanärä *et al.* (2001) have noted that the increased feeding rates will also increase rates of intraspecific competition such that individuals of lowest competitive ability may be unable to increase their feeding rate. Given that predators can be an important component in structuring an aquatic community (Werner *et al.*, 1983), those individuals unable to meet their increasing energetic demands may be the most likely to risk exposure to predators as a way to gain additional food (Abrahams and Cartar, 2000). However, it should be noted that even though gross rates of energetic intake can be achieved by feeding in more dangerous locations (Abrahams and Dill, 1989), the net effect may be reduced because such locations can generate an increased metabolic cost through fear bradycardia (Cooke *et al.*, 2003).

Assuming that size will impact relative competitive abilities, Hughes and Grand (2000) have developed a model that predicts temperature-based size segregation between habitats. The relation between competitive ability and maximum ration size will determine whether smaller or larger fish preferentially occupy cooler waters. Currently there are no tests of this model, although these data are consistent with experimental work by Krause *et al.* (1998).

4.2. Hypoxic Environments

As long as there has been water, there has been considerable variation in the levels of dissolved oxygen. This can occur through variation in temperature or salinity that alters the solubility of different gasses in water (known as Henry's Law). Any body of water will also have processes that add and remove dissolved oxygen. These include agitation at the water's surface that allows atmospheric oxygen to dissolve into the water, and photosynthesis that produces oxygen. Respiration by any aerobic organism, including decomposition, will remove dissolved oxygen. As a consequence, aquatic ecosystems that have low light and reduced mixing such as heavily vegetated wetlands, flooded forests, and floodplain lakes will tend to have low levels of dissolved oxygen. Similarly, benthic regions of lakes, water subjected to extreme solar or geothermic heating, and the salinity gradients around estuaries will also be hypoxic. The consistent availability of hypoxic environments means that adaptation to such conditions has been a major evolutionary force affecting fish design. Such adaptations include increased gill ventilation rates, (Johansen, 1982; Randall, 1982; Smith and Jones, 1982), increased gill perfusion and consequent increase in functional gill surface area (Booth, 1979; Johansen, 1982), increased blood oxygen carrying capacity and affinity (Powers, 1980), and the behaviour and morphology necessary for aquatic surface respiration. Comparative studies indicate that chronic exposure to hypoxic conditions result in enlarged gill surface area (Hughes and Morgan, 1973; Palzenberger and Pohla, 1992; Chapman *et al.*, 1999, 2000, 2002; Chapman and Hulen, 2001; Schaack and Chapman, 2003) and the use of anaerobic metabolism (Blažka, 1958; Holeton, 1980; Hochachka, 1986).

One mechanism by which fish can reduce aerobic metabolic pathways and activate anaerobic pathways is modification of lactate dehydrogenase isozyme distribution within major organs in response to long-term chronic hypoxia (Almeida-Val *et al.*, 1995). More recently, Almeida-Val *et al.* (2000)

analyzed the levels of lactate dehydrogenase and malate dehydrogenase within some of the major organs of one of the most hypoxia tolerant fish of the Amazon, *Astronotus ocellatus*. They found a positive relationship between hypoxia tolerance and body size, suggesting that the mechanism may be through the action of these enzymes increasing the anaerobic potential of these organs. But it is not clear whether the result from this species will apply generally to all fish species, because relationships between body size and the presence of these and related enzymes are not always observed (Somero and Childress, 1980; Pelletier *et al.*, 1993).

As individuals become larger, so too are they influenced by the forces that affect patterns of growth with increasing size (Schmidt-Nielson, 1984). A larger individual is not simply a scaled-up version of the smaller individual. Some parts of the body grow at rates different than other parts. One example is the negative allometric relationship for mass-specific gill-surface area (Muir, 1969; Hughes, 1984; Figure 3.4), suggesting that

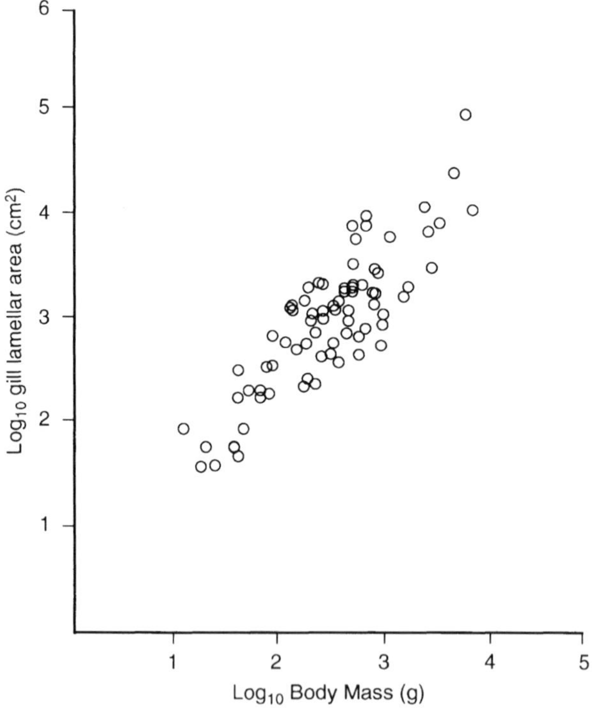

Fig. 3.4. The relation between $\log_{10}$ gill lamellar area and $\log_{10}$ body mass for several different fish species. (Reprinted from Muir, B. S. (1969). Gill dimensions as a function of fish size. *J. Fish. Res. Bd. Canada* **26,** 165–170, with permission from NRC Research Press.)

smaller individuals have more efficient gas exchange with their environment. Likewise, as individuals become larger, there is a well-known reduction in the respiratory rate per unit biomass (Schmidt-Nielson, 1984). When considering tolerance to hypoxic environments, a key component is whether factors that affect the rate of oxygen availability (i.e., gill surface-to-body volume ratios) are more than compensated by allometric scaling that reduces oxygen requirements.

A key parameter that may tip the balance in favor of the smaller fish is that respiratory rates can also be affected by fractal scaling. In its very simplest sense, allometric scaling relates to how two- and three-dimensional variables associated with areas and volume are affected as an animal becomes larger, measured in some one-dimensional parameter, usually either mass or length. Fractal scaling can describe how internal branching structures such as the circulatory system relate to animal size (West *et al.*, 1997, 1999). Given that the situation is significantly more complex, so too are the mathematics. However, there are really only two things that are important for this argument. First, for fractal scaling, there must be some component fundamental to the system under study that is relatively invariant as size changes. For gas transport, such a component is the red blood cell. The second is that the structure must somehow be a limiting factor affecting the physiology and survival of the organism in question. West *et al.* (1997, 1999) believe that because metabolic rates scale as three-quarters of the power of mass, and that such an exponent is inconsistent with allometric scaling, that the functional explanation associated with a fractal scaling model is the mechanistic argument from which this three-quarter power relationship is derived.

More recently, Darveau *et al.* (2002) have argued that metabolic rate is a complex process based upon many biological parameters, and that attempting to explain allometric scaling using a single parameter is folly. Their argument is compelling and may provide useful insight into understanding the processes that affect scaling relationships of metabolic rate. But hypoxia tolerance within fish may benefit from the approach advocated by West *et al.* (1997, 1999). A key distinction is that hypoxia tolerance represents an extreme challenge to the respiratory system and is determined by the physiological limits of the animal. As argued by Darveau *et al.* (2002), a problem with trying to develop a simple model to understand the scaling of basal metabolic rates is that they do not represent any physiological constraint, but reflect the combined energy requirements of all the cells within an organism. This rate can increase by many multiples to the maximum metabolic rate, which is dominated by the energy demand of the muscles, no longer all the cells of the body (Darveau *et al.*, 2002).

Regardless of the mechanism, there is evidence that small fish tend to be more tolerant of hypoxic environments than larger individuals. Robb and

Abrahams (2003) placed fathead minnows (*Pimephales promelas*) and their predator the yellow perch (*Perca flavescens*) in one of three different hypoxic environments. Within these hypoxic environments fathead minnows were capable of feeding and behaving normally for almost the entire duration of the three hour challenge, whereas yellow perch would lose equilibrium and have to be removed from the experiment after only 30 minutes (Figure 3.5). In addition, use of juvenile yellow perch that were the same size as the fathead minnows demonstrated an intermediate result that indicated that tolerance to hypoxia appears to be driven both by body size and species-specific parameters (Figure 3.5). Robb and Abrahams (2002) also demonstrated that fathead minnows were capable of visually detecting hypoxia stress in yellow perch at higher dissolved oxygen levels when these fish appeared outwardly healthy. These data suggest that if there is a size-based variation in hypoxia tolerance, then small fish may intentionally seek such locations to avoid attacks by their predators.

There is already some evidence in support of this hypothesis. Many lakes in temperate climates freeze over during the winter. When snow covers the

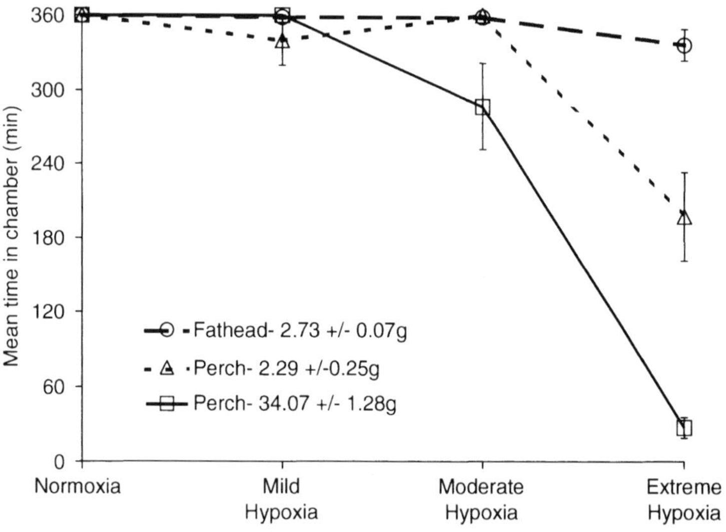

Fig. 3.5. The effect of variation in fish size and species on tolerance to a hypoxic environment. Fish were placed in an apparatus that had dissolved oxygen concentrations (measured in mg/l) of 7.2 (normoxic), 3.5 (mild hypoxic), 2.5 (hypoxic) and 1.8 (extreme hypoxic). Results measure the mean tolerance time (± 1 SE) spent in the chamber for all fish types (fathead minnow, small yellow perch or large yellow perch). (Reprinted from Robb, T., and Abrahams, M. V. (2003). Variation in tolerance to hypoxia in a predator and prey species: An ecological advantage to being small? *J. Fish Biol.* **62,** 1067–1081, with permission from Blackwell Publishing.)

ice, there is no longer any solar input into these lakes, stopping photosynthesis and hence the production of any additional dissolved oxygen to the lake. As a consequence, winter means a progressive decline in dissolved oxygen concentration. The longer the winter, the lower dissolved oxygen levels become. Klinger *et al.* (1982) and Fox and Keast (1990) have reported higher mortality of larger individuals within a species, suggesting that smaller individuals are more tolerant of hypoxic conditions. In Lake Victoria, the Nile Perch was introduced to provide a large species that could be harvested for food by fishers in the region. The Nile Perch did exceedingly well in this lake, consuming species within the highly diverse haplochromine assemblage and driving many species extinct. This change in the species composition of Lake Victoria fundamentally changed the lake's limnology. What was once a mesotrophic lake dominated by diatoms became a eutrophic lake dominated by blue-green algae (Hecky, 1993). Primary productivity doubled and ultimately the physical characteristics of the lake changed (Balirwa *et al.*, 2003). These included the development of hypolimnetic anoxia, almost year-round stratification of the deeper parts of the lake, and increasing hypoxia in shallower waters (Wanink *et al.*, 2001). Research by Chapman *et al.* (1996a,b, 1999) found that several species of haplochromines that were believed to be extinct were in fact extant within wetlands adjacent to the lake. These environments not only provide the structural complexity to assist fish in evading predators, but also contain hypoxic water that can not be tolerated by the Nile perch.

The most direct evidence of how hypoxic environments may affect predator–prey relationships comes from Rahel and Nutzman (1994). They captured mudminnows (*Umbra limi*) foraging for prey in the deep hypoxic waters of a bog lake. Restraining the mudminnows within the hypoxic regions in which they were feeding was lethal. This result indicates that the predatory mudminnows have modified their behaviour to allow them to forage in a lethal environment for short periods. Prey movement to this location does not provide a complete refuge from predators, but presumably the requirement to "dive" into this hypoxic region must constrain the predator's foraging ability.

Likewise, predatory fish may be able to take advantage of areas that have a horizontal distribution of hypoxic water. In particular, piscivorous fish that are capable of aquatic surface respiration will be able to enter such locations. The interesting caveat with such an approach is that the predators will be required to spend a considerable amount of their time near the surface of the water, rendering them particularly vulnerable to aerial predators (Kramer, 1987). Should such a situation exist within shallow hypoxic waters, then vulnerable prey species may effectively be protected by their predator's predators—a tritrophic predator refuge.

It should be noted that a variety of mechanisms will affect the ability of small fish to tolerate hypoxic environments and hence not every situation will fit the size-based model described above.

4.3. Morphology

Of more importance to prey are tactics to allow them to escape predation, and specifically to defeat predators by employing the most effective technique of all—not fitting in their mouth. Relationships between gape limitation (the biggest possible prey that can be consumed) and body size have been characterised for several species (Damsgård, 1995; Persson *et al.*, 1996; Nilsson and Brönmark, 2000; Magnhagen and Heibo, 2001; see Figure 3.1). What all of these studies tell us is that for a very large predator to be capable of consuming a prey that has increased its body depth, it must compensate by increasing its body mass as a cubic function of this change in length—or a considerable change in body mass is required to exploit larger size classes of prey. This relation has not been lost on prey species. Crucian carp are known to alter their pattern of growth when exposed to predators (Brönmark and Miner, 1992) and more recent experiments have shown that this altered growth is stimulated by chemical cues from predators that have consumed prey containing alarm substance (Petterson *et al.*, 2000). Their response is to alter their normal allometric pattern of growth by becoming deep bodied and hence escaping predation by larger pike more rapidly than retaining their normal growth trajectory. The antipredator benefits of larger body size are amplified further when recognising that predators also have predators. The increased handling time associated with consuming larger prey can result in pike preferring prey smaller than would be predicted when considering only energetics (Nilsson and Brönmark, 1999). Because this is an inducible defense, there presumably must be some cost; otherwise, carp should adopt this morphology as a normal growth pattern. Pettersson and Brönmark (1999) theorised that this deep-bodied morphology must have a higher drag and therefore must be more energetically demanding when moving. Their laboratory experiments showed that this explanation was not quite that simple. Deep-bodied carp actually had a lower standard metabolic rate, and when moving at the velocity associated with the minimum cost of transport did not exhibit the increased energy consumption predicted from the increased drag hypothesis. However, when they increased their velocity there was a significant increase in energy requirements beyond what was observed for the normal morph, suggesting that they would fare poorly under intense intraspecific competition with the normal morph. This result provides a functional explanation for their earlier field experiments, in which they demonstrated that the deep-bodied morph had an impaired ability to compete for food (Pettersson and Brönmark, 1997).

Changing overall body shape is not the only potential response to predators. Sticklebacks appear to develop lateral plates and armour in habitats that have greater densities of predators (Gross, 1978; Bell and Richkind, 1981). The combination of erect spines and lateral plates can render these fish immune to smaller predators, although this morphology may make them more vulnerable to invertebrate predators (Reimchen, 1980, 1983; Reist, 1980; Ziuganov and Zotin, 1995; Vamosi, 2002). Populations of *Galaxias platei* within Andean lakes also have a reduced caudal fin where they are associated with predators (Milano *et al.*, 2002). This morphology is hypothesised to make these fish more maneuverable and hence better able to escape attack by a predator. With this reduction in susceptibility to predation, these fish may then reap a competitive benefit when foraging for food in the presence of predators. The modification in behaviour in response to the risk of predation should be in proportion to the threat. If the threat is diminished by an altered morphology, these animals will not only enjoy reduced predation, but can also obtain a competitive advantage over other species that lack this morphological advantage (Grand, 2000; Abrahams, 1994, 1995). Antipredator morphology will allow these individuals to exploit the niche created by the risk of predation.

One additional option that prey have after encountering and successfully avoiding a predator is to hide within a refuge. Once within the refuge the prey is safe, but how long should prey stay there? Dill and Gillett (1991) provide an idea based upon economic theory, but the solution to this problem will be dependent upon the costs of remaining within a refuge. In particular, some species vary in their metabolic rates, meaning that the energetic costs associated with hiding are not equal for all species or individuals within a species. Krause *et al.* (2000) show that for two sympatric freshwater fishes, three-spine sticklebacks lose weight when in a nonfeeding refuge at a rate greater than do European minnow (*Phoxinus phoxinus*). They also observed that minnows tended to remain within a refuge for longer than the sticklebacks. They interpret this variation in refuge use as being consistent with the variation in cost associated with its use.

5. SO WHAT DO YOU DO WITH WHAT YOU'VE GOT?

There is an old joke about two field biologists who are preparing to work in a location that contains large numbers of grizzly bears. One biologist spots his colleague training on a local outdoor track. He asks why he's training and the answer is that if they are attacked by a grizzly bear, he wants to be prepared. The first biologist scoffs, "That's ridiculous; you can't outrun a grizzly bear." To that his colleague responds, "I don't need to outrun the bear, I just need to outrun you."

The point of this story is that when examining what antipredator options are available, you must consider both the physiology of the animal and that of the other individuals and species that are likely to be subject to predation by the same predators. There has been considerable attention devoted to understanding which prey should be selected by predators to allow the predators to maximise their net rate of energy intake (see Stephens and Krebs, 1986). One specific component of this research is determining the characteristics of individual potential prey and whether they should or should not be included in a predator's diet. The detail of this theory is unnecessary here, but the main point is that three major components affect inclusion within a diet: the rate at which prey are encountered within the environment, the time it takes upon encounter to capture the prey, and the net energetic content of the prey. All that has been presented in this Chapter falls generally into these three categories. Prey physiology can influence their likelihood of encountering a predator by winning the detection war. Investment in sensory modalities will allow them to detect a predator before it detects them. If prey are able to do this, they will never be encountered and hence never consumed. This is unlikely to happen but also unnecessary. They need only win more of these detection games than other species within their environment. Likewise, unique features of a small fishes' physiology may allow them to occupy habitats that are inaccessible to their predators. Although such a situation likely has energetic costs, it does provide the ultimate opportunity to drive encounter rates with predators to zero.

However, prey cannot devote their entire lives to avoiding predation. If they did, they would be unable to gather resources, find mates, and pass this trait on to new generations. Rather, prey must ultimately risk exposure to predators to gain access to important resources, and they will have some finite encounter rate with predators. But even under these conditions, their physiology is partially designed to render them less preferred as a diet item. Some prey may contain potent toxins that render them poisonous to predators, or they may mimic such species to get off the menu (Gittleman and Harvey, 1980). However, even in the absence of this ultimate strategy, prey physiology possesses other tactics. These include the benefit of small size and increased maneuverability. This feature in conjunction with the Mauthner system means that predators that pursue their prey will often require considerable effort to capture them.

The validity of the "grow big and do it fast" rule can now be tested directly through hormonal and genetic manipulation of fish that elevates and sustains growth hormone levels through all seasons (Devlin *et al.*, 1994; Du *et al.*, 1992). The impact of such manipulations upon the fishes' phenotypes has been profound and includes fish that can be up to ten times larger by 16 months of age and achieve sexual maturity in approximately one-half the time (Devlin *et al.*, 1995). The intriguing feature of genetically manipulated fish is that they are heterozygous for the transgene, meaning that a sib group will contain individuals that do and do not express this altered phenotype. Challenging genetically modified and wild-type Atlantic salmon with foraging decisions that involve a risk of predation also reveals a variation in behaviour that matches the variation in morphology. Size-matched individuals differ markedly in their willingness to risk exposure to a predator to gain access to additional food, whether this predator is restrained behind a transparent Plexiglas partition or if the fish must physically enter a location containing a predator (Abrahams and Sutterlin, 1999). Under both circumstances, salmon that have artificially elevated growth rates are much more willing to risk exposure to a predator and hence gain access to additional food.

Of course, to determine whether the modification in behavioural rules is associated with this altered growth pattern must ultimately involve a determination of whether the transgenic fish pay a much higher cost: are they more likely to be killed by predators under natural conditions? For obvious reasons, the simple experiment of releasing genetically modified salmon into the wild to assess their survival rates cannot be performed. To adequately understand the potential costs associated with transgenic manipulations, it is necessary to challenge these animals with environmental conditions that mirror those they will encounter in the field. This has now been done with newly hatched salmon and these data demonstrate that a mortality cost is associated with growth acceleration (Sundström *et al.*, 2004). Similar experiments have been attempted with transgenic smolts and so far have shown no mortality cost associated with the greater willingness of transgenic salmon to risk exposure to a predator (Abrahams and Devlin, unpublished data). However, challenging these larger-size classes of individuals with environmental conditions that mirror those they will face in the wild, while simultaneously meeting guidelines for the containment of genetically modified fish, make these experiments very difficult to perform. It is therefore not clear whether a failure to identify mortality costs for larger-size classes truly reflects less cost or an inability to adequately challenge these fish.

Almost all small fish species will have to confront some risk of predation associated with their evolutionary success. Whether feeding, finding a mate, obtaining and defending a spawning site or territory, or provisioning their young, there is some likelihood they will encounter a predator. The physiology of most small species is adapted to dealing with this problem, and not simply by following the "grow fast and quickly" rule presented at the opening of this chapter. Their options are much more diverse and often benefit from the size difference between themselves and their predators.

ACKNOWLEDGEMENTS

I greatly appreciate the generosity of Geoff Eales, Ted Wiens, Robert MacArthur, Jennifer McLeese, and Kevin Campbell for lending an ear and providing advice on some of the topics discussed in this chapter. This chapter also benefited from the critical comments of Lauren Chapman, Gail Davoren, Larry Dill, Anne Magurran and one anonymous referee.

REFERENCES

Abrahams, M. V. (1995). The interactions between antipredator behaviour and antipredator morphology: Experiments with fathead minnows and brook sticklebacks. *Can. J. Zool.* **73,** 2209–2215.

Abrahams, M. V. (1994). Risk of predation and its influence on the relative competitive abilities of two species of freshwater fishes. *Can. J. Fish. Aquat. Sci.* **51,** 1629–1633.

Abrahams, M. V., and Cartar, R. V. (2000). Within group variation in the willingness to risk exposure to a predator: The influence of species and size. *Oikos* **89,** 340–344.

Abrahams, M. V., and Dill, L. M. (1989). A determination of the energetic equivalence of the risk of predation. *Ecology* **70,** 999–1007.

Abrahams, M. V., and Sutterlin, A. (1999). The foraging and antipredator behaviour of growth enhanced transgenic Atlantic Salmon. *Anim. Behav.* **58,** 933–942.

Alanärä, A., Burns, M. D., and Metcalfe, N. B. (2001). Intraspecific resource partitioning in brown trout: The temporal distribution of foraging is determined by rank. *J. Anim. Ecol.* **76,** 980–986.

Alderdice, D. F., Brett, J. R., Idler, D. R., and Fagerlund, U. (1954). Further observations on olfactory perception in migrating adult coho and sping salmon: Properties of the repellent in mammalian skin. *Fish. Res. Bd. Canada* **98,** 10–12.

Almeida-Val, V. M. F., Val, A. L., Duncan, W. P., Souza, F. C. A., Paula-Silva, M. N., and Land, S. (2000). Scaling effects on hypoxia tolerance in the Amazon fish *Astronotus ocellatus* (Perciformes: Cichlidae): Contribution of tissue enzyme levels. *Comp. Biochem. Phys. B* **125,** 219–226.

Almeida-Val, V. M. F., Farias, I. P., Silva, M. N. P., Duncan, W. P., and Val, A. L. (1995). Biochemical adjustments to hypoxia by Amazon cichlids. *Braz. J. Med. Biol. Res.* **28,** 1257–1263.

Anderson, M. T., Kiesecker, J. M., Chivers, D. P., and Blaustein, A. R. (2001). The direct and indirect effects of temperature on a predator-prey relationship. *Can. J. Zool.* **79,** 1834–1841.

Bell, M. A., and Richkind, K. E. (1981). Clinal variation of lateral plates in threespine stickleback fish. *Am. Nat.* **117,** 113–132.

Balirwa, J. S., Chapman, C. A., Chapman, L. J., Cowx, I. G., Geheb, K., Kaufman, L., Lowe-McConnell, R. H., Seehausen, O., Wanink, J. H., Welcomme, R. L., and Witte, F. (2003). Biodiversity and Fishery Sustainability in the Lake Victoria Basin: An Unexpected Marriage? *Bioscience* **53,** 703–715.

Barimo, J. F., and Fine, M. L. (1998). Relationship of swim-bladder shape to the directionality pattern of underwater sound in the oyster toadfish. *Can. J. Zool.* **76,** 134–143.

Blažka, P. (1958). The anaerobic metabolism of fish. *Phys. Zool.* **31,** 117–128.

Blaxter, J. H. S. (1988). Sensory performance, behaviour, and ecology of fish. *In* "Sensory Biology of Aquatic Animals" (Atema, J., Fay, R. R., Popper, A. N., and Tavolga, W. N., Eds.), pp. 203–232. Springer-Verlag, New York.

Booth, J. H. (1979). The effect of oxygen supply, epinephrine and acetycholine on the distribution of blood flow in trout gills. *J. Exp. Biol.* **83,** 31–39.

Brandstätter, R., and Kotrschal, K. (1990). Brain growth patterns in four European cyprinid fish species (Cyprinidae, Teleostei); roach (*Rutilus rutilus*), bream (*Abramis brama*), common carp (*Cyprinus carpio*) and sabre carp (*Pelecus cultratus*). *Brain Behav. Evol.* **35,** 195–211.

Brönmark, C., and Miner, J. G. (1992). Predator-induced phenotypical change in body morphology in crucian carp. *Science* **258,** 1348–1350.

Brown, G. E., Adrian, J. C., Jr., Naderi, N. T., Harvey, M. C., and Kelly, J. M. (2003). Nitrogen-oxides elicit antipredator responses in juvenile channel catfish, but not convict cichlids or rainbow trout: Conservation of the Ostariophysan alarm pheromone. *J. Chem. Ecol.* **29,** 1781–1796.

Brown, G. E., and Magnavacca, G. (2003). Predator inspection behaviour in a characin fish: An interaction between chemical and visual information? *Ethology* **109,** 739–750.

Brown, G. E., Adrian, J. C., Jr., Smyth, E., Leet, H., and Brennan, S. (2000). Ostariophysan alarm pheromones: Laboratory and field tests of the functional significance of nitrogen-oxides. *J. Chem. Ecol.* **26,** 139–154.

Brown, G. E., and Godin, J.-G. J. (1997). Anti-predator responses to conspecific and heterospecific skin extract by threespine sticklebacks: Alarm pheromones revisited. *Behaviour* **134,** 1123–1134.

Brown, G. E., and Smith, R. J. F. (1998). Acquired predator recognition in juvenile rainbow trout (*Oncorhynchus mykiss*): Conditioning hatchery reared fish to recognise chemical cues of predator. *Can. J. Fish. Aquat. Sci.* **55,** 611–617.

Brown, G. E., Chivers, D. P., and Smith, R. J. F. (1995a). Fathead minnows avoid conspecific and heterospecific alarm pheromones in the faeces of northern pike. *J. Fish Biol.* **47,** 387–393.

Brown, G. E., Chivers, D. P., and Smith, R. J. F. (1995b). Localised defecation by pike: A response to labelling by cyprinid alarm pheromone? *Behav. Ecol. Sociobiol.* **36,** 105–110.

Canfield, J. G., and Rose, G. J. (1996). Hierarchical sensory guidance of Mauthner-mediated escape responses in goldfish (*Carassius auratus*) and cichlids (*Haplochromis burtoni*). *Brain Behav. Evol.* **48,** 137–156.

Chapman, L. J., Chapman, C. A., Nordlie, F. G., and Rosenberger, A. E. (2002). Physiological refugia: Swamps, hypoxia tolerance, and maintenance of fish biodiversity in the Lake Victoria Region. *Comp. Biochem. Phys.* **133(A),** 421–437.

Chapman, L. J., Chapman, C. A., Brazeau, D., McGlaughlin, B., and Jordan, M. (1999). Papyrus swamps and faunal diversification: Geographical variation among populations of the African cyprinid *Barbus neumayeri*. *J. Fish Biol.* **54,** 310–327.

Chapman, L. J., Chapman, C. A., and Chandler, M. (1996a). Wetland ecotones as refugia for endangered fishes. *Biol. Conserv.* **78,** 263–270.

Chapman, L. J., Chapman, C. A., Ogutu-Ohwayo, R., Chandler, M., Kaufman, L., and Keiter, A. E. (1996b). Refugia for endangered fishes from an introduced predator in Lake Nabugabo, Uganda. *Cons. Biol.* **10,** 554–561.

Chapman, L. J., Galis, F., and Shinn, J. (2000). Phenotypic plasticity and the possible role of genetic assimilation: Hypoxia-induced trade-offs in the morphological traits of an African cichlid. *Ecol. Lett.* **3,** 388–393.

Chapman, L. J., and Hulen, K. (2001). Implications of hypoxia for the brain size and gill surface area of mormyrid fishes. *J. Zool.* **254,** 461–472.

Chivers, D. P., Brown, G. E., and Smith, R. J. F. (1996). Evolution of chemical alarm signals: Attracting predators benefits alarm signal senders. *Am. Nat.* **148,** 649–659.

Chivers, D. P., Mirza, R. S., Bryer, P. J., and Kiesecker, J. M. (2001). Threat-sensitive predator avoidance by slimy sculpins: Understanding the role of visual versus chemical information. *Can. J. Zool.* **79,** 867–873.

Chivers, D. P., and Smith, R. J. F. (1994). Intra- and interspecific avoidance of areas marked with skin extract from brook sticklebacks (*Culaea inconstans*) in a natural habitat. *Env. Biol. Fishes* **49,** 89–96.

Cooke, S. J., Steinmetz, J., Degner, J. F., Grant, E. C., and Philipp, D. P. (2003). Metabolic fright responses of different-sized largemouth bass (*Micropterus salmoides*) to two avian predators show variations in nonlethal energetic costs. *Can. J. Zool.* **81,** 699–709.

Coombs, S., Janssen, J., and Webb, J. C. (1998). Diversity of lateral line systems: Evolutionary and functional considerations. *In* "Sensory Biology of Aquatic Animals" (Atema, J., Fay, R.R., Popper, A. N., and Tavolga, W. N., Eds.), pp. 553–594. Springer-Verlag, New York.

Damsgård, B. (1995). Arctic charr, *Salvelinus alpinus* (L.), as prey for piscivorous fish: A model to predict prey vulnerabilities and prey size refuges. *Nord. J. Fresh. Res.* **71,** 190–196.

Darveau, C.-A., Suarez, R. K., Andrews, R. D., and Hochachka, P. W. (2002). Allometric cascade as a unifying principle of body mass effects on metabolism. *Nature* **417,** 166–170.

Devlin, R. H., Yesaki, T. Y., Donaldson, E. M., Du, S. J., and Hew, C.-L. (1995). Production of germline transgenic Pacific salmonids with dramatically increased growth performance. *Can. J. Fish. Aquat. Sci.* **52,** 1376–1384.

Devlin, R. H., Yesaki, T. Y., Biagi, C. A., Donaldson, E. M., Swanson, P., and Chan, W.-K. (1994). Extraordinary salmon growth. *Nature* **371,** 209–210.

Domenici, P. (2002). The visually mediated escape response in fish: Predicting prey responsiveness and the locomotor behaviour of predators and prey. *Mar. Fresh. Behav. Physiol.* **35,** 87–110.

Domenici, P., and Blake, R. W. (1993). Escape trajectories in angelfish (*Pterophyllum eimekei*). *J. Exp. Biol.* **177,** 253–272.

Dill, L. M., and Gillett, J. F. (1991). The economic logic of barnacle *Balanus glandula* (Darwin) hiding behaviour. *J. Exp. Mar. Biol. Ecol.* **153,** 115–127.

Du, S. J., Gong, Z., Fletcher, G. L., Shears, M. A., King, M. J., Idler, D. R., and Hew, C.-L. (1992). Growth enhancement in transgenic Atlantic salmon by the use of an "all fish" chimeric growth hormone gene construct. *Biotechnology* **10,** 176–181.

Dusenbery, D. B. (1992). "Sensory Ecology: How organisms acquire and respond to information." W.H. Freeman and Co., New York.

Eaton, R. C., and Emberley, D. S. (1991). How stimulus direction determines the trajectory angle of the Mauthner initiated escape response in a teleost fish. *J. Exp. Biol.* **161,** 469–487.

Eaton, R. C., Guzik, A. L., and Casagrand, J. L. (1997). Mauthner system discrimination of stimulus direction from the acceleration and pressure components at sound onset. *Biol. Bull.* **192,** 146–149.

Elliott, J. M. (1976). The energetics of feeding, metabolism and growth of brown trout (*Salmo trutta* L.) in relation to body weight, water temperature, and ration size. *J. Anim. Ecol.* **45,** 923–948.

Fox, M. G., and Keast, A. (1990). Effects of winterkill on population structure, body size and prey consumption patterns of pumpkinseed in isolated beaver ponds. *Can. J. Zool.* **68,** 2489–2498.

Fritzsch, B. (1999). Hearing in two worlds: Theoretical and actual adaptive changes of the aquatic and terrestrial ear for sound reception. *In* "Comparative Hearing: Fish and Amphibians" (Fay, R. R., and Popper, A. N., Eds.), pp. 15–42. Springer-Verlag, New York.

Ganong, W. F. (2001). "Review of Medical Physiology" 20th ed., Lange Medical Books/McGraw-Hill Medical Publishing Division, New York.

Gittleman, J. L., and Harvey, P. H. (1980). Why are distasteful prey not cryptic? *Nature* **286,** 149–150.

Godin, J.-G..J., Classon, L. J., and Abrahams, M. V. (1988). Group vigilance and shoal size in a small characin fish. *Behaviour* **104,** 29–40.

Goldsmith, T. H. (1990). Optimisation, constraint, and history in the evolution of eyes. *Quart. Rev. Biol.* **65,** 281–322.

Grand, T. C. (2000). Risk-taking by threespine stickleback (*Gasterosteus aculeatus*) pelvic phenotypes: Does morphology predict behaviour? *Behaviour* **137,** 889–906.

Greenwood, M. F. D., and Metcalfe, N. B. (1998). Minnows become nocturnal at low temperatures. *J. Fish Biol.* **53,** 25–32.

Gross, H. P. (1978). Natural selection by predators on the defensive apparatus of the three-spined stickleback, *Gasterosteus aculeatus* L,. *Can. J. Zool.* **56,** 398–413.

Hartman, E. J., and Abrahams, M. V. (2000) Sensory compensation and the detection of predators: The interaction between chemical and visual information. *Proc. Roy. Soc. B* **267,** 571–575.

Hasson, O. (1991). Pursuit-deterrent signals: Communication between prey and predator. *TREE* **6,** 325–329.

Hecky, R. E. (1993). The eutrophication of Lake Victoria. *Berhandlungen der-* Internationale Vereinigung für Theoretische und Angewandte Limnologie **25,** 39–48.

Hochachka, P. W. (1986). Defence strategies against hypoxia and hypothermia. *Science* **231,** 234–241.

Holeton, G. F. (1980). Oxygen as an environmental factor of fishes. *In* "Environmental Physiology of Fishes" (Ali, M. A., Ed.), pp. 7–32. Plenum, New York.

Huber, R., and Rylander, M. K. (1992). Quantitative histological study of the optic nerve in species of minnows (Cyprinidae, Teleostei) inhabiting clear and turbid water. *Brain Behav. Evol.* **40,** 250–255.

Huey, R. B., and Kingsolver, J. G. (1989). Evolution of thermal sensitivity of ectotherm performance. *TREE* **4,** 131–135.

Hughes, G. M. (1984). Scaling of respiratory areas in relation to oxygen consumption of vertebrates. *Experientia* **40,** 519–652.

Hughes, G. M., and Morgan, M. (1973). The structure of fish gills in relation to their respiratory function. *Biol. Rev.* **48,** 419–475.

Hughes, N. F., and Grand, T. C. (2000). Physiological ecology meets the ideal free distribution: Predicting the distribution of size-structured fish populations across temperature gradients. *Env. Biol. Fishes* **59,** 285–298.

Idler, J. R., Fagerlund, U. H. M., and Mayoh, H. (1956). Olfactory perception in migrating salmon. I. L-serine, a salmon repellent in mammalian skin. *J. Gen. Phys.* **39,** 889–892.

Irving, P. W., and Magurran, A. E. (1997). Context-dependent fright reactions in captive European minnows: The importance of naturalness in laboratory experiments. *Anim. Behav.* **53,** 1193–1201.

Isaacs, R. (1965). "Differential games: A mathematical theory with applications to warfare and pursuit, control and optimisation." Wiley, New York.

Johansen, K. (1982). Respiratory gas exchange of vertebrate gills. *In* "Gills" (Houlihan, D. F., Rankin, J. C., and Shuttleworth, T. J., Eds.), pp. 99–128. Cambridge University Press, Cambridge, MA.

Kalmijn, A. J. (1988). Detection of weak electric fields. *In* "Sensory Biology of Aquatic Animals." (Atema, J., Fay, R. R., Popper, A. N., and Tavolga, W. N., Eds.), pp. 151–186. Springer-Verlag, New York.

Kalmijn, A. J. (1966). Electro-perception in sharks and rays. *Nature* **212,** 1232–1233.

Kats, L. B., and Dill, L. M. (1998). The scent of death: Chemosensory assessment of predation risk by prey animals. *Ecoscience* **5,** 361–394.

Kiltie, R. A. (2000). Scaling of visual acuity with body size in mammals and birds. *Funct. Ecol.* **14,** 226–234.

Kirschfield, K. (1976). The resolution of lens and compound eyes. *In* "Neural Principles of Vision" (Zettler, F., and Weiler, R., Eds.), pp. 354–369. Springer-Verlag, Berlin.

Klinger, S. A., Magnuson, J. J., and Gallepp, G. W. (1982). Survival mechanisms of the central mudminnow (*Umbra limi*), fathead minnow (*Pimephales promelas*) and brook stickleback (*Culaea inconstans*) for low oxygen in winter. *Env. Biol. Fishes* **7,** 113–120.

Knouft, J. H., and Page, L. M. (2003). The evolution of body size in extant groups of North American freshwater fishes: Speciation, size distributions, and Cope's Rule. *Am. Nat.* **161,** 413–421.

Koops, M. A., and Abrahams, M. V. (2003). Integrating the roles of information and competitive ability on the spatial distribution of social foragers. *Am. Nat.* **161,** 586–600.

Kotrschal, K., van Staaden, M. J., and Huber, R. (1998). Fish brains: Evolution and functional relationships. *Rev. Fish Biol. Fisheries* **8,** 373–408.

Kramer, D. L. (1987). Dissolved oxygen and fish behaviour. *Env. Biol. Fishes* **18,** 81–90.

Krause, J., Cheng, D. J.-S., Kirkman, E., and Ruxton, G. D. (2000). Species-specific patterns of refuge use in fish: The role of metabolic expenditure and body length. *Behaviour* **137,** 1113–1127.

Krause, J., Staaks, G., and Mehner, T. (1998). Habitat choice in shoals of roach as a function of water temperature and feeding rate. *J. Fish Biol.* **53,** 377–386.

Lampert, W. H. (1989). The adaptive significance of diel vertical migration of zooplankton. *Funct. Ecol.* **3,** 21–27.

Leopold, L. B., Wolman, M. G., and Miller, J. P. (1964). "Fluvial processes in geomorphology." W. H. Freeman, San Francisco.

Lima, S. L., and Dill, L. M. (1990). Behavioural decisions made under the risk of predation: A review and prospectus. *Can. J. Zool.* **68,** 619–640.

Magnuson, J. J., and De Stasio, B. T. (1996). Thermal niche of fishes and global warming. *In* "Global Warming—Implications for Freshwater and Marine Fish" (Wood, C. M., and McDonald, D. G., Eds.), pp. 377–408. Society for Experimental Biology Seminar Series 61, Cambridge University Press, Cambridge, UK.

Magnhagen, C., and Heibo, E. (2001). Gape size allometry in pike reflects variation between lakes in prey availability and relative body depth. *Funct. Ecol.* **15,** 754–762.

Magurran, A. E., Irving, P. W., and Henderson, P. A. (1996). Is there a fish alarm pheromone? A wild study and critique. *Proc. Roy. Soc. London B* **263,** 1551–1556.

Martin, J., and López, P. (2001). Are fleeing "noisy" lizards signalling to predators? *Acta Ethologia* **3,** 95–100.

Mathis, A., and Smith, R. J. F. (1992). Avoidance of areas marked with a chemical alarm substance by fathead minnows (*Pimephales promelas*) in a natural habitat. *Can. J. Zool.* **70,** 1473–1476.

Milano, D., Cussac, V. E., Macchi, P. J., Ruzzante, D. E., Alonso, M. F., Vigliano, P. H., and Denegri, M. A. (2002). Predator associated morphology in *Galaxias platei* in Patagonian lakes. *J. Fish Biol.* **61,** 138–156.

Muir, B. S. (1969). Gill dimensions as a function of fish size. *J. Fish. Res. Bd. Canada* **26,** 165–170.

Neilson, J. D., and Perry, R. I. (1990). Diel vertical migrations of marine fishes: An obligate or facultative process? *Adv. Mar. Biol.* **26,** 115–168.

Nelson, M. E., and Mac Iver, M. A. (1999). Prey capture in the weakly electric fish *Apteronotus albifrons*: Sensory acquisition strategies and electrosensory consequences. *J. Exp. Biol.* **202,** 1195–1203.

Nilsson, P. A., and Brönmark, C. (2000). Prey vulnerability to a gape-size limited predator: Behavioural and morphological impacts on northern pike piscivory. *Oikos* **88,** 539–546.

Nilsson, P. A., and Brönmark, C. (1999). Foraging among cannibals and kleptoparasites: Effects of prey size on pike behaviour. *Behav. Ecol.* **10,** 557–566.

Odell, J. P., Chappell, M. A., and Dickson, K. A. (2003). Morphological and enzymatic correlates of aerobic and burst swimming performance in different populations of Trinidadian guppies, *Poecilia reticulata*. *J. Exp. Biol.* **206,** 3707–3718.

Page, L. M., and Burr, B. M. (1991). "A field guide to freshwater fishes: North America north of Mexico." Houghton Mifflin, Boston.

Persson, L., Andersson, J., Wahlström, E., and Eklov, P. (1996). Size-specific interactions in lake systems: Predator game limitation and prey growth rate and mortality. *Ecology* **77,** 900–911.

Petterson, L. B., Nilsson, P. A., and Brönmark, C. (2000). Predator recognition and defence strategies in crucian carp, *Carassius carassius*. *Oikos* **88,** 200–212.

Pfeiffer, W., Riegelbauer, G., Meier, G., and Scheibler, B. (1985). Effect of hypoxanthine-3-*N* oxide and hypoxanthine-1-*N*-oxide on central nervous excitation of the black tetra, *Gymnocorymbus ternetzi* (Characaidae, Ostariophysi, Pisces) indicated by dorsal light response. *J. Chem. Ecol.* **11,** 507–523.

Palzenberger, M., and Pohla, H. (1992). Gill surface area of water-breathing freshwater fish. *Rev. Fish. Biol. Fisheries,* **2,** 187–216.

Pelletier, D., Guderley, H., and Dutil, J. D. (1993). Effects of growth rate, temperature, season, and body size on glycolytic enzyme activities in the white muscle of Atlantic Cod (*Gadus morhua*). *J. Exp. Zool.* **265,** 477–487.

Pocklington, R., and Dill, L. M. (1995). Predation on females or males: Who pays for bright male traits? *Anim. Behav.* **49,** 1122–1124.

Powers, D. A. (1980). Molecular ecology of teleost fish hemoglobins: Strategies for adapting to changing environments. *Am. Zool.* **20,** 139–162.

Rahel, F. J., and Nutzman, J. W. (1994). Foraging in a lethal environment: Fish predation in hypoxic waters of a stratified lake. *Ecology* **75,** 1246–1253.

Randall, D. (1982). The control of respiration and circulation in fish during exercise and hypoxia. *J. Exp. Biol.* **100,** 275–288.

Reimchen, T. E. (1983). Structural relationships between spines and lateral plates in threespine stickleback (*Gasterosteus aculeatus*). *Evolution* **37,** 931–946.

Reimchen, T. E. (1980). Spine deficiency and polymorphism in a population of *Gasterosteus aculeatus*: An adaptation to predators? *Can. J. Zool.* **68,** 1232–1244.

Reist, J. (1980). Predation upon pelvic phenotypes of brook stickleback, *Culea inconstans*, by selected invertebrates. *Can. J. Zool.* **58,** 1253–1258.

Robb, T., and Abrahams, M. V. (2003). Variation in tolerance to hypoxia in a predator and prey species: An ecological advantage to being small? *J. Fish Biol.* **62,** 1067–1081.

Robb, T., and Abrahams, M. V. (2002). The influence of hypoxia on risk of predation and habitat choice by the fathead minnow, *Pimephales promelas*. *Behav. Ecol. Sociobiol.* **52,** 25–30.

Schaack, S. R., and Chapman, L. J. (2003). Interdemic variation in the African cyprinid *Barbus neumayeri*: Correlations among hypoxia, morphology, and feeding performance. *Can. J. Zool.* **81,** 430–440.

Schellart, N. A. M. (1992). Interactions between the auditory, the visual and the lateral line systems of teleosts; a mini-review of modelling sensory capabilities. *Neth. J. Zool.* **42,** 459–477.

Schmidt-Nielson, K. (1984). "Scaling: Why is animal size so important?". Cambridge University Press, New York.

Smith, F. M., and Jones, D. R. (1982). The effect of changes in blood oxygen-carrying capacity on ventilation volume in the rainbow trout (*Salmo gairdneri*). *J. Exp. Biol.* **97,** 325–334.

Smith, M. E., and Belk, M. C. (2001). Risk assessment in western mosquitofish (*Gambusia affinis*): Do multiple cues have additive effects? *Behav. Ecol. Sociobiol.* **51,** 101–107.

Smith, R. J. F. (1997). Does one result trump all others? A response to Magurran, Irving, and Henderson. *Proc. Roy. Soc. London B* **264,** 445–450.

Smith, R. J. F. (1992). Alarm signals in fishes. *Rev. Fish. Biol. Fisheries* **2,** 33–63.

Smith, R. J. F. (1986). Evolution of alarm signals: Role of benefits of retaining group members or territorial neighbours. *Am. Nat.* **128,** 33–63.

Somero, G. N., and Childress, J. J. (1980). A violation of the metabolism size scaling paradigm: Activities of glycolytic enzymes in muscle increase in large-size fish. *Phys. Zool.* **53,** 322–337.

Stephens, D. W., and Krebs, J. R. (1986). "Foraging Theory." Princeton University Press, Princeton, NJ.

Sundström, L. F., Lõhmus, M., Johnsson, J. I., and Devlin, R. H. (2004). Growth hormone transgenic salmon pay for growth potential with increased predation mortality. *Proc. Roy. Soc. Lond. B* **271,** S350–S352.

Sweatman, H. P. A. (1984). A field study of the predatory behaviour and feeding rate of piscivorous coral reef fish, the lizardfish, *Synodus englemani. Copeia* **1984,** 187–194.

van Staaden, M. J., Huber, R., Kaufman, L. S., and Liem, K. F. (1995). Brain evolution in cichlids of the African Great Lakes: Brain and body size, general patterns, and evolutionary trends. *Zoology* **98,** 165–178.

Vamosi, S. (2002). Predation sharpens the adaptive peaks: Survival trade-offs in sympatric sticklebacks. *Ann. Zool. Fenn.* **39,** 1–28.

Wanink, J. H., Kashindye, J. J., Goudswaard, P. C., and Witte, F. (2001). Dwelling at the oxycline: Does increased stratification provide a predation refugium for the Lake Victoria sardine *Rastrineobola argentea*? *Fresh. Biol.* **46,** 75–86.

Webb, P. W., and Skadsen, J. M. (1980). Strike tactics of. *Esox. Can. J. Zool.* **58,** 1462–1469.

Werner, E. E., and Anholt, B. R. (1993). Ecological consequences of the trade-off between growth and mortality rates mediated by foraging activity. *Am. Nat.* **142,** 242–272.

Werner, E. E., Mittelbach, G. G., Hall, D. J., and Gilliam, J. F. (1983). Experimental tests of optimal habitat use in fish: The role of relative habitat profitability. *Ecology* **64,** 1525–1539.

West, G. B., Brown, J. H., and Enquist, B. J. (1999). The fourth dimension of life: Fractal geometry and allometric scaling of organisms. *Science* **284,** 1677–1679.

West, G. B., Brown, J. H., and Enquist, B. J. (1997). A general model for the origin of allometric scaling laws in biology. *Science* **276,** 122–126.

Wisenden, B. D., Vollbrecht, K. A., and Brown, J. L. (2004). Is there a fish alarm cue? Affirming evidence from a wild study. *Anim. Behav.* **67,** 59–67.

Wisenden, B. D., and Thiel, T. A. (2002). Field verification of predator attraction to minnow alarm substance. *J. Chem. Ecol.* **28,** 433–438.

Wurtsbaugh, W. A., and Neverman, D. (1988). Post-feeding thermotaxis and daily vertical migration in a larval fish. *Nature* **333,** 846–848.

Ydenberg, R. C., and Dill, L. M. (1986). The economics of fleeing from predators. *Adv. Study Behav.* **16,** 229–249.

Ziuganov, V. V., and Zotin, A. A. (1995). Pelvic girdle polymorphism and reproductive barriers in the ninespine stickleback *Pungitius pungitius* (L.) from northwest Russia. *Behaviour* **132,** 1095–1105.

Zottoli, S. J., and Faber, D. S. (2000). The Mauthner Cell: What has it taught us? *Neuroscientist* **6,** 25–37.

4

EFFECTS OF PARASITES ON FISH BEHAVIOUR: INTERACTIONS WITH HOST PHYSIOLOGY

IAIN BARBER
HAZEL A. WRIGHT

Behaviour and Physiology of Fish: Volume 24
FISH PHYSIOLOGY

DOI: 10.1016/S1546-5098(05)24004-9

1. INTRODUCTION

1.1. Parasites, Physiology, and Behaviour

Being infected with parasites is the normal state for the vast majority of animals in natural, as well as in most managed populations. Consequently, developing an understanding of how infections impact on the biology of host organisms has considerable fundamental and applied value. Given the intimate associations between parasites and their hosts, infections are expected to impact on their hosts at a range of organisational levels, from gene expression to population dynamics. Parasites that cause mortality through direct pathogenic effects on their hosts have obvious importance because they impose a direct selection pressure on host populations. However, in recent years there has been recognition that sublethal physiological effects on hosts—which may lead to alterations in the behaviour of infected animals—may also play an important role in regulating populations through demographic effects (Dobson, 1988; Hudson and Dobson, 1997; Finley and Forrester, 2003). The rise of behavioural ecology as a discipline devoted to the study of the adaptive function of behaviour (Krebs and Davies, 1997) has facilitated our understanding of the role of sublethal parasite infections as selective agents in evolution. Elegant experimental tests of key hypotheses (e.g., Moore, 1983; Milinski and Bakker, 1990; Lafferty and Morris, 1996) strongly suggest that parasite infections can impact significantly on the natural and sexual selection of host populations through their effects on behaviour.

Although behavioural and evolutionary ecologists have focused on examining the function and evolutionary consequences of behavioural changes associated with infection, there has also been interest in the proximate (physiological) mechanisms of host behavioural modification. A significant body of research has focused on the mechanisms of behavioural change in parasitised invertebrates (reviewed by Hurd, 1990; Adamo, 1997; Moore, 2002), which offer scientists an opportunity to study the mechanisms of manipulation in biological systems with relatively simple nervous systems (Adamo, 1997). However, research into the mechanisms by which parasites alter the behaviour of vertebrate hosts has been limited by a lack of general knowledge regarding the physiological control of behaviour in vertebrates. Writing over a decade ago, primarily from a mammal research perspective, Thompson and Kavaliers (1994) pointed out that:

> "While it is clear that parasitism often brings about dramatic changes in host behaviour, little is understood of the physiological bases for these

changes and this in large measure reflects our primitive knowledge of the interaction of behaviour and physiology."

In the intervening years, considerable research effort by those authors and others has begun to shed light on the physiological control of behaviour in mammals, and the mechanisms by which parasites can exert their influence.

Coordinated studies examining the physiological mechanisms of infection-associated behavioural change in fish have been similarly slow to emerge, and there are very few studies available that unambiguously link parasite-induced behavioural change to a physiological mechanism in fish. However, both physiologists and behaviourists have realised the value of fish as models with which to study the effects of parasites, and there is a wealth of information on the (separate) behavioural and the physiological effects of a wide range of parasitic infections. As a consequence, this Chapter often uses evidence from studies demonstrating a link between behaviour and physiology in noninfected fish to identify potential mechanisms by which parasites may induce their observed behavioural effects on hosts. Clearly, more investigations are required that examine contemporal physiological and behavioural effects in the same host–parasite systems (Figure 4.1). As more of the mechanistic bases of behaviour are being demystified by fish physiologists, there are an increasing number of opportunities for coordinated investigations into how parasites exert their influences on host behaviour.

1.2. Why Study the Physiological Basis of Behavioural Changes Associated with Infections?

There are two fundamental reasons for studying the physiological bases of infection-associated behavioural change in fish. First, there is still very little data in support of a key hypothesis in evolutionary biology, which states that behavioural changes associated with infection in hosts have arisen as parasite adaptations to maximise transmission success (the Manipulation Hypothesis; Moore and Gotelli, 1990; see reviews by Barnard and Behnke, 1990; Barber *et al.*, 2000; Moore, 2002). Studies that demonstrate evolutionarily relevant changes in host ecology that are mediated through behavioural changes associated with infection have proven difficult to undertake and interpret (but for notable exceptions see Moore, 1983; Lafferty and Morris, 1996). Given the difficulty of acquiring data on the fitness effects of behavioural changes, studies that uncover the complexity of the physiological mechanism of behavioural change potentially provide alternative indirect evidence on the likely evolution of infection-associated traits (Poulin, 1998). Altered host behaviours that result from "simple" mechanisms, such

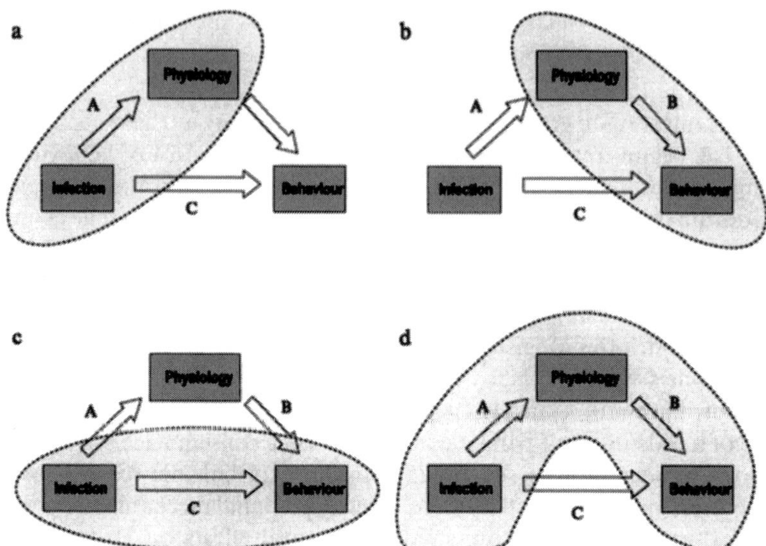

Fig. 4.1. Schematic illustrating the types of studies that have examined interactions between parasite infection, behaviour and physiology in fish. (a) Differences in the physiology of parasitised fish, compared with uninfected individuals, are often recorded by physiologists or parasitologists studying either naturally infected or experimentally exposed individuals. (b) Physiological parameters are also commonly associated with behavioural traits, for example, reduced swimming performance associated with anemia. (c) In many cases, parasite infections may be demonstrated to influence host behaviour. (d) More studies are needed to provide causal links between infection, host physiology, and behaviour in the same host–parasite system.

as occupying a specific tissue to be in the required place for development or egg production, may arise as unavoidable side effects of infection that have neutral, or even detrimental, consequences for parasite transmission. On the other hand, behavioural effects of parasites that are mediated through complex mechanisms, such as parasite secretions that interfere with the neurochemical control of host behaviour, are more likely to be true adaptations because they require investment on the behalf of the parasite (Poulin, 1998).

Second, because behavioural effects of parasites ultimately arise from physiological changes in the host, understanding the mechanisms by which parasites manipulate host behaviour can yield important general information on the physiological processes responsible for control and modulation of behaviour in the healthy organism (Helluy and Holmes, 1990; Thompson and Kavaliers, 1994).

There may also be applied value in examining the physiological basis of infection-associated behavioural change, because some of the traits exhibited by infected fish may ultimately prove to be exploitable. For example, fish infected with fast-growing plerocercoids of pseudophyllidean cestodes feed more readily and as a consequence grow more quickly under particular feeding regimes than nonparasitised conspecifics (see Section 3.1.1). Such traits may be desirable in aquaculture or commercial fishery species, and understanding the physiological mechanisms by which parasites alter fish behaviour (for example, by increasing the fish's appetite) may provide future opportunities to maximise productivity.

1.3. Biological Diversity of Fish Parasites

If we define parasites as organisms that live in or on another organism (termed the *host*) for at least some of their lifecycle and as a result cause harm to the host, then parasitic organisms include all viruses, many bacteria, and some fungi. However, the majority of parasitologists focus on protozoan and metazoan parasites (Bush *et al.*, 2001), as will this Chapter. Fish parasites are a diverse assemblage of organisms, both in terms of their taxonomy and in the ways in which they exploit fish as hosts. Lifecycle variation between parasite groups is substantial and important in understanding the ecological consequences of—and therefore the evolutionary pressures acting on—infection-associated changes in host behaviour. From this perspective, perhaps the most important consideration is the type of lifecycle exhibited by the parasite. Some taxa (*directly-transmitted parasites*) are capable of completing their lifecycle on one individual fish, whereas others (*indirectly-transmitted parasites*) utilise the fish as one of a number of essential, sequential hosts. If more than one host is involved in the parasite's lifecycle, then fish may harbour the sexually active adult stage—in which case the fish is termed the *definitive host*—or they may serve as an intermediate host for a nonreproductive, developmental, or dormant stage of parasites. Because of their abundance and their typically intermediate position in aquatic food webs, fish are utilised as intermediate hosts by many indirectly transmitted parasites, with transmission to definitive hosts (which are often birds, mammals, or predatory fish) occurring through the food chain (*trophic transmission*). However, it is important to recognise that in many cases—particularly for marine parasites, and many of the myxozoans and microsporidians—lifecycles are not yet fully resolved, and so inferring likely ecological or transmission outcomes of infection or their evolutionary significance is not always possible.

This biological diversity of fish parasites—in terms of their taxonomy, life cycle details, the variety of ways in which they acquire their nutrients from host fish, their sites of infection and the extent to which they are nutritionally demanding—means that there are both mechanistic and evolutionary reasons why we should not expect all parasites to have the same kinds of effects on the physiology or behaviour of hosts.

1.4. Parasites and Fish Behaviour

Parasites can impact on normal patterns of fish behaviour in a number of ways. First, because animals are expected to have evolved behavioural mechanisms to limit contact with infective stages to reduce the demand placed on the immune system (Hart, 1990), the threat posed by the presence of parasites in an environment may influence the behaviour of fish even before they become infected. For example, fish have been shown to avoid particular types of habitat associated with infection risk (e.g., Poulin and FitzGerald, 1989), select against joining shoals containing parasitised individuals (e.g., Krause and Godin, 1996; Barber *et al.*, 1998), and reject parasitised sexual partners (e.g., Kennedy *et al.*, 1987; Milinski and Bakker, 1990; Rosenqvist and Johansson, 1995).

Second, parasitised fish may perform behaviours that reduce levels of infection with already-acquired parasites. Such behaviours range from simple "flashing" against the substratum or rubbing against other structural components of their environments to dislodge ectoparasites (e.g., Urawa, 1992) to complex interspecific social behaviours such as visiting "cleaning stations" on coral reefs (see Losey, 1987; Poulin and Grutter, 1996 for reviews). These first two types of behavioural effects of parasites constitute the "behavioural resistance" repertoire of a host (Hart, 1990), and these behaviours of hosts or potential hosts are generally regarded as host adaptations (e.g., Grutter, 1999). An in-depth coverage of behaviours that limit contact with parasites or reduce infection levels is outside the scope of this review, but further discussion can be found in Barber *et al.* (2000).

Third, parasite infections may cause host behaviours to alter in ways that serve to mediate the detrimental effects of infection. For example, the altered prey preferences and foraging behaviour of sticklebacks infected with *Schistocephalus solidus* plerocercoids (e.g., Milinski, 1990; Ranta, 1995) increase food intake rates, and as such compensate to some extent for the nutritional demands of the parasites. Such behaviours are therefore readily explained as host adaptations to infection.

However, not all infection-induced behavioural changes are likely to be host adaptations, because parasite infections often cause behavioural changes in infected hosts that have no obvious direct benefit to their hosts. In some cases, these behavioural changes may reflect parasite adaptations, increasing the probability of successful transmission; in other cases, the behavioural changes may be inevitable side effects of infection that benefit neither parasite nor host (Poulin, 1998). However, phenotypic correlations between infection status and behaviour can also be generated if behaviour influences infection status rather than vice versa, because fish that exhibit pre-existing atypical behaviour may be more exposed, or less resistant, to infections. For this reason, studies that demonstrate naturally infected hosts to display atypical behaviours cannot definitively identify causality; conversely, studies that demonstrate behavioural changes in experimentally infected fish may provide convincing evidence that behavioural changes are caused by infections. Fortunately, there are an increasing number of fish parasite systems for which experimental infections are possible.

1.5. How do Parasites Alter the Behaviour of Fish Hosts?

Parasite infections that alter the behaviour of their hosts have traditionally been proposed to do so through direct or indirect physiological mechanisms (Milinski, 1990). Direct mechanisms of behavioural manipulation are those in which the parasites themselves, or their biochemical secretions, act directly on the host's system of behavioural control. Examples of these kinds of mechanisms would include parasites that alter host behaviour by releasing substances with neurotransmitter or neuromodulation capabilities, or which locate in and damage specific lobes of the brain. Alternatively, host behaviour may be altered indirectly if parasites impose a constraint on some other aspect of the host's physiology. An example of indirect behavioural modification would be the increased foraging behaviour of fish infected with nutritionally demanding parasites such as larval cestodes.

However, although separating indirect from direct mechanisms of behavioural manipulation may have heuristic value, it requires a level of detail regarding the host–parasite interaction that is generally not available (Milinski, 1990; Barber *et al.*, 2000). For instance, it is not known whether the altered swimming behaviour of cyprinid fishes harboring heavy infections of brain-dwelling diplostomatid trematodes (see Section 4.1) results from physical damage to the brain itself, a build-up of metabolic compounds from the parasites affecting neural or muscle function, or the secretion of behaviour-modifying chemicals. In this Chapter, a different approach to understanding the physiological basis of behavioural change in parasitised fish is used.

Behaviour can be broadly described as an animal's motor responses to multiple perceived external stimuli, detected and transduced into afferent nerve signals by a number of different sensory systems that are integrated by the central nervous system (CNS) and modulated by the physiological status of the animal. This Chapter highlights the impact of parasitic infections on normal patterns of host behaviour by interference at each of four different physiological levels (Figure 4.2). First, parasites may influence behaviour by

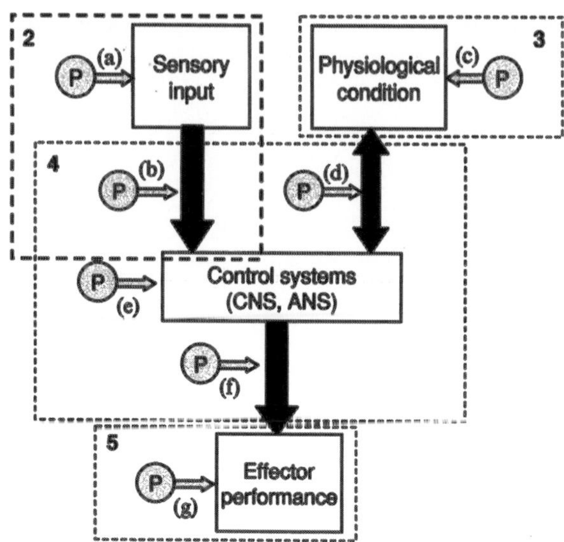

Fig. 4.2. Schematic illustrating the different ways in which parasite infections potentially influence host behaviour. Parasites (P) can directly influence the quality of information received by a host by interfering with sensory function (a) or with the transmission of input signals by affecting neural performance (b). Parasites can also impact directly on host physiological condition (c) and on how host physiology and control systems (such as the CNS and ANS) interact (d). Furthermore, parasites can affect control systems if they cause either physical or neurochemical disruption (e) and they may also affect efferent neural function (f). Finally, parasites can impact directly on the functioning of effectors, such as muscles (g). Numbers relate to sections of this review dealing each type of effect.

affecting the quality and/or quantity of information obtained by hosts as they sample their environments, by affecting the functioning of peripheral sense organs (Section 2). Second, parasites may change the internal nutritional status of fish hosts—and thus the motivational basis to respond to external stimuli—if infection has significant energetic consequences (Section 3). Third, parasite infections may interfere directly with the control of host behaviours by physically damaging the CNS by their site selection, by manipulating levels of hormones or neurotransmitters, or by having neuromodulatory effects (Section 4). Finally, many of the effects that parasites have on host respiration, circulation, locomotion, or stamina can impact on the host's capacity to perform normal patterns of behaviour in response to perceived stimuli (Section 5).

The primary intention of this Chapter is to stimulate others to provide much-needed studies linking behaviour and physiology in parasitised fish. With this in mind, the conclusion presents major gaps in knowledge and highlights new opportunities to exploit postgenomic technologies.

2. EFFECTS OF PARASITES ON THE SENSORY PERFORMANCE OF FISH HOSTS

Because parasites cause local pathology to host tissues by their attachment, movements, growth, or development, the specific sites they occupy may have important consequences for the type and extent of host behavioural change (Holmes and Zohar, 1990). Many endoparasites have a predilection for occupying sensory organs within their hosts, and damage to sensory tissues or occlusion of sensory organs may be sufficient to bring about changes to host behaviour. It is thought that the evolution of parasite preferences for such sites has arisen primarily as a mechanism for evading the host's immune system (Szidat, 1969; Ratanarat-Brockelman, 1974; Cox, 1994), with any associated behavioural changes arising as (potentially fortuitous) side effects of immune avoidance (O'Connor, 1976). In fish, the eyes, nares, inner ear, and lateral line are frequently used as sites of infection by parasites (Williams and Jones, 1994). Here we examine the sensory systems of fish that are used as infection sites by parasites and review studies that have investigated their effects on host behaviour and/or physiology. The sensory systems of fish are described in detail in Chapter 2.

2.1. Behavioural Effects of Visual Impairment

A number of parasites utilise the fish eye as an infection site. Particularly common are the metacercariae of diplostomatid trematodes, including *Diplostomum* and *Tylodelphys* spp., which locate in the lens, retina, or vitreous humour. A range of freshwater species become infected with these and other "eyeflukes" when free-swimming cercariae, released from aquatic snails, penetrate the skin and travel to the eye in the host's circulatory system (Erasmus, 1959). In the case of *D. spathaceum*, the parasites invade the lens tissue, causing parasitic cataract disease in heavy infections (Chappell *et al.*, 1994). Intact fish lenses are optically complex tissues with remarkable resolving capacity (Fernald, 1993), largely as a consequence of variable refractive index throughout their depth, which counteracts spherical aberration (Sivak, 1990). Infected fish display behavioural changes that suggest the physical presence of parasites in the eye, and the damage caused during tissue migration is sufficient to impair vision. Infection with *D. spathaceum* is associated with reduced visual acuity in the three-spined stickleback *Gasterosteus aculeatus*, with infected fish initiating approaches toward motile prey at significantly reduced distances compared to noninfected fish (Owen *et al.*, 1993; Figure 4.3). Similar effects suggestive of reduced visual acuity have been demonstrated in infected dace (*Leuciscus leuciscus*), which feed less

a

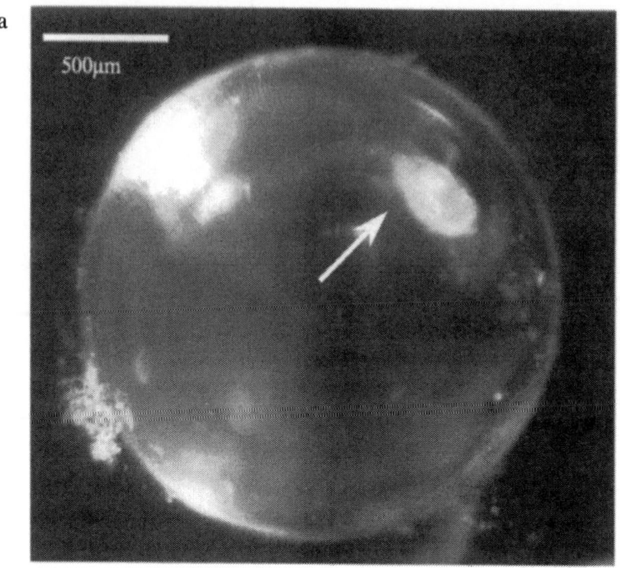

b

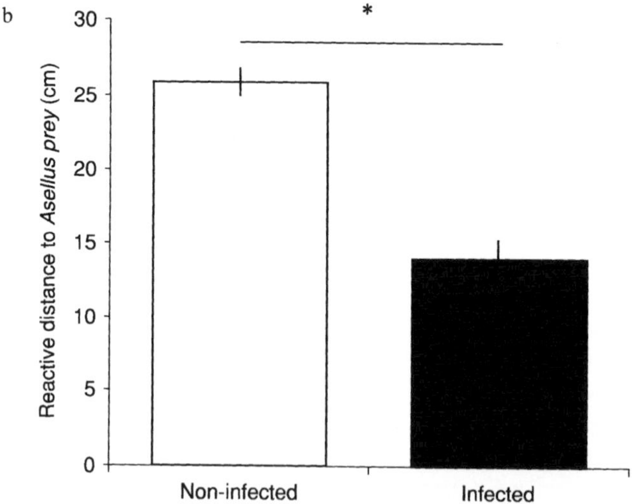

Fig. 4.3. (a) A single *Diplostomum spathaceum* metacercaria infecting the lens of a three-spined stickleback *Gasterosteus aculeatus*. (b) The effect of *D. spathaceum* infection on the reactive distance of three-spined sticklebacks to aquatic prey (*Asellus* sp.). Bar heights are means; error bars represent 1 standard deviation. $*P < 0.05$. (Data from Owen *et al.*, 1993.)

successfully on surface prey items and as a consequence spend a greater proportion of time feeding than noninfected fish (Crowden, 1976; Crowden and Broom, 1980), and in trout, which are caught less frequently by anglers in rod-and-line fisheries than in gill nets when infected with eyeflukes (Moody and Gaten, 1982). The reduced visual performance of infected fish, and the altered time budgets and spatial distributions that result, apparently increase their predation risk (Brassard *et al.*, 1982), with consequences for parasite transmission.

Other parasites, which are not restricted to infecting optical tissues, may nonetheless impair vision if they locate in the cornea or elsewhere in the head region. For example, xenomae of the microsporidian *Glugea anomala* often form in the head, causing the eye to swell (I. Barber, personal observation) and metacercariae of the digenean trematodes *Bucephaloides gracilescens* and *Cryptocotyle lingua* may encyst in the cornea (Karlsbakk, 1995), with likely but untested consequences for vision and behaviour. Parasites that locate in lobes of the brain associated with vision or in the optical nerves may additionally impact on vision and behaviour (see Section 4.1).

2.2. Behavioural Effects of Impaired Chemoreception

The chemosensory systems of fish are extremely well developed and mediate many behaviours of fundamental importance (Hara, 1992; Sorenson and Caprio, 1998) including the location of food (Jones, 1992) and predators (Chivers and Smith, 1993). Fish in the order Ostariophysi have the additional ability to detect alarm substance ('Schreckstoff') released from the damaged skin of conspecifics and heterospecifics, and thereby can acquire immediate information on predator activity (Chapter 3, Section 3.4; von Frisch, 1938; Wisenden *et al.*, 1994). Chemoreception is also known to play a major role in the return migrations of anadromous fish returning to spawn in natal rivers (Chapter 7, Section 12; Hara, 1993; Lucas and Barras, 2001) and in chemical communication (Brønmark and Hansson, 2000), with recent studies demonstrating that individuals can identify kin (Winberg and Olsén, 1992) and even nonrelated familiar individuals (Brown and Smith, 1994; Griffiths, 2003) by their odors. Moreover, recent research on sticklebacks suggests that mate choice decisions may be mediated by odours that relate to major histocompatibility complex allelic diversity (Reusch *et al.*, 2001; Milinski, 2003).

Chemosensation is achieved via at least two different channels in fish: olfaction (smell) and gustation (taste), but because both involve the detection of waterborne molecules, the distinction is not always clear (Hara, 1993). The primary sites of chemosensation in fishes are the olfactory epithelium situated within the narial canals (nares), taste buds located in the mouth, and taste cells on the skin (Sorenson and Caprio, 1998), all of which may be invaded by parasites (Williams and Jones, 1994). Parasites locating in these organs may affect the sensitivity of their hosts to waterborne odours if they either restrict the passage of water through olfactory organs or damage the olfactory epithelium. Because fish use chemosensory detection to detect predators, there are reasons to expect that the chemosensory abilities of hosts may be a target of behavioural manipulation by trophically transmitted parasites. Given the current level of interest in the role of chemoreception in fish behaviour, it is perhaps surprising that as yet there have been no studies that we are aware of examining the influence of parasite infections on the chemosensory abilities of fish and subsequent consequences for their behaviour, and we highlight this as an area of significant research potential (Section 6.1).

As well as impacting on the chemosensory abilities of their fish hosts, parasite infections may also alter host odours, allowing them to be discriminated by conspecifics in social or sexual contexts. In laboratory trials, mice are capable of detecting conspecifics infected with the nematode *Heligmosomoides polygyrus* based on odour cues alone (Kavaliers and Colwell, 1995; Kavaliers *et al.*, 1998a). Parasitised fish are frequently demonstrated to be unattractive companions or mates to individuals making shoaling or mate choice decisions (e.g., Rosenqvist and Johansson, 1995; Krause and Godin, 1996; Barber *et al.*, 1998) and although visual cues can be used to identify some infections (e.g., Krause and Godin, 1996), infection-associated odour variation may be a potential, though untested, mechanism allowing determination of the infection status of individuals by conspecifics. Given the importance of chemosensation to fish, it is not unrealistic to expect fish to be capable of identifying parasitised individuals by their odours, particularly if there are fitness costs of associating with them.

2.3. Behavioural Effects of Impaired Octavolateral Functioning

The octavolateral system of fish comprises three separate mechanosensory systems—the auditory, the vestibular (or equilibrium), and the lateral-line systems—that are traditionally studied together because they have a common sensory cell type as their major receptor (Schellart and Wubbels, 1998). The lateral-line and auditory systems detect mechanical vibrations, whereas the major roles of the vestibular apparatus of fishes are in the control of balance and posture and in visual stabilisation (Popper and Platt, 1993; Schellart and Wubbels, 1998).

The organisation of the auditory systems of fish has been reviewed in detail by Schellart and Wubbels (1998), and the importance of the production and detection of underwater sound in fish behaviour was reviewed by Hawkins (1993) and in Chapter 2, Section 6 of this volume. The auditory system comprises three semicircular canals and three calcareous otolith organs and is located in the inner ear. Underwater sound waves are detected when movements of the fish's body, caused by molecular vibration of the water, cause the hair cells to oscillate with respect to the otoliths. The orthogonally arranged semicircular canals of the inner ear form the equilibrium system, providing information on the three-dimensional orientation of the fish (Bone *et al.*, 1995). The inner-ear labyrinth, containing both systems, is calcified to form a single massive structure in teleosts and is filled with a fluid (endolymph). Parasites that inhabit the inner ear, or cause a change in the structure of the semicircular canals or the endolymph contained within, have the potential to impact significantly on the ability to achieve balance and postural control and to detect underwater sounds.

The most common parasite infections of the inner ear of teleost fish are various species of myxosporean, including *Myxobolus cerebralis*, which destroy the cartilage of the inner ear of salmonids (Markiw, 1992). The main behavioural symptom of infection is uncontrolled erratic circular swimming at the water surface termed *whirling* (Uspenskaya, 1957; Kreirer and Baker, 1987; Markiw, 1992), as would be expected for a parasite that damaged the major balance organs. However, other pathomechanisms generating the observed whirling behaviours have been suggested (see Section 4.1) and it is possible that a range of physiological effects may generate phenotypically similar behaviour effects in parasitised fish.

An additional component of the auditory system of fishes is the swimbladder, which functions as a pressure-to-displacement transducer and lowers the auditory threshold (Schellart and Wubbels, 1998). Parasites that inhabit the swimbladder (see also Section 5.3) therefore potentially interfere with sound detection, but we are aware of no studies examining this.

The lateral-line system of fishes serves primarily as a hydrodynamic receiver, detecting pressure changes in the water immediately surrounding the fish (Popper and Platt, 1993; Schellart and Wubbels, 1998), and is utilised by a range of parasites as an infection site. Metacercariae of the digenean *Ribeiroia marini* are found in the lateral-line scale canals of goldfish *Carassius auratus* (Huizinga and Nadakavukaren, 1997) and philichthyid copepods inhabit the lateral-line sensory canals of tropical and temperate marine fishes (Kabata, 1979; Grabda, 1991; Hayward, 1996). The latter may be common in fish populations: 31% of corkwing wrasse (*Crenilabrus melops*: Labridae) off the coast of Ireland were infected with *Leposphilus labrei* (Donnelly and Reynolds, 1994), with heavy infections being associated with tumor formation. The behavioural functions of the lateral line include the localisation of moving prey, the detection of approaching conspecifics and predators, obstacle detection (particularly in low-visibility conditions), and schooling behaviour (Bleckmann, 1993).

Studies in which lateral-line function is experimentally impaired demonstrate the potential for such parasites to impact on host behaviour. For example, mechanical blocking of lateral-line canal pores eliminates the feeding response of mottled sculpin *Cottus bairdi* (Hoekstra and Janssen, 1985), and cauterisation of the posterior lateral line nerves is sufficient to impair the ability of saithe *Pollachius virens* to maintain accurate spatial positions within dynamic swimming schools (Partridge and Pitcher, 1980). Infections with parasites such as those detailed above are likely to impact on water flow in the lateral-line canals and reduce sensitivity to pressure changes; however, despite the potential for infection-associated behavioural change, the authors are aware of no studies that have examined behavioural effects of lateral-line–dwelling parasites.

Endoparasites such as *Ligula intestinalis* or *Schistocephalus solidus* that grow to a large size and cause gross swelling of their hosts' bodies also have the potential to impact on lateral-line function. Although there are as yet no physiological data available, the schooling behaviour of European minnows infected with *Ligula* is strikingly similar to that of saithe deprived of lateral-line function. In mixed infection schools, infected minnows exhibit larger nearest-neighbour distances (NNDs) and a greater propensity to occupy peripheral positions within schools than uninfected conspecifics; although they respond to a simulated avian strike by reducing their NND, they continue to take up spatially inappropriate positions after attack (Barber and Huntingford, 1996). Sticklebacks infected with *Schistocephalus* plerocercoids also exhibit impaired escape responses, including a reduced propensity to react to a striking model heron (Giles, 1983; Godin and Sproul, 1988; Tierney *et al.*,1993; Ness and Foster, 1999; Barber *et al.*, 2005), as do roach infected with *Ligula intestinalis* (Loot *et al.*, 2002a). Because the model predators may strike the water surface in these trials, one explanation for the absence of a response from infected fish is that infected fish fail to sense water pressure changes via their lateral line. Future studies of escape response using model predators should be designed more carefully to allow a detailed interpretation of the results and ascertain which sensory modalities are impaired by infection.

2.4. Behavioural Effects of Impaired Electrosensation

Some fish have the capacity to produce electrical pulses that are used actively in foraging, object detection, or communication (Bleckmann, 1993). However, the ability to detect electric fields (*electroreception*) is not just limited to these electric fish; the majority of nonteleost fish, as well as some teleosts including the Ostariophysans and Osteoglossiformes, also have low-frequency electroreceptors and use these passively in prey detection, long-distance orientation, and migration (Bleckmann, 1993). Despite considerable research interest in electrosensation and electrobiology of fishes (comprehensively reviewed by Moller, 1995), the authors are not aware of any studies that have examined the impact of parasites on active or passive electrosensory function or behaviour.

3. BEHAVIOURAL CONSEQUENCES OF PARASITE-IMPOSED CONSTRAINTS ON HOST PHYSIOLOGY

The internal physiological state of animals may have a considerable effect on their behaviour. For example, following a period of food deprivation, fish will take more risks while foraging (Damsgård and Dill, 1998) and ingest more food in a single meal than fish in better nutritional condition (Ali *et al.*, 2003), whereas fish that are under osmoregulatory stress often exhibit altered locomotory and antipredator behaviour (e.g., Handeland *et al.*, 1996). Hence, parasites that impose constraints on the physiological status of their hosts potentially impact indirectly on the behaviour of their hosts. This Section focuses on the observed and potential behavioural consequences of infections resulting from their effects on host energetic status and capacity for osmoregulation.

3.1. Behavioural Consequences of Nutritionally Demanding Parasites

3.1.1. Effects on Foraging Behaviour

All parasites utilise host-derived resources for their growth and development, and as such impose an energetic drain on host organisms. As a consequence, energetic effects of infection are commonly reported, particularly when the parasites involved are large, rapidly-growing, numerous, or highly pathogenic. In addition, if infections reduce the competitive ability or foraging success of hosts (for example, as a consequence of infection-induced damage to sense organs; see Section 2), then parasites may limit the nutrient intake of their hosts. Fish infected with parasites that impose a significant energetic cost may therefore be forced to increase their foraging effort or alter their foraging strategies to compensate.

The most detailed studies of behavioural changes induced by energetically demanding parasites have examined fish harboring the large plerocercoid larvae of the closely related pseudophyllidean cestodes *Schistocephalus solidus* and *Ligula intestinalis*, which infect sticklebacks and cyprinids respectively. Infections are acquired when fish hosts feed on parasitised copepods, and the plerocercoids grow rapidly to a large size in the peritoneal cavity of host fish. Evidence from laboratory and field studies shows that, during the parasite growth phase, these parasites impose a considerable energetic drain on their hosts. Infected wild caught fish typically have lower somatic body condition and/or liver energy reserves (Arme and Owen, 1967; Pennycuick, 1971; Tierney, 1994; Tierney *et al.*, 1996) as do experimentally infected fish held under a fixed ration of 8% body weight per day over the parasite growth phase (Barber and Svensson, 2003). Furthermore, data on the amino acid composition of *Ligula*-infected roach are consistent with those of starved fish (Soutter *et al.*, 1980). The energetic problems faced by infected fish are compounded by two further factors: (1) the growing plerocercoid reduces the space available for stomach expansion, limiting meal size (Cunningham *et al.*, 1994; Wright *et al.*, in press); and (2) infected fish are poor competitors for food (Milinski, 1990; Tierney, 1994; Barber and Ruxton, 1998) as a result of impaired foraging performance (Cunningham *et al.*, 1994).

Fish under nutritional stress, such as after a period of food deprivation, typically exhibit hyperphagia on refeeding to maximise their food intake (Ali *et al.*, 2003). Although increasing meal size is not an option for fish infected with cestode plerocercoids, they are able to partially alleviate the combined effects of infection by altering their foraging behaviour in other ways (Milinski, 1990). First, infected sticklebacks switch foraging strategies, taking advantage of "risky" food sources such as those available close to potential predators (Giles, 1983; Milinski, 1985; Godin and Sproul, 1988; Barber *et al.*, 2005) and altering food selection strategies to focus on prey types for which there is less competition (Milinski, 1984; Tierney, 1994; Ranta, 1995; see discussion in Barber and Huntingford, 1995). Second, infected fish may spend a greater proportion of their time foraging (e.g., Giles, 1987), possibly at the expense of shoal membership (Barber *et al.*, 1995).

Intriguingly, these strategies employed by infected sticklebacks to counter the energetic costs of infection are remarkably similar to those documented for salmonids injected with a growth hormone (GH) supplement (e.g., Jönsson *et al.*, 1998). By stimulating growth, GH increases metabolic demands, which must largely be met by increased nutrient intake, and one possibility is that energetically demanding parasites may exert their behavioural effects on hosts through the same physiological mechanisms. The control of food intake in fish is under the influence of a complex suite of factors including stomach distension, the level of circulating hormones (including insulin), and blood nutrients (including glucose) (see reviews by Le Bail and Boeuf, 1997; Donald, 1998). Indirect evidence supporting the hypothesis that parasites can exploit the growth hormone pathway to induce behavioural change in their hosts comes from two separate studies. First, mice infected with plerocercoids of *Spirometra mansonoides* (a close relative of *Schistocephalus* and *Ligula*) show almost identical changes in growth and physiology to those given GH supplementation (see review by Phares, 1997). Subsequent investigations have revealed that plerocercoids secrete a 27.5-kD protein known as plerocercoid growth factor (PGF), which has growth hormone activity (Phares and Kubik, 1996). Second, Arnott *et al.* (2000) demonstrated that three-spined sticklebacks experimentally infected with *Schistocephalus solidus*, when allowed access to unlimited food resources, outgrew sham-infected controls, suggesting that infections increased feeding motivation and total food intake. As yet there is no direct evidence that a parasite-produced growth factor is involved in mediating foraging behaviour in sticklebacks infected with *Schistocephalus solidus*, but one explanation for their enhanced growth in laboratory studies is that they are more willing than uninfected fish to feed under unfamiliar conditions. Future studies should focus on whether this readiness to feed is also observed under natural conditions, and whether it may be mediated through infection-induced changes to GH pathways.

Although we have focused primarily on infections that are associated with an increase in host foraging behaviour, fish infected with highly pathogenic infections may more typically show a reduction in food intake. Reductions in voluntary meal size have been demonstrated in rainbow trout (*Oncorhynchus mykiss*) infected with the kinetoplast *Cryptobia salmositica* (Lowe-Jinde and Zimmerman, 1991) and in Atlantic salmon (*Salmo salar*) infected with sea lice *Lepeophtheirus salmonis* (Dawson *et al.*, 1999). The physiological mechanisms reducing appetite in these groups have not been explicitly investigated.

3.1.2. Effects on Reproductive Behaviour

Gonadogenesis is typically delayed, impaired, or reversed in fish that are under nutritional stress, so it is not unexpected that parasite infections are often associated with reduced gonadogenesis or fecundity (e.g., Chen and Power, 1972; Wiklund *et al.*, 1996). Female common gobies (*Pomatoschistus microps*) harboring the adult stage of the trematode *Aphalloïdes coelomicola* in their body cavities exhibit reduced gonad weight as a consequence of reduced mass and energy content (but not diameter or number) of individual ova (Pampoulie *et al.*, 1999). The myxozoan *Kudoa paniformis*, which infects fibers of the skeletal muscle system of Pacific hake *Merluccius productus*, is associated with reduced fecundity of female hosts, with increased infection intensity having more severe effects (Adlerstein and Dorn, 1998). Furthermore, because reproductive behaviour in fish is initiated by the release of hormones from mature gonads (see Chapter 9) then parasite infections that impair gonadogenesis through their nutritional effects are likely to have consequences for the sexual behaviour of hosts.

However, it is difficult to attribute parasite-associated changes in sexual development solely to the energetic consequences of infection, as gonadogenesis in fishes is under the control of pituitary gonadotropins (GTH-I and GTH-II) as well as a multitude of hormones and growth factors (see Van Der Kraak *et al.*, 1998 for a detailed review). Yet the fact that *Schistocephalus solidus* infections impair gonadogenesis in female three-spined sticklebacks in natural populations—with infected individuals having smaller ovaries (McPhail and Peacock, 1983; Heins *et al.*, 1999) that contain fewer and smaller eggs (Heins and Baker, 2003) despite pituitary function being apparently unaffected in this system (Arme and Owen, 1967)—suggests that the energetic effects of infection per se may be important. In male sticklebacks, the finding that *S. solidus* infection does not impair nest building and courtship behaviour in lab trials (with freely-available food), despite the fact that few infected breeding males are ever located in surveys of natural populations (Candolin and Voigt, 2001), also suggests that infection reduces sexual behaviour solely through its nutritional effects.

Parasites that localise inside the host gonads may also considerably reduce the reproductive capacity of host fish through direct nutrient depletion. *Proteocephalus ambloplitis* plerocercoids penetrate the ovaries of host bass *Micropterus salmoides*, destroying individual oocytes and possibly utilising nutrients in the yolk for their own growth (McCormick and Stokes, 1982). Not all parasite infections reduce gonad size, though. Infections with *Kudoa ovivora*—a recently discovered myxozoan that lives inside the ova of Caribbean labroid fishes—are associated with increased egg mass as the parasite channels nutrient resources from the host into gonad development. This benefits the parasites developing within individual ova, but as infected ova are nonviable, infections impact negatively on host fecundity and sexual activity (Swearer and Robertson, 1999). Roach infected with *Ligula* do not engage in species-typical spawning behaviour (Dogiel *et al.*, 1961), but it is not clear whether this is a consequence of direct manipulation (via pituitary action of the worm; see Section 4.4) or an indirect effect of the failure of gonadogenesis.

3.2. Behavioural Consequences of Parasites with Homeostatic Effects

Although the intact teleost epidermis is relatively impervious to water, all fishes have large areas of permeable epithelia (including the gill lamellae, oral and narial mucosae) in contact with their aquatic medium (Bone *et al.*, 1995). As very few fish are isosmotic with the water they live in, they must cope with diffusion gradients that build up across these permeable surfaces by investing in active osmoregulation (Karnaky, 1998).

Fish ectoparasites that damage water-resistant epidermal tissues or permeable epithelia of their hosts, or which impair the functioning of osmoregulatory systems or tissues such as the gills, thus have the potential to impose osmoregulatory stress on their hosts. The effects of salmon lice (*Lepeophtheirus salmonis*) on the osmoregulatory capacity of host fish have been particularly well studied. Heavy infections with these parasites, which feed by sloughing off skin cells and damage the epidermis, result in osmoregulatory failure (Birkeland, 1996; Birkeland and Jakobsen, 1997), which leads to a reduction in swimming and cardiovascular performance (Wagner *et al.*, 2003) and ultimately results in the cessation of swimming activity (Grimnes and Jakobsen, 1996). As active osmoregulation is energetically expensive, locating ideal (isosmotic) aquatic environments may have significant benefits even to healthy fish, because emancipation from osmoregulation means that more energy can be made available for growth (e.g., Woo and Kelly, 1995; Riley *et al.*, 2003). Although this option is unavailable to marine or landlocked freshwater species, infected diadromous or estuarine fishes may be able to locate to osmotically favourable environments to facilitate survival and recuperation from infections. Infection-induced osmotic intervention may therefore be expected to impact on the migratory strategies of diadromous fish, and data exists to suggest this may be the case (see Chapter 7 for more details on fish migration).

Bjørn *et al.* (2001) found that anadromous sea trout *Salmo trutta* and Arctic char *Salvelinus alpinus* returning earliest to freshwaters were those most heavily infected with *L. salmonis*, suggesting that early return migrations are triggered to reduce the physiological (osmoregulatory) consequences of infection (see also Birkeland and Jakobsen, 1997). Boyce and Clarke (1983) demonstrated that experimentally induced *Eubothrium salvelini* infections reduce the capacity of migratory sockeye salmon to adapt physiologically to seawater, and suggest that, as this inability would be likely to reduce ocean survival, infection may delay or prevent seaward migrations. Sproston and Hartley (1941) recorded disproportionately high levels of infection with the ectoparasitic copepod *Lernaeocerca branchialis* among whiting (*Merlangius merlangus*, Gadidae) and pollack (*Pollachius pollachius*, Gadidae) delaying annual seaward migrations in an English estuary at the end of the winter. The authors suggest that the delay in migration could result from chronic anemia or a physiological response to ionic imbalance and subsequent avoidance of higher salinity water to which migrations normally are directed. However, as it has also been suggested that heavily infected individuals may also alter the timing of migrations to maximise foraging success (e.g., Bean and Winfield, 1992), experimental studies examining the salinity preferences of experimentally infected fish would be valuable in separating these factors.

4. EFFECTS OF PARASITES ON HOST CONTROL SYSTEMS

4.1. Behavioural Effects of Parasites that Invade the Central Nervous System

Like the peripheral sense organs, the CNS offers a refuge from the host's immune system and therefore is attractive to parasites. Locating in the CNS also provides parasites with the potential to directly affect behaviour and there is evidence that, in some cases, fine-scale site selection may have evolved to influence host behaviour in ways that benefit the parasites. Some of the most commonly recorded parasites of the CNS are the metacercariae of diplostomatid trematodes, including *Diplostomum phoxini* (Figure 4.4) and *Ornithodiplostomum ptychocheilus*, which infect old and new world minnows respectively. Histological studies have demonstrated that metacercariae are not randomly dispersed throughout the brains of infected fish, but are aggregated in lobes of brain concerned with vision and motor control (Barber and Crompton, 1997; Shirakashi and Goater, 2002). Heavy infections are associated with impaired optomotor responses (Shirakashi and Goater, 2002), altered shoaling behaviour (Radabaugh, 1980), and altered swimming behaviour of host fish (Ashworth and Bannerman, 1927; Rees, 1955; Lafferty and Morris, 1996), suggesting that damage caused to specific

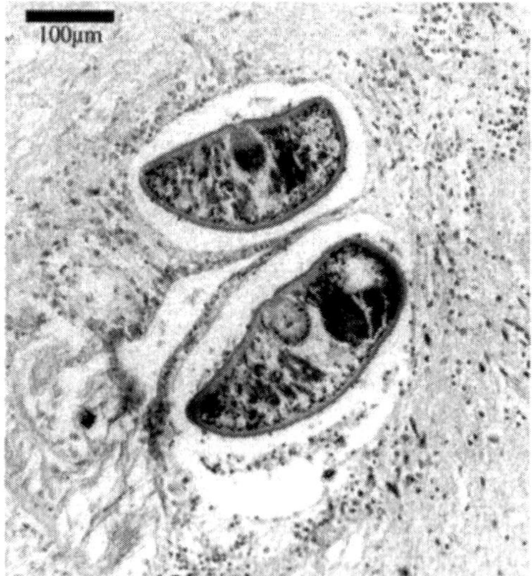

Fig. 4.4. Light micrograph of a sagittal section of two metacercariae of the digenean trematode *Diplostomum phoxini* in the medulla oblongata of the brain of a European minnow *Phoxinus phoxinus*.

brain regions by large aggregations of parasites is likely to be responsible. Such infections may have a considerable ecological impact through their effects on host behaviour. In a landmark study, Lafferty and Morris (1996) found that California killifish (*Fundulus parvipinnis*) infected with the brain-encysting metacercariae of the trematode *Euhaplorchis californiensis* performed more "conspicuous" swimming movements than noninfected conspecifics, and in field enclosure experiments were 30 times more likely than noninfected conspecifics to be eaten by piscivorous birds—the definitive hosts of the parasites.

Other parasites that do not live in the brain tissues may nonetheless damage the CNS and impact directly on host behaviour if they occupy the cranial cavity and/or vertebral column and exert pressure on the CNS tissues. The myxozoan *Myxobolus arcticus* inhabits the cranial cavity of sockeye salmon, and infections are associated with a significant reduction in swimming speed (Moles and Heifetz, 1998). Salmonids infected with *Myxobolus cerebralis* display a behavioural syndrome consisting of sequences of tight turns (whirling), interspersed with periods at rest with the tail elevated above the head, and episodes of postural collapse and immobility. Although the behavioural syndrome has previously been thought to result from damage to the vestibular apparatus (see Section 2.3), evidence from recent studies suggests a neurophysiological basis of behavioural changes associated with infection. Rose *et al.* (2000) argue that the behavioural syndrome is inconsistent with impaired vestibular function, alternatively proposing that behavioural changes result from granulomatous inflammation (associated with parasite invasion of the skull and vertebral column), which constricts the upper spinal cord, compressing and deforming the lower brainstem. Supportive evidence comes from studies of other myxozoans, for which infection is associated with both swimming disorders and similar host pathology (Grossel *et al.*, 2003; Longshaw *et al.*, 2003), and further studies are required to determine the precise physiological basis of whirling behaviour in infected fish.

4.2. Behavioural Effects of Parasites Impacting on the Autonomic Nervous System

The autonomic nervous system (ANS) of fish regulates the cardiovascular system, gastrointestinal system, swimbladder function, spleen and urogenital system, as well as controlling the expansion and contraction of melanophores and iris colouration (Donald, 1998). The ANS therefore plays an important role in regulating aspects of behaviour, and studies on non-fish taxa have demonstrated that parasite infections can impact on host ANS function. For example, in the rat (*Rattus rattus*), L3 larvae of the parasitic nematode *Anisakis simplex* (a fish parasite) induce cholinergic hyperactivity and adrenergic blockade in the whole of the small intestine, causing gastrointestinal symptoms that influence foraging behaviour and appetite (Sanchez-Monsalvez *et al.*, 2003). In humans, although the mechanism is not well understood, some trypanosome infections are associated with autonomic dysfunction, including neurogenic cardiomyopathy and digestive damage (Sterin-Borda and Borda, 2000; Pinto *et al.*, 2002). Recent studies on brown trout *Salmo trutta* have shown that infections with the intestinal acanthocephalan *Pomphorhynchus laevis* and the cestode *Cyathocephalus truncatus* are associated with changes in the numbers of endocrine cells in the gastrointestinal tract showing immunoreactivity to a range of neurotransmitters (including bombesin, cholecystokinin-8 [CCK-8], leu-enkephalin and 5-hydroxytryptamine [5-HT], serotonin; Dezfuli *et al.*, 2000, 2002) known to control gut motility and digestive/absorptive processes in fish (Donald, 1998). Behavioural studies allied to these investigations of ANS and gut function infection would be invaluable in generating a more complete understanding of how parasite infections impact on the foraging ecology and growth of infected fish.

4.3. Neurochemical Interference with Host Behaviour

Brain monoamine neurotransmitters are involved in the control of feeding behaviour (De Pedro *et al.*, 1998), social behaviour and aggression in fish (Winberg and Nilsson, 1993), and are sensitive to a range of stressors (Chapter 5, Section 3; Winberg and Nilsson, 1993), immunological factors (Lacosta *et al.*, 2000) and nutritional status (Levin and Routh, 1996). Monoamine neurotransmitters are also essential in pathways associated with sensory and motor performance, in addition to higher brain function (at least in mammals). Moreover, monoamines such as serotonin are also now known to act as neuromodulators, playing an active role in the recruitment of neurones to neuronal assemblies, resculpting neural circuits and giving animals the flexibility to shape their behaviour in response to the changing demands of their internal state and the external environment (Thompson and Kavaliers, 1994; Adamo, 2002). This feature of monoamines potentially provides parasites with a ready-made mechanism for manipulating host behaviour.

A number of studies have examined the role of neurotransmitters in the control of behaviour and the influence of parasites in invertebrate host models (see reviews by Adamo, 1997; Moore, 2002). Helluy and Holmes (1990) demonstrated that the peculiar impaired escape responses of gammarids infected with *Polymorphus paradoxus* (Bethel and Holmes, 1973) could be precisely replicated in healthy specimens by the injection of serotonin;

moreover, the altered behaviours could be reversed by injecting with octopamine, a serotonin antagonist. More recently, Helluy and Thomas (2003) used immunocytochemical methods to examine changes in the serotonin-immunoreactivity of brain regions in gammarids infected with another behaviour-changing parasite, *Microphallus papillorobustus*. Infection was found to be associated with altered serotonergic sensitivity in specific brain regions and altered architecture of the serotonergic tracts and neurons, suggesting that parasites may have exerted their influence through neuromodulatory mechanisms.

In mammals, the most detailed studies of the neurochemical basis of parasite-induced behavioural change have been carried out by Kavaliers and coworkers (reviewed by Kavaliers *et al.*, 1998b; Moore, 2002) and have suggested a role for serotonin. In noninfected mice, brief exposure to predatory threat typically induces nonopioid-mediated, serotonin-sensitive analgesia (reduced pain sensitivity). However, Kavaliers and Colwell (1994) found that mice infected with the coccidian *Eimeria vermiformis* failed to exhibit serotonin-sensitive analgesia when presented with a cat stimulus. In addition, although general olfactory functioning appeared unaffected, infected mice failed to avoid cat odor. Although the parasite would not benefit from increased predation, because transmission occurs directly between mice via fecal contamination, the observed "fearlessness" of infected mice may also increase mouse–mouse social contact, which would have consequences for transmission. As with infected gammarids, one physiological explanation for this fearlessness of infected mice is that the parasite influences host behaviour by interfering with serotonergic activity.

Although there is considerable research interest in the neurochemical regulation of behaviour in fish, there is very little information on parasite modulation. In the only study of parasite-associated neuromodulation in fish, Øverli *et al.* (2001) investigated the effects of *Schistocephalus solidus* infection status on the concentrations of neuromodulators, including norepinephrine (NE) and serotonin, in the brains of three-spined sticklebacks. Infected fish had elevated concentrations of NE and serotonin in the telencephalons, but the clearest effect of infection was on the serotonin metabolite, 5-hydroxy-indoleacetic acid (5-HIAA), which increased in the brainstem of infected sticklebacks. As a consequence, the ratio of 5-HIAA to serotonin (an indicator of stress in fish) was elevated in both the hypothalamus and brainstem regions of infected sticklebacks (Figure 4.5). One explanation for these patterns is that the parasite induces chronic stress, but it is also possible that the parasite manipulates the neuroendocrine status of its host directly. Behavioural changes in sticklebacks infected with the plerocercoid larvae of *Schistocephalus solidus* have been extremely well documented, and infected fish are known to show altered risk-taking and antipredator behaviour (see

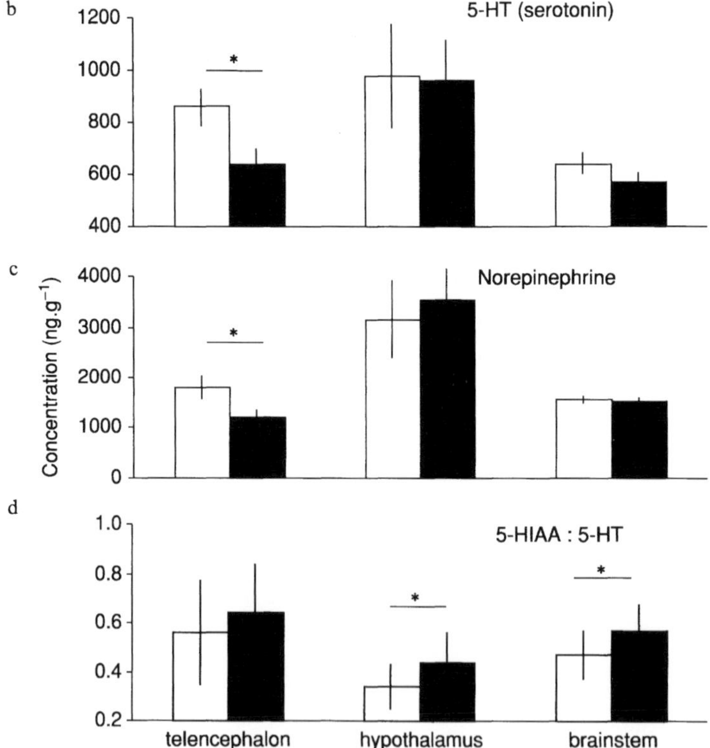

Fig. 4.5. (a) A dissected three-spined stickleback *Gasterosteus aculeatus*, and four plerocercoids of the pseudophyllidean cestode *Schistocephalus solidus* recovered from its body cavity. (b-d) Effects of *S. solidus* infection on the concentration of 5-hydroxytryptamine (5-HT, serotonin),

Section 3.1.1; Milinski, 1990; Barber *et al.*, 2000 for reviews). These behaviours closely match the fearlessness shown by parasitised mice and gammarids, and because *Schistocephalus solidus* is transmitted when birds ingest host sticklebacks (Smyth, 1985), it is possible that adaptive host behavioural change in this system is mediated through neurochemical manipulation. Further research is required to separate the potential direct and indirect mechanisms of infection-induced behavioural change in sticklebacks infected with *Schistocephalus solidus*.

4.4. Endocrinological Interference with Reproductive Behaviour

As well as impacting on host reproduction by reducing the amount of energy available for gonadogenesis (see Section 3.1.2), parasites may also interfere directly with the reproductive endocrinology of their fish hosts, with consequences for the reproductive physiology and behaviour of infected fish. The tapeworm *Ligula intestinalis* achieves host castration by interfering with the pituitary-gonadal axis of fish hosts (Arme, 1968). Although the precise details of the mechanism are still to be discovered (Arme, 1997), the parasite appears to act at the level of the host's hypothalamus, restricting gonadotropin-releasing hormone (GnRH) secretion. This results in poor gonad development (Williams *et al.*, 1998), which in turn is likely to be responsible for the impaired spawning behaviour (Dogiel *et al.*, 1961). With increasing interest in the mechanisms causing the increased levels of endocrine disruption in freshwater fish, particularly resulting from anthropogenic pollution events, the value of further research into the mechanistic basis by which *Ligula* alters host sexual development and behaviour is clear (Jobling and Tyler, 2003).

5. PHYSIOLOGICAL EFFECTS OF INFECTION THAT IMPAIR THE HOST'S BEHAVIOURAL CAPACITY

5.1. Cardiovascular Effects of Parasites on Host Swimming Performance

Parasites that inhabit the heart muscle, live in the lumen of blood vessels, or reduce the oxygen-carrying capacity of the blood may reduce the efficiency of the cardiovascular system, with likely consequences for behaviour

---

norepinephrine (NE), and the ratio of 5-hydroxy-indoleacetic acid (5-HIAA) to 5-HT in various brain regions of infected fish. Open bars represent values for control (noninfected) fish and filled bars represent values for infected fish. Bar heights are means; error bars represent 1 standard deviation. $^*P < 0.05$. (Data from Øverli *et al.*, 2001).

and stamina of infected fish. Sanguilicolid and heterophyid digenean flukes are understood to be the most pathogenic of the helminths that inhabit the heart and vascular system of fishes (Williams and Jones, 1994). Infections reduce the capacity of the blood to carry and exchange gases by causing mechanical obstruction, by altering the number and type of blood cells, and by causing hemorrhage (Smith, 1972). *Ascocotyle pachycystis*, a heterophyid trematode, locates in and occludes the bulbus arteriosus of sheepshead minnows (*Cyprinodon variegatus*, Cyprinodontidae), physically obstructing blood flow and reducing the time infected fish are able to swim at their maximum sustainable velocity before becoming exhausted (Coleman, 1993). A number of species of philometrid nematodes are also known to occupy the cardiovascular system of the fish hosts, with similar pathological effects.

Anemia is a commonly reported symptom of infection, particularly for fish that are infected with parasites that feed on the blood of their hosts, such as ectoparasitic lice and copepods. Mann (1952, 1953; cited in Grabda, 1991) reports that the haemoglobin levels of whiting infected with the copepod *Lernaeocera* were reduced from 30–40% (normal) to 20–22%. Anemia typically decreases the swimming performance of fish (Jones, 1971; Gallaugher *et al.*, 1995). Wagner *et al.* (2003) demonstrated that sublethal *Lepeophtheirus salmonis* infections caused a reduction in critical swimming speed (U_{crit}) of rainbow trout. Although the behaviourally significant sublethal infections were not associated with anemia in that study, a subsequent study showed that controlled blood loss from noninfected fish could generate similar negative effects on U_{crit} (Wagner and McKinley, 2004), suggesting that as infection levels increase a change in swimming performance would accompany other morbidity effects of infection. Parasitic blood-dwelling haematozoans, including leech-vectored kinetoplastids such as *Trypanoplasma borreli*, may also impact on host swimming behaviour as a consequence of their effects on renal function, osmoregulation and hematocrit (Bunnajirakul *et al.*, 2000).

5.2. Habitat Selection Consequences of Parasites that Affect Host Metabolism and Respiration

Fish infected with metabolically demanding parasites, such as rapidly growing cestode plerocercoids, may have elevated oxygen requirements with consequences for habitat selection and swimming performance. Lester (1971) demonstrated that *Schistocephalus*-infected three-spined sticklebacks had increased oxygen consumption and as a result selected shallower water habitats, with potential consequences for predation risk. Nine-spined sticklebacks (*Pungitius pungitius*) infected with the same parasite exhibited an increased frequency of aquatic surface respiration (gasping at the surface), and when exposed to constantly decreasing oxygen tensions in experimental studies were found to have a higher lethal oxygen level (Smith and Kramer, 1987).

Infections may also reduce the level of oxygen available for host respiration if they interfere with ventilation or gill function. Parasites that are ectoparasitic on the gills damage filaments and epithelia of the secondary lamellae and cause an increase in host mucous secretion (e.g., Ishimatsu *et al.*, 1996; Dezfuli *et al.*, 2003) and potentially restrict the flow of water over these tissues. Eggs of *Sanguinicola* spp. flukes, described in Section 5.1. above, block gill capillaries causing thrombosis and necrosis of gill tissue (Smith, 1972). Such changes are likely to impair gas exchange efficiency, with the consequence that infected fish may be unable to withstand low oxygen levels and be forced into seeking oxygen-rich waters.

5.3. Buoyancy Regulation Effects of Parasites

Teleost fish rely on the swimbladder, which acts as a hydrostatic organ by replacing parts of the fish body with gas to maintain vertical position in the water column, so parasites that impact on swimbladder function and/or inflation might be expected to impair the ability of fish to achieve buoyancy regulation. Largemouth bass (*Micropterus salminoides*, Centrarchidae) infected with the swimbladder-dwelling nematode *Eustrongylides ignotus* exhibit buoyancy abnormalities (Coyner *et al.*, 2001), as do *Eimeria*-infected haddock (*Melanogrammus aeglefinus*, Gadidae; Odense and Logan, 1976). *Anguillicola crassus* is an introduced and economically important swimbladder parasite of European eels (*Anguilla anguilla*, Anguillidae), which has significant impact on commercial and natural populations (Kirk, 2003). Histological studies show that *Anguillicola* infections cause swimbladder pathology (Nimeth *et al.*, 2000), and infection is associated with a change in the composition of swimbladder gases (Wurtz *et al.*, 1996). Although the consequences of infection for the migratory performance of eels is not yet known, the potential for an effect would appear to be substantial, because *Anguillicola* infection in eels is associated with a reduction in maximum swimming speed (Sprengel and Lüchtenberg, 1991). However, whether this is due to buoyancy effects or to general morbidity is as yet unknown.

Buoyancy control may also be impaired if parasite infections alter the specific gravity of fish hosts. This is most likely for fish infected with the large plerocercoid larvae of pseudophyllidean cestodes, which have a different specific gravity than their fish hosts and can thus potentially interfere with swimbladder inflation (Ness and Foster, 1990; LoBue and Bell, 1993).

5.4. Effects of Parasites on Swimming Performance

Techniques for quantifying fish swimming performance are well established, and a number of studies have documented quantitative effects of parasite infection. In some cases, it is likely that the effects of the parasites are related to energetic consequences of infection. Smith and Margolis (1970) demonstrated that sockeye salmon smolts infected with adult *Eubothrium salvelini* cestodes, which inhabit the intestine and reduce host growth rates (Saksvik *et al.*, 2001), fatigued more quickly than uninfected smolts. Such effects may have consequences for the migratory performance of anadromous fish. Boyce (1979) demonstrated significant impacts of experimentally-induced *Eubothrium salvelini* infections on the swimming performance of sockeye and Smith (1973) found that numbers of infected fish were concentrated towards the end of the run, probably as a result of their impaired swimming ability.

In most cases, there is little information available on the physiological basis of impaired swimming performance, but the location of the parasite may often be informative. Parasites encysting in the musculature, or otherwise impacting on muscle development or function, clearly have the potential to interfere directly with host swimming performance. Metacercariae of the trematode *Nanophyetus salmonicola* encyst throughout the musculature and organ systems of salmonid fish, with effects on burst swimming performance of coho salmon (*Oncorhynchus kisutch*) and steelhead rainbow trout (*Oncorhynchus mykiss*) smolts (Butler and Millemann, 1971). Anisakid nematodes encysted in the musculature of host gadoids secrete metabolic products (alcohol and ketones) which anesthetise host musculature (Ackman and Gjelstad, 1975), and McClelland (1995) speculates that this mechanism is likely to be responsible for the reported slow swimming speed of smelt *Osmerus eperlanus*, Osmeridae infected with *Pseudoterranova decipiens* (Sprengel and Lüchtenberg, 1991). Because the definitive hosts of these indirectly transmitted parasites are phocid seals, which acquire the worms after eating infected fish, reduced swimming performance could conceivably enhance transmission.

Parasites that have significant nutritional effects also potentially reduce muscle mass of fish: *Ligula* infections in roach are associated with atrophy of the body wall musculature (Sweeting, 1977), and in common bream *Abramis brama* the same parasite reduces the mass of the body wall muscle (Richards and Arme, 1981).

Changes in the body shape of fish infected with cestodes that grow to a large size in the body cavity or intestine (e.g., Smith, 1973; Barber, 1997; Loot *et al.*, 2002b) are likely to increase the flow resistance and frictional drag of hosts (Rodewald and Foster, 1998) and reduce the body flexibility necessary for fast starts, as well as being costly in terms of locomotive speed and efficiency (Blake, 1983; Videler, 1993). The "jerky" swimming movements characterised by increased lateral amplitude relative to swimming speed reported from bleak (*Alburnus alburnus*, Cyprinidae) infected with *Ligula intestinalis* (Harris and Wheeler, 1974), and the "sluggish" movements of common shiners (*Luxilus cornutus*, Cyprinidae; Dence, 1958) probably result from such mechanisms. However, although infection-associated modifications in the swimming movements (kinematics) of host fish are likely to be important mechanisms in determining detection and selection by predators, this area has so far attracted little attention from researchers, and few quantitative studies have been undertaken.

5.5. Effects of Parasites on Sound Production

The production and detection of underwater sounds is an important component of fish ecology (Chapter 2, Section 6; Myrberg and Spires, 1980; Hawkins, 1993) and plays a significant role in many aspects of fish behaviour including aggressive interaction, courtship, and intraspecific communication. Fish are capable of using three types of mechanisms to produce sound: by stridulation (the grinding or rubbing together of body parts such as fins or pharyngeal teeth), by percussive action (typically involving the swimbladder), and by performing rapid turns (which result in hydrodynamic sounds) (Hawkins, 1993). During all types of sound production, the swimbladder plays a key role in amplifying the noises produced, and so parasites that inhabit the swimbladder have the potential to interfere with the sound production capabilities of host fish.

As discussed in Section 5.3, the swimbladder is a common site for parasites and pathological infections may have consequences for sound production. The coccidian protozoan *Eimeria gadi* infects the swimbladders of a number of gadoid species. Odense and Logan (1976) reported that among haddock from Nova Scotia fishing banks, prevalence of *Eimeria* ranged between 4% and 58% and the swimbladders of infected fish were filled with "a creamy viscous to yellow semisolid material, [consisting] of various parasite stages, fibrous and cellular debris and lipid material." The authors speculate that infections would impact sound production. As courtship in haddock and other gadoids is a complex process, requiring subtle posturing and the production of drumming sounds made by the vibrations of specially adapted muscles surrounding the swimbladder (Hawkins and Amorim, 2000; Bremner *et al.*, 2002), it seems likely that parasites have the potential to impact significantly on spawning success, with potential consequences for the conservation of these threatened species (Rowe and Hutchings, 2003).

6. FUTURE DIRECTIONS

6.1. Parasites and the Sensory Ecology of Hosts

With significant developments in our understanding of the physiology of the nonvisual senses of fish including chemosensation, sound perception, and electrosensation, and an improved recognition of the pivotal role such senses play in behaviour, the scarcity of studies focused on examining the physiological and behavioural effects of parasites on these systems is surprising. It seems apparent to us that further research on the effects of parasites on both the odor profiles and chemosensation ability of fishes would be highly valuable, and we highlight this as a potentially fruitful area for collaborations between behavioural ecologists, sensory ecologists, and parasitologists.

6.2. Extending the Range of Systems under Study

As is evident from this review, there has been a bias towards studying the behavioural and physiological consequences of parasite infections in freshwater and anadromous hosts. This bias is understandable given the commercial value of many of these species and the relative ease with which they are kept in the laboratory, compared to marine species. Furthermore, our most detailed knowledge of the physiological basis of behavioural change associated with infection stems from studies into a small number of conveniently studied host–parasite systems, such as the stickleback–*Schistocephalus* system. Although there is considerable value in developing model systems such as this, the fact that parasite infections have such diverse effects on hosts means that the number of model systems developed should be increased. Studies examining behavioural and physiological impacts of parasites of marine species are relatively scarce, and yet ecologists are beginning to realise that parasites can have a major effect on the demography and population dynamics of marine host fish species (Finley and Forrester, 2003). The development of studies investigating behavioural consequences of infection in marine species, and their physiological bases, are therefore to be encouraged.

6.3. Studying Host Behavioural Change in the Postgenomic Era

Biology is moving into an exciting era, with tools developed as part of large-scale genome sequencing projects, such as transcriptomics, proteomics, and metabolomics, now being routinely used in a large number of molecular biology laboratories. Fish physiologists are beginning to utilise these technologies to investigate the environmental and ontogenetic control of gene expression and metabolism in their study organisms (Parrington and Coward, 2002). These analytical tools, which allow infection-associated changes in gene regulation, protein expression, and metabolic activity to be qualified and in some cases quantified (Barrett *et al.*, 2000), can be used to investigate physiological aspects of host–parasite interactions in systems for which experimental infection protocols are well-established. By combining the various-omic technologies with carefully designed behavioural examination of experimentally-infected fish hosts, it is now possible to track changes in the behaviour of parasitised fish and relate these to concurrent changes in the expression of host genes and proteins. These approaches clearly offer a significant opportunity to further our understanding of how parasites, through their physiological effects on hosts, impact on patterns of host behaviour.

7. SUMMARY

There is an increased interest from fisheries biologists, aquaculturists, evolutionary, and behavioural ecologists in the behavioural changes in fish hosts that are associated with parasite infections. This Chapter introduced the various ways in which parasites may influence the behaviour of teleost fishes, focusing particularly on behavioural changes that are induced by parasites following infection. We systematically reviewed each of the major physiological systems of fish (e.g., ionic balance, neurochemistry, endocrine function, and nutritional status), the effects of parasites on them, and examined how infection-associated changes in functioning may impact on normal patterns of host behaviour. There are few host–parasite systems for which physiological and behavioural effects of infection have been quantified experimentally, but where possible those studies for which both types of data are available were reviewed. Major gaps in knowledge were also highlighted for further research. This Chapter ended by emphasising the value of a mechanistic approach for understanding the evolution and likely fitness consequences of infection-associated host behaviour modification, and highlighting opportunities to exploit postgenomic technologies to further elucidate the physiological basis of infection-associated changes in host behaviour.

ACKNOWLEDGEMENTS

We thank Katherine Sloman, Rod Wilson, and Sigal Balshine for giving us the opportunity to write this review. I.B. was in receipt of NERC fellowship funding during much of the writing of this chapter (GT/5/98/6/FS and NER/I/S/2000/00971). H.W. is funded by a UWA Ph.D. studentship.

REFERENCES

Ackman, R. G., and Gjelstad, R. T. (1975). Gas-chromatographic resolution of isomeric pentanols and pentanones in the identification of volatile alcohols and ketones in the codworm *Terranova decipiens* (Krabbe, 1878). *Anal. Biochem.* **67,** 684–687.

Adamo, S. A. (1997). How parasites alter the behaviour of their insect hosts. *In* "Parasites and pathogens: Effects on host hormones and behaviour" (Beckage, N. E., Ed.), pp. 231–245. Chapman and Hall, New York.

Adamo, S. A. (2002). Modulating the modulators: Parasites, neuromodulators and host behavioural change. *Brain Behav. Evolut.* **60,** 370–377.

Adlerstein, S. A., and Dorn, M. W. (1998). The effect of *Kudoa paniformis* infection on the reproductive effort of female Pacific hake. *Can. J. Zool.* **76,** 2285–2289.

Ali, M., Nicieza, A., and Wootton, R. J. (2003). Compensatory growth in fishes: A response to growth depression. *Fish Fisheries* **4,** 147–190.

Arme, C. (1968). Effects of the plerocercoid larvae of a pseudophyllidean cestode, *Ligula intestinalis*, on the pituitary gland and gonads of its host. *Biol. Bull.* **134,** 15–25.

Arme, C. (1997). *Ligula intestinalis*: Interactions with the pituitary-gonadal axis of its fish host. *J. Helminthol.* **71,** 83–84.

Arme, C., and Owen, R. W. (1967). Infections of the three-spined stickleback, *Gasterosteus aculeatus* L., with the plerocercoid larvae of *Schistocephalus solidus* with special reference to pathological effects. *Parasitology* **57,** 301–304.

Arnott., S. A., Barber, I., and Huntingford, F. A. (2000). Parasite-associated growth enhancement in a fish-cestode system. *Proc. Roy. Soc. Lond. B. Bio.* **267,** 657–663.

Ashworth, J. H., and Bannerman, J. C. W. (1927). On a tetracotyle (*T. phoxini*) in the brain of the minnow. *Trans. Roy. Soc. Edin.* **55,** 159–171.

Barber, I. (1997). A non-invasive morphometric technique to estimate the cestode plerocercoid load of small freshwater fishes. *J. Fish Biol.* **51,** 654–658.

Barber, I., and Huntingford, F. A. (1995). The effect of *Schistocephalus solidus* (Cestoda: Pseudophyllidea) on the foraging and shoaling behaviour of three-spined sticklebacks, *Gasterosteus aculeatus*. *Behaviour* **132,** 1223–1240.

Barber, I., and Huntingford, F. A. (1996). Parasite infection impairs schooling behaviour: Spatial deviance of helminth-infected minnows in conspecific groups. *Proc. R. Soc. Lond. Ser. B* **263,** 1095–1102.

Barber, I., and Crompton, D. W. T. (1997). The distribution of *Diplostomum phoxini* metacercariae (Trematoda) in the brain of minnows (*Phoxinus phoxinus*). *Folia Parasit.* **44,** 19–25.

Barber, I., and Ruxton, G. D. (1998). Temporal prey distribution affects the competitive ability of parasitised sticklebacks. *Anim. Behav.* **58,** 1477–1483.

Barber, I., and Svensson, P. A. (2003). Effects of experimental *Schistocephalus solidus* infections on growth, morphology and sexual development of female three-spined sticklebacks, *Gasterosteus aculeatus*. *Parasitology* **126,** 359–367.

Barber, I., Huntingford, F. A., and Crompton, D. W. T. (1995). The effect of hunger and cestode parasitism on the shoaling decisions of small freshwater fish. *J. Fish Biol.* **47,** 524–536.

Barber, I., Downey, L. C., and Braithwaite, V. A. (1998). Parasitism, oddity and the mechanism of shoal choice. *J. Fish Biol.* **52,** 1365–1368.

Barber, I., Hoare, D., and Krause, J. (2000). The effects of parasites on fish behaviour: An evolutionary perspective and review. *Rev. Fish Biol. Fisher.* **10,** 131–165.

Barber, I., Walker, P., and Svensson, P. A. (2004). Behavioural responses to simulated avian predation in female three-spined sticklebacks: The effect of experimental *Schistocephalus* infections. *Behaviour* **141,** 1425–1440.

Barnard, C. J., and Behnke, J. M. (1990). "Parasitism and Host Behaviour." Taylor and Francis, London.

Barrett, J., Jeffries, J. R., and Brophy, P. M. (2000). Parasite proteomics. *Parasitol. Today* **16,** 400–403.

Bean, C. W., and Winfield, I. J. (1992). Influences of the tapeworm *Ligula intestinalis* (L.) on the spatial distributions of juvenile roach *Rutilus rutilus* (L.) and Gudgeon *Gobio gobio* (L.) in Lough Neagh, Northern Ireland. *Neth. Jnl. Zool.* **42,** 416–423.

Bethel, W. M., and Holmes, J. C. (1973). Altered evasive behaviour and responses to light in amphipods harbouring acanthocephalan cystacanths. *J. Parasitol.* **59,** 945–956.

Birkeland, K. (1996). Consequences of premature return by sea trout (*Salmo trutta*) infested with the salmon louse (*Lepeophtheirus salmonis* Kroyer): Migration, growth, and mortality. *Can. J. Fish. Aq. Sci.* **53,** 2808–2813.

Birkeland, K., and Jakobsen, P. J. (1997). Salmon lice, *Lepeophtheirus salmonis*, infestation as a causal agent of premature return to rivers and estuaries by sea trout, *Salmo trutta*, juveniles. *Environ. Biol. Fish.* **49,** 129–137.

Bjørn, P. A., Finstad, B., and Kristoffersen, R. (2001). Salmon lice infection of wild sea trout and Arctic char in marine and freshwaters: The effects of salmon farms. *Aquaculture Res.* **32,** 947–962.

Blake, R. W. (1983). "Fish Locomotion." Cambridge University Press: Cambridge, UK.

Bleckmann, H. (1993). Role of the lateral line in fish behaviour. *In* "Behaviour of Teleost Fishes" (Pitcher, T. J., Ed.), pp. 201–246. Chapman and Hall, London.

Bone, Q., Marshall, N. B., and Blaxter, J. H. S. (1995). "Biology of Fishes" (2nd ed.). Chapman and Hall, London.

Boyce, N. P. (1979). Effects of *Eubothrium salvelini* (Cestoda: Pseudophyllidea) on the growth and vitality of sockeye salmon, *Oncorhynchus nerka*. *Can. J. Zool.* **57,** 597–602.

Boyce, N. P., and Clarke, W. C. (1983). *Eubothrium salvelini* (Cestoda, Pseudophyllidea) impairs seawater adaptation of migrant sockeye salmon yearlings (*Oncorhynchus nerka*) from Babine Lake, British Columbia. *Can. J. Fish. Aq. Sci.* **40,** 821–824.

Brassard, P., Rau, M. E., and Curtis., M. A. (1982). Parasite-induced susceptibility to predation in diplostomiasis. *Parasitology* **85,** 495–501.

Bremner, A. A., Trippel, E. A., and Terhune, J. M. (2002). Sound production by adult haddock, *Melanogrammus aeglefinus*, in isolation, pairs and trios. *Env. Biol. Fish* **65,** 359–362.

Brønmark, C., and Hansson, L. A. (2000). Chemical communication in aquatic systems: An introduction. *Oikos* **88,** 103–109.

Brown, G. E., and Smith, R. J. F. (1994). Fathead minnows use chemical cues to discriminate natural shoalmates from unfamiliar conspecifics. *J. Chem. Ecol.* **20,** 3051–3061.

Bunnajirakul, S., Steinhagen, D., Hetzel, U., Korting, W., and Drommer, W. (2000). A study of sequential histopathology of *Trypanoplasma borreli* (Protozoa: Kinetoplastida) in susceptible common carp *Cyprinus carpio*. *Dis. Aq. Org.* **39,** 221–229.

Bush, A. O., Fernández, J. C., Esch., G. W., and Seed, R. (2001). "Parasitism: The diversity and ecology of animal parasites." CUP, Cambridge, UK.

Butler, J. A., and Millemann, R. E. (1971). Effect of the "salmon poisoning" trematode, *Nanophyetus salmincola* on the swimming ability of juvenile salmonid fishes. *J. Parasitol.* **57,** 860–865.

Candolin, U., and Voigt, H.-R. (2001). No effect of a parasite on reproduction in stickleback males: A laboratory artefact? *Parasitology* **122,** 457–464.

Chappell, L. H., Hardie, L. J., and Secombes, C. J. (1994). Diplostomiasis: The disease and host-parasite interactions. *In* "Parasitic Diseases of Fish" (Pike, A. W., and Lewis, J. W., Eds.), pp. 59–86. Samara Publishing, Dyfed. UK.

Chen, M., and Power, G. (1972). Infection of American smelt in Lake Ontario and Lake Erie with the microsporidian parasite *Glugea hertwigi*. *Can. J. Zool.* **50,** 1183–1188.

Chivers, D. P., and Smith, R. J. R. (1993). The role of olfaction in chemosensory-based predator recognition in the fathead minnow, *Pimephales promelas*. *J. Chem. Ecol.* **19,** 623–633.

Coleman, F. C. (1993). Morphological and physiological consequences of parasites encysted in the bulbus arteriosis of an estuarine fish, the sheepshead minnow, *Cyprinodon variegatus*. *J. Parasitol.* **79,** 247–254.

Cox, F. E. G. (1994). Immunology. *In* "Modern Parasitology" (Cox, F. E. G., Ed.), 3rd edn., pp. 193–218. Blackwell Scientific Publications, Oxford.

Coyner, D. F., Schaack, S. R., Spalding, M. G., and Forrester, D. J. (2001). Altered predation susceptibility of mosquitofish infected with *Eustrongylides ignotus*. *J. Wildl. Dis.* **37,** 556–560.

Crowden, A. E. (1976). *Diplostomum spathaceum* in the Thames; occurrence and effects on fish behaviour. *Parasitology* **73,** vii.

Crowden, A. E., and Broom, D. M. (1980). Effects of the eyefluke, *Diplostomum-spathaceum*, on the behaviour of dace (*Leuciscus leuciscus*). *Anim. Behav.* **28,** 287–294.

Cunningham, E. J., Tierney, J. F., and Huntingford, F. A. (1994). Effects of the cestode *Schistocephalus solidus* on food intake and foraging decisions in the three-spined stickleback *Gasterosteus aculeatus*. *Ethology* **79,** 65–75.

Damsgård, B., and Dill, L. M. (1998). Risk-taking behaviour in weight-compensating coho salmon, *Oncorhynchus kisutch*. *Behav. Ecol.* **9,** 26–32.

Dawson, L. H. J., Pike, A. W., Houlihan, D. F., and McVicar, A. H. (1999). Changes in physiological parameters and feeding behaviour of Atlantic salmon *Salmo salar* infected with sea lice *Lepeophtheirus salmonis*. *Dis. Aquat. Org.* **35,** 89–99.

Dence, W. A. (1958). Studies on *Ligula*-infected common shiners (*Notropis cornutus frontinalis* Agassiz) in the Adirondacks. *J. Parasitol.* **44,** 334–338.

De Pedro, N., Pinillos, M. L., Valenciano, A. I., Alonso-Bedate, M., and Delgado, M. J. (1998). Inhibitory effect of serotonin on feeding behaviour in goldfish: Involvement of CRF. *Peptides* **19,** 505–511.

Dezfuli, B. S., Arrighi, S., Domeneghini, C., and Bosi, G. (2000). Immunohistochemical detection of neuromodulators in the intestine of *Salmo trutta* L. naturally infected with *Cyathocephalus truncatus* Pallas (Cestoda). *J. Fish Dis.* **23,** 265–273.

Dezfuli, B. S., Pironi, F., Giari, L., Domeneghini, C., and Bosi, G. (2002). Effect of *Pomphorhynchus laevis* (Acanthocephala) on putative neuromodulators in the intestine of naturally infected *Salmo trutta*. *Dis. Aquat. Org.* **51,** 27–35.

Dezfuli, B. S., Giari, L., Konecny, R., Jaeger, P., and Manera, M. (2003). Immunohistochemistry, ultrastructure and pathology of gills of *Abramis brama* from Lake Mondsee, Austria, infected with *Ergasilus sieboldi* (Copepoda). *Dis. Aquat. Org.* **53,** 257–262.

Dogiel, V. A., Petrushevski, G. K., and Polyanski, Y. I. (1961). "Parasitology of Fishes." (Translation from Russian: "Basic Problems of the Parasitology of Fishes" (1958). Izdatel'stvo Leningradskogo Universiteta, Leningrad), pp. 384. Oliver and Boyd, Edinburgh and London.

Dobson, A. P. (1988). The population biology of parasite-induced changes in host behaviour. *Quart. Rev. Biol.* **63,** 139–165.

Donald, J. A. (1998). Autonomic Nervous System. *In* "The Physiology of Fishes," (Evans, D. H., Ed.), 2nd edn., pp. 407–439. CRC Press, Boca Raton.

Donnelly, R. E., and Reynolds, J. D. (1994). Occurrence and distribution of the parasitic copepod *Leposphilus labrei* on corkwing wrasse (*Crenilabrus melops*) from Mulroy Bay, Ireland. *J. Parasitol.* **80,** 331–332.

Erasmus, D. A. (1959). The migration of *Cercaria X* Baylis (Strigeida) within the fish intermediate host. *Parasitology* **48,** 312–335.

Fernald, R. D. (1993). Vision. *In* "The Physiology of Fishes" (Evans, D. H., Ed.), pp. 161–190. CRC Press, Boca Raton.

Finley, R. J., and Forrester, G. E. (2003). Impact of ectoparasites on the demography of a small reef fish. *Mar. Ecol. Prog. Ser.* **248,** 305–309.

Frisch, K. Von (1938). Zur Psychologie des Fische-Schwarmes. *Naturwissenschaften* **26,** 601–606.

Gallaugher, P., Thorarensen, H., and Farrell, A. P. (1995). Haematocrit in oxygen transport and swimming in rainbow trout (*Oncorhynchus mykiss*). *Resp. Phys.* **102,** 279–292.

Giles, N. (1983). Behavioural effects of the parasite *Schistocephalus solidus* (Cestoda) on an intermediate host, the three-spined stickleback, *Gasterosteus aculeatus*. *Anim. Behav.* **31,** 1192–1194.

Giles, N. (1987). A comparison of the behavioural responses of parasitised and non-parasitised three-spined sticklebacks, *Gasterosteus aculeatus* L., to progressive hypoxia. *J. Fish Biol.* **30,** 631–638.

Godin, J. G. J., and Sproul, C. D. (1988). Risk taking in parasitised sticklebacks under threat of predation: Effects of energetic need and food availability. *Can. J. Zool.* **66,** 2360–2367.

Grabda, J. (1991). Marine fish parasitology: An outline. PWN-Polish Scientific Publishers, Warsaw.

Griffiths, S. W. (2003). Learned recognition of conspecifics by fishes. *Fish Fish.* **4,** 256–268.

Grimnes, A., and Jakobsen, P. J. (1996). The physiological effects of salmon lice infection on post-smolt of Atlantic salmon. *J. Fish Biol.* **48,** 1179–1194.

Grossel, G. W., Dykova, I., Handlinger, J., and Munday, B. L. (2003). *Pentacapsula neurophila* sp. (Multivalvulida) from the central nervous system of striped trumpeter, *Latris lineata* (Forster). *J. Fish Dis.* **26,** 315–320.

Grutter, A. S. (1999). Cleaner fish really do clean. *Nature* **398,** 672–673.

Handeland, S. O., Järvi, T., Fernö, A., and Stefansson, S. O. (1996). Osmotic stress, antipredator behaviour, and mortality of Atlantic salmon (*Salmo salar*) smolts. *Can. J. Fish. Aq. Sci.* **53,** 2673–2680.

Hara, T. J. (1992). "*Fish Chemoreception.*" Chapman and Hall, London.

Hara, T. J. (1993). Role of olfaction in fish behaviour. *In* "Behaviour of Teleost Fishes" (Pitcher, T. J., Ed.), pp. 171–200. Chapman and Hall, London.

Harris, M. T., and Wheeler, A. (1974). *Ligula* infestation of bleak *Alburnus alburnus* (L.) in the tidal Thames. *J. Fish Biol.* **6,** 181–188.

Hart, B. (1990). Behavioural adaptations to pathogens and parasites: Five strategies. *Neuro. Biobehav. Rev.* **14,** 273–294.

Hawkins, A. D. (1993). Underwater sound and fish behaviour. *In* "Behaviour of Teleost Fishes" (Pitcher, T. J., Ed.), pp. 129–169. Chapman and Hall, London.

Hawkins, A. D., and Amorim, M. C. P. (2000). Spawning sounds of the male haddock, *Melanogrammus aeglefinus*. *Env. Biol. Fish.* **59,** 29–41.

Hayward, C. J. (1996). Copepods of the genus *Colobomatus* (Poecilostomatoida: Philichthyidae) from fishes of the family Sillaginidae (Tereostei: Perciformes). *J. Nat. Hist.* **30,** 1779–1798.

Heins, D. C., and Baker, J. A. (2003). Reduction in egg size in natural populations of threespine stickleback infected with a cestode macroparasite. *J. Parasitol.* **89,** 1–6.

Heins, D. C., Singer, S. S., and Baker, J. A. (1999). Virulence of the cestode *Schistocephalus solidus* and reproduction in infected threespine stickleback, *Gasterosteus aculeatus*. *Can. J. Zool.* **77,** 1967–1974.

Helluy, S. M., and Holmes, J. C. (1990). Serotonin, octopamine and the clinging behaviour induced by the parasite *Polymorphus paradoxus* (Acanthocephala) in *Gammarus lacustris* (Crustacea). *Can. J. Zool.* **68,** 1214–1220.

Helluy, S., and Thomas, F. (2003). Effects of *Microphallus papillorobustus* (Platyhelminthes: Trematoda) on serotonergic immunoreactivity and neuronal architecture in the brain of *Gammarus insensibilis* (Crustacea: Amphipoda). *Proc. Roy. Soc. Lond. B* **270,** 563–568.

Hoekstra, D., and Janssen, J. (1985). Non-visual feeding behaviour of the mottled sculpin, *Cottus bairdi*, in Lake Michigan. *Env. Biol. Fishes* **12,** 111–117.

Holmes, J. C., and Zohar, S. (1990). Pathology and host behaviour. *In* "Parasitism and Host Behaviour" (Barnard, C. J., and Behnke, J. M., Eds.), pp. 193–229. Taylor and Francis, London, UK.

Hudson, P. J., and Dobson, A. P. (1997). Host-parasite processes and demographic consequences. *In* "Host-parasite evolution: General principles and avian models" (Clayton, D. H., and Moore, J., Eds.), pp. 128–154. Oxford University Press, Oxford.

Huizinga, H. W., and Nadakavukaren, M. J. (1997). Cellular responses of goldfish, *Carassius auratus* (L.), to metacercariae of *Ribeiroia marini* (Faust and Hoffman, 1934). *J. Fish Dis.* **20,** 401–408.

Hurd, H. (1990). Physiological and behavioural interactions between parasites and invertebrate hosts. *Adv. Parasitol.* **29,** 271–318.

Ishimatsu, A., Sameshima, M., Tamura, A., and Oda, T. (1996). Histological analysis of the mechanisms of *Chatonella*-induced hypoxemia in yellowtail. *Fish. Sci.* **62,** 50–58.

Jobling, S., and Tyler, C. R. (2003). Endocrine disruption, parasites and pollutants in wild freshwater fish. *Parasitology* **126,** S103–S108.

Jones, D. R. (1971). The effect of hypoxia and anaemia on the swimming performance of rainbow trout (*Salmo gairdneri*). *J. Exp. Biol.* **55,** 541–551.

Jones, K. A. (1992). Food search behaviour in fish and the use of chemical lures in commercial and sports fishing. *In* "Fish Chemoreception" (Hara, T. J., Ed.), pp. 288–319. Chapman and Hall, London.

Jönsson, E., Johnsson, J. I., and Thrandur Bjornsson, B. (1998). Growth hormone increases risk-taking in foraging rainbow trout. *Ann. New York Acad. Sci.* **839,** 636–638.

Kabata, Z. (1979). "Parasitic Copepoda of British Fishes." Ray Society, London.

Karlsbakk, E. (1995). The occurrence of metacercariae of *Bucephaloides gracilescens (Digenea, Gasterostomata)* in an intermediate host, the 4-bearded rockling, *Enchelyopus cimbrius (Gadidae)*. *J. Fish Biol.* **46,** 18–27.

Karnaky, K. J., Jr. (1998). Osmotic and ionic regulation. *In* "The Physiology of Fishes" (Evans, D. H., Ed.), 2nd edn., pp. 157–176. CRC Press, Boca Raton.

Kavaliers, M., and Colwell, D. D. (1994). Parasite infection attenuates non-opioid mediated predator-induced analgesia in mice. *Phys. Behav.* **55,** 505–510.

Kavaliers, M., and Colwell, D. D. (1995). Odours of parasitised males induce adverse responses in female mice. *Anim. Behav.* **50,** 1161–1169.

Kavaliers, M., Colwell, D. D., and Choleris, E. (1998a). Analgesic responses of male mice exposed to the odours of parasitised females: Effects of male sexual experience and infection status. *Behav. Neurosci.* **112,** 1001–1011.

Kavaliers, M., Colwell, D. D., and Choleris, E. (1998b). Parasites and behaviour: An ethopharmacological analysis and biomedical implications. *Neurosci. Biobehav. Rev.* **23,** 1037–1045.

Kennedy, C. E. J., Endler, J. A., Poynton, S. L., and McMinn, H. (1987). Parasite load predicts mate choice in guppies. *Behav. Ecol. Sociobiol.* **21,** 291–295.

Kirk, R. S. (2003). The impact of *Anguillicola crassus* on European eels. *Fish. Man. Ecol.* **10,** 385–394.

Krause, J., and Godin, J.-G. J. (1996). Influence of parasitism on shoal choice in the banded killifish (*Fundulus diaphanus*, Teleostei, Cyprinodontidae). *Ethology* **102,** 40–49.

Krebs, J. R., and Davies, N. B. (1997). The evolution of behavioural ecology. *In* "Behavioural Ecology: An Evolutionary Approach," (Krebs, J. R., and Davies, N. B., Eds.), 4th edn., pp. 3–14. Blackwell Science, Oxford.

Kreirer, J. P., and Baker, J. R. (1987). "Parasitic Protozoa." Allen and Unwin, Boston.

Lacosta, S., Merali, Z., and Anisman, H. (2000). Central monoamine activity following acute and repeated systemic interleukin-2 administration. *Neuroimmunomodulation* **8,** 83–90.

Lafferty, K. D., and Morris, A. K. (1996). Altered behaviour of parasitised killifish increases susceptibility to predation by bird final hosts. *Ecology* **77,** 1390–1397.

Le Bail, P. Y., and Boeuf, G. (1997). What hormones may regulate food intake in fish? *Aq. Liv. Res.* **10,** 371–379.

Lester, R. J. G. (1971). The influence of *Schistocephalus* plerocercoids on the respiration of *Gasterosteus* and a possible resulting effect on the behaviour of the fish. *Can. J. Zool.* **49,** 361–366.

Levin, B. E., and Routh, V. H. (1996). Role of the brain in energy balance and obesity. *Am. J. Physiol. Reg. Int. Comp. Physiol.* **271,** R491–R500.

LoBue, C. P., and Bell, M. A. (1993). Phenotypic manipulation by the cestode parasite *Schistocephalus solidus* of its intermediate host, *Gasterosteus aculeatus*, the threespine stickleback. *Am. Nat.* **142,** 725–735.

Longshaw, M., Frear, P., and Feist, S. W. (2003). *Myxobolus buckei* sp n. (Myxozoa), a new pathogenic parasite from the spinal column of three cyprinid fishes from the United Kingdom. *Folia Parasitol.* **50,** 251–262.

Loot, G., Aulagnier, S., Lek, S., Thomas, F., and Guegan, J. F. (2002a). Experimental demonstration of a behavioural modification in a cyprinid fish, *Rutilus rutilus* (L.), induced by a parasite, *Ligula intestinalis* (L.). *Can. J. Zool.* **80,** 738–744.

Loot, G., Giraudel, J. L., and Lek, S. (2002b). A non-destructive morphometric technique to predict *Ligula intestinalis* L. plerocercoid load in roach (*Rutilus rutilus* L.) abdominal cavity. *Ecological Modelling* **156,** 1–11.

Losey, G. S. (1987). Cleaning symbiosis. *Symbiosis* **4,** 229–258.

Lowe-Jinde, L., and Zimmerman, A. M. (1991). Influence of Cryptobia salmositica on feeding, body composition and growth in rainbow trout. *Can. J. Zool.* **69,** 1397–1401.

Lucas, M. C., and Barras, E. (2001). "Migration of Freshwater Fishes." Blackwell Science, Oxford.

Markiw, M. E. (1992). Salmonid whirling disease. U.S. Fisheries and Wildlife Service Leaflet 17. Washington, DC.

McClelland, G. (1995). Experimental infection of fish with larval sealworm, *Pseudoterranova decipiens* (Nematoda, Aniskinae), transmitted by amphipods. *Can. J. Fish. Aquat. Sci.* **52,** 140–155.

McCormick, J. H., and Stokes, G. N. (1982). Intra-ovarian invasion of smallmouth bass oocytes by *Proteocephalus ambloplitis* (Cestoda). *J. Parasitol.* **68,** 973–975.

McPhail, J. D., and Peacock, S. D. (1983). Some effects of the cestode (*Schistocephalus solidus*) on reproduction in the threespine stickleback (*Gasterosteus aculeatus*): Evolutionary aspects of a host-parasite interaction. *Can. J. Zool.* **61,** 901–908.

Milinski, M. (1984). Parasites determine a predator's optimal feeding strategy. *Behav. Ecol. Sociobiol.* **15,** 35–37.

Milinski, M. (1985). Risk of predation of parasitised sticklebacks (*Gasterosteus aculeatus* L.) under competition for food. *Behaviour* **93,** 203–216.

Milinski, M. (1990). Parasites and host decision-making. *In* "Parasitism and Host Behaviour" (Barnard, C. J., and Behnke, J. M., Eds.), pp. 95–116. Taylor and Francis, London.

Milinski, M. (2003). The function of mate choice in sticklebacks: Optimizing Mhc genetics. *J. Fish. Biol.* **63** (Suppl. A), 1–16.

Milinski, M., and Bakker, T. C. M. (1990). Female sticklebacks use male colouration in mate choice and hence avoid parasitised males. *Nature* **344,** 330–333.

Moles, A., and Heifetz, J. (1998). Effects of the brain parasite *Myxobolus arcticus* on sockeye salmon. *J. Fish Biol.* **52,** 146–151.

Moller, P. (1995). "Electric Fishes: History and Behaviour." Chapman and Hall, London.

Moody, J., and Gaten, E. (1982). The population dynamics of eyeflukes *Diplostomum spathaceum* and *Tylodelphys clavata* (Digenea: Diplostomatidae) in rainbow and brown trout in Rutland Water: 1974–1978. *Hydrobiologia* **88,** 207–209.

Moore, J. (1983). Responses of an avian predator and its isopod prey to an acanthocephalan parasite. *Ecology* **64,** 1000–1015.

Moore, J. (2002). "Parasites and the behaviour of animals." OUP, Oxford.

Moore, J., and Gotelli, N. J. (1990). A phylogenetic perspective on the evolution of altered host behaviours: A critical look at the manipulation hypothesis. *In* "Parasitism and Host Behaviour" (Barnard, C. J., and Behnke, J. M., Eds.), pp. 193–229. Taylor and Francis, London, UK.

Myrberg, A. A., and Spires, J. Y. (1980). Hearing in damselfishes – an analysis of signal-detection among closely related species. *J. Comp. Physiol. A* **140,** 135–144.

Ness, J. H., and Foster, S. A. (1999). Parasite-associated phenotype modifications in three-spined stickleback. *Oikos* **85,** 127–134.

Nimeth, K., Zwerger, P., Wurtz, J., Salvenmoser, W., and Pelster, B. (2000). Infection of the glass-eel swimbladder with the nematode *Anguillicola crassus. Parasitology* **121,** 75–83.

O' Connor, G. R. (1976). Parasites of the eye and brain. *In* "Ecological aspects of parasitology" (Kennedy, C. R., Ed.), pp. 327–374. North Holland Publishing, Amsterdam.

Odense, P. H., and Logan, V. H. (1976). Prevalence and morphology of *Eimeria gadi* (Fiebiger, 1913) in the haddock. *J. Protozool.* **23,** 564–571.

Owen, S. F., Barber, I., and Hart, P. J. B. (1993). Low level infection by eyefluke, *Diplostomum* spp. affects the vision of three-spined sticklebacks, *Gasterosteus aculeatus. J. Fish Biol.* **42,** 803–806.

Øverli, Ø., Pall, M., Borg, B., Jobling, M., and Winberg, S. (2001). Effects of *Schistocephalus solidus* infection on brain monoaminergic activity in female three-spined sticklebacks, *Gasterosteus aculeatus. Proc. Roy. Soc. B. Lond.* **268,** 1411–1415.

Pampoulie, C., Morand, S., Lambert, A., Rosecchi, E., Bouchereau, J. L., and Crivelli, A. J. (1999). Influence of the trematode *Aphalloïdes coelomicola* Dollfus, Chabaud and Golvan, 1957 on the fecundity and survival of *Pomatoschistus microps* (Kroyer, 1838) (Teleostei : Gobiidae). *Parasitology* **119,** 61–67.

Parrington, J., and Coward, K. (2002). Use of emerging genomic and proteomic technologies in fish physiology. *Aq. Liv. Res.* **15,** 193–196.

Partridge, B. L., and Pitcher, T. J. (1980). The sensory basis of fish schools: Relative roles of lateral line and vision. *J. Comp. Physiol.* **135A,** 315–325.

Pennycuick, L. (1971). Quantitative effects of three species of parasites on a population of three-spined sticklebacks (*Gasterosteus aculeatus* L.). *J. Zool, Lond.* **165,** 143–162.

Phares, C. K. (1997). The growth hormone-like factor from plerocercoids of the tapeworm Spirometra mansonoides a multifunctional protein. *In* "Parasites and pathogens: Effects on host hormones and behaviour" (Beckage, N. E., Ed.), pp. 99–112. Chapman and Hall, New York.

Phares, C. K., and Kubik, J. (1996). The growth factor from plerocercoids of *Spirometra mansonoides* is both a growth hormone agonist and a cysteine proteinase. *J. Parasitol.* **82,** 210–215.

Pinto, N. X., Torres-Hillera, M. A., Mendoza, E., and Leon-Sarmiento, F. E. (2002). Immune response, nitric oxide, autonomic dysfunction and stroke: A puzzling linkage on *Trypanosoma cruzi* infection. *Med. Hyptoh.* **58,** 374–377.

Popper, A. N., and Platt, C. (1993). Inner Ear and Lateral Line. *In* "The Physiology of Fishes" (Evans, D. H., Ed.), pp. 99–136. CRC Press, Boca Raton.

Poulin, R. (1998). "Evolutionary Ecology of Parasites: From Individuals to Communities." Chapman and Hall, London.

Poulin, R., and Fitzgerald, G. J. (1989). Risk of parasitism and microhabitat selection in juvenile sticklebacks. *Can. J. Zool.* **67,** 14–18.

Poulin, R., and Grutter, A. S. (1996). Cleaning symbioses: Proximate and adaptive explanations. *Bioscience* **46,** 512–517.

Radabaugh, D. C. (1980). Changes in minnow, *Pimephales promelas* Rafinesque, schooling behaviour associated with infections of brain-encysted larvae of the fluke, *Ornithodiplostomum ptychocheilus*. *J. Fish Biol.* **16,** 621–628.

Ranta, E. (1995). *Schistocephalus* infestation improves prey size selection by three-spined sticklebacks, *Gasterosteus aculeatus*. *J. Fish Biol.* **46,** 156–158.

Ratanarat-Brockelman, C. (1974). Migration of *Diplostomum spathaceum* (Trematoda) in the fish intermediate host. *Z. Parasitenkunde* **43,** 123–134.

Rees, G. (1955). The adult and Diplostomulum stage (*Diplostomulum phoxini* (Faust)) of *Diplostomum pelmatoides* Dubois and an experimental demonstration of part of the life cycle. *Parasitology* **45,** 295–311.

Reusch, T. B. H., Häberli, M. A., Aeschlimann, P. B., and Milinski, M. (2001). Female sticklebacks count alleles in a strategy of sexual selection explaining MHC polymorphism. *Nature* **414,** 300–302.

Richards, K. S., and Arme, C. (1981). The effects of the pseudophyllidean cestode *Ligula intestinalis* on the musculature of bream, *Abramis brama*. *Z. für Parasitkde* **65,** 207–215.

Riley, L. G., Hirano, T., and Grau, E. G. (2003). Effects of transfer from seawater to fresh water on the growth hormone/insulin-like growth factor-I axis and prolactin in the Tilapia, *Oreochromis mossambicus*. *Comp. Biochem. Physiol. B* **136,** 647–655.

Rodewald, A. D., and Foster, S. A. (1998). Effects of gravidity on habitat use and anti-predator behaviour in three-spined sticklebacks. *J. Fish Biol.* **52,** 973–984.

Rose, J. D., Marrs, G. S., Lewis, C., and Schisler, G. (2000). Whirling disease behaviour and its relation to pathology of brain stem and spinal cord in rainbow trout. *J. Aquat. Anim. Health* **12,** 107–118.

Rosenqvist, G., and Johansson, K. (1995). Male avoidance of parasitised females explained by direct benefits in a pipefish. *Anim. Behav.* **49,** 1039–1045.

Rowe, S., and Hutchings, J. A. (2003). Mating systems and the conservation of commercially exploited marine fish. *Trends Ecol. Evol.* **18,** 567–572.

Saksvik, M., Nilsen, F., Nylund, A., and Berland, B. (2001). Effect of marine *Eubothrium* sp (Cestoda: Pseudophyllidea) on the growth of Atlantic salmon, *Salmo salar* L. *J. Fish Dis.* **24,** 111–119.

Sanchez-Monsalvez, I., De, Armas-Serra, C., Bernadina, W., and Rodriguez-Caabeiro, F. (2003). Altered autonomic control in rat intestine due to both infection with *Anisakis simplex* and incubation with the parasite's crude extract. *Dig. Dis. Sci.* **48,** 2342–2352.

Schellart, N. A. M., and Wubbels, R. J. (1998). The auditory and mechanosensory lateral line system. *In* "The Physiology of Fishes" (Evans, D. H., Ed.), 2nd edn., pp. 283–312. CRC Press, Boca Raton.

Shirakashi, S., and Goater, C. P. (2002). Intensity-dependent alteration of minnow (*Pimephales promelas*) behaviour by a brain-encysting trematode. *J. Parasitol.* **88,** 1071–1074.

Sivak, J. G. (1990). Optical variability of the fish lens. *In* "The Visual System of Fish" (Douglas, R. H., and Djamgoz, M. B. A., Eds.), pp. 63–80. Chapman and Hall, London.

Smith, H. D. (1973). Observations on the cestode *Eubothrium salvelini* in juvenile sockeye salmon (*Oncorhynchus nerka*) at Babine Lake, British Columbia. *J. Fish. Res. Bd. Can.* **30,** 947–964.

Smith, H. D., and Margolis, L. (1970). Some effects of *Eubothrium salvelini* (Schrank, 1790) on sockeye salmon, *Oncorhynchus nerka* (Walbaum), in Babine Lake, British Columbia. *J. Parasitol.* **56,** 321–322.

Smith, J. D. (1972). The blood flukes (Digenea: Sanguinicolidae and Spirorchidae) of cold-blooded vertebrates and some comparisons with schistosomes. *Helminth. Abstr.* **41,** 161–204.

Smith, R. S., and Kramer, D. L. (1987). Effects of a cestode (*Schistocephalus sp.*) on the response of ninespine sticklebacks (*Pungitius pungitius*) to aquatic hypoxia. *Can. J. Zool.* **65,** 1862–1865.

Smyth, J. D. (1985). "An Introduction to Animal Parasitology." Hodder and Stoughton, London.

Sorenson, P. W., and Caprio, J. (1998). Chemoreception. *In* "Physiology of Fishes" (Evans, D. H., Ed.), 2nd edn, pp. 375–405. CRC Press, Boca Raton.

Soutter, A. M., Walkey, M., and Arme, C. (1980). Amino acids in the plerocercoid of *Ligula intestinalis* (Cestoda, Pseudophyllidea) and its fish host, *Rutilus rutilus*. *Z. für Parasitkde* **63,** 151–158.

Sprengel, G., and Lüchtenberg, H. (1991). Infection by endoparasites reduces maximum swimming speed of European smelt *Osmerus eperlanus* and European eel *Anguilla anguilla*. *Dis. Aq. Org.* **11,** 31–35.

Sproston, N. G., and Hartley, P. H. T. (1941). The ecology of some parasitic copepods of gadoids and other fishes. *J. Mar. Biol. Assn. UK* **25,** 361–392.

Sterin-Borda, L., and Borda, E. (2000). Role of neurotransmitter autoantibodies in the pathogenesis of chagasic peripheral dysautonomia. *Neuroimmunomodulat. Ann. NY. Acad. Sci.* **917,** 273–280.

Swearer, S. E., and Robertson, D. R. (1999). Life history, pathology, and description of *Kudoa ovivora* sp. (Myxozoa, Myxosporea): An ovarian parasite of Caribbean labroid fishes. *J. Parasitol.* **85,** 337–353.

Sweeting, R. A. (1977). Studies on *Ligula intestinalis*: Some aspects of the pathology in the second intermediate host. *J. Fish Biol.* **10,** 43–50.

Szidat, L. (1969). Structure, development and behaviour of new strigeatoid metacercariae from subtropical fishes of South America. *J. Fish. Res. Bd. Can.* **26,** 753–786.

Thompson, S. N., and Kavaliers, M. (1994). Physiological basis for parasite-induced alterations of host behaviour. *Parasitology* **109,** S119–S138.

Tierney, J. F. (1994). Effects of *Schistocephalus solidus* (Cestoda) on the food intake and diet of the three-spined stickleback, *Gasterosteus aculeatus*. *J. Fish Biol.* **44,** 731–735.

Tierney, J. F., Huntingford, F. A., and Crompton, D. W. T. (1996). Body condition and reproductive status in sticklebacks exposed to a single wave of *Schistocephalus solidus* infection. *J. Fish Biol.* **49,** 483–493.

Tierney, J. F., Huntingford, F. A., and Crompton, D. W. T. (1993). The relationship between infectivity of *Schistocephalus solidus* (Cestoda) and anti-predator behaviour of its intermediate host, the three-spined stickleback, *Gasterosteus aculeatus*. *Anim. Behav.* **46,** 603–605.

Urawa, S. (1992). *Trichodina truttae* Mueller 1937 (Ciliophora, Peritrichida) on juvenile chum salmon (*Oncorhynchus keta*) – pathogenicity and host-parasite interactions. *Fish Pathol.* **27,** 29–37.

Uspenskaya, A. V. (1957). The ecology and spreading of the pathogen of trout whirling disease – *Myxosoma cerebralis* (Hofer, 1903, Plehn, 1905) in the fish ponds of the Soviet Union. *In* "Izvestiya Vsesoyuznogo naucho-issledovatel'skogo instituta ozernogo i rechnogo rybnogo khozyaistva," Vol. XLII.

Van Der Kraak, G., Chang, J. P., and Janz, D. M. (1998). Reproduction. *In* "Physiology of Fishes" (Evans, D. H., Ed.), 2nd edn., pp. 465–490. CRC Press, Boca Raton.

Videler, J. J. (1993). "Fish Swimming." Chapman and Hall, London.

Wagner, G. N., McKinley, R. S., Bjorn, P. A., and Finstad, B. (2003). Physiological impact of sea lice on swimming performance of Atlantic salmon. *J. Fish Biol.* **62,** 1000–1009.

Wagner, G. N., and McKinlay, R. S. (2004). Anaemia and salmonid swimming performance: Potential effects of sub-lethal sea lice infection. *J. Fish Biol.* **64,** 1027–1038.

Wiklund, T., Lounashelmo, L., Lom, J., and Bylund, G. (1996). Gonadal impairment in roach *Rutilus rutilus* from Finnish coastal areas of the northern Baltic sea. *Dis. Aquat. Org.* **26,** 163–171.

Williams, H. H., and Jones, A. (1994). "Parasitic worms of fish." Taylor and Francis, London.

Williams, M. A., Penlington, M. C., King, J. A., Hoole, D., and Arme, C. (1998). *Ligula intestinalis* (Cestoda) infections of roach (*Rutilus rutilus*) (Cyprinidae): Immunocytochemical investigations into the salmon- and chicken-II type gonadotrophin-releasing hormone (GnRH) systems in host brains. *Acta Parasitol.* **43,** 232–235.

Winberg, S., and Olsén, K. H. (1992). The influence of rearing conditions on the sibling odour preferences of juvenile arctic charr, *Salvelinus alpinus* L. *Anim. Behav.* **44,** 157–164.

Winberg, S., and Nilsson, G. E. (1993). Roles of brain monoamine neurotransmitters in agonistic behaviour and stress reactions, with particular reference to fish. *Comp. Biochem. Physiol. C* **106,** 597–614.

Wisenden, B. D., Chivers, D. P., and Smith, R. J. F. (1994). Risk sensitive habitat use by brook stickleback (*Culaea* inconstans) in areas associated with minnow alarm pheromone. *J. Chem. Ecol.* **20,** 2975–2983.

Woo, N. Y. S., and Kelly, S. P. (1995). Effects of salinity and nutritional status on growth and metabolism of *Sparus sarba* in a closed seawater system. *Aquaculture* **135,** 229–238.

Wright, H. A., Wooton, R. J., and Barber, I. (in press). The effect of *Schistocephalus solidus* infection on meal size of three-spined stickleback. *J. Fish Biol.*

Wurtz, J., Taraschewski, H., and Pelster, B. (1996). Changes in gas composition in the swimbladder of the European eel (*Anguilla anguilla*) infected with *Anguillicola crassus* (Nematoda). *Parasitology* **112,** 233–238.

5

SOCIAL INTERACTIONS

JÖRGEN I. JOHNSSON
SVANTE WINBERG
KATHERINE A. SLOMAN

1. INTRODUCTION

This Chapter summarises recent research on the interrelationships between social behaviour and physiology in fish, discussing the physiological factors that determine social status, the physiological consequences of social status, and the environmental and genetic effects influencing these interactions. So far, most of the work in this area has been concerned with competitive and aggressive, rather than cooperative or mutualistic, interactions. Thus, this Chapter will mainly cover links between physiology, competitive behaviour, and social status in fish. Although we will often treat environmental and genetic effects separately to simplify discussion, it should

Behaviour and Physiology of Fish: Volume 24
FISH PHYSIOLOGY

DOI: 10.1016/S1546-5098(05)24005-0

be emphasised that the phenotype is determined by the often complex interactions between genotype and environment (Pigliucci, 2001).

1.1. Why, When, and How Do Fish Compete?

Most fish species produce a large number of offspring in a resource-limited environment, commonly without parental care (Sargent, 1997; Wootton, 1998). Consequently, juvenile mortality rates are often high and only the most efficient competitors survive to pass their genes on to future generations (Darwin, 1859). Competition over important resources such as food, territories, and mates occurs in various forms depending on species, developmental stage, and environmental conditions.

In exploitative (or scramble) competition, all individuals have access to the resource and all try to exploit it first, but individuals do not interact aggressively. In contrast, interference competition occurs through aggressive exclusion of competitors (Milinski and Parker, 1991). Interference competition is theoretically expected to be favoured in situations when a resource is economically defendable, that is, when the costs of defending the resource are lower than the benefits of monopolising it (Brown, 1964). The economic defensibility of a resource is dependent on the density of the competitors as well as on the density and distribution of the resource in time and space (for an excellent review, see Grant, 1997). In reality, modes of competition in fish vary continuously from extreme interference competition leading to territorial structures, to pure exploitative competition such as in a pelagic shoal of herring (*Clupea harengus*). Resource defense is more common in structured habitats such as coral reefs and streams than in unstructured areas like the pelagic zone of lakes and oceans (Grant, 1997), which may be partly explained by the fact that structured habitats often are associated with a patchy distribution of resources and, thus, a high potential for resource defense. Interference competition is generally thought to lead to a higher degree of resource monopolisation (unequal sharing of resources) than exploitative competition (Lomnicki, 1988; Davies, 1991) but recent research suggests that this relationship does not always hold true for fish (Weir and Grant, 2004).

When population density increases, resource defense becomes too costly and many fish species form groups. Group living may confer both benefits and costs and the probability of joining a group theoretically depends on the balance between these (see review by Krause and Ruxton, 2002). One benefit of joining a group is that predators can be detected earlier (Godin *et al.*, 1988) and the risk of predation for each individual may decrease through dilution (Turner and Pitcher, 1986). A large homogenous group may also confuse a predator, making it difficult to single out an individual for pursuit

(Neill and Cullen, 1974). Foraging in groups may increase hunting efficiency; for example, predatory jacks (*Caranx ignobilis*) are more efficient at capturing anchovies (*Stolephorus purpureus*) when they hunt in groups (Major, 1978). Fish may also be able to make hydrodynamic savings through schooling behaviour (e.g., Herskin and Steffensen, 1998).

One obvious cost of group living is that resources have to be shared with others. In fish groups, social hierarchies often develop where the most dominant individuals get prior access to resources. Laboratory studies suggest that subordinate individuals in the group often have a lower food intake and growth rate than dominants (Metcalfe *et al.*, 1989). In field studies, however, this relationship is less clear (Section 4; Höjesjö *et al.*, 2002; Martin-Smith and Armstrong, 2002). The modes of competition in hierarchically structured groups can include both interference behaviour from dominants and scrambling among subordinate individuals. Several studies suggest that fish are capable of recognising kin (Brown and Brown, 1996) as well as specific individuals (Griffiths, 2003), which tends to reduce aggression in groups as social hierarchies develop. Dominance relations in fish are generally assumed to be linear and relatively stable (Abbott *et al.*, 1985; Elliott, 1994), although reversals sometimes occur (Jenkins, 1969; Johnsson, 1997).

1.2. Measuring Dominance in Fish

Dominance has been defined many times, each definition reflecting the methods used for identifying social rank. For example, Drews (1993) defined dominance as "an attribute of the pattern of repeated, agonistic interactions between two individuals, characterised by a consistent outcome in favor of the same dyad member and a default yielding response of its opponent rather than escalation. The status of the consistent winner is dominant and that of the loser subordinate." This definition fits well with dominance measured in its simplest form between a dyad of fish in which one member will yield following repeated agonistic encounters and the other will become subordinate (Peters *et al.*, 1980; Abbott and Dill, 1989; Øverli *et al.*, 1999). The simplicity of this experimental design means that behavioural interactions can be easily observed and the dominant animal determined.

Among groups of fish, dominance status can be rapidly assessed by a serial removal technique. Fish are observed to see which individual monopolises a food source and that individual is removed and termed the dominant. The individual taking the place of the removed fish in monopolising the food source is then removed as the second dominant and so on until the whole group has been ranked (Metcalfe *et al.*, 1989; Huntingford and Garcia de Leaniz, 1997; Maclean and Metcalfe, 2001). However, although aggression may be recorded during the serial removal technique, focus is placed more upon the ability to monopolise a food resource. Thus, here a more appropriate definition of dominance may be where a central requisite is priority access to resources in competitive situations (Clutton-Brock and Harvey, 1976). Among groups of fish, dominance may also be ascertained by group observation. Usually both resource monopolisation and aggression are measured (Li and Brocksen, 1977; Harwood *et al.*, 2002) and observations may be carried out in simple laboratory tanks, stream tanks designed to simulate the natural environment of the fish (Sloman *et al.*, 2000a; Harwood *et al.*, 2002), or in the natural environment (Bachman, 1984; Nakano, 1995). One of the problems of observing larger groups of fish is that some animals may remain inactive or absent from aggressive interactions, making it necessary to consider more behavioural parameters when assigning social status to lower-ranking fish. However, here a linear hierarchical structure is not assumed as it is with the serial removal technique, and there are methods for assigning ranks under these conditions (Boyd and Silk, 1983).

Dominance relations are often fluid and capable of rapid change (Martin and Bateson, 1993). Variations in both group size and environmental setting will also affect levels of social stress. Zayan (1991) highlights that social stress is a physiological response, reflecting both physical (aggression) and psychological stressors. Simple dyad interactions may therefore be influenced mainly by physical encounters, whereas group interactions can involve a more complex level of sociobiological stress. This difference in social complexity is important when considering the relevance of social stress in the natural environment.

2. COMPETITIVE INTERACTIONS

2.1. Factors Affecting their Structure

Escalated fights over resources can impose a number of costs. Prolonged fighting increases energetic expenditure in coho salmon, *Oncorhynchus kisutch* (Puckett and Dill, 1985) and sticklebacks, *Gasterosteus aculeatus* (Chellappa and Huntingford, 1989). Fighting may also lead to accumulation of harmful byproducts (Sneddon *et al.*, 1999), increased risk of injury (Abbott and Dill, 1985; Enquist *et al.*, 1990), and reduced vigilance, increasing the risk of being preyed upon (Jakobsson *et al.*, 1995). Because of these costs, selection has promoted the ability to assess the relative strength of an opponent.

The development of game theoretical models during the 1970s and 1980s (Maynard Smith, 1974) led to dramatically new insights into how animal conflicts are structured and resolved. In the sequential assessment model (Enquist and Leimar, 1983), opponents are assumed to obtain increasingly accurate information about their relative strength as the contest unfolds. Thus, when asymmetries are large, contests are expected to be resolved quickly without escalation. When asymmetries are small, however, escalation becomes increasingly likely (Leimar and Enquist, 1984; Figure 5.1). Recent studies on rainbow trout, *Oncorhynchus mykiss*, (Johnsson and Åkerman, 1998) and male siamese fighting fish, *Betta splendens* (Oliveira *et al.*, 1998), show that fish also gain information about the fighting ability of opponents by observing their contest success against other individuals (i.e., eavesdropping; see also Chapter 1, Section 5.2). This information is then used to reduce costs in subsequent contests against these opponents.

Traits such as body size and fighting experience affect the fighting ability (or resource holding potential [RHP]) of an animal, thereby influencing its chances of winning a fight, whereas variation in prior residence duration and energetic status can affect the value associated with a resource (i.e., payoff or value asymmetries), and thereby the energy the contestant invests in a conflict (for a review, see Bradbury and Vehrencamp, 1998). Table 5.1 summarises traits that have been found to affect the outcome of competitive interactions in fish. It is important to acknowledge that several traits can affect a conflict simultaneously. For example, the outcomes of territorial contests between brown trout fry (*Salmo trutta*) depend on asymmetries in size as well as prior residence (Johnsson *et al.*, 1999; Figure 5.2).

2.2. Physiological Predictors of Outcome

Energetic status, metabolic rate, growth hormone, androgen levels, and cortisol are interrelated physiological factors that affect the structure and/or outcome of competitive interactions in fish. These effects will be discussed briefly in this section. Physiological status may affect both fighting ability and/or motivation, depending on the state of the individual. The motivation to fight for food increases with hunger level because the relative value of the contested food increases as energy status decreases (Symons, 1968; Dill *et al.*, 1981; Dill, 1983). However, after prolonged food deprivation, this effect will be outbalanced by reduced physical strength due to nutrition deficiency (Johnsson *et al.*, 1996).

Metabolic rate has been found to correlate positively with dominance status in a number of studies (Table 5.1), which may be explained by a higher metabolic scope (Priede, 1985) resulting in higher capacity for energetically costly activities like aggression. Alternatively, the elevated aggression results from increased hunger levels associated with increased metabolic demands (Symons, 1968; Metcalfe *et al.*, 1995). Similarly, behavioural studies

Fig. 5.1. Fighting sequence in the cichlid *Nannacara anomala*, with increasingly escalating interactions from (a) to (f). Generally, the more similar the contestants are in fighting ability and/or motivation, the longer and more escalated the fight. (a) Broadside display. (b) Tail beating. (c) Frontal orientation. (d) Biting. (e) Mouthwrestling. (f) Termination. The loser adopts midline darkening and lowered fins. (Drawing by Bibbi Mayrhofer; redrawn with permission from Jakobsson, 1987.)

Table 5.1
Studies Identifying Traits that Affect the Outcome of Competitive Social Interactions in Fish

Trait	Species	Reference	Comment
Body size	*Nannacara anomala*	Enquist *et al.*, 1990	
	Oncorhynchus kisutch	Rhodes and Quinn, 1998	
	Oncorhynchus mykiss	Abbott *et al.*, 1985	
		Johnsson, 1993	
	Oreochromis mossambica	Turner and Huntingford, 1986	
	Salmo salar	Wankowski and Thorpe, 1979	
		Huntingford *et al.*, 1990	Effect only in socially experienced fish
		Cutts *et al.*, 1999a	
	Salmo trutta	Newman, 1956	
		Jenkins, 1969	
		Johnsson *et al.*, 1999	
	Salvelinus fontinalis	Newman, 1956	
	Trichogaster trichopterus	Frey and Miller, 1972	
Prior residence	*Salmo salar*	Cutts *et al.*, 1999a, b	
		O'Connor *et al.*, 2000	
	Salmo trutta	Johnsson *et al.*, 1999, 2000, 2004	
		Johnsson and Forser, 2002	
		Sundström *et al.*, 2003	
Innate aggression	*Oncorhynchus kisutch*	Holtby *et al.*, 1993	In these studies fry aggression towards a mirror (MIS-test) predicted social dominance
	Oncorhynchus mykiss	Berejikian *et al.*, 1996	
	Salmo trutta	Höjesjö *et al.*, 2004	

(*continued*)

Table 5.1 (*continued*)

Trait	Species	Reference	Comment
Winning experience	*Macropodus opercularis*	Francis, 1983	
	Rivulus marmoratus	Hsu and Wolf, 1999	
	Tricogaster trichopterus	Hollis *et al.*, 1995	
Nutritional status	*Oncorhynchus mykiss*	Johnsson *et al.*, 1996	Effect status dependent, negative at moderate food deprivation
Metabolic rate	*Oncorhynchus masou*	Yamamoto *et al.*, 1998	
	Oncorhynchus mykiss	McCarthy, 2001	
	Salmo salar	Metcalfe *et al.*, 1995	
		Cutts *et al.*, 1998	
	Salmo trutta	Lahti *et al.*, 2002	
	Salvelinus alpinus	Cutts *et al.*, 2001	
Growth hormone	*Oncorhynchus kisutch*	Devlin *et al.*, 1999	GH-transgene
	Oncorhynchus mykiss	Johnsson and Björnsson, 1994	GH-injection
	Oreochromis hornorum	Guillén *et al.*, 1999	GH-transgene, effect negative
	Salmo salar	Abrahams and Sutterlin, 1999	GH-transgene
Reproductive state	*Tilapia mariae*	Schwank, 1980	Measure: size of genital papilla
	Tilapia zillii	Neat and Mayer, 1999	Measure: gonadosomatic index
Cortisol	*Oncorhynchus mykiss*	Pottinger and Carrick, 2001	Effect negative
		Sloman *et al.*, 2001a	Effect negative
L-dopa	*Salvelinus alpinus*	Winberg and Nilsson, 1992	

If not otherwise stated, the traits were found to be positively correlated with the probability of becoming dominant.

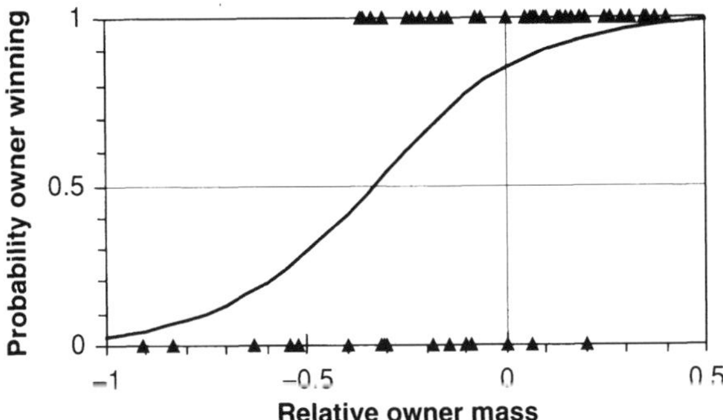

Fig. 5.2. Logistic regression of the probability of the territory owner winning over the intruder as a function of their relative mass difference, calculated as ([owner–intruder]/owner). Data from 72 dyadic contests between wild brown trout fry (*Salmo trutta*). When there is no size difference, owners win about 85% of the contests. Black triangles show the relative mass difference and outcome of each observed contest. (Reprinted from Johnsson *et al.*, 1999, with permission from Blackwell Publishing.)

on growth hormone (GH) treated and GH-transgenic salmonids suggest that GH increases feeding motivation, in turn elevating aggression and promoting competitive exclusion (Johnsson and Björnsson, 1994; Abrahams and Sutterlin, 1999; Devlin *et al.*, 1999). In a study on rainbow trout, GH-treatment increased aggression levels but not the probability of winning contests, suggesting that GH increases feeding motivation rather than fighting ability (Jönsson *et al.*, 1998). Interestingly, GH injections did elevate activity levels but not the resting metabolic rate of Atlantic salmon, *Salmo salar* (Herbert *et al.*, 2001), further suggesting that GH increases feeding motivation rather than metabolic scope (Sloman and Armstrong, 2002).

In a recent study, seventh-generation farm salmon grew faster and had higher plasma GH levels and pituitary GH content compared with wild fish (Fleming *et al.*, 2002). This suggests a possible link between growth selection, endocrine growth regulation, and competitive behaviour, as GH increases feeding motivation and aggression, as discussed above. Hatchery-selected salmonid juveniles are generally more aggressive than their wild counterparts, although the results from different studies are highly variable (review by Weber and Fausch, 2003).

Androgens (sex steroids) have been shown to influence aggression in a number of studies, but their association with social status is less clear-cut, which will be discussed further in Section 3. Treatment with the antiandrogen cyproterone acetate reduces aggression and displacement fanning in presence of another male in mature male sticklebacks (Rouse *et al.*, 1977). Similarly, androgen treatment increases aggression and territorial defense in the cichlid *Oreochromis mossambicus* (Billy and Liley, 1985). When androgen levels in male cichlids (*Haplochromis burtoni*) were lowered by castration, aggression levels were reduced, whereas dominance status was not affected (Francis *et al.*, 1992). In *Tilapia zillii*, the state of reproductive readiness, indicated by gonadosomatic index (Neat and Mayer, 1999) or the size of the male genital papilla (Schwank, 1980), influences the probability of becoming dominant (Table 5.1).

Cortisol level, an endocrine indicator of stress, is negatively correlated with the likelihood of becoming dominant in rainbow trout (Pottinger and Carrick, 2001; Sloman *et al.*, 2001a). Neurotransmitters also have the potential to influence dominance, as shown by oral administration of L-dopa (an immediate precursor of dopamine), which induced dominant behaviour in Arctic charr, *Salvelinus alpinus* (Winberg and Nilsson, 1992).

3. NEUROENDOCRINE RESPONSES: SHORT-TERM EFFECTS OF SOCIAL STATUS

Social interaction has been connected with stress in both dominant and subordinate animals (Blanchard *et al.*, 1993; Creel *et al.*, 1996). Stress in dominant individuals may be related to the fact that they must defend their dominant position, especially during periods of social instability (Sapolsky, 1992). The formation of a dominance hierarchy is highly stressful even for the winner, the future dominant individual. However, subordinate animals often continue to show signs of chronic stress, such as elevated plasma levels of cortisol, even in stable dominance hierarchies where the level of overt aggressive behaviour is low (Winberg and Lepage, 1998; Øverli *et al.*, 1999). This chronic stress experienced by subordinates in a stable dominance hierarchy is likely to be very different from the stress experienced during hierarchy formation, and is probably more related to being constantly threatened and a general lack of control than to receiving overt aggressive acts. Numerous studies indicate that loss of environmental control is a major factor inducing stress responses and in a social setting this could mean loss of social control (i.e., social defeat; Blanchard *et al.*, 1993).

3.1. Central Nervous Control of Stress Responses and Agonistic Behaviour

In fish as well as in other vertebrates, autonomic, neuroendocrine, and behavioural stress responses are controlled by the central nervous system. However, the brain is not just the source but also a target of the stress response. Our knowledge of the central control of stress responses and agonistic behaviour in fish is still very rudimentary. In the mammalian brain, the norepinephric and the corticotrophin-releasing hormone (CRH) systems are rapidly activated when the animal is challenged by a stressor, and these two systems act as the principal effectors of the generalised stress response. Descending CRH-containing fibers project to the medullary norepinephric cell groups and CRH stimulates norepinephric activity, which in turn will provoke further release of CRH, creating a positive feedback loop that escalates the release of CRH and norepinephrine in the course of the stress response (Huether, 1996). Brain dopaminergic and serotonergic systems are also activated as part of the mammalian central stress response and the serotonergic system is particularly important in the control of agonistic behaviour (e.g., Nelson and Chiavegatto, 2001).

3.1.1. Effects of Stress and Agonistic Interactions on Brain Monoaminergic Activity

Brain monoaminergic systems (i.e., dopamine, norepinephrine, and serotonin) appear to act mainly as neuromodulators, adjusting the responsiveness of neurones to other excitatory and inhibitory inputs (reviewed by Huether, 1996); the brain dopaminergic and norepinephric systems might to some extent have behavioural effects opposite to those of serotonin (Eichelman, 1987; Winberg and Nilsson, 1992; Arregui *et al.*, 1993; Höglund *et al.*, 2001). In mammals, the stress-induced activation of norepinephric neurons is correlated with increasing levels of arousal. Activation of the brain norepinephric system seems to facilitate reactivity to stimuli, processing of sensory information, and active behavioural output, a fight-or-flight reaction serving to regain control (Aston-Jones, 1986; Jacobs *et al.*, 1991; Huether, 1996). Similarly, the dopaminergic system appears to promote sensory processing and active behavioural responses (Sawaguchi *et al.*, 1990; Huether, 1996), whereas the serotonergic system seems to inhibit active behavioural responses and promote a conservation/withdrawal response, or what has also been described as passive or reactive stress coping (see Section 6).

Agonistic interactions result in a rapid activation of brain monoaminergic neurotransmission in several teleost species, as well as in other vertebrates (Winberg and Nilsson, 1993a). Following fights for social dominance in pairs of juvenile rainbow trout, subordinate fish show elevated norepinephric, dopaminergic, and serotonergic activity in selected brain areas (Øverli *et al.*, 1999). Brain norepinephric, dopaminergic, and serotonergic activity are often estimated by the ratio of the catabolite to the parent monoaminergic neurotransmitter concentrations. For example, norepinephric activity is estimated by the ratio of 3-methoxy-4-hydroxyphenylglycol (MHPG) to norepinephrine (NE) concentrations, dopaminergic activity is estimated by the ratio of 3,4-dihydroxyphenylacetic acid (DOPAC) to dopamine (DA) concentrations, and serotonergic activity is estimated by the ratio of 5-hydroxyindoleacetic acid (5-HIAA) to serotonin (5-hydroxytryptamine, 5-HT) concentrations. Brain levels of the monoamines themselves often remain relatively stable; when changes in MHPG/NE, DOPAC/DA, and 5-HIAA/5-HT ratios are observed, they are often mirrored by changes in brain concentrations of their respective catabolites such as MHPG, DOPAC, and 5-HIAA. However, the ratio is a more direct index of neuronal activity than catabolite levels per se, because variance related to tissue sampling and total levels of monoamines and monoamine catabolites are reduced (Shannon *et al.*, 1986).

In the study by Øverli *et al.* (1999), telencephalic 5-HIAA/5-HT ratios were elevated in both dominant and subordinate rainbow trout 3 hours after the termination of fights for social dominance. This effect was reversed in dominant fish within 24 hours of social interaction, whereas in subordinates elevated 5-HIAA/5-HT ratios were observed in all brain regions after 24 hours. Similarly, substantial activation of brain catecholaminergic systems (DA and NE), as indicated by elevated MHPG/NE and DOPAC/DA ratios, was observed after 24 hours in subordinates but not in dominant trout (Øverli *et al.*, 1999). The duration of fights for social dominance in pairs of rainbow trout is highly variable; in the study by Øverli *et al.* (1999), correlations were found between fight duration and brain levels of 5-HIAA (in all brain regions), MHPG (in the brainstem), and DOPAC (in the optic tectum) in both dominant and subordinate fish (Figure 5.3). Thus, brain monoaminergic systems are probably activated very early during dyadic agonistic interactions in both winners and losers, even though in the winner's brain monoaminergic activity rapidly returns to baseline levels.

Subordinate fish continue to show elevated brain serotonergic activity even after long-term social interaction in stable dominance hierarchies. In Arctic charr interacting in pairs over a 21-day period, 5-HIAA/5-HT ratios were elevated in brain regions (telencephalon, hypothalamus, and brainstem) of subordinates after 1 day of social interaction and remained equally high in subordinate fish sampled after 21 days of interaction (Winberg and Nilsson, 1993b). When interacting in groups of four for up to 11 weeks, subordinate juvenile Arctic charr of social rank 2, 3, and 4 displayed elevated brain 5-HIAA/5-HT ratios compared to dominant fish, with an inverse correlation between social rank and brain 5HIAA/5-HT ratios (Winberg *et al.*, 1991, 1992a). In rainbow trout interacting in groups of three, plasma cortisol declined in subordinate fish after 7 days of interaction, whereas brain 5-HIAA/5-HT remained elevated (Winberg and Lepage, 1998). Moreover, there was a positive correlation between the number of aggressive acts received and telencephalic 5-HIAA/5-HT ratios.

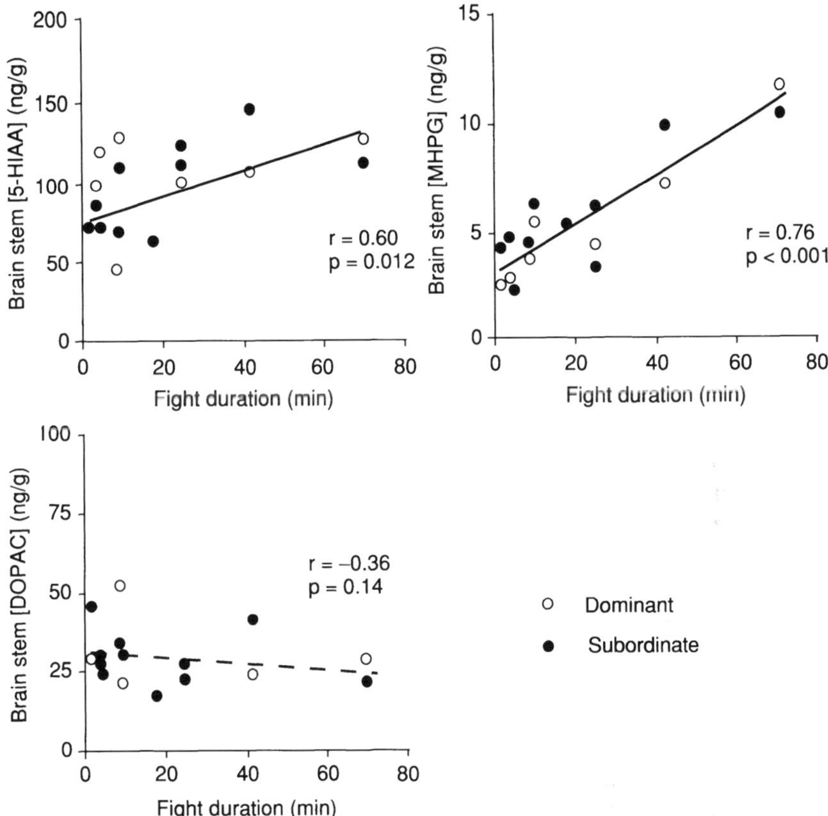

Fig. 5.3. The relationship between fight duration and the concentrations of 5-HIAA, DOPAC, and MHPG in the brainstem in rainbow trout sampled 5 minutes following the termination of fights for social dominance. Note that data from dominant and subordinate fish are pooled. Spearman r and p values are given. (Reprinted from Øverli *et al.*, 1999, with permission from S. Karger AG.)

Stressors other than social interaction—such as handling, confinement, and predator exposure—also result in elevated brain 5-HIAA/5-HT ratios in fish (Winberg *et al.*, 1992b; 1993a; Lepage *et al.*, 2000; Øverli *et al.*, 2001), whereas feed restriction does not (Winberg *et al.*, 1992b). Thus, the brain 5-HIAA/5-HT ratio appears to provide a remarkably sensitive index of chronic social stress in salmonid fish in pairs or small groups.

As stated above, acute social stress during the establishment of a dominance hierarchy results in activation of the brain catecholaminergic (NE and DA) systems in both dominant and subordinate fish, but the effects of long-term social interaction on the activity of these neurotransmitter systems is less clear. Effects of social interaction on catecholamine levels in fish brains were first reported by McIntyre *et al.* (1979). They found that following 2 weeks of social interaction in groups of six juvenile rainbow trout, dominant fish displayed lower NE and higher DA concentrations in the brain than subordinate fish. Also, they reported that the subordinate fish that were most frequently attacked displayed the largest reduction in brain DA levels. However, the fact that McIntyre *et al.* (1979) did not measure catecholamine catabolite levels makes their results hard to interpret. A decreased steady-state level of a monoamine could indicate that the monoamine is being more extensively used, or the opposite, that the neurotransmitter synthesis is being downregulated due to decreased utilisation of the system. Höglund *et al.* (2000) reported that MHPG/NE is elevated in the optic tectum of subordinate Arctic charr as compared to isolated controls following 5 days of social interaction in groups of three fish. Winberg *et al.* (1991) found that dominant Arctic charr had higher concentrations of the DA metabolite, homovanillic acid, in the telencephalon than subordinate fish following 2–11 weeks of social interaction in groups consisting of four fish.

3.1.2. Brain Monoamines and the Control of the Hypothalamo-Pituitary-Interrenal Axis

The brain monoaminergic systems are important in the control of neuroendocrine release factors at the level of the hypothalamus and pituitary. Brain 5-HIAA/5-HT ratios have been found to correlate with plasma levels of cortisol (Winberg and Lepage, 1998; Øverli *et al.*, 1999) and adrenocorticotropic hormone (ACTH; Höglund *et al.*, 2000) in rainbow trout and Arctic charr, suggesting that brain 5-HT has a stimulatory action on the hypothalamo-pituitary-interrenal (HPI) axis. However, the role of the brain 5-HT system in the control of the HPI axis is still not clear. For instance, 8-hydroxy-2-(di-*N*-propylamino) tetralin (8-OH-DPAT), a selective 5-HT_{1A} receptor agonist, may have either stimulatory or inhibitory effects on HPI axis activity in rainbow trout depending on the dose and context. In undisturbed fish, 8-OH-DPAT stimulates the HPI axis (Winberg *et al.*, 1997; Höglund *et al.*, 2002), whereas if administrated at low doses to stressed fish, 8-OH-DPAT has the opposite effect, suppressing the stress-induced elevation of plasma ACTH and cortisol concentrations (Höglund *et al.*, 2002). Similarly, stimulation of brain 5-HT activity in rainbow trout by feeding the fish a diet supplemented with L-tryptophan, the amino-acid precursor of 5-HT, for 7 days results in slightly elevated basal plasma levels of cortisol. At the same time, the stress-induced elevation of plasma cortisol concentrations is reduced (Lepage *et al.*, 2002, 2003).

It has been suggested that the central NE system stimulates HPI axis activity in salmonids (Øverli *et al.*, 1999; Höglund *et al.*, 2000). The brain DA system, on the other hand, may have an inhibitory effect, and L-3,4-dihydroxyphenylalanine (levodopa, L-DOPA) treatment, which elevates brain DA activity, has been reported to counteract the stress-induced elevation of plasma cortisol concentrations and brain 5-HT activity in Arctic charr (Höglund *et al.*, 2001).

3.1.3. Candidate Mechanisms Mediating Behavioural Inhibition in Subordinate Animals

Social subordination results in a general behavioural inhibition including a suppression of aggressive behaviour (Winberg *et al.*, 1992a; Höglund *et al.*, 2001), an inhibition of appetite and food intake (Øverli *et al.*, 1998), and lowered locomotor activity (Winberg *et al.*, 1993b; Øverli *et al.*, 1998). An antiaggressive effect of the brain 5-HT system has been reported in a number of vertebrates (Blanchard *et al.*, 1991, 1993; Raleigh *et al.*, 1991; Deckel, 1996; Deckel and Jevitts, 1997; Edwards and Kravitz, 1997; Larson and Summers, 2001) including teleost fish (Winberg and Nilsson, 1993a; Adams *et al.*, 1996), and overall the behavioural effects observed following stimulation of brain 5-HT activity are strikingly similar to those displayed by subordinate fish. Stimulation of brain 5-HT activity, using 5-HT precursors or selective serotonin reuptake inhibitors (SSRIs; substances that elevate synaptic 5-HT levels by inhibiting synaptic clearance by reuptake of 5-HT, e.g., prozac) inhibits aggressive behaviour (Winberg *et al.*, 2001) and locomotor activity (Winberg *et al.*, 1993b). Conversely, inhibition of 5-HT activity, using the 5-HT synthesis inhibitor para-chlorophenylalanine, stimulates aggressive behaviour (Adams *et al.*, 1996) and locomotor activity (Winberg *et al.*, 1993b) in teleost fish.

However, only long-term activation of the brain 5-HT system appears to result in behavioural inhibition. During fights for social dominance, the winners show a rapid but transient activation of the brain 5-HT system but no signs of behavioural inhibition (Øverli *et al.*, 1999). In fact, even a short-term social defeat resulting in elevated brain 5-HT activity increases aggression in rainbow trout (Øverli *et al.*, 2004a). In this case, an activation of brain DA systems may counteract the behavioural effects of an early activation of the brain 5-HT system (Winberg *et al.*, 1991; Winberg and Nilsson, 1993a).

The hypothesis that behavioural effects of transient increases in brain 5-HT activity observed in dominant fish are counteracted by a rapid activation of brain catecholaminergic systems is supported by the observation that L-DOPA treatment, which elevates brain DA activity, counteracts the stress-induced activation of the brain 5-HT system (Höglund *et al.*, 2001) and increases the chance of juvenile Arctic charr winning fights for social dominance over size-matched conspecifics (Winberg and Nilsson, 1992).

Another explanation for the divergent behavioural effects of short- and long-term activation of the brain 5-HT system could be that chronic activation of the 5-HT system has effects on the 5-HT receptor expression and/or sensitivity. Long-term elevation of 5-HT release could result in a downregulation of inhibitory 5-HT_{1A}-receptors at 5-HT cell bodies (somatodendritic autoreceptors) and thus in a decrease in 5-HT negative feedback, which in turn could cause an upregulation of 5-HT neurotransmission in terminal fields (Mongeau *et al.*, 1997). Stimulation of brain 5-HT activity, by feeding rainbow trout L-tryptophan supplemented feed, results in suppression of aggressive behaviour, but only after feeding for 7 days. Feeding rainbow trout L-tryptophan supplemented feed for 3 days has no effect on aggressive behaviour (Winberg *et al.*, 2001). Moreover, a 1-week treatment with the SSRI citalopram reduces aggressive behaviour in dominant rainbow trout (Lepage *et al.*, 2005). Larson and Summers (2001) obtained similar results when treating lizards (*Anolis carolinensis*) with setralin, another SSRI.

3.1.4. Autonomic and Behavioural Effects of CRH and Other Neuropeptides: Possible Interactions with 5-HT

Neuropeptides, such as CRH and argininevasotocin (AVT), involved in the control of the HPI axis may also have behavioural effects. The CRH distribution in the teleost brain is widespread (Peter, 1986; Pepels *et al.*, 2002). Stress (Ando *et al.*, 1999), including social subordination (Doyon *et al.*, 2003), has been reported to increase brain levels of CRH mRNA in rainbow trout. In fish as in mammals (Arborelius *et al.*, 1999; Lovejoy and Balment, 1999), CRH appears to have autonomic and behavioural effects. For instance, Mimassi *et al.* (2003) reported that intracerebroventricular (i.c.v.) injections of low doses of CRH induce bradycardia in rainbow trout, an effect that is probably mediated by an increase in parasympathetic drive to the heart. Moreover, in juvenile chinook salmon (*Oncorhynchus tshawytscha*), central administration of CRH stimulates locomotor activity in a dose-dependent manner, an effect that was blocked by concomitant administration of a synthetic CRH antagonist, α-helical CRH_{9-41} (ahCRH) (Clements *et al.*, 2002). In addition to its effects on locomotor activity, CRH administration increased the amount of time the fish spent in the center of the test arena. Moreover, the ability to find cover was reduced in fish treated with the CRH antagonist, ahCRH. In mammals, CRH is known to have anxiogenic effects, but it is not known if this can account for the increase in the amount of time taken for fish treated with ahCRH to seek cover, or if it is just a result of lower locomotor activity in the ahCRH-treated fish (Clements *et al.*, 2002).

Interestingly, the stimulatory effect of CRH on locomotor activity in juvenile chinook salmon was potentiated by acute treatment with the SSRI, fluoxetine (Clements *et al.*, 2003). The 5-HT system is believed to act in a stimulatory manner on CRH (Dinan, 1996). However, in the study by Clements *et al.* (2003), fluoxetine by itself had no effects on locomotor activity, suggesting that the effect on locomotor activity is not mediated by a 5-HT stimulation of CRH. Instead, the interaction between 5-HT and CRH seems to be in the other direction, with CRH stimulating serotonergic activity, which in turn stimulates locomotor activity. This suggestion is supported by the observation that the stimulatory effect of CRH on locomotor activity was abolished if CRH was administrated together with a 5-HT_{1A} receptor antagonist (Clements *et al.*, 2003).

Social subordination often results in a drastic decrease in food intake (Koebele, 1985). In pairs of juvenile Arctic charr, the subordinate continued to show reduced food intake for up to a week following the removal of the dominant fish, suggesting that social subordination results in a stress-induced anorexia (Abbott *et al.*, 1985; Øverli *et al.*, 1998). Similarly, Griffiths and Armstrong (2002) found that feeding rates of subordinate Atlantic salmon remained suppressed several days after the dominant salmon was removed. In goldfish, *Carassius auratus*, CRH as well as 5-HT has anorexic effects, which seem to be mediated by direct action of 5-HT on appetite as well as by a 5-HT stimulation of CRH (Bernier and Peter, 2001). Thus, interactions between the brain 5-HT and CRH systems appear to be important in mediating behavioural stress responses, and these two systems seem to regulate each other by reciprocal connections.

The neurohypophysial peptide AVT, which is a potent ACTH secretagogue in fish (Fryer *et al.*, 1985; Baker *et al.*, 1996), is synthesised by both parvocellular and magnacellular neurons within the hypothalamic preoptic area. AVT produced by parvocelluar preoptic neurons is mainly transported via their axons to specific brain regions, whereas AVT synthesised by the magnocellular portion of the preoptic area is transported to the posterior pituitary gland for release into the circulation. Stress has been reported to increase AVT mRNA levels of parvocellular but not magnocellular neurons in the hypothalamic preoptic area of rainbow trout (Gilchriest *et al.*, 2000). Moreover, AVT released within the central nervous system alters the activity of selective neuronal circuitry affecting behaviour. Thus, in addition to its role as an ACTH secretagogue and a hormone with suggested effects on osmoregulation and circulation, AVT also acts as a neuromodulator/neurotransmitter with effects on courtship and/or sexual behaviour in various nonmammalian vertebrates (e.g., Moore and Zoeller, 1979; Boyd, 1994;

Castagna *et al.*, 1998). In teleosts, exogenous AVT has, for example, been found to induce spawning reflex in hypophysectomized killifish, *Fundulus heteroclitus* (Pickford, 1952; Wilhelmi *et al.*, 1955; Pickford and Strecker, 1977), and closely related species of Cyprinodontids (Macey *et al.*, 1974; Peter, 1977; Liley and Stacey, 1983); to stimulate courtship behaviour in male white perch, *Morone americana* (Salek *et al.*, 2002); and to affect the activity of the neuronal circuits subserving vocalisation in the plainfin midshipman, *Porichthys notatus* (Goodson and Bass, 2000a,b). Similarly, the mammalian homologoue arginine-vasopressin (AVP) has been reported to affect communicative, affiliative, and aggressive behaviour in voles, mice, and hamsters (e.g., Ferris *et al.*, 1984, 1997; Hennessey *et al.*, 1992; Bluthe and Dantzer, 1993; Winslow *et al.*, 1993). Behavioural effects of AVT will be further discussed in Chapter 8, Section 5.2.

3.2. The Endocrine Stress-Response

What follows in this section is an overview of the endocrine stress response in teleost fish, mainly focusing on the HPI axis. For more detailed surveys on the effect of stress on teleost endocrine systems and secondary physiological effects, see recent reviews by Pickering and Pottinger (1995), Sumpter (1997), Wendelaar-Bonga (1997) and Mommsen *et al.* (1999).

Stress in teleost fish results in an almost instant activation of the sympathetic nervous system and an immediate release of catecholamines, primarily epinephrine and norepinephrine, from chromaffin cells found along the posterior cardinal vein in the region of the head kidney (Nilsson, 1984; Reid *et al.*, 1995; Fabbri *et al.*, 1998). A stress-mediated increase in sympathetic tone and circulating levels of catecholamines will optimise cardiovascular and respiratory functions and mobilise stored energy to meet the higher metabolic demands associated with stressful challenges, such as agonistic interactions.

The almost instant release of catecholamines in response to stress makes sampling difficult, as techniques for collecting blood are themselves invasive and stressful, making it hard to untangle effects of experimental treatment from those of sampling. Cannulation techniques have been used (Gamperl *et al.*, 1994), but these are not suitable for studies on fish interacting in groups, and there are no studies on the effect of social interaction on plasma catecholamine concentrations in fish.

On the other hand, the effect of social interaction on the other major stress axis, the HPI axis, is well documented. During or immediately following social interactions, plasma levels of cortisol are elevated in both dominant and subordinate fish (Øverli *et al.*, 1999; Sloman *et al.*, 2001a); however, although dominants readily recover from this social encounter

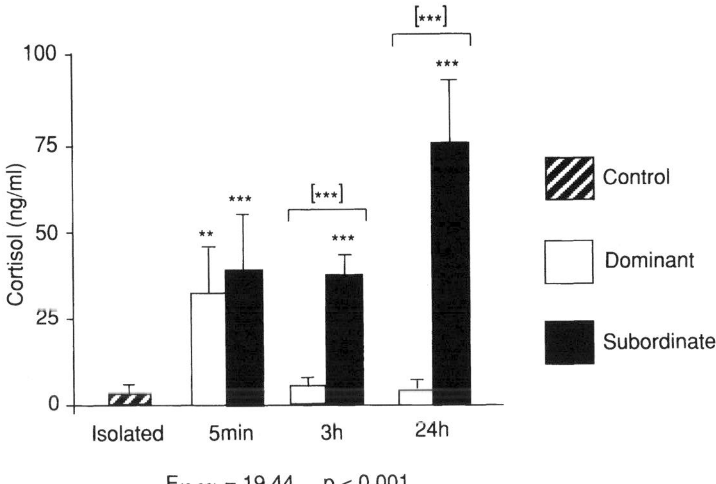

Fig. 5.4. Effects of fights for social dominance and continued interaction in established dominant-subordinate relationships on blood plasma concentrations of cortisol in dominant and subordinate rainbow trout, as compared to nonstressed controls. Post hoc significance levels are indicated by asterisks, where * is used to indicate a difference to controls and [*] indicates a difference between social ranks at a given time point. $^{**}p < 0.01$, $^{***}p < 0.001$. (Reprinted from Øverli *et al.*, 1999, with permission from S. Karger AG.)

with cortisol levels returning to normal, subordinate concentrations of plasma cortisol generally remain elevated (Figure 5.4). The elevation of plasma cortisol in subordinates has been widely documented in a variety of fish species including rainbow trout (Laidley and Leatherland, 1988; Pottinger and Pickering, 1992), coho salmon (Ejike and Schreck, 1980), European eels, *Anguilla anguilla* (Peters *et al.*, 1980), Arctic charr (Höglund *et al.*, 2000), and the cichlid *Haplochromis burtoni* (Fox *et al.*, 1997).

An increase in interrenal cell diameter has also been documented in subordinate fish (Ejike and Schreck, 1980; Peters *et al.*, 1980) and is likely to be associated with the increased synthesis and release of cortisol. In straight-line hierarchies formed among groups of 12 swordtails (*Xiphophorus helleri*), the highest-ranking fish showed the least adrenocortical activity (Scott and Currie, 1980).

Not only does social stress result in the release of cortisol and its subsequent physiological effects, but it also has effects at multiple levels of the HPI axis. Sloman *et al.* (2002a) demonstrated that when the interrenal cells were stimulated in situ with ACTH, subordinate rainbow trout displayed a lower rate of cortisol secretion than dominants, suggesting a reduced ability of subordinates to respond to additional stressors. Social subordination also results in elevated CRH mRNA levels in the hypothalamic preoptic area of rainbow trout, suggesting increased synthesis and release of CRH, the main ACTH secretagogue (Doyon *et al.*, 2003). At the pituitary level, Boddingius (1976) found that the activity of type IV cells, associated with producing ACTH, was significantly higher in subordinate trout compared with dominants. Moreover, Höglund *et al.* (2000) showed that subordinate Arctic charr display elevated plasma levels of ACTH, compared to dominant fish and isolated controls.

Several other biologically active peptides, including α-melanocyte-stimulating hormone (α-MSH) and β-endorphin are derived from the large preprohormone, preopiomelanocortin, which gives rise to ACTH. This preprohormone is processed in a tissue-specific manner, forming ACTH in the corticotrops of the pituitary *pars distalis*, and mainly α-MSH and β-endorphin in the melanotrops of the pituitary neurointermediate lobe. Social subordination results in a sustained elevation of pituitary preopiomelanocortin mRNA levels in juvenile rainbow trout, an effect mainly related to an increase in melanotropic preopiomelanocortin expression (Winberg and Lepage, 1998). Activation of the pituitary corticotrops and an elevation of circulating plasma levels of ACTH seem to be a general response to all stressors (Sumpter, 1997), whereas effects on the pituitary melanotrops and circulating plasma levels of α-MSH and β-endorphin seem to depend on the nature and/or the intensity of the stressor (Wendelaar Bonga *et al.*, 1997).

3.2.1. Behavioural Effects of Cortisol

In addition to their physiological effects, glucocorticoids may also have direct effects on behaviour. Øverli *et al.* (2002a) reported that short-term (1.5-hour) cortisol treatment resulted in an increase in locomotor activity of isolated rainbow trout challenged by a small conspecific intruder, whereas long-term (48-hour) cortisol treatment had the opposite effect, suppressing the increase in locomotor activity induced by the presence of the intruder. These behavioural effects of cortisol seem to be context-dependent because similar cortisol treatments had no effects on locomotor activity in isolated trout not challenged by an intruder. The effects on aggressive behaviour showed a similar pattern: long-term cortisol treatment had an inhibitory effect, whereas short-term treatment appeared to stimulate aggressive behaviour (Øverli *et al.*, 2002a).

These results suggest that cortisol could be a factor contributing to the divergent behavioural effects of short- and long-term social stress (discussed in Section 3.1.3). Moreover, rainbow trout strains selectively bred for high- (HR) and low-plasma cortisol levels (LR) in response to confinement stress, show divergent behavioural profiles and cortisol could not be excluded as a factor mediating these divergent behavioural profiles (Section 6;

Pottinger and Carrick, 1999, 2001; Øverli *et al.*, 2002b). Behavioural effects of cortisol could be mediated through interactions with other systems such as central serotonergic (Chaouloff, 2000; Summers *et al.*, 2003), dopaminergic (Marinelli *et al.*, 1998), or CRH systems (Schulkin *et al.*, 1998), and/or they may reflect behavioural effects of brain glucocorticoid and/or mineralocorticoid receptors. Glucocorticoid and mineralocorticoid receptors have been cloned in rainbow trout (Ducouret *et al.*, 1995; Colombe *et al.*, 2000) and both of these receptor types are expressed in the brain (Teitsma *et al.*, 1997; Sturm *et al.*, 2005). Moreover, rapid behavioural effects of cortisol could be mediated by membrane-bound steroid receptors, as has been described in amphibians (Orchinik *et al.*, 1991; Moore and Orchinik, 1994).

3.3. Modulation of Skin Colouration

In juvenile Arctic charr, social subordination also results in elevated plasma levels of α-MSH (Höglund *et al.*, 2000). Because this hormone is well known to cause skin darkening in poikilotherm animals, the skin darkening observed in subordinate salmonids could well be mediated by a stress-induced elevation of plasma α-MSH levels. In fact, in the study by Höglund *et al.* (2000), plasma α-MSH concentrations were positively correlated with skin darkness (Figure 5.5).

It has long been acknowledged that subordinate salmonids may take on a darkened appearance (Keenleyside and Yamamoto, 1962; Boddingius, 1976) and the positive correlation between plasma α-MSH concentrations and skin darkening in subordinate salmonids suggests a physiological role of α-MSH in what is postulated to be a signal during competitive interactions. O'Connor *et al.* (1999) demonstrated that the sclera of the eye and overall body colouration darkened in Atlantic salmon that became subordinate in a paired social encounter. Following the change in colour of the subordinate fish, aggression level decreased, suggesting that this may act as a signal modifying social behaviour. The dark colouration of subordinate Arctic charr takes days to develop and probably reflects the chronic social stress experienced by the subordinate fish, being persistent as long as the stressor (i.e., the dominant fish) is present (Höglund *et al.*, 2000). Rapid changes in body colouration, signaling intent, or motivational state during agonistic interactions are often mediated by neural mechanisms (e.g., Demski, 1992) or rapid changes in circulating plasma catecholamine levels (Bentley, 1998). O'Connor *et al.* (2000) showed that fish familiar with their opponents darken their body colouration at lower contest intensities than those that are unfamiliar with their opponents, reducing the likelihood of fight escalation with familiar opponents. The relationship between eye colour and social status

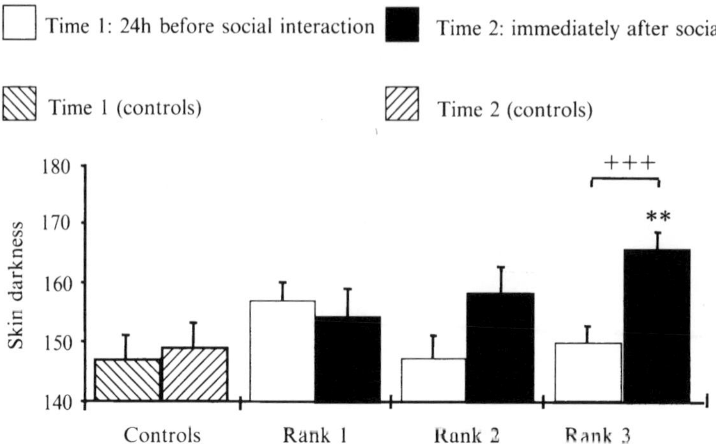

Fig. 5.5. Skin darkness of Arctic charr occupying different positions in a dominance hierarchy, social rank 1 being the dominant and social rank 3 the most subordinate fish in a group. Measurements were performed on a linear grey scale, on which 0 is white and 255 is black, before and after 5 days of social interaction. Controls are fish that were kept visually isolated. *Significant difference from controls at time 2. +Significant difference between times 1 and 2. $^{**}p < 0.01$, $+++\ p < 0.001$. (Reprinted from Höglund *et al.*, 2000.)

was investigated further by Suter and Huntingford (2002) in groups of Atlantic salmon held in a seminatural environment. Dominant fish showed a stable pale colouration of their sclera, whereas the sclera colouration of subordinates was darker and fluctuated from day to day. They conclude that colour change in salmonids is a complex response to local social events. Many colour changes in other teleosts, for example reddening of the eye and earflaps in male pumpkinseed sunfish, *Lepomis gibbosus* (see review by Guthrie and Muntz, 1993), may be associated with signaling dominance and interindividual communication, a topic covered in greater depth in Chapter 2.

3.4. Alterations in Gonadal Hormones

Social modulation of androgen levels is widespread among teleosts (for a review, see Oliveira *et al.*, 2002) and social dominance and/or territoriality is often associated with elevated levels of circulating androgens. For instance, Cardwell *et al.* (1996) found that in small groups of male rainbow trout spawning in the laboratory, dominant fish had higher plasma levels of testosterone and 17α,20β-dihydroxy-4-pregnen-3-one (17,20β-P) than subordinates. Similarly, in Arctic charr following 4 days of dyadic interaction in the absence of females, dominant males showed elevated plasma levels of testosterone and 11-ketotestosterone as compared to subordinate males (Elofsson *et al.*, 2000). In teleost fish, 11-ketotestosterone is the dominant androgen, whereas testosterone often occurs at high levels also in females (Borg, 1994). In *Brienomyrus brachyistius*, a weakly electric fish of the family Mormyridae, males becoming dominant show an increase in the duration of pulsatile electric organ discharges, along with an increase in plasma levels of 11-ketotestosterone but not testosterone (Carlson *et al.*, 2000). In male *Tilapia zilli*, on the other hand, dyadic interactions had no effects on plasma levels of either testosterone or 11-ketotestosterone (Neat and Mayer, 1999). Thus, the effect of social interaction on plasma androgen levels may vary depending on species. Other factors that may affect plasma androgen levels in socially interacting males are the intensity of the social stress experienced by subordinate fish and the access of subordinate males to female sexual cues. In the field, the greater complexity of the natural environment could reduce stress in subordinates by providing refuges and opportunities for escape. Moreover, the more complex natural environment would make it difficult for dominants to exclude subordinate males completely from female sexual cues (Cardwell *et al.*, 1996).

In general, elevated plasma levels of androgens have been associated with aggression. However, a stimulatory role of androgens on aggressive behaviour has been questioned (Monaghan and Glickman, 1992) and it is often not clear if differences in plasma androgen levels between dominant and subordinate males are the cause or the consequence of a specific social position. In the study by Cardwell *et al.* (1996), elevated 11-ketotestosterone levels in dominant rainbow trout males appeared to be a result and not a cause of social dominance, because plasma levels of 11-ketotestosterone increased in subordinate males that became dominant after the dominant males were experimentally removed. Moreover, in the cichlid fish *Oreochromis mossambicus*, there was a strong correlation between androgen levels (both testosterone and 11-ketotestosterone) measured after group formation and the social status acquired, but no correlation between androgen levels (both testosterone and 11-ketotestosterone) measured prior to group formation and the status subsequently achieved (Oliveira *et al.*, 1996). This further supports the suggestion that differences in androgen levels are the consequence and not the cause of a specific social position.

Hannes (1986) reported that there was an association between aggressive behaviour and androgen levels in dyadic interactions of swordtails, an association that was absent in males sampled from a socially stable community tank (Hannes, 1984). Thus, an association between androgen levels and social status may occur only at times of social instability (Oliveira *et al.*, 2002). In line with this suggestion, Cardwell and Liley (1991) found that peaks in androgen levels could be induced by experimental territorial intrusions in a natural population of the stoplight parrotfish (*Sparisoma viride*). A recent experiment on the cichlid *Oreochromis mossambicus* showed that androgen levels increased in males observing fights between other conspecific males, whereas no increase occurred in controls. These results indicate that androgens may act as competition hormones that respond to the social environment, preparing the individual for competitive situations (Oliveira *et al.*, 2001).

4. LONGER-TERM EFFECTS OF SOCIAL STATUS

The short-term physiological effects of social status have been well documented, but long-term physiological effects have received less attention. We know very little about the prolonged physiological effects of social status (i.e., over months to years), if indeed they occur. The examples below illustrate long-term differences in growth, life-history strategy, and reproductive function that suggest potential for long-term changes in underlying physiologies, but this is clearly an area that warrants future research.

Appetite suppression associated with stress-induced anorexia (see Section 3.1.3) and decreased nutrient digestibility in subordinate fish (Olsen and Ringø, 1999) can lead to decreased growth in the short- and long-term. Whether 5-HT and CRH systems are involved in longer-term differences in growth rates between dominants and subordinates is unclear. Growth is one of the few indicators of physiological condition that has been measured over long periods under both laboratory and natural conditions, and is likely to be influenced by both physiological and behavioural factors. Studies on growth rates in the field have yielded conflicting results. For example, Nakano (1995) demonstrated that within groups of red-spotted masu salmon (*Oncorhynchus masou ishikawai*) living in a mountain stream, dominant fish had higher daily growth increments. Other field studies have found that growth is not related to social status (Martin Smith and Armstrong, 2002; Harwood *et al.*, 2003). Although it is likely that interrelations between physiology (e.g., brain monoaminergic systems) and behaviour (e.g., competitive interference) are responsible for long-term effects of social status on growth, the specific physiological mechanisms are yet to be addressed.

The growth rates of juvenile salmonids can exert an important influence on life history strategy (Thorpe, 1994). Dominant Atlantic salmon achieving high growth rates in the first few months of life are more likely to metamorphose into smolts the following spring, compared to fish with lower growth rates that stop growing and enter a period of anorexia for 4–6 months over their first winter (Metcalfe and Thorpe, 1992; Thorpe *et al.*, 1992). This phenomenon has been well documented in the laboratory and is supported by the bimodal size distribution of Atlantic salmon found in wild populations (Heggenes and Metcalfe, 1991; Nicieza *et al.*, 1991). A change in life history may be an indirect effect of social status with associated long-term implications for both behaviour and physiology, but it is likely to be brought about by short-term correlates of social status. Any physiological factor affecting the ability of a fish to achieve dominant status or any physiological consequence of social status depressing growth rates has the potential to influence subsequent life history strategy. One of the major criticisms of many experiments documenting the physiological effects of social stress is the inability to demonstrate these effects in the natural environment, as illustrated by the varying effects of social status on growth. Competitive ability influences the decision of when to migrate in Atlantic salmon through its effects on growth (Metcalfe, 1994). Therefore, variation in salmon life history strategy may provide a good example of how short-term, transient effects of social status at crucial points in life history, perhaps only accessible to measure in the laboratory environment, may have longer-term effects.

Long-term effects on growth rates also occur within groups of fish where position in a dominance hierarchy is equivalent to place in line for a breeding position. Groups of clownfish (*Amphiprion percula*) live within their anemones and form a perfect queue for obtaining breeding positions. Each group consists of a breeding pair and zero to four nonbreeders. Individuals adjust their size and growth rates, and by maintaining size differences the queue can remain stable and uncontested (Buston, 2004). The physiological mechanisms that allow fish of different social rank to maintain relative size differences have not been investigated. However, it appears likely that socially-mediated changes in growth rates over long periods of time are underpinned by physiological systems; one candidate is brain monoamines, as demonstrated by short-term effects on appetite suppression. In the African cichlid, *Haplochromis burtoni*, social status influences the size of neurons containing somatostatin, a neuropeptide that inhibits the release of growth hormone, which in turn causes reduced somatic growth in mature territorial males. This suggests a possible mechanism responsible for socially induced physiological plasticity allowing resources to be shifted between growth and reproduction (Hoffmann and Fernald, 2000).

There may also be long-term effects of social status on reproductive physiology. Social encounters are known to affect reproductive hormones in the short-term (see Section 3.4) but the long-term effects of social stress on reproduction have not been well studied. Chronic confinement stress reduces levels of reproductive hormones (Pickering *et al.*, 1987) and quality of gametes (Campbell *et al.*, 1994) in trout. Stress-related impairment of reproductive function is usually attributed to elevations of plasma cortisol;

indeed, cortisol implantation has deleterious effects on trout reproductive function (Carragher *et al.*, 1989). However, there is a large gap in our knowledge as to whether social status has long-term effects on circulating concentrations of plasma cortisol, but if socially-induced elevations in cortisol (see Section 3.2) persist for long periods of time, then reproductive function could be disrupted. Some evidence exists for differences in reproductive hormones in relation to social status in the natural environment, where dominant male brown trout show elevated plasma 11-ketotestosterone levels compared to subordinates (Cardwell *et al.*, 1996). However, in brook trout (*Salvelinus fontinalis*) sampled at the same field location, dominant and subordinate males did not differ in plasma levels of either 11-ketotestosterone or testosterone. Additionally, socially-mediated alterations in plasma hormones could result in altered development of subsequent offspring as maternal transfer of hormones to oocytes has been documented (Schreck *et al.*, 1991; Hwang *et al.*, 1992).

5. ENVIRONMENTAL FACTORS AND SOCIAL RELATIONS

Fish inhabit a dynamic environment that has profound effects on both their social behaviour and associated physiological consequences. The effects of environmental factors such as food availability, habitat complexity, predation, chemical cues, water flow rates, and pollutants on social behaviour are therefore a key consideration when looking at behaviour/physiology interactions. One of the major parameters influencing the degree of social interaction, and therefore the ensuing effects of social stress, is the spatial and temporal prediction of food availability. A spatially and temporally predictable food source is easy to defend and monopolise. In convict cichlids, *Cichlasoma nigrofasciatum*, increasing resource predictability results in dominant fish becoming more aggressive, more sedentary, and monopolising a greater share of the food (Grand and Grant, 1994). Under these conditions, food consumption will generally correlate with dominance rank and the benefits of a higher growth rate are often associated with the dominant position. If a food source is monopolised, size differences occur within a population and reduced growth becomes a physiological consequence of low social status. Under these conditions, physiological effects of subordination such as elevated cortisol concentrations are more likely to occur. However, a reduction in predictability of food supply (e.g., by distributing food evenly over the water surface of a tank) can reduce or eliminate the growth benefits of dominance (McCarthy *et al.*, 1999).

Rearing of fish in environments of different physical complexity can also affect competitive ability. Steelhead trout (*Oncorhynchus mykiss*) raised in habitat-enriched rearing tanks socially dominated size-matched competitors raised in conventional bare tanks. When fish from both rearing environments were placed together in a stream tank, fish from the enriched tanks displayed higher growth rates, suggesting a physiological advantage of their competitive dominance (Berejikian *et al.*, 2000). Environmental enrichment promotes neural growth and proliferation in the brain (Chapter 1, Section 5.1, Jacobs *et al.*, 2000). Recent work on rainbow trout indicates that conventional hatchery rearing may impair the development of brain areas linked to aggression and feeding behaviour, such as the optic tectum and the telencephalon (Marchetti and Newitt, 2003).

The presence of an immediate predation threat can directly influence aggressive social behaviour; social behaviour in itself may influence susceptibility to predation. A visual cue to simulate predation was used by Jakobsson *et al.* (1995) to investigate the predation risk associated with fighting in mature male *Nannacara anomala*. A predator model was moved towards competing fish, and the distance between the fish and the model predator was recorded at the point where predator detection occurred. Males held singly or involved in paired tail-beating contests noticed the approach of the predator significantly earlier than males involved in mouth fighting (Figure 5.1), demonstrating that escalating fighting behaviour incurs an increased risk of predation. Martel and Dill (1993), using predator-conditioned water as a chemically-mediated predation threat, also found a reduced amount of aggressive behaviour performed by coho salmon, *Oncorhynchus kisutch*, towards a mirror stimulus. The clear relationship between intensity of social encounter and short-term physiological effects of social status (Øverli *et al.*, 1999; Sloman *et al.*, 2000b) suggests that both habitat complexity and predation threat can potentially influence physiological correlates of social status.

Olfactory cues can alter fish social behaviour. Many fish species preferentially associate with familiar individuals, and this can aid in stabilising social structures such as shoals and dominance hierarchies (e.g., Barber and Ruxton, 2000). Salmonid fish are able to distinguish kin from nonkin according to olfactory cues, and aggressive interactions can be higher among nonkin groups of both Atlantic salmon and rainbow trout (Brown and Brown, 1993). Griffiths and Armstrong (2000) demonstrate that the presence of pheromonal cues in the water can influence levels of aggression between kin and nonkin and that recirculation of water, which is likely to concentrate pheromones, increases aggression in dominant fish towards nonkin subordinates. As well as recognising a conspecific as kin or nonkin, olfactory cues can be used to identify a conspecific's social status. This recognition can lead to acceleration of the process of individual recognition and hierarchy stabilisation as seen in pairs of Nile tilapia, *Oreochromis niloticus* (Giaquinto and

Volpato, 1997). In general, the specific pheromones involved in individual recognition have not been identified, but pheromone-induced behaviours rely upon the physiological ability to detect pheromones by olfaction (Chapter 2) and to process the information received (Chapter 1). Physical environmental conditions, such as backcurrents and local water flow patterns that can carry olfactory cues, can therefore strongly affect social interaction.

Dominance hierarchies formed within stream and river environments are subject to seasonal fluctuations (e.g., associated with both water speed and depth) such as a reduction in water current speed increasing intraspecific aggression of Arctic charr (Damsgård *et al.*, 1997). Sloman *et al.* (2001b) found a nonsignificant increase in aggression among groups of brown trout following reduced water levels. In control groups, dominance was correlated with higher growth rates and liver condition (a measure of both hepatosomatic index and glycogen content) but in groups of fish exposed to lowered water levels the dominance/high growth association was lost, although dominant fish still maintained higher liver conditions. Drought affected the physiological condition of the fish indirectly through its effect on the behaviour and structure of the dominance hierarchy; those fish that were dominant under control conditions were no longer able to successfully compete and maintain their growth advantage. Similar results were found with increased flow rates simulating spate conditions (Sloman *et al.*, 2002b).

Increased flow rates may increase the necessity for swimming and associated energy expenditure, thus reducing the amount available for competitive interactions. Enforced swimming of Arctic charr at a moderate speed can reduce aggression, although not necessarily the degree of food monopolisation shown by aggressive individuals (Adams *et al.*, 1995). Swimming ability (including fast-start performance) of poikilotherms is reduced at lower temperatures, potentially increasing vulnerability to homeothermic predators (e.g., birds) during winter (Fraser *et al.*, 1993). In addition, switches from diurnal to nocturnal behaviour have been documented in Atlantic salmon (Valdimarsson *et al.*, 1997). It is possible that seasonally-reduced aggressive behaviour could be stimulated by an increased predation risk. Whether physiological mechanisms are also linked to seasonal changes in aggression in fish is not completely known, but there is some evidence in mammals that melatonin is associated with aggression (Jasnow *et al.*, 2002), which could be linked in to circadian rhythms (Chapter 6).

Some studies have considered the effects of parasites on competitive ability. Hamilton and Poulin (1995) found no effect of the dignean trematode parasite *Telogaster opithorchison* on the aggression and competitive ability of male upland bullies (*Gobiomorphus breviceps*). However, there is some evidence that parasites can affect the competitive ability of sticklebacks (Chapter 4, Section 3, Barber and Ruxton, 1998). Anthropogenic alterations of an environment can affect the formation and maintenance of dominance hierarchies, for example, contamination of aquatic habitats with trace metals (Sloman *et al.*, 2003a,b). Some trace metals (e.g., cadmium) interfere with the olfactory ability of trout and will impair the detection of pheromones, altering competitive interactions and hierarchy formation (Sloman *et al.*, 2003a). Moreover, some of the physiological differences that exist between ranks of fish as a result of their social position can influence susceptibility to aquatic pollutants. Subordinate rainbow trout take up more copper and silver across their gills than their dominant counterparts due to socially mediated differences in ionoregulation (Chapter 10, Section 5; Sloman *et al.*, 2002c, 2003c).

6. GENETIC FACTORS AFFECTING SOCIAL RELATIONS AND STRESS

Studies on animal behaviour have demonstrated that individuals differ in behavioural traits in much the same way as humans differ in personality traits, and that such behavioural differences between individuals (even within populations) appear to be adaptive (Clark and Ehlinger, 1987; Wilson *et al.*, 1994; Wilson, 1998; Gosling, 2001). Some individuals, in ecological studies often referred to as "bold," are highly aggressive and approach unfamiliar objects quickly. Such individuals are also prone to develop routines and display very low behavioural flexibility when presented with a challenge. In contrast, other individuals often referred to as "shy" or "timid" behave quite differently. They very rarely start a fight, approach new objects with marked caution, do not develop routines, and are much more flexible in their behaviour (Huntingford, 1976; McLeod and Huntingford, 1994; Verbeek *et al.*, 1994; Wilson *et al.*, 1994).

Various behavioural and neuroendocrine studies, predominantly on rodents, have stated the existence of two distinct "personality types," each associated with a set of behavioural and physiological characteristics. In these studies, two divergent personalities usually referred to as proactive (or active) and reactive (or passive) coping styles are recognised (Bohus *et al.*, 1987; Koolhaas *et al.*, 1999, 2001). When challenged, proactive animals are characterised in terms of their behaviour by being more aggressive, showing higher general activity, and, in terms of their physiology, by displaying a predominant sympathetic activation (fight/flight reaction; Bohus *et al.*, 1987; Koolhaas *et al.*, 1999, 2001). On the other hand, animals with a reactive coping style respond to the same challenge with immobility and lack of initiative, and physiologically by a predominantly parasympathetic/hypothalamic activation (a conservation/withdrawal reaction; Bohus *et al.*,

1987; Koolhaas *et al.*, 1999, 2001). Thus, compared to proactive animals, reactive animals show higher HPA/HPI axis reactivity and therefore a larger increase in plasma glucocorticoid levels in response to stress (Koolhaas *et al.*, 1999, 2001). As discussed above, animals referred to as bold and shy in ecological studies differ in behavioural traits in a way resembling the stress-coping styles described in rodents or personality traits in humans. However, because these studies usually do not include physiological variables, the physiological traits associated with shy and bold behavioural profiles are not known.

Still, several studies on rainbow trout suggest the presence of differential coping styles akin to those described in mammals. For instance, van Raaij *et al.* (1996) observed differential behavioural strategies in rainbow trout subjected to hypoxia. The behavioural strategy appeared to be highly related to survival, because nonsurviving fish displayed strenuous avoidance behaviour but surviving fish remained calm. These behavioural differences were associated with marked differences in plasma catecholamine levels, which were four- to fivefold higher in the fish that panicked and showed low survival, compared to fish that remained calm during hypoxia (van Raaij *et al.*, 1996). In contrast, the elevation of cortisol tended to be higher in quiet fish (van Raaij *et al.*, 1996).

When rainbow trout are transferred from a holding tank to visual isolation in an experimental aquarium, it usually takes some days for the fish to resume feeding. Øverli *et al.* (2004b) showed that within pairs of rainbow trout, the fish that resumed feeding first nearly always won the subsequent fights for social dominance. Similarly, when challenged by a smaller conspecific intruder, rainbow trout that most rapidly resumed feeding after transfer to social isolation in experimental aquaria showed shorter attack latencies and a higher number of attacks towards the intruder than fish that took longer to resume feeding (Schjolden *et al.*, 2005).

Furthermore, HPI axis reactivity, as indicated by changes in plasma cortisol concentrations in response to transfer to a novel environment, appears to be a stable individual trait in rainbow trout. In size-matched pairs of rainbow trout consisting of one fish showing consistently low and one fish showing consistently high plasma cortisol levels in response to experience of a novel environment, the fish showing low plasma cortisol became dominant in 10 out of 12 cases (Schjolden *et al.*, 2005). By contrast, in the study by Øverli *et al.* (2004b), there was no indication of divergent HPI axis reactivity in fish subsequently winning or losing dyadic fights for social dominance. However, among fish becoming dominant, confinement-induced changes in plasma cortisol levels were negatively correlated with aggressive behaviour (Øverli *et al.*, 2004a). These results suggest that divergent stress coping strategies akin to the proactive and reactive coping style described in mammals may exist in rainbow trout.

Stress coping patterns of individuals appear more stable within rather than between contexts, and part of the variation in coping patterns is probably related to previous experiences and life history. However, at least in rodents, different stable coping strategies appear to coexist within populations, and these strategies appear to be largely innate. In rainbow trout, HPI axis reactivity, as indicated by the elevation of plasma cortisol concentrations induced by confinement, appears to be a heritable trait (Pottinger and Carrick, 1999).

Strains of rainbow trout, one showing consistently high plasma cortisol levels (HR) and the other consistently low plasma cortisol levels (LR) in response to confinement stress, have been created through selective breeding (Pottinger and Carrick, 1999). In addition to divergent HPI axis reactivity, these two strains of rainbow trout also show divergent behavioural profiles, suggesting that they have been selected for divergent stress coping strategies. In staged fights for social dominance in size-matched pairs of HR and LR rainbow trout, the LR trout became dominant in 43 out of 46 cases (Pottinger and Carrick, 2001). Moreover, LR trout grow faster than the HR fish when reared in mixed groups (although growth rates are similar when reared apart), further suggesting that LR trout are more aggressive and thus competitively superior in the highly constant and predictable hatchery environment (T.G. Pottinger, unpublished observation). Thus, the hatchery environment may select for an aggressive LR-like strategy. When transferred to social isolation, LR trout resume feeding faster than HR trout (Øverli *et al.*, 2002b). Moreover, Øverli *et al.* (2001) reported that brain 5-HT turnover is higher in LR fish than in HR trout. Together, these results suggest that LR and HR trout are displaying divergent stress coping styles, LR fish being proactive and HR fish reactive. As stated above, in rodents, proactive animals are more aggressive, rely more on environmental cues, and show a smaller elevation in plasma glucocorticoid levels in response to a stressor than reactive animals (Koolhaas *et al.*, 1999, 2001). Furthermore, as in LR rainbow trout, proactive male rats show higher brain 5-HT turnover than reactive male rats (Koolhaas *et al.*, 2001).

Thus, the concept of coping strategies seems to address a fundamental issue in evolutionary biology that can be observed in many species. Studies on feral populations suggest that the individual differentiation in coping style may be highly functional in population dynamics, with divergent phenotypes having differential fitness depending on environmental conditions (Clark and Ehlinger, 1987; Wilson, 1998; Koolhaas *et al.*, 1999, 2001). Relative fitness of divergent stress coping strategies is also likely to be affected by frequency- and density-dependent selection.

7. CONCLUSIONS

The aim of this Chapter was to provide insight into the complex interrelationships between social relations and physiology in fish, summarising the current state of knowledge. Section 2 showed how a number of interacting physiological factors influence the structure of contests and the outcome of competition for social dominance. Section 3 went on to discuss the complex neuroendocrinological stress response associated with alterations of social status in hierarchical systems and the resulting behavioural effects. The long-term physiological effects of social status are less well known, but Section 4 provided some examples of the interrelationships between social status and fitness-related traits such as growth rate and reproduction. Section 5 explored how biotic and abiotic environmental variation influences social relations and associated physiological consequences. Finally, Section 6 discussed how genetic factors (i.e., selection) can generate stress-coping phenotypes with distinct physiological and behavioural characteristics. Although this chapter reflects considerable progress in a flourishing field of interdisciplinary research, knowledge is still lacking in many important areas.

First, knowledge of the central control of stress responses and social behaviour is still very rudimentary, and the specific physiological mechanisms responsible for long-term effects of social status on growth and reproduction are yet to be addressed. Secondly, the physiological traits associated with behavioural profiles, such as shy or bold personality types, are not known. Perhaps most importantly, the present knowledge is almost exclusively based on laboratory studies in simple environments. Given the considerable influences of environmental conditions on physiology and behaviour, can we really extrapolate laboratory findings to the wild? Although growth rates have been measured in wild populations in relation to dominance, few other physiological parameters have been considered. To further understanding of how social behaviour interacts with physiology, more physiological correlates must be measured in the field. Admittedly this is difficult to achieve with the more traditional, often terminal, laboratory measures such as plasma cortisol and brain monoamines. There is now a need to develop noninvasive or minimally invasive techniques for measuring physiological parameters as correlates of social relations in the natural environment (e.g., Goddard *et al.*, 1998; Höjesjö *et al.*, 1999). Such development, in conjunction with continuing to further our knowledge of laboratory-measurable indicators of social relations, will help us to improve our understanding of how physiology and behaviour relate during social interactions.

ACKNOWLEDGMENTS

We thank Rod Wilson, Sigal Balshine, and two anonymous referees for helpful comments.

REFERENCES

Abbott, J. C., and Dill, L. M. (1985). Patterns of aggressive attack in juvenile steelhead trout (*Salmo gairdneri*). *Can. J. Fish. Aquat. Sci.* **42,** 1702–1706.

Abbott, J. C., and Dill, L. M. (1989). The relative growth of dominant and subordinate juvenile steelhead trout (*Salmo gairdneri*) fed equal rations. *Behaviour* **108,** 104–113.

Abbott, J. C., Dunbrack, R. L., and Orr, C. D. (1985). The interaction of size and experience in dominance relationship of juvenile steelhead trout (*Salmo gairdneri*). *Behaviour* **92,** 241–253.

Abrahams, M. V., and Sutterlin, A. (1999). The foraging and antipredator behaviour of growth-enhanced transgenic Atlantic salmon. *Anim. Behav.* **58,** 933–942.

Adams, C. E., Huntingford, F. A., Krpal, J., Jobling, M., and Burnett, S. J. (1995). Exercise, agonistic behaviour and food acquisition in Arctic charr, *Salvelinus alpinus*. *Environ. Biol. Fish.* **43,** 213–218.

Adams, C. F., Liley, N. R., and Gorzalka, B. B. (1996). PCPA increases aggression in male firemouth cichlids. *Pharmacology* **53,** 328–330.

Ando, H., Hasegawa, M., Ando, J., and Urano, A. (1999). Expression of salmon corticotrophin-releasing hormone precursor gene in the preoptic nucleus in stressed rainbow trout. *Gen. Comp. Endocrinol.* **113,** 87–95.

Arborelius, L., Owens, M. J., Plotsky, P. M., and Nemeroff, C. B. (1999). The role of corticotropin-releasing factor in depression and anxiety disorders. *J. Endocrinol.* **160,** 1–12.

Arregui, A., Azpiroz, A., Brain, P. F., and Simon, V. (1993). Effects of two selective dopaminergic antagonists on ethologically assessed encounters in male mice. *Gen. Pharmacol.* **24,** 353–356.

Aston-Jones, G. (1986). Behavioural functions of locus coeruleus derived from cellular attributes. *Physiol. Psychol.* **13,** 118–126.

Bachman, R. A. (1984). Foraging behaviour of free-ranging wild and hatchery brown trout in a stream. *Trans. Am. Fish. Soc.* **113,** 1–32.

Baker, B. I., Bird, D. J., and Buckingham, J. C. (1996). In the trout, CRH and AVT synergize to stimulate ACTH release. *Reg. Peptides* **67,** 207–210.

Barber, I., and Ruxton, G. D. (1998). Temporal prey distribution affects the competitive ability of parasitised sticklebacks. *Anim. Behav.* **56,** 1477–1483.

Barber, I., and Ruxton, G. D. (2000). The importance of stable schooling: Do familiar sticklebacks stick together? *Proc. Biol. Sci.* **267,** 151–157.

Bentley, P. J. (1998). "Comparative Vertebrate Endocrinology," pp. 313–315. Cambridge University Press, Cambridge.

Berejikian, B. A., Mathews, S. B., and Quinn, T. P. (1996). Effects of hatchery and wild ancestry and rearing environments on the development of agonistic behaviour in steelhead trout (*Oncorhynchus mykiss*) fry. *Can. J. Fish. Aquat. Sci.* **53,** 2004–2014.

Berejikian, B. A., Tezak, E. P., Flagg, T. A., LaRae, A. L., Kummerow, E., and Mahnken, C. V. W. (2000). Social dominance, growth, and habitat use of age-0 steelhead (*Oncorhynchus mykiss*) grown in enriched and conventional hatchery reared environments. *Can. J. Fish. Aquat. Sci.* **57,** 628–636.

Bernier, N. J., and Peter, R. E. (2001). The hypothalamic-pituitary-interrenal axis and the control of food intake in teleost fish. *Comp. Biochem. Physiol. B* **129,** 639–644.

Billy, A. J., and Liley, N. R. (1985). The effects of early and late androgen treatments on the behaviour of *Sarotherodon Mossambicus* (Pisces: Cichlidae). *Horm. Behav.* **19,** 311–330.

Blanchard, D. C., Cholvanich, P., Blanchard, R. J., Clow, D. W., Hammer, R. P., Rowlett, J. K., and Bardo, M. T. (1991). Serotonin, but not dopamine, metabolites are increased in selected brain regions of subordinate male rats in a colony environment. *Brain Res.* **568,** 61–66.

Blanchard, D. C., Sakai, R. R., McEwen, B., Weis, S. M., and Blanchard, R. J. (1993). Subordination stress: Behavioural, brain, and neuroendocrine correlates. *Behav. Brain Res.* **58,** 113–121.

Bluthe, R. M., and Dantzer, R. (1993). Role of the vomeronasal system in vasopressinergic modulation of social recognition in rats. *Brain Res.* **604,** 205–210.

Boddingius, J. (1976). The influence of social rank on adenohypophysial cell activity in *Salmo irideus. Cell Tiss. Res.* **170,** 383–414.

Bohus, B., Benus, R. F., Fokkema, D. S., Koolhaas, J. M., Nyakas, C., van Oortmerssen, G. A., Prins, A. J. A., de Ruiter, A. J. H., Scheurink, A. J. W., and Steffens, A. B. (1987). *In* "Progress in Brain Research, Vol. 72" (De Kloet, E. R., Wiegant, V. M., and De Wied, D., Eds.), pp. 57–70. Elsevier, Amsterdam.

Borg, B. (1994). Androgens in teleost fishes. *Comp. Biochem. Physiol. C* **109,** 219–245.

Boyd, S. K. (1994). Arginine vasotocin facilitation of advertisement calling and call phonotaxis in bullfrogs. *Horm. Behav.* **28,** 232–240.

Boyd, R., and Silk, J. B. (1983). A method for assigning cardinal dominance ranks. *Anim. Behav.* **31,** 45–58.

Bradbury, J. W., and Vehrencamp, S. L. (1998). "Principles of animal communication." Sinauer, Sunderland.

Brown, G. E., and Brown, J. A. (1993). Social dynamics in salmonid fishes: Do kin make better neighbours? *Anim. Behav.* **45,** 863–871.

Brown, G. E., and Brown, J. A. (1996). Kin discrimination in salmonids. *Rev. Fish Biol. Fisheries* **6,** 201–219.

Brown, J. L. (1964). The evolution of diversity in avian territorial systems. *Wilson. Bull.* **76,** 160–169.

Buston, P. (2004). Territory inheritance in clownfish. *Proc. R. Soc. Lond. B (Suppl.)* **271,** S252–S254.

Campbell, P. M., Pottinger, T. P., and Sumpter, J. P. (1994). Preliminary evidence that chronic confinement stress reduces the quality of gametes produced by brown and rainbow trout. *Aquaculture* **120,** 151–169.

Carragher, J. F., Sumpter, J. P., Pottinger, T. G., and Pickering, A. D. (1989). The deleterious effects of cortisol implantation on reproductive function in two species of trout, *Salmo trutta* L. and *Salmo gairdneri* Richardson. *Gen. Comp. Endo.* **76,** 310–321.

Cardwell, J. R., and Liley, N. R. (1991). Androgen control of social status in males of a wild population of stoplight parrotfish, *Sparisoma viride* (Scaridae). *Horm. Behav.* **25,** 1–18.

Cardwell, J. R., Sorensen, P. W., Van der Kraak, G. J., and Liley, N. R. (1996). Effect of dominance status on sex hormone levels in laboratory and wild-spawning male trout. *Gen. Comp. Endocrinol.* **101,** 333–341.

Carlson, B. A., Hoplons, C. D., and Thomas, P. (2000). Androgen correlates of socially induced changes in the electric organ discharge waveform of a mormyrid fish. *Horm. Behav.* **38,** 177–186.

Chaouloff, F. (2000). Serotonin, stress and corticoids. *J. Psychopharmacol.* **14,** 139–151.

Chellappa, S., and Huntingford, F. A. (1989). Depletion of energy reserves during reproductive aggression in male three-spined stickleback, *Gasterosteus aculeatus* L. *J. Fish Biol.* **35,** 315–316.

Clark, A. B., and Ehlinger, T. J. (1987). Pattern and adaptation in individual behavioural differences. *In* "Perspectives in Ethology" (Bateson, P. P. G., and Klopfer, P. H., Eds.), pp. 1–47. Plenum, New York.

Clements, S., Schreck, C. B., Larsen, D. A., and Dickhoff, W. W. (2002). Central administration of corticotrophin-releasing hormone stimulates locomotor activity in juvenile Chinook salmon (*Oncorhynchus tshawytscha*). *Gen. Comp. Endocrinol.* **125,** 319–327.

Clements, S., Moore, F. L., and Schreck, C. B. (2003). Evidence that acute serotonergic activation potentiates the locomotor-stimulating effects of corticotropin-releasing hormone in juvenile chinook salmon (*Oncorhynchus tshawytscha*). *Horm. Behav.* **43,** 214–221.

Clutton-Brock, T. H., and Harvey, P. H. (1976). Evolutionary rules and primate societies. *In* "Growing Points in Ethology" (Bateson, P. P. G., and Hinde, R. A., Eds.), pp. 195–237. The University Press, Cambridge.

Colombe, L., Fostier, A., Bury, N., Pakdel, F., and Guiguen, Y. (2000). A mineralcorticoid-like receptor in the rainbow trout, *Oncorhynchus mykiss*: Cloning and characterisation of its steroid binding domain. *Steroids* **65,** 319–328.

Creel, S., Creel, N. M., and Monfort, S. L. (1996). Social stress and dominance. *Nature* **379,** 212.

Cutts, C. J., Metcalfe, N. B., and Taylor, A. C. (1998). Aggression and growth depression in juvenile Atlantic salmon: The consequences of individual variation in standard metabolic rate. *J. Fish Biol.* **52,** 1026–1037.

Cutts, C. J., Metcalfe, N. B., and Taylor, A. C. (1999a). Competitive asymmetries in territorial juvenile Atlantic salmon, *Salmo salar*. *Oikos* **86,** 479–486.

Cutts, C. J., Brembs, B., Metcalfe, N. B., and Taylor, A. C. (1999b). Prior residence, territory quality and life-history strategies in juvenile Atlantic salmon (*Salmo salar* L.). *J. Fish Biol.* **55,** 784–794.

Cutts, C. J., Adams, C. E., and Campbell, A. (2001). Stability of physiological and behvioural determinants of performance in Arctic charr (*Salvelinus alpinus*). *Can. J. Fish. Aquat. Sci.* **58,** 961–968.

Damsgård, B., Arnesen, A. M., Baardvik, B. M., and Jobling, M. (1997). State-dependent feed acquisition among two strains of hatchery-reared artic charr. *J. Fish Biol.* **50,** 859–869.

Darwin, C. R. (1859). "On the Origin of Species by Means of Natural Selection." Murray, London.

Davies, N. B. (1991). Mating systems. *In* "Behavioural Ecology" (Krebs, J. R., and Davies, N. B., Eds.), 4th ed., pp. 263–294. Blackwell Scientific, Oxford.

Deckel, A. W. (1996). Behavioural changes in *Anolis carolinensis* following injection with fluoxetine. *Behav. Brain Res.* **78,** 175–182.

Deckel, A. W., and Jevitts, E. (1997). Left vs. right-hemisphere regulation of aggressive behaviours in *Anolis carolinensis*: Effects of eye-patching and fluoxetine administration. *J. Exp. Zool.* **278,** 9–21.

Demski, L. S. (1992). Chromatophore systems in teleosts and cephalopods: A level oriented analysis of convergent systems. *Brain Behav. Evol.* **40,** 141–156.

Devlin, R. H., Johnsson, J. I., Smailus, D. E., Biagi, C. A., Jönsson, E., and Björnsson, B. Th. (1999). Increased ability to compete for food by growth hormone-transgenic coho salmon *Oncorhynchus kisutch* (Walbaum). *Aquacult. Res.* **30,** 479–482.

Dill, L. M. (1983). Adaptive flexibility in the foraging behaviour of fishes. *Can. J. Fish. Aquat. Sci.* **40,** 398–408.

Dill, L. M., Ydenberg, R. C., and Fraser, A. H. G. (1981). Food abundance and territory size in juvenile coho salmon (*Oncorhynchus kisutch*). *Can. J. Zool.* **59,** 1801–1809.

Dinan, T. G. (1996). Serotonin and the regulation of the hypothalamic-pituitary-adrenal axis function. *Life Sci.* **58,** 1683–1694.

Doyon, C., Gilmour, K. M., Trudeau, V. L., and Moon, T. W. (2003). Corticotropin-releasing factor and neuropeptide Y mRNA levels are elevated in the preoptic area of socially subordinate rainbow trout. *Gen. Comp. Endocrinol.* **133,** 260–271.

Drews, C. (1993). The concept and definition of dominance in animal behaviour. *Behaviour* **125,** 283–313.

Ducouret, B., Tujague, M., Ashraf, J., Mouchel, N., Servel, N., Valotaire, Y., and Thompson, E. B. (1995). Cloning of a teleost fish glucocorticoid receptor shows that it contains a deoxyribonucleic acid-binding domain different from that of mammals. *Endocrinol.* **136,** 3774–3783.

Edwards, D. H., and Kravitz, E. A. (1997). Serotonin, social status and aggression. *Curr. Opin. Neurobiol.* **7,** 812–819.

Eichelman, B. (1987). Neurochemical and psychopharmacologic aspects of aggressive behaviour. *In* "Psychopharmacology: The Third Generation of Progress" (Meltzer, H. Y., Ed.), pp. 697–704. Raven Press, NewYork.

Ejike, C., and Schreck, C. B. (1980). Stress and social hierarchy rank in coho salmon. *Trans. Am. Fish. Soc.* **109,** 423–426.

Elliott, J. M. (1994). "Quantitative Ecology and the Brown Trout." Oxford University Press, Oxford.

Elofsson, U. O. E., Mayer, I., Damsgård, B., and Winberg, S. (2000). Intermale competition in sexually mature Arctic charr: Effects of brain monoamines, endocrine stress responses, sex hormone levels, and behaviour. *Gen. Comp. Endocrinol.* **118,** 450–460.

Enquist, M., and Leimar, O. (1983). Evolution of fighting behaviour: Decision rules and assessment of relative strength. *J. Theor. Biol.* **102,** 387–410.

Enquist, M., Leimar, O., Ljungberg, T., Mallner, Y., and Segerdahl, N. (1990). A test of sequential assessment game: Fighting in the cichlid fish *Nannacara anomala. Anim. Behav.* **40,** 1–14.

Fabbri, E., Capuzzo, A., and Moon, T. W. (1998). The role of circulating catecholamines in the regulation of fish metabolism: An overview. *Comp. Biochem. Physiol.* **C 120,** 177–192.

Ferris, C. F., Albers, S. M., Wesolowski, S. M., and Goldman, B. D. (1984). Vasopressin injected into the hypothalamus triggers a stereotypic behaviour in golden hamsters. *Science* **224,** 521–523.

Ferris, C. F., Melloni, R. H., Koppel, G., Perry, K. W., Fuller, R. W., and Delville, Y. (1997). Vasopressin/serotonin interactions in the anterior hypothalamus control aggressive behaviour in golden hamsters. *J. Neurosci.* **17,** 4331–4340.

Fleming, I. A., Agustsson, T., Finstad, B., Johnsson, J. I., and Björnsson, B. Th. (2002). Effects of domestication on growth physiology and endocrinology of Atlantic salmon (Salmo salar). *Can. J. Fish. Aquat.Sci.* **59,** 1323–1330.

Fox, H. E., White, S. A., Kao, M. H. F., and Fernald, R. D. (1997). Stress and dominance in a social fish. *J. Neurosci.* **17,** 6463–6469.

Francis, R. C. (1983). Experiential effects on agonistic behaviour in the paradise fish, *Macropodus opercularis. Behaviour* **85,** 292–313.

Francis, R. C., Jacobson, B., Wingfield, J. C., and Fernald, R. D. (1992). Castration lowers aggression but not social dominance in male *Haplochromis burtoni* (Cichlidae). *Ethology* **90,** 247–255.

Fraser, N. H. C., Metcalfe, N. B., and Thorpe, J. E. (1993). Temperature-dependent switch between diurnal and nocturnal foraging in salmon. *Proc. R. Soc. Lond. B* **252,** 135–139.

Frey, D. F., and Miller, R. J. (1972). The establishment of dominance relationships in the blue gourami. *Behaviour* **42,** 8–62.

Fryer, J., Lederis, K., and Rivier, J. (1985). ACTH-releasing activity of urotensin I and ovine CRF, interactions with arginine vasotocin, isotocin and arginine vasopressin. *Reg. Peptides* **11,** 11–15.

Gamperl, A. K., Vijayan, M. M., and Boutilier, R. G. (1994). Epinephrine, norepinephrine, and cortisol concentrations in cannulated seawater–acclimated rainbow trout (*Oncorhynchus mykiss*) following black-box confinement and epinephrine injection. *J. Fish Biol.* **45,** 313–324.

Giaquinto, P. C., and Volpato, G. L. (1997). Chemical communication, aggression, and conspecific recognition in the Nile Tilapia. *Physiol. Behav.* **62,** 1333–1338.

Gilchriest, B. J., Tipping, D. R., Hake, L., Levy, A., and Baker, B. I. (2000). The effects of acute and chronic stresses on vasotocin gene transcripts in the brain of the rainbow trout (*Oncorhynchus mykiss*). *J. Neuroendocrinol.* **12,** 795–801.

Goddard, P. J., Gaskin, G. J., and Macdonald, A. J. (1998). Automatic blood sampling equipment for use in studies of animal physiology. *Anim. Sci.* **66,** 769–775.

Godin, J.-G. J., Classon, L. J., and Abrahams, M. V. (1988). Group vigilance and shoal size in a small characin fish. *Behaviour* **104,** 29–40.

Goodson, J. L., and Bass, A. (2000a). Forebrain peptides modulate sexually polymorphic vocal circuitry. *Nature* **403,** 769–772.

Goodson, J. L., and Bass, A. (2000b). Vasotocin innervation and modulation of vocal-acoustic circuitry in the teleost *Porichthys notatus*. *J. Comp. Neurol.* **422,** 363–379.

Gosling, S. D. (2001). From mice to men: What can we learn about personality from animal research? *Psychol. Bul.* **127,** 45–86.

Grand, T. C., and Grant, J. W. A. (1994). Spatial predictability of food influences its monopolisation and defence by juvenile convict cichlids. *Anim. Behav.* **47,** 91–100.

Grant, J. W. A. (1997). Territoriality. *In* "Behavioural Ecology of Teleost Fishes" (Godin, J.-G. J., Ed.), pp. 81–103. Oxford University Press, Oxford.

Griffiths, S. W. (2003). Learned recognition of conspecifics by fishes. *Fish Fish* **4,** 256–268.

Griffiths, S. W., and Armstrong, J. D. (2000). Differential responses of kin and nonkin salmon to patterns of water flow: Does recirculation influence aggression? *Anim. Behav.* **59,** 1019–1023.

Griffiths, S. W., and Armstrong, J. D. (2002). Kin-based territory overlap and food sharing among Atlantic salmon juveniles. *J. Anim. Ecol.* **71,** 408–486.

Guillén, I., Berlanga, J., Valenzuela, C. M., Morales, A., Toledo, J., Estrada, M. P., Puentes, P., Hayes, O., and de la Fuente, J. (1999). Safety evaluation of transgenic tilapia with accelerated growth. *Mar. Biotechnol.* **1,** 2–14.

Guthrie, D. M., and Muntz, W. R. A. (1993). Role of vision in fish behaviour. *In* "Behaviour of Teleost Fishes" (Pitcher, T., Ed.), pp. 89–128. Chapman and Hall, London.

Hamilton, W. J., and Poulin, R. (1995). Parasites, aggression and dominance in male upland bullies. *J. Fish. Biol.* **47,** 302–307.

Hannes, R. P. (1984). Androgen and corticoid levels in blood and body extracts of high and low ranking swordtail males (Xiphophorus helleri) before and after social isolation. *Z. Tierpsychol.* **66,** 70–76.

Hannes, R. P. (1986). Blood and whole-body androgen levels of male swordtails correlated with aggression measures in a standard-opponent test. *Aggress. Behav.* **12,** 249–254.

Harwood, A. J., Armstrong, J. D., Griffiths, S. W., and Metcalfe, N. B. (2002). Sympatric association influences within-species dominance relations among juvenile Atlantic salmon and brown trout. *Anim. Behav.* **64,** 85–95.

Harwood, A. J., Armstrong, J. D., Metcalfe, N. B., and Griffiths, S. W. (2003). Does dominance status correlate with growth in wild stream-dwelling Atlantic salmon (*Salmo salar*?). *Behav. Ecol.* **14,** 902–908.

Heggenes, J., and Metcalfe, N. B. (1991). Bimodal size distributions in wild juvenile Atlantic salmon populations and their relationship with age at smolt migration. *J. Fish Biol.* **39,** 905–907.

Hennessey, A. C., Whitman, D. C., and Albers, H. E. (1992). Microinjection of arginine-vasopressin into the periaqueductal gray stimulates flank marking in Syrian hamsters (*Mesocricetus auratus*). *Brain Res.* **569,** 136–140.

Herbert, N. A., Armstrong, J. D., and Björnsson, B. Th. (2001). Evidence that growth hormone-induced elevation in routine metabolism of juvenile Atlantic salmon is a result of increased spontaneous activity. *J. Fish Biol.* **59,** 754–757.

Herskin, J., and Steffensen, J. F. (1998). Energy savings in sea bass swimming in a school: Measurements of tail beat frequency and oxygen consumption at different swimming speeds. *J. Fish Biol.* **53,** 366–376.

Hoffmann, H. A., and Fernald, R. D. (2000). Social status controls somastotatin neuron size and growth. *J. Neurosci.* **20,** 4740–4744.

Höglund, E., Balm, P. H. M., and Winberg, S. (2000). Skin darkening, a potential social signal in subordinate Arctic charr (*Salvelinus alpinus*): The role of brain monoamines and pro-opiomelanocortin-derived peptides. *J. Exp. Biol.* **203,** 1711–1721.

Höglund, E., Kolm, N., and Winberg, S. (2001). Stress-induced changes in brain serotonergic activity, plasma cortisol and aggressive behaviour in Arctic charr (*Salvelinus alpinus*) is counteracted by L-DOPA. *Physiol. Behav.* **74,** 381–389.

Höglund, E., Balm, P. H. M., and Winberg, S. (2002). Stimulatory and inhibitory effects of 5-HT1A receptors on ACTH and cortisol secretion in a teleost fish, the Arctic charr (*Salvelinus alpinus*). *Neurosci. Lett.* **324,** 193–196.

Hollis, K. L., Dumas, M., Singh, P., and Fackelman, P. (1995). Pavlovian conditioning of aggressive behaviour in blue gourami fish (Trichogaster trichopterus): Winners become winners and losers stay losers. *J. Comp. Psychol.* **109,** 123–133.

Holtby, L. B., Swain, D. P., and Allan, G. M. (1993). Mirror-elicited agonistic behaviour and body morphology as predictors of dominance status in juvenile coho salmon (*Oncorhynchus kisutch*). *Can. J. Fish. Aquat. Sci.* **50,** 676–684.

Höjesjö, J., Johnsson, J. I., and Axelsson, M. (1999). Behavioural and heart rate responses to food limitation and predation risk: An experimental study on rainbow trout. *J. Fish Biol.* **55,** 1009–1019.

Höjesjö, J., Johnsson, J. I., and Bohlin, T. (2002). Can laboratory studies on dominance predict fitness of young brown trout in the wild? *Behav. Ecol. Sociobiol.* **52,** 102–108.

Höjesjö, J., Johnsson, J. I., and Bohlin, T. (2004). Habitat complexity selects against aggressive behaviour. *Behav. Ecol. Sociobiol.* **56,** 286–289.

Hsu, Y., and Wolf, L. (1999). The winner and loser effect: Integrating multiple experiences. *Anim. Behav.* **57,** 903–910.

Huether, G. (1996). The central adaptation syndrome: Psychosocial stress as a trigger for adaptive modifications of brain structure and brain function. *Prog. Neurobiol.* **48,** 569–612.

Huntingford, F. A. (1976). The relationship between anti-predator behaviour and aggression among conspecifics in the three-spined stickleback, *Gasterosteus aculeatus*. *Anim. Behav.* **24,** 245–260.

Huntingford, F. A., and Garcia de Leaniz, C. (1997). Social dominance, prior residence and the acquisition of profitable feeding sites in juvenile Atlantic salmon. *J. Fish Biol.* **51,** 1009–1014.

Huntingford, F. A., Metcalfe, N. B., Thorpe, J. E., Graham, W. D., and Adams, C. E. (1990). Social dominance and body size in Atlantic salmon parr, *Salmo salar* L. *J. Fish Biol.* **36,** 877–881.

Hwang, P. P., Wu, S. M., Lin, J. H., and Wu, L. S. (1992). Cortisol content of eggs and larvae of teleosts. *Gen. Comp. Endocrinol.* **86,** 189–196.

Jacobs, B. L., Abercrombie, K. E. D., Fornal, C. A., Levine, E. S., Morilak, D. A., and Stafford, I. L. (1991). Single-unit and physiological analyses of brain norepinephrine function in behaving animals. *Progr. Brain Res.* **88,** 159–165.

Jacobs, B. L., Van Praag, H., and Gage, F. H. (2000). Depression and the birth and death of brain cells. *Am. Sci.* **88,** 340–345.

Jakobsson, S. (1987). Male behaviour in conflicts over mates and territories. Doctoral dissertation at the University of Stockholm, pp. 1–124.

Jakobsson, S., Brick, O., and Kullberg, C. (1995). Escalated fighting behaviour incurs increase predation risk. *Anim. Behav.* **49,** 235–239.

Jasnow, A. M., Huhman, K. L., Bartness, T. J., and Demas, G. E. (2002). Short days and exogenous melatonin increase aggression of male Syrian hamsters (*Mesocricetus auratus*). *Horm. Behav.* **42,** 13–20.

Jenkins, T. M. (1969). Social structure position choice and microdistribution of two trout species (Salmo trutta and Salmo gairdneri) resident in mountain streams. *Anim. Behav. Monogr.* **2,** 57–123.

Johnsson, J. I. (1993). Big and brave: Selection affects foraging under risk of predation in juvenile rainbow trout, *Oncorhynchus mykiss. Anim. Behav.* **45,** 1219–1225.

Johnsson, J. I. (1997). Individual recognition affects aggression and dominance relations in rainbow trout, *Oncorhynchus mykiss. Ethology* **103,** 267–282.

Johnsson, J. I., and Björnsson, B. Th. (1994). Growth hormone increases growth rate, appetite and dominance in juvenile rainbow trout, *Oncorhynchus mykiss. Anim. Behav.* **48,** 177–186.

Johnsson, J. I., and Åkerman, A. (1998). Watch and learn: Preview of the fighting ability of opponents alters contest behaviour in rainbow trout. *Anim. Behav.* **56,** 771–776.

Johnsson, J. I., and Forser, A. (2002). Residence duration influences the outcome of territorial conflicts in brown trout (*Salmo trutta*). *Behav. Ecol. Sociobiol.* **51,** 282–286.

Johnsson, J. I., Jönsson, E., and Björnsson, B. Th. (1996). Dominance, nutritional state, and growth hormone levels in rainbow trout (*Oncorhynchus mykiss*). *Horm. Behav.* **30,** 13–21.

Johnsson, J. I., Nöbbelin, F., and Bohlin, T. (1999). Territorial competition among wild brown trout fry: Effects of ownership and body size. *J. Fish Biol.* **54,** 469–472.

Johnsson, J. I., Carlsson, M., and Sundström, L. F. (2000). Habitat preference increases territorial defence in brown trout (*Salmo trutta*). *Behav. Ecol. Sociobiol.* **48,** 373–377.

Johnsson, J. I., Rydeborg, A., and Sundström, L. F. (2004). Predation risk and the territory value of cover: An experimental study. *Behav. Ecol. Sociobiol.* **56,** 388–392.

Jönsson, E., Johnsson, J. I., and Björnsson, B. Th. (1998). Growth hormone increases aggressive behaviour in juvenile rainbow trout. *Horm. Behav.* **33,** 9–15.

Keenleyside, M. H. A., and Yamamoto, F. T. (1962). Territorial behaviour of juvenile Atlantic salmon (*Salmo salar* L.). *Behaviour* **19,** 138–169.

Koebele, B. P. (1985). Growth and size hierarchy effect: An experimental assessment of three proposed mechanisms; activity differences, disproportional food acquisition, physiological stress. *Environ. Biol. Fish.* **12,** 181–188.

Koolhaas, J. M., Korte, S. M., De Boer, S. F., Van Der Vegt, B. J., Van Reenen, C. G., Hopster, H., De Jong, I. C., Ruis, M. A. W., and Blokhuis, H. J. (1999). Coping styles in animals: Current status in behaviour and stress physiology. *Neurosci. Biobehav. Rev.* **23,** 925–935.

Koolhaas, J. M., de Boer, S. F., Buwalda, B., van der Vegt, B. J., Carere, C., and Groothuis, A. G. G. (2001). How and why coping systems vary among individuals. *In* "Coping with Challenge: Welfare in Animals including Humans" (Broom, D. M., Ed.), pp. 197–209. Dahlem University Press, Berlin.

Krause, J., and Ruxton, G. D. (2002). "Living in Groups." Oxford University Press, Oxford.

Lahti, K., Huuskonen, H., Laurila, A., and Piironen, J. (2002). Metabolic rate and aggressiveness between Brown Trout populations. *Funct. Ecol.* **16,** 167–174.

Laidley, C. W., and Leatherland, J. F. (1988). Cohort sampling, anaesthesia and stocking-density effects on plasma cortisol, thyroid hormone, metabolite and ion levels in rainbow trout, *Salmo gairdneri* Richardson. *J. Fish Biol.* **33,** 73–88.

Larson, E. T., and Summers, C. H. (2001). Serotonin reverses dominant social status. *Behav. Brain Res.* **121,** 95–102.

Leimar, O., and Enquist, M. (1984). Effects of asymmetries in owner-intruder conflicts. *J. Theor. Biol.* **111,** 475–491.

Lepage, O., Øverli, Ø., Petersson, E., Järvi, T., and Winberg, S. (2000). Differential neuroendocrine stress responses in wild and domesticated sea trout (*Salmo trutta*). *Brain Behav. Evol.* **56,** 259–268.

Lepage, O., Tottmar, O., and Winberg, S (2002). Elevated dietary intake of L-tryptophan counteracts the stress-induced elevation of plasma cortisol in rainbow trout (*Oncorhynchus mykiss*). *J. Exp. Biol.* **205,** 3679–3687.

Lepage, O., Vilchez, I. M., Pottinger, T. G., and Winberg, S. (2003). Time-course of the effect of dietary TRP on plasma cortisol in rainbow trout (*Oncorhyncus mykiss*). *J. Exp. Biol.* **206,** 3589–3599.

Lepage, O., Larson, E. T., Mayer, I., and Winberg, S. (2005). Serotonin, but not melatonin, plays a role in shaping dominant-subordinate relationships and aggression in rainbow trout. *Horm. Behav.* **48,** 233–242.

Li, H. W., and Brocksen, R. W. (1977). Approaches to the analysis of energetic costs of intraspecific competition for space by rainbow trout (*Salmo gairdneri*). *J. Fish Biol.* **11,** 329–341.

Liley, N. R., and Stacey, N. E. (1983). Hormones, pheromones and reproductive behaviour. *In* "Fish Physiology" (Hoar, W. S., and Randall, D. J., Eds.), Vol. 9B, pp. 1–63. Academic Press, New York.

Lomnicki, A. (1988). "Population Ecology of Individuals." Princeton University Press, Princeton.

Lovejoy, D. A., and Balment, R. J. (1999). Evolution and physiology of the corticotrophin-releasing factor (CRF) family of neuropetides in vertebrates. *Gen. Comp. Endocrinol.* **115,** 1–22.

Macey, M. J., Pickford, G. E., and Peter, R. E. (1974). Forebrain localisation of the spawning reflex response to exogenous neurohypophysial hormones in the killifish *Fundulus heteroclitus*. *J. Exp. Zool.* **190,** 269–280.

Mac Lean, A., and Metcalfe, N. B. (2001). Social status, access to food, and compensatory growth in juvenile Atlantic salmon. *J. Fish Biol.* **58,** 1331–1346.

Major, P. F. (1978). Predator-prey interactions in two schooling fishes, *Caranx ignobilis* and *Stolephorus purpureus*. *Anim. Behav.* **26,** 760–777.

Marchetti, M. P., and Nevitt, G. A. (2003). Effects of hatchery rearing on brain structures of rainbow trout. *Env. Biol. Fish* **66,** 9–14.

Marinelli, M., Aouizerate, B., Barrot, M., Le Moal, M., and Piazza, P. V. (1998). Dopamine-dependent responses to morphine depend on glucocorticoid receptors. *Proc. Natl. Acad. Sci. USA* **95,** 7742–7747.

Martel, G., and Dill, L. M. (1993). Feeding and aggressive behaviours in juvenile coho salmon (*Oncorhynchus kisutch*) under chemically-mediated risk of predation. *Behav. Ecol. Sociobiol.* **32,** 365–370.

Martin, P., and Bateson, P. (1993). "Measuring Behaviour: An Introductory Guide." Cambridge University Press.

Martin-Smith, K. M., and Armstrong, J. D. (2002). Growth rates of wild stream-dwelling Atlantic salmon correlate with activity and sex but not dominance. *J. Anim. Ecol.* **71,** 413–423.

Maynard Smith, J. (1974). The theory of games and the evolution of animal conflicts. *J. Theor. Biol.* **47,** 209–221.

McCarthy, I. D. (2001). Competitive ability is related to metabolic asymmetry in juvenile rainbow trout. *J. Fish Biol.* **59,** 1002–1014.

McCarthy, I. D., Gair, D. J., and Houlihan, D. F. (1999). Feeding rank and dominance in *Tilapia rendalli* under defensible and indefensible patterns of food distribution. *J. Fish. Biol.* **55,** 854–867.

McIntyre, D. C., Healy, I. M., and Saari, M. (1979). Intraspecies aggression and monoamine levels in rainbow trout (*Salmo gairdneri*) fingerlings. *Behav. Neural. Biol.* **25,** 90–98.

McLeod, P. G., and Huntingford, F. A. (1994). Social rank and predator inspection in sticklebacks. *Anim. Behav.* **47,** 1238–1240.

Metcalfe, N. B. (1994). The role of behaviour in determining salmon growth and development. Aquacult. *Fish. Manag.* **25,** 67–76.

Metcalfe, N. B., and Thorpe, J. E. (1992). Anorexia and defended energy levels in over-wintering juvenile salmon. *J. Anim. Ecol.* **61,** 175–181.

Metcalfe, N. B., Huntingford, F. A., Graham, W. D., and Thorpe, J. E. (1989). Early social status and the development of life-history strategies in Atlantic salmon. *Proc. R. Soc. Lond. B.* **236,** 7–19.

Metcalfe, N. B., Taylor, A., and Thorpe, J. E. (1995). Metabolic rate, social status and life-history strategies in Atlantic salmon. *Anim. Behav.* **49,** 431–436.

Milinski, M., and Parker, G. A. (1991). Competition for resources. *In* "Behavioural Ecology an Evolutionary Approach." (Krebs, J. R., and Davies, N. B., Eds.), pp. 137–168. Blackwell Scientific Publications, Oxford.

Mimassi, N., Lancien, F., Mabin, D., Delarue, C., Conlon, J. M., and Le Mével, J.-C. (2003). Induction of bradycardia in trout by centrally administered corticotrophin-releasing-hormone (CRH). *Brain Res.* **982,** 211–218.

Mommsen, T. P., Vijayan, M. M., and Moon, T. W. (1999). Cortisol in teleosts: Dynamics, mechanisms of action, and metabolic regulation. *Rev. Fish Biol. Fish.* **9,** 211–268.

Monaghan, E. P., and Glickman, S. E. (1992). Hormones and aggressive behaviour. *In* "Behavioural Endocrinology" (Becker, J. B., Breedlove, S. M., and Crews, D., Eds.), pp. 261–285. MIT Press, Cambridge, MA.

Mongeau, R., Blier, P., and de Montigny, C. (1997). The serotonergic and noradrenergic systems of the hippocampus: Their interactions and the effects of antidepressant treatments. *Brain Res. Rev.* **23,** 145–195.

Moore, F. L., and Zoeller, R. T. (1979). Endocrine control of amphibian sexual behaviour: Evidence for a neurohormone-androgen interaction. *Horm. Behav.* **13,** 207–213.

Moore, F. L., and Orchinik, M. (1994). Membrane-receptors for corticosterone – a mechanism for rapid behavioural-responses in an amphibian. *Horm. Behav.* **28,** 512–519.

Nakano, S. (1995). Individual differences in resource use, growth and emigration under the influence of a dominance hierarchy in fluvial red-spotted masu salmon in a natural habitat. *J. Anim. Ecol.* **64,** 75–84.

Neat, F. C., and Mayer, I. (1999). Plasma concentrations of sex steroids and fighting in male *Tilapia zillii*. *J. Fish Biol.* **54,** 695–697.

Neill, S. R. St. J., and Cullen, J. M. (1974). Experiments on whether schooloing of prey affects hunting behaviour of cephalopods and fish predators. *J. Zool.* **172,** 549–569.

Nelson, R. J., and Chiavegatto, S. (2001). Molecular basis of aggression. *Trends Neurosci.* **24,** 713–719.

Newman, M. A. (1956). Social behaviour and inter-specific competition in two trout species. *Physiol. Zool.* **29,** 64–81.

Nicieza, A. G., Braña, F., and Toledo, M. M. (1991). Development of length-bimodality and smolting in wild stocks of Atlantic salmon, *Salmo salar* L., under different growth conditions. *J. Fish Biol.* **38,** 509–523.

Nilsson, S. (1984). Adrenergic control systems in fish. *Mar. Biol. Lett.* **5,** 127–146.

O'Connor, K. I., Metcalfe, N. B., and Taylor, A. C. (1999). Does darkening signal submission in territorial contests between juvenile Atlantic salmon, *Salmo salar*? *Anim. Behav.* **58,** 1269–1276.

O'Connor, K. I., Metcalfe, N. B., and Taylor, A. C. (2000). Familiarity influences body darkening in territorial disputes between juvenile salmon. *Anim. Behav.* **59,** 1095–1101.

Oliveira, R. F., Almanda, V. C., and Carnario, A. V. M. (1996). Social modulation of sex steroid concentrations in the urine of male cichlid fish *Oreochromis mossambicus*. *Horm. Behav.* **30,** 2–12.

Oliveira, R. F., McGregor, P. K., and Latruffe, C. (1998). Know thine enemy: Fighting fish gather information from observing conspecific interactions. *Proc. R. Soc. Lond. B* **265,** 1045–1049.

Oliveira, R. F., Lopes, M., Carneiro, L. A., and Canário, A. V. M. (2001). Watching fights raises fish hormone levels. *Nature* **409,** 475.

Oliveira, R. F., Hirschenhauser, K., Carneiro, L. A., and Carnario, A. V. M. (2002). Social modulation of androgen levels in male teleost fish. *Comp. Biochem. Physiol. B* **132,** 203–215.

Olsen, R. E., and Ringø, E. (1999). Dominance hierarchy formation in Arctic charr *Salvelinus alpinus* (L.): Nutrient digestibility of subordinate and dominant fish. *Aquacult. Res.* **30,** 667–671

Orchinik, M., Murray, T. F., and Moore, F. L. (1991). A corticosteroid receptor in neuronal membranes. *Science* **252,** 1848–1851.

Øverli, Ø., Winberg, S., Damsgård, B., and Jobling, M. (1998). Food intake and spontaneous swimming activity in Arctic charr (*Salvelinus alpinus* L.): Role of brain serotonergic activity and social interaction. *Can. J. Zool.* **76,** 1366–1370.

Øverli, Ø., Harris, C. A., and Winberg, S. (1999). Short-term effects of fights for social dominance and the establishment of dominant-subordinate relationships on brain monoamines and cortisol in rainbow trout. *Brain Behav. Evol.* **54,** 263–275.

Øverli, Ø., Pottinger, T. G., Carrick, T. R., Øverli, E., and Winberg, S. (2001). Brain monoaminergic activity in rainbow trout selected for high and low stress responsiveness. *Brain Behav. Evol.* **57,** 214–224.

Øverli, Ø., Kotzian, S., and Winberg, S. (2002a). Effects of cortisol on aggression and locomotor activity in rainbow trout. *Horm. Behav.* **42,** 53–61.

Øverli, Ø., Pottinger, T. G., Carrick, T. R., Øverli, E., and Winberg, S. (2002b). Differences in behaviour between rainbow trout selected for high- and low-stress responsiveness. *J. Exp. Biol.* **205,** 391–395.

Øverli, Ø., Korzan, W. J., Larson, E. T., Winberg, S., Lepage, O., Pottinger, T. G., Renner, K. J., and Summers, C. H. (2004a). Behavioural and neuroendocrine correlates of displaced aggression in trout. *Horm. Behav.* **45,** 324–329.

Øverli, Ø, Korzan, W. J., Höglund, E., Winberg;, S., Bollig, H., Watt, M., Forster, G. L., Barton, B., Øverli, E., Renner, K. J., and Summers, C. H. (2004b). Stress coping style predicts aggression and social dominance in rainbow trout. *Horm. Behav.* **45,** 235–241.

Pepels, P. P. L. M., Meek, J., Wendelaar Bonga, S. E., and Balm, P. H. M. (2002). Distribution and quantification of corticotrophin-releasing hormone (CRH) in the brain of the teleost fish *Oreochromis mossambicus* (Tilapia). *J. Comp. Neurol.* **453,** 247–268.

Peter, R. E. (1977). Preoptic nucleus in fishes-comparative discussion of function – activity relationships. *Am. Zool.* **17,** 775–785.

Peter, R. E. (1986)."Vertebrate Neurohormonal systems." Vol. 1, pp. 57–104. Academic Press, Orlando, FL.

Peters, G., Delventhal, H., and Klinger, H. (1980). Physiological and morphological effects of social stress in the eel (*Anguilla anguilla* L.). *Arch. Fischereiwissenschaft* **307,** 157–180.

Pickering, A. D., and Pottinger, T. G. (1995). Biochemical effects of stress. *In* "Biochemistry and Molecular Biology of Fishes" (Hochachka, P. W., and Pottinger, T. G., Eds.), Vol. 5, pp. 349–379. Elsevier, Amsterdam.

Pickering, A. D., Pottinger, T. G., Carragher, J., and Sumpter, J. P. (1987). The effects of acute and chronic stress on the levels of reproductive hormones in the plasma of mature male brown trout, *Salmo trutta* L. *Gen. Comp. Endo.* **68,** 249–259.

Pickford, G. E. (1952). Induction of a spawning reflex in a hypophysectomized killifish. *Nature* **107,** 807–808.

Pickford, G. E., and Strecker, E. L. (1977). The spawning reflex response of the killifish *Fundulus heteroclitus*; isotocin is relatively inactive in comparison with arginine vasotocin. *Gen. Comp. Endocrinol.* **32,** 132–137.

Pigliucci, M. (2001). "Phenotypic Plasticity: Beyond Nature and Nurture." John Hopkins Press, Baltimore, MD.

Pottinger, T. G., and Pickering, A. D. (1992). The influence of social interaction on the acclimation of rainbow trout, *Oncorhynchus mykiss* (Walbaum) to chronic stress. *J. Fish Biol.* **41,** 435–447.

Pottinger, T. G., and Carrick, T. R. (1999). Modification of plasma cortisol response to stress in rainbow trout by selective breeding. *Gen. Comp. Endocrinol.* **116,** 122–132.

Pottinger, T. G., and Carrick, T. R. (2001). Stress responsiveness affects dominant-subordinate relationships in rainbow trout. *Horm. Behav.* **40,** 419–427.

Priede, I. G. (1985). Metabolic scope in fishes. *In* "Fish Energetics: New Perspectives" (Tytler, P., and Calow, P., Eds.), pp. 33–64. Croom Helm, London.

Puckett, K. J., and Dill, L. M. (1985). The energetics of feeding territoriality in juvenile coho salmon (Oncorhynchus kisutch). *Behaviour* **92,** 97–111.

Raleigh, M. J., McGuire, M. T., Brammer, G. L., Pollack, D. B., and Yuwiler, A. (1991). Serotonergic mechanisms promote dominance acquisition in adult male vervet monkeys. *Brain Res.* **559,** 181–190.

Reid, S. G., Fritsche, R., and Jonsson, A. C. (1995). Immunohistochemical localisation of bioactive peptides and amines associated with the chromaffin tissue of five species of fish. *Cell Tissue Res.* **280,** 499–512.

Rhodes, J. S., and Quinn, T. P. (1998). Factors affecting the outcome of territorial contests between hatchery and naturally reared coho salmon parr in the laboratory. *J. Fish Biol.* **53,** 1220–1230.

Rouse, E. F., Coppenger, C. J., and Barnes, P. R. (1977). The effect of an androgen inhibitor on behaviour and testicular morphology in the stickleback *Gasterosteus aculeatus*. *Horm. Behav.* **9,** 8–18.

Salek, S. J., Sullivan, C. V., and Godwin, J. (2002). Arginine vasotocin effects on courtship behaviour in male white perch (*Morone americana*). *Behav. Brain Res.* **133,** 177–183.

Sapolsky, R. M. (1992). Cortisol concentrations and the social significance of rank instability among wild baboons. *Psychoneuroendocrinology* **17,** 701–709.

Sargent, R. C. (1997). Parental care. *In* "Behavioural Ecology of Teleost Fishes" (Godin, J.-G. J., Ed.), pp. 292–315. Oxford University Press, Oxford.

Sawaguchi, T., Matsumura, M., and Kubota, K. (1990). Catecholaminergic effects on neuronal activity related to a delayed response task in monkey frontal cortex. *J. Neurophysiol.* **63,** 1401–1411.

Schjolden, J., Stoskus, A., and Winberg, S. (2005). Does intraspecific variation in stress responses and agonistic behaviour reflect divergent stress coping strategies in juvenile rainbow trout? *Physiol. Biochem. Zool.* **78,** 715–723.

Schreck, C. B., Fitzpatrick, M. S., Feist, G. W., and Yeoh, C. G. (1991). Steroids: Developmental continuum between mother and offspring. *In* "Reproductive Physiology of Fish. Proceedings of the 4th International Symposium on the Reproductive Physiology of Fish, University of East Anglia, Norwich, 7–12 July 1991." (Scott, A. P., Sumpter, J. P., Kinne, D. E., and Rolfe, M. S., Eds.), pp. 256–258. Published by Fish Symposium 91, Sheffield.

Schulkin, J., Gold, P. W., and McEwen, B. S. (1998). Induction of corticotrophin-releasing hormone gene expression by glucocorticoids: Implications for understanding the states of fear and anxiety and allostatic load. *Psychoneuroendocrinology* **23,** 219–243.

Schwank, E. (1980). The effect of size and hormonal state on the establishment of dominance in young males of *Tilapia Mariae* (Pisces: Cihlidae). *Behav. Proc.* **5,** 45–54.

Scott, D. B. C., and Currie, C. E. (1980). Social hierarchy in relation to adrenocortical activity in *Xiphophorus helleri* Heckel. *J. Fish Biol.* **16,** 265–277.

Shannon, N. J., Gunnet, J. W., and More, K. E. (1986). A comparison of biochemical indices of 5-hydroxytryptaminergic activity following electrical stimulation of dorsal raphe nucleus. *J. Neurochem.* **47,** 958–965.

Sloman, K. A., Gilmour, K. M., Taylor, A. C., and Metcalfe, N. B. (2000a). Physiological effects of dominance hierarchies within groups of brown trout, *Salmo trutta*, held under simulated natural conditions. *Fish Physiol. Biochem.* **22,** 11–20.

Sloman, K. A., Gilmour, K. M., Metcalfe, N. B., and Taylor, A. C. (2000b). Does socially induced stress in rainbow trout cause chloride cell proliferation? *J. Fish Biol.* **56,** 725–738.

Sloman, K. A., and Armstrong, J. D. (2002). Physiological effects of dominance hierarchies: Laboratory artefacts or natural phenomena? *J. Fish Biol.* **61,** 1–23.

Sloman, K. A., Metcalfe, N. B., Taylor, A. C., and Gilmour, K. M. (2001a). Plasma cortisol concentrations before and after social stress in rainbow trout and brown trout. *Physiol. Biochem. Zool.* **74,** 383–389.

Sloman, K. A., Taylor, A. C., Metcalfe, N. B., and Gilmour, K. M. (2001b). Effects of an environmental perturbation on the social behaviour and physiological function of brown trout. *Anim. Behav.* **61,** 325–333.

Sloman, K. A., Montpetit, C. J., and Gilmour, K. M. (2002a). Modulation of catecholamine release and cortisol secretion by social interactions in the rainbow trout, *Oncorhynchus mykiss.*. *Gen. Comp. Endocrinol.* **127,** 136–146.

Sloman, K. A., Wilson, L., Freel, J. A., Taylor, A. C., Metcalfe, N. B., and Gilmour, K. M. (2002b). The effects of increased flow rates on linear dominance hierarchies and physiological function in brown trout, *Salmo trutta*. *Can. J. Zool.* **80,** 1221–1227.

Sloman, K. A., Baker, D. W., Wood, C. M., and McDonald, D. G. (2002c). Social interactions affect physiological consequences of sublethal copper exposure in rainbow trout, *Oncorhynchus mykiss*. *Environ. Toxicol. Chem.* **21,** 1255–1263.

Sloman, K. A., Scott, G. R., Diao, Z., Rouleau, C., Wood, C. M., and McDonald, D. G. (2003a). Cadmium affects the social behaviour of rainbow trout, *Oncorhynchus mykiss*. *Aquat. Toxicol.* **65,** 171–185.

Sloman, K. A., Baker, D. W., Ho, C. G., McDonald, D. G., and Wood, C. M. (2003b). The effects of trace metal exposure on agonistic encounters in juvenile rainbow trout, *Oncorhynchus mykiss*. *Aquat. Toxicol.* **63,** 187–196.

Sloman, K. A., Morgan, T. P., McDonald, D. G., and Wood, C. M. (2003c). Socially-induced changes in sodium regulation affect the uptake of water-borne copper and silver in the rainbow trout, *Oncorhynchus mykiss*. *Comp. Biochem. Physiol. C.* **135,** 393–403.

Sneddon, L. U., Taylor, A. C., and Huntingford, F. A. (1999). Metabolic consequences of agonistic behaviour: Crab fights in declining oxygen tensions. *Anim. Behav.* **57,** 353–363.

Sturm, A., Bury, N., Dengreville, L., Fagart, J., Flouriot, G., Rafestin-Oblin, M. E., and Prunet, P. (2005). 11-deoxycorticosterone is a potent agonist of the rainbow trout (*Oncorhynchus mykiss*) mineralocorticoid receptor. *Endocrinol.* **146,** 47–55.

Summers, T. R., Matter, J. M., McKay, J. M., Ronan, P. J., Larson, E. T., Renner, K. J., and Summers, C. H. (2003). Rapid glucocorticoid stimulation and GABAergic inhibition of hippocampal serotonergic response: *In vivo* dialysis in the lizard *Anolis carolinensis*. *Horm. Behav.* **43,** 245–253.

Sumpter, J. P. (1997). The endocrinology of stress. *In* "Fish Stress and Health in Aquaculture" (Iwama, G. K., Pickering, A. D., Sumpter, J. P., and Schreck, C. B., Eds.), pp. 95–118. Cambridge University Press, Cambridge.

Sundström, L. F., Löhmus, M., and Johnsson, J. I. (2003). Investment in territorial defence depends on rearing environment in brown trout (*Salmo trutta*). *Behav. Ecol.*

Suter, H. C., and Huntingford, F. A. (2002). Eye colour in juvenile Atlantic salmon: Effects of social status, aggression and foraging success. *J. Fish Biol.* **61,** 606–614.

Symons, P. E. K. (1968). Increased aggression and in strength of the social hierarchy among juvenile Atlantic salmon deprived of food. *J. Fish. Res. Board Can.* **25,** 2387–2401.

Teitsma, C. A., Bailhache, T., Tujague, M., Balment, R. J., Ducouret, B., and Kah, O. (1997). Distribution and expression of glucocorticoid receptor mRNA in the forebrain of the rainbow trout. *Neuroendocrinology* **66,** 294–304.

Thorpe, J. E., Metcalfe, N. B., and Huntingford, F. A. (1992). Behavioural influences on life-history variation in juvenile Atlantic salmon, *Salmo salar*. *Env. Biol. Fish.* **33,** 331–340.

Thorpe, J. E. (1994). An alternative view of smolting in salmonids. *Aquaculture* **121,** 105–113.

Turner, G. F., and Pitcher, T. J. (1986). Attack abatement: A model for group protection by combined avoidance and dilution. *Am. Nat.* **128,** 228–240.

Valdimarsson, S. K., Metcalfe, N. B., Thorpe, J. E., and Huntingford, F. A. (1997). Seasonal changes in sheltering: Effect of light and temperature on diel activity in juvenile salmon. *Anim. Behav.* **54,** 1405–1412.

van Raaij, M. T. M., Pit, D. S. S., Balm, P. H. M., Steffens, A. B., and van den Thillart, G. E. E. M. (1996). Behavioural strategy and the physiological stress response in rainbow trout exposed to severe hypoxia. *Horm. Behav.* **30,** 85–92.

Verbeek, M. E. M., Drent, P. J., and Wiepkema, P. R. (1994). Consistent individual differences in early exploratory behaviour of male great tits. *Anim. Behav.* **48,** 1113–1121.

Wankowski, J. W. J., and Thorpe, J. E. (1979). Spatial distribution and feeding in Atlantic salmon, *Salmo salar* juveniles. *J. Fish Biol.* **14,** 239–247.

Weber, E. D., and Fausch, K. D. (2003). Interactions between hatchery and wild salmonids in streams: Differences in biology and evidence for competition. *Can. J. Fish. Aquat. Sci.* **60,** 1018–1036.

Weir, L. K., and Grant, J. W. A. (2004). The causes of resource monopolisation: Interaction between resource dispersion and mode of competition. *Ethology* **110,** 63–74.

Wendelaar-Bonga, S. E. (1997). The stress response in fish. *Physiol. Rev.* **77,** 591–625.

Wilhelmi, A. E., Pickford, G. E., and Sawyer, W. H. (1955). Initiation of the spawning reflex response in *Fundulus* by the administration of fish and mammalian neurohypophysial preparations and synthetic oxytocin. *Endocrinology* **57,** 243–252.

Wilson, D. S., Clark, A. B., Coleman, K., and Dearstyne, T. (1994). Shyness and boldness in humans and other animals. *Trends Ecol. Evol.* **9,** 442–446.

Wilson, S. D. (1998). Adaptive individual differences within single populations. *Phil. Trans. R. Soc. Lond. B* **353,** 199–205.

Winberg, S., and Nilsson, G. E. (1992). Induction of social dominance by L-DOPA treatment in Arctic charr. *NeuroReport* **3,** 243–246.

Winberg, S., and Lepage, O. (1998). Elevation of brain 5-HT activity, POMC expression and plasma cortisol in socially subordinate rainbow trout. *Am. J. Physiol.* **274,** R645–R654.

Winberg, S., and Nilsson, G. E. (1993a). Roles of brain monoamine neurotransmitters in agonistic behaviour and stress reactions, with particular reference to fish. *Comp. Biochem. Physiol.* **106,** 597–614.

Winberg, S., and Nilsson, G. E. (1993b). Time course of changes in brain serotonergic activity and brain tryptophan levels in dominant and subordinate juvenile Arctic charr. *J. Exp. Biol.* **179,** 181–195.

Winberg, S., Nilsson, G. E., and Olsén, K. H. (1991). Social rank and brain levels of monoamines and monoamine metabolites in Arctic charr, *Salvelinus alpinus* (L.). *J. Comp. Physiol. A* **168,** 241–246.

Winberg, S., Nilsson, G. E., and Olsén, K. H. (1992a). Changes in brain serotonergic activity during hierarchic behaviour in Arctic charr (*Salvelinus alpinus* L.) are socially induced. *J. Comp. Physiol. A* **170,** 93–99.

Winberg, S., Nilsson, G. E., and Olsén, K. H. (1992b). The effect of stress and starvation on brain serotonin utilisation in Arctic charr (*Salvelinus alpinus* L.). *J. Exp. Biol.* **165,** 229–239.

Winberg, S., Myrberg, A. A., and Nilsson, G. E. (1993a). Predator exposure alters brain serotonin metabolism in bicolor damselfish. *NeuroReport* **4,** 399–402.

Winberg, S., Nilsson, G. E., Spruijt, B. M., and Höglund, U. (1993b). Spontaneous locomotor activity in Arctic charr measured by computerized imaging technique: Role of brain serotonergic activity. *J. Exp. Biol.* **179,** 213–232.

Winberg, S., Nilsson, A., Hylland, P., Söderström, V., and Nilsson, G. E. (1997). Serotonin as a regulator of hypothalamic-pituitary-interrenal activity in teleost fish. *Neurosci. Lett.* **230,** 113–116.

Winberg, S., Øverli, Ø., and Lepage, O. (2001). Suppression of aggression in rainbow trout (*Oncorhyncus mykiss*) by dietary L-tryptophan. *J. Exp. Biol.* **204,** 3867–3886.

Winslow, J. T., Hastings, N., Carter, C. S., Harbaugh, C. R., and Insel, R. I. (1993). A role for central vasopressin in pair bonding in monogamous prairie voles. *Nature* **365,** 545–548.

Wootton, R. J. (1998). "Ecology of Teleost Fishes." Kluwer Academic Publishers, Dordrecht.

Yamamoto, T., Ueda, H., and Higashi, S. (1998). Correlation among dominance status, metabolic rate and otolith size in masu salmon. *J. Fish Biol.* **52,** 281–290.

Zayan, R. (1991). The specificity of social stress. *Behav. Proc.* **25,** 81–93.

6

CIRCADIAN RHYTHMS IN FISH

IRINA V. ZHDANOVA
STÉPHAN G. REEBS

1. INTRODUCTION

Circadian rhythmicity in behaviour and metabolism is a ubiquitous phenomenon in biology, documented in such diverse species as unicellular organisms, plants, invertebrates, and vertebrates. The reason is quite clear: the survival of all of these organisms depends on adapting to the regular changes of their environment, defined mostly by the 24-hour period of earth's rotation relative to the sun. Importantly, the organisms can immensely benefit from predicting when the day or night comes and, with it, changes in illumination, temperature, or food availability. Anticipating these environmental changes allows organisms to adjust all of their metabolic and behavioural processes in advance and to do everything "on time." Hence, biological clocks have evolved, and even in the absence of any environmental

Behaviour and Physiology of Fish: Volume 24
FISH PHYSIOLOGY

DOI: 10.1016/S1546-5098(05)24006-2

cues they autonomously oscillate with a circadian (*circa* = about, *dia* = day) period.

The intrinsic period of circadian clocks can be somewhat shorter or longer than 24 hours and can be adjusted (reset) to the environment by gradually shifting the oscillation phase until it coincides with the environmental cycle and then stabilises (entrains). Such entrainment can be induced by different environmental cues, called *zeitgebers* ("time givers") or *synchronisers*. Light appears to be the strongest of these synchronisers and, as a result, a large amount of data characterising light-dependent intrinsic oscillations has been collected over the years. In addition, other critical environmental parameters changing daily or annually (such as environmental temperature, food availability, and predation risk) can also affect the phase of intrinsic clocks and, in some species, may even play a more important role than light.

Conceivably, every cell in the body can contain an intrinsic clock mechanism. Although some cells may initiate oscillations only in response to specific internal or external factors, others constantly express this function. The continuously oscillating cells and structures specialising in providing circadian signals to the entire organism are called *central oscillators*. The most renowned of them is the suprachiasmatic nucleus (SCN) of the hypothalamus in mammals, a neuronal structure defining the majority (if not all) of the circadian rhythms in this group The autonomous oscillations displayed by peripheral cells and tissues may remain independent or can synchronise with each other and with the central oscillators, organising complex networks and affecting multiple physiological functions in a species-specific manner.

The circadian system of fish follows the same general design as in other vertebrates and invertebrates; they show circadian rhythms of activity, food intake, and some physiological parameters (Table 6.1). The existence of an elaborate network of coexisting or competing central and peripheral circadian oscillators and a large number of circadian clock genes in fish, when compared to other vertebrates, may explain some specific characteristics such as instability in their circadian rhythms, spontaneous shifts between diurnal and nocturnal activity, and dramatic seasonal changes in activity patterns.

A principal circadian hormone, melatonin, and melatonin-producing organs, the pineal gland and retina, also play a central role in the circadian rhythms of fish and their entrainment to changing environments. Rest in fish is under both circadian and homeostatic control, showing distinct behavioural and pharmacological features of sleep. A sleeplike state can be induced in fish by different hypnotic agents and melatonin, further suggesting that rest in fish is analogous to sleep in higher vertebrates.

Table 6.1
Self-Sustained Circadian Rhythms (and Attempts to Find Such) in Fishes

Parameter	Common name (or family)	Latin name	Lighting and tau	Reference	Comment
Body colour change	Mummichog	*Fundulus heteroclitus*	LL and dimLL: 22.5–26.5 h	Kavaliers and Abbott, 1977	Hypophysectomy abolishes Rhythm
Cell proliferation in retina	(Cichlid)	*Haplochromis burtoni*	DD: ca. 24 h	Chiu *et al.*, 1995	
Demand-feeding	European sea bass	*Dicentrarchus labrax*	LL: 21.3–26 h	Boujard *et al.*, 2000	Groups of 50 fish
	European sea bass	*Dicentrarchus labrax*	LD 40:40 min: 22.5–24.5 h	Sánchez-Vázquez *et al.*, 1995a	Singles and groups of four
	European sea bass	*Dicentrarchus labrax*	DD: no rhythm detected	Sánchez-Vázquez *et al.*, 1995b	After many days under a restricted feeding schedule
	Rainbow trout	*Oncorhynchus mykiss*	LL: 24.6–26.0 h	Chen and Tabata 2002	After many days under a restricted feeding schedule
	Rainbow trout	*Oncorhynchus mykiss*	LL: >24 h	Sánchez-Vázquez *et al.*, 2000	Pinealectomy does not abolish rhythm
	Rainbow trout European catfish	*Oncorhynchus mykiss* *Silurus glanis*	LL (trout): 23–30 h DD (catfish): 20–26 h	Bolliet *et al.*, 2001	Groups and individuals were tested, but no difference detected
	Rainbow trout	*Oncorhynchus mykiss*	LL: 16–32 h	Chen *et al.*, 2002a	Longer tau in groups than in individuals

(*continued*)

Table 6.1 (*continued*)

Parameter	Common name (or family)	Latin name	Lighting and tau	Reference	Comment
Demand-feeding (ambulatory)	Bluegill sunfish Largemouth bass	*Lepomis macrochirus* *Micropterus salmoides*	DD: ca. 24 h	Davis, 1964	Rhythm lasts for only 2 days after a restricted feeding schedule
Electric discharge	(Gymnotid)	*Gymnorhamphichthys* hypostomus	DD and LL: 23.5–25.4 h	Schwassmann, 1971	Well-defined rhythm; also a review of the older literature
Locomotion, demand-feeding	Goldfish	*Carassius auratus*	DD: 22.2–27.5 h LD 45:45 min: 23.0–24.8 h	Sánchez-Vázquez *et al.*, 1996	Rhythms lacked constancy and damped out within a few days
	Brown bullhead	*Ictalurus nebulosus*	LD 45:15 min: 21.5–23.8 h	Eriksson and Van Veen, 1980	No rhythm found in DD, LL, or dimLL
	Rainbow trout	*Oncorhynchus mykiss*	LD 45:45 min: 21.9 h (feeding) LL: 26.2 h (feeding) and 25.8 h (locomotion)	Sánchez-Vázquez and Tabata, 1998	More fish expressed rhythm under LL than LD 45:45 min or dimLL
Locomotion	American shad (Characin)	*Alosa sapidissima*	dimLL: no rhythm found	Katz, 1978	
		Anoptichthys jordani	DD: ca. 24 h LL: ca. 24 h	Erckens and Weber, 1976	Cave-living population; rhythm damps out after 1–4 day
	Sea catfish	*Arius felis*	DD and LL: ca. 24 h	Steele, 1984	

Mexican Tetra	*Astyanax mexicanus*	DD: 21.9–25.9 h LL: ca. 24 h	Erckens and Martin, 1982a	River-living population
Mexican tetra	*Astyanax mexicanus*	DD: no rhythm found	Erckens and Martin, 1982b	Cave-living population
Shanny	*Blennius pholis*	DD: 23–25 h	Gibson, 1971	Only after many days of LD in the lab
Goldfish	*Carassius auratus*	DD: 24.4 dimLL: 26.0 LL: 25.2 h	Iigo and Tabata, 1996	Only half of all individuals were rhythmic
Goldfish	*Carassius auratus*	LL: >24 h	Kavaliers, 1981a	More stable rhythms in groups than in individuals
Goldfish	*Carassius auratus*	DD: 21.5–28.8 h	Sánchez-Vázquez *et al.*, 1997	Rhythm better detected at bottom than at top of tank
Goldfish	*Carassius auratus*	DD and LL: ca. 24 h	Spieler and Clougherty, 1989	After many days under a restricted feeding schedule
White sucker	*Catostomus commersoni*	DD: <24 h	Kavaliers, 1981b	Long-lasting rhythm
White sucker	*Catostomus commersoni*	DD: 22.4–23.5 h (single) and 24.6–26.0 (groups of 25)	Kavaliers, 1980a	Better defined rhythm in groups of 25 than in individuals
Herring	*Clupea harengus*	DD: ca. 24 h	Stickney, 1972	Rhythm damps out after 2 days
Lake chub	*Couesius plumbeus*	DD: 24.8–28.1 h	Kavaliers and Ross, 1981	Seasonal variation in tau
Lake chub	*Couesius plumbeus*	DD: 24.8–28.1 h	Kavaliers, 1978	Seasonal variation in tau
Lake chub	*Couesius plumbeus*	DD: ca. 25 h	Kavaliers, 1980b	In both intact and pinealectomized individuals

(*continued*)

Table 6.1 (*continued*)

Parameter	Common name (or family)	Latin name	Lighting and tau	Reference	Comment
	Lake chub	*Couesius plumbeus*	DD: 24.6–26.9 h	Kavaliers, 1979	In both intact and pinealectomized individuals
	Lake chub	*Couesius plumbeus*	DD: >24 h	Kavaliers, 1980c	In both intact and pinealectomized individuals
	Zebrafish	*Danio rerio*	DD: many averages, all >24 h	Hurd and Cahill, 2002	Larvae
	Inshore hagfish	*Eptatetrus burgeri*	DD: <24 h	Kabasawa and Ooka-Souda, 1989	Well defined and long-lasting rhythm
	Inshore hagfish	*Eptatretus burgeri*	DD: 22.8–25.1 h	Ooka-Souda and Kabasawa, 1995	Well defined and long-lasting rhythm; phase-response curve to light
	Inshore hagfish	*Eptatretus burgeri*	DD: 23.0–24.7 h	Ooka-Souda *et al.*, 1985	Nice transient cycles after LD shift
	Northern pike	*Esox lucius*	DD: 24 h LL: 25.25 h	Beauchamp *et al.*, 1993	Actograms not shown
	Mummichog	*Fundulus heteroclitus*	LL: >24 h DD: >24 h	Kavaliers, 1980d	Better defined rhythm in groups of 25 than in groups of 5 or in individuals
	Yellow wrasse	*Halichoeres chrysus*	dimLL: 24.2 h on average	Gerkema *et al.*, 2000	16 tested individually; well defined rhythms; phase-response curve to light

(Wrasse)	*Halichoeres tenuispinnis*	DD: 23.6–24.3 h LL: 23.5–23.7 h	Nishi, 1989	Well defined rhythm
Horn shark Swell shark	*Heterodontus francisci*	DD and LL: no rhythm for horn shark	Nelson and Johnson, 1970	Only one individual tested; relation between presence of sleep and of rhythm
	Cephaloscyllium ventriosum	DD and LL: 23.4 h and 24.4 h for swell shark		
Stinging catfish	*Heteropneustes fossilis*	DD: <24 h	Garg and Sundararaj, 1986	Pinealectomy abolishes rhythm
Rosy tetra	*Hyphessobrycon rosaceus*	LL: no rhythm DD: ca. 24 h	Thines, 1967	Groups of 12 fish; two cycles only
Channel catfish	*Ictalurus punctatus*	DD and LL: no rhythm found	Goudie *et al.*, 1983	
Arctic lamprey	*Lampetra japonica*	DD: 20.0–23.5 h	Morita *et al.*, 1992	Pinealectomy abolishes rhythm
Burbot	*Lota lota*	DD: 21.2–23.5 h	Kavaliers, 1980e	Seasonal variation in tau; well defined and long-lasting rhythms
Oriental weatherfish	*Misgurnus anguillicaudatus*	DD: 22.0–28.2 h	Naruse and Oishi, 1994	No more than 50% of fish showed rhythm
(River loach)	*Nemacheilus evezardi*	DD: ca. 24 h	Pati, 2001	In both cave-dwelling and surface-dwelling populations
(River loach)	*Noemacheilus barbatulus*	dimLL: ca. 24 h	Burdeyron and Buisson, 1982	Groups of 9 fish; individuals are arhythmic

(continued)

Table 6.1 (*continued*)

Parameter	Common name (or family)	Latin name	Lighting and tau	Reference	Comment
	Pink salmon	*Oncorhynchus gorbuscha*	LL: 20.9–26.4 h dimLL: 20.0–35.6 h	Godin, 1981	Juveniles; only half the fish were rhythmic
	Sea lamprey	*Petromyzon marinus*	dimLL: ca. 23 h	Kleerekoper *et al.*, 1961	Only one individual
	European minnow	*Phoxinus phoxinus*	LL: no rhythm found	Harden Jones, 1956	
	Bluefish	*Pomatomus saltatrix*	dimLL: ca. 24 h	Olla and Studholme, 1972	One group of 6 fish
	(Bagrid catfish)	*Pseudobagrus aurantiacus*	DD and LL: no rhythm found	Mashiko, 1979	
	Longnose dace	*Rhinichthys cataractae*	DD: 21.6–23.6 on average	Kavaliers, 1981c	Seasonal variation in tau
	Amur catfish	*Silurus asotus*	DD: 22.5–27.3 h LL: 20.1–27.0 h LD 45:15 min: 20.9–27.5 h	Tabata *et al.*, 1989	
	Amur catfish	*Silurus asotus*	DD: 24.3–26.2 h LL: 21.2–25.6 h dimLL: 25.9–26.3 h	Tabata *et al.*, 1991	Neither pinealectomy nor blinding abolish rhythm
	Slender wrasse Cleaner wrasse	*Suezichthys gracilis*	LL: depends on species, but generally <24h	Nishi, 1990	Well-defined rhythm

	Threadfin wrasse	*Thalassoma cupido* *Labroides dimidiatus* *Cirrhilabrus temminckii*	DD: activity often suppressed		
	Slender wrasse	*Suezichthys gracilis*	LL: >24 h DD: activity suppressed	Nishi, 1991	In Japanese with English summary; rhythm dependent on presence of refuge (sand)
	(Heptapterid)	*Taunayia sp.*	DD: no rhythm found	Trajano and Menna-Barreto, 2000	Cave-living
	Tench	*Tinca tinca*	LD 40:40 min: 20.8–28.6 h DD: 21.0–26.4 h	Herrero *et al.*, 2003	Rhythm detected in only half the fish
	Dark chub	*Zacco temmincki*	DD: 23.9 h dimLL: 26.1 h	Minh-Nyo *et al.*, 1991	
	Viviparous blenny	*Zoarces viviparus*	DD: ca. 24 h	Cummings and Morgan, 2001	Mixed with a stronger circatidal rhythm
	Bluegill sunfish	*Lepomis macrochirus*	DD: ca. 24 h	Reynolds and Casterlin, 1976a	Scant details of methods and results
	Largemouth bass	*Micropterus salmoides*	LL: ca. 24 h	Reynolds and Casterlin, 1976b	Scant details of methods and results
	Atlantic salmon	*Salmo salar*	DD: 23.5–24.2 h LL: no rhythm found	Richardson and McCleave, 1974	Only 5 out of 30 individuals were rhythmic in DD
Locomotion, oxygen consumption	Lemon shark	*Negaprion brevirostris*	Dim LL: ca. 24 h	Nixon and Gruber, 1988	One individual tested

(continued)

Table 6.1 (*continued*)

Parameter	Common name (or family)	Latin name	Lighting and tau	Reference	Comment
Oxygen consumption	Nile tilapia	*Oreochromis niloticus*	LL: ca. 24 h DD: ca. 24 h	Ross and McKinney, 1988	
	River puffer fish	*Takifugu obscurus*	DD: 22.7–24.8 h	Kim *et al.*, 1997	In singles as well as in groups
Parental behaviour	Convict cichlid Rainbow cichlid	*Cichlasoma nigrofasciatum* *Herotilapia multispinosa*	DD: ca. 24 h	Reebs and Colgan, 1991	Rhythm only 2 cycles long
	Threespine stickleback	*Gasterosteus aculeatus*	LL: no rhythm found	Sevenster *et al.*, 1995	Also no rhythm of self-selection of light
Response to light	Lake chub	*Couesius plumbeus*	DD: ca. 24 h	Kavaliers, 1981d	In blind and pinealectomized individuals
Self-selection of light	Pumpkinseed sunfish	*Lepomis gibbosus*	Self-selected: no rhythm found	Colgan, 1975	
	(River loach)	*Nemacheilus evezardi*	Self-selected: only one cycle measured	Pradhan *et al.*, 1989	Tested in groups of 16 fish
Melatonin production	Northern pike	*Esox lucius*	DD: 24–27 h	Bolliet *et al.*, 1997	Pineal cell culture
	Ayu	*Plecoglossus altivelis*	DD: averages 25.7 26.1 h	Iigo *et al.*, 2003a	Pineal organ culture
	Ayu	*Plecoglossus altivelis*	DD: 26.1 h	Iigo *et al*, 2004	Pineal organ culture

Many of these rhythms lasted no longer than 10 days. Many were also hard to distinguish visually on actograms and required periodogram analysis for detection.

Tau, period of free-run; DD, constant darkness; LL, constant light; LD, light–dark cycle.

Remarkable intra- and interspecies variability in the design and function of the circadian systems in fish (Table 6.1) add challenge to the study of their circadian rhythmicity, but also could provide unique insights into the general mechanisms of biological timekeeping in vertebrates.

2. CIRCADIAN CLOCK GENES

The discovery in insects and mammals of a network of circadian clock genes and the proteins they encode, with a possibility to study them *in vivo* and *in vitro*, provided a remarkable breakthrough in our understanding of the circadian system and its development. Some of these genes define intrinsic circadian oscillations, whereas others influence the distribution of the circadian message. The genetic regulation of circadian rhythm generation in fish appears to be similar to the general scheme described in *Drosophila* and in mice (Dunlap, 1999). It is based on a self-sustaining transcriptional–translational feedback loop, in which the expression of the clock genes is periodically suppressed by their protein products. The oscillation involves heterodimerisation of CLOCK and BMAL proteins, members of the basic helix-loop-helix PAS superfamily of transcription factors, which bind to the E-box DNA motif. This complex activates expression of the circadian clock genes *Period* (*Per*) and *Cryptochrome* (*Cry*). Then, PER and CRY proteins form complexes that enter into the nucleus and repress the CLOCK-BMAL dependent transcription of their own genes, thereby generating approximately a 24-hour period molecular oscillator (see Okamura, 2003 for a review). This loop also controls the rhythmic expression of the repressor REV-ERBα, which is required for rhythmic Bmal1 transcription. Such a second loop appears to promote an overall robustness of the circadian oscillator. The negative feedback loop, reinforced by a positive transcriptional loop, can be reset by external time cues (predominantly a light–dark cycle) synchronising the intrinsic processes with the periodically changing environment (Glossop *et al.*, 1999).

Our knowledge of the molecular mechanisms of the fish circadian system is based mainly on studies conducted in zebrafish (Cahill, 2002). A distinct feature of the zebrafish circadian system is a large number of circadian clock genes, although all of them appear to be homologs of the circadian clock genes known in other species. This phenomenon is likely to stem from whole or partial genome duplication in some teleosts (Postlethwait *et al.*, 1998; Leggatt and Iwama, 2003; Le Comber and Smith, 2004).

So far, three zebrafish *Clock* genes (*zfClock1, 2,* and *3*) and three *Bmal* genes (*Bmal1*, *2,* and *3*) have been cloned (Whitmore *et al.*, 1998; Ishikawa *et al.*, 2002). In mammals, *Clock* expression does not oscillate but *Bmal* expression does (Oishi *et al.*, 1998; Shearman *et al.*, 2000). In contrast, in zebrafish, all of the *Clock* and *Bmal* genes show robust rhythmic expression under a light–dark cycle or in constant darkness (Cermakian *et al.*, 2000; Ishikawa *et al.*, 2002).

Three zebrafish *Period* genes are homologs of the mammalian *Per1*, *Per2*, and *Per3*. At least two of them, *Per2* and *Per3*, are expressed early in zebrafish development. Their differential spatiotemporal expression suggests that they play distinct roles in the establishment of the embryonic circadian system (Delaunay *et al.*, 2000, 2003).

Zebrafish have six *Cry* genes (*zCry1a*, *1b*, *2a*, *2b*, *3*, and *4*) (Kobayashi *et al.*, 2000). Four of them (*zCry1a*, *1b*, *2a*, *2b*) have high sequence similarity to mammalian *Cry* genes. They are rhythmically expressed in the retina, pineal, brain, and other zebrafish tissues and their products can inhibit CLOCK-BMAL mediated transcription. *zCry3 and zCry4* have higher sequence similarity to the Drosophila *Cry* gene and their protein products do not affect CLOCK-BMAL mediated transcription.

Further studies in other fish species would substantiate the data collected in zebrafish and help determine the basis for interspecies variability in circadian rhythms and the circadian plasticity exhibited by these vertebrates.

3. ENTRAINMENT TO LIGHT

Three major properties of a light-dependent circadian system include photoreception, intrinsic oscillation, and ability to communicate circadian phase information to the local environment (to peripheral oscillators, for example). Apparently, such a "trio" can be present within one cell because unicellular organisms can show robust intrinsic circadian rhythms. Moreover, all these properties are retained by specialised photosensitive cells in phylogenetically advanced species. Both retinal photoreceptors and photosensitive cells of the pineal gland contain circadian clocks and secrete melatonin, a bioamine that can convey circadian messages either to the nearby surroundings (retinal melatonin) or to the entire organism (pineal melatonin).

Furthermore, it has recently been shown that multiple tissues in vertebrates, including peripheral ones, can directly detect light and entrain to it, and can express circadian clock genes (Whitmore *et al.*, 2000). This suggests that cell-autonomous circadian oscillation modulated by environmental light is a more widespread phenomenon than previously thought.

It is also becoming increasingly clear that the circadian system does not rely only on the cone- or rod-based opsins, although they are likely to play some role in light entrainment. Several new photopigments have been identified in fish retinas, pineal glands, brain and peripheral tissues (see Foster *et al.*, 2003 for a review). Such photopigments include three isoforms of vertebrate-ancient (VA) opsin (Soni and Foster, 1997; Soni *et al.*, 1998; Kojima *et al.*, 2000; Moutsaki *et al.*, 2000; Minamoto and Shimizu, 2002; Jenkins *et al.*, 2003), melanopsin (Bellingham *et al.*, 2002; Jenkins *et al.*, 2003; Drivenes *et al.*, 2003), cryptochromes (Kobayashi *et al.*, 2000; Cermakian and Sassone-Corsi, 2002), and *tmt*-opsin (Moutsaki *et al.*, 2003). These molecules became candidates for a putative "circadian photoreceptor." It remains to be elucidated which of them play critical roles in circadian systems. So far, it appears that several rather than one photopigment might be involved in photic entrainment.

3.1. Centralised Circadian Oscillators

In fish, the existence of a centralised neuronal central clock structure, analogous to the mammalian SCN, has not yet been documented. However, another evolutionary conserved circadian structure, the pineal gland (*epiphysis cerebri*) plays an important role in fish circadian rhythmicity via its strictly periodic secretion of melatonin. Melatonin is also produced by the retina, providing a local paracrine signal. Analysis of the temporal parameters of melatonin secretion by the pineal gland and retina and their role in different physiological functions allows interesting comparisons between the centralised and peripheral circadian clocks.

The pineal gland is one of the first brain structures to develop in fish. For example, in zebrafish, the pineal gland develops by 19 hours postfertilisation and, almost immediately, becomes responsive to light and starts secreting melatonin (Kazimi and Cahill, 1999; Danilova *et al.*, 2004). If the embryos are exposed to a light–dark cycle during these early stages of development, their pineal glands secrete melatonin in a distinct circadian pattern, persisting after the embryos are moved to constant conditions. However, if the embryos are kept in constant darkness from the start, the circadian rhythm of melatonin production does not develop and melatonin is secreted continuously at a relatively low level (Kazimi and Cahill, 1999). Thus, the pineal gland in zebrafish has intrinsic circadian oscillators but these may need to be "turned on" by the environmental light–dark cycle.

Fish show a surprising variety of tissue-specific circadian adaptations, which are well reflected in peculiarities of melatonin secretion and function. This might be explained by fish having more genes for encoding the same or similar proteins than mammals or birds, presumably due to a duplication of the teleost genome early in the course of evolution. For example, as opposed to mammals having one gene that encodes a limiting melatonin-synthesising enzyme, arylalkylamine-N-acetyltransferase (AANAT), fish have at least two of them (Begay *et al.*, 1998; Mizusawa *et al.*, 1998, 2000; Coon *et al.*, 1999;

Benyassi *et al.*, 2000). Similarly, mammals appear to have two genes for melatonin receptor proteins, but zebrafish have at least five of them (Reppert *et al.*, 1995). Such an abundance of proteins with similar functions may be at the root of the high functional and tissue specialisation shown by fish, resulting in more complex and often unpredictable patterns of circadian adaptations in cells and organs of different fish species.

In the vast majority of vertebrates, including mammals, the pineal gland and the retina secrete melatonin only at night, and in both structures melatonin synthesis is inhibited by bright light. This is not necessarily true in fish, at least for the retina.

Pineal glands in teleosts (e.g., zebrafish and pike) are directly photosensitive and contain intrinsic circadian clocks. Similarly to mammalian pineals, they secrete melatonin exclusively at night under a light–dark cycle, or exclusively during what would be nighttime (subjective night) under constant darkness. Pinealectomy can significantly disrupt circadian rhythmicity in these fish (as has been reported in white suckers, burbot, and catfish), presumably because their pineal glands serve as central circadian clocks (Kavaliers, 1980e; Garg and Sundararaj, 1986).

In contrast, the pineal glands of salmonids, though photosensitive, lack an intrinsic circadian oscillator (Gern and Greenhouse, 1988; Max and Menaker, 1992; Iigo *et al.*, 1997a; Falcon, 1999). In the absence of an SCN-like "master" clock or an intrinsic clock in the pineal gland, melatonin production in these species is under exclusive light control, occurring at any dark period, whether it is during their subjective day or at night (Masuda *et al.*, 2003). Consequently, pinealectomy does not dramatically alter their circadian rhythmicity, as shown by studies conducted in trout (Sánchez-Vázquez *et al.*, 2000).

Even more remarkable are the differences between melatonin rhythms in fish retinas. In some fish (e.g., sea bass, trout, or pike), the retina produces melatonin during the day, contrary to the "rule." In these species, nighttime light pulses stimulate retinal melatonin production and at the same time inhibit melatonin secretion by the pineal gland (Zachmann *et al.*, 1992; Iigo *et al.*, 1997b; Garcia-Allegue *et al.*, 2001; Falcon *et al.*, 2003). Studies in European sea bass show other differences in the pineal-retinal relationship; for example, stable circadian rhythms of nighttime pineal melatonin production can be associated with seasonally-adjustable retinal melatonin synthesis (Garcia-Allegue *et al.*, 2001) and alterations in retinal melatonin production can attenuate the pineal secretory activity, despite independent photosensitivity in this species (Bayarri *et al.*, 2003). Such remarkable variability in melatonin rhythms underscores the overall plasticity of the fish circadian system reflected in circadian patterns of behaviour, as discussed in detail below.

3.2. Peripheral Circadian Oscillators

Along with centralised clock structures, such as the pineal gland, fish have multiple peripheral oscillators in a variety of tissues, if not in all of them. This has been shown by documenting periodic expression of zebrafish circadian clock genes in isolated tissues, such as the heart or the kidney (Whitmore *et al.*, 1998). Importantly, these peripheral tissue pacemakers, at least in zebrafish, are responsive to light and thus can be reset and entrained independently of the central circadian structures or photic input via the retina (Whitmore *et al.*, 2000). In addition, the peripheral clocks appear to be able to entrain to other zeitgebers, such as feeding time (Tamai *et al.*, 2003).

Central and peripheral oscillators can retain a similar circadian phase or be out of phase with one another by up to 12 hours. However, peripheral oscillators are likely to have a high degree of independence from central circadian signals, because there appears to be no time lag between pacemakers in the brain and the periphery in responding to a zeitgeber. Such a time lag would be expected in hierarchical structures that feature a leading "master" clock (Figure 6.1).

A complex network of coexisting central and peripheral clocks, coordinated to maintain physiological integrity, poses an interesting and challenging system to decipher. Depending on environmental or internal processes, the relative contribution of centralised circadian signals (such as circulating melatonin) and local clocks in peripheral organs can probably vary. Moreover, some oscillators may stay dormant, allowing a central oscillator to define the major rhythm, until the central clock activity is altered or special events "wake up" the local clocks.

3.3. Melatonin: Functional Significance

Conveying a circadian message to the immediate surrounding or to faraway structures and organs requires some delivery mechanisms. Presumably, there are multiple mechanisms, both neuronal and hormonal, that allow central clocks to achieve this. However, we still know very little about those pathways, with the exception of one of them involving the bioamine melatonin. Documenting changes in melatonin secretion proved to be an excellent way to monitor circadian phase and its changes. Similarly, studying the effects of increased or reduced melatonin levels during certain hours of the day provided important data on some of the physiological effects of the circadian clock (for a review, see Zhdanova and Tucci, 2003).

Melatonin produced by the pineal gland and melatonin produced by the retina have different physiological roles. The pineal gland provides its

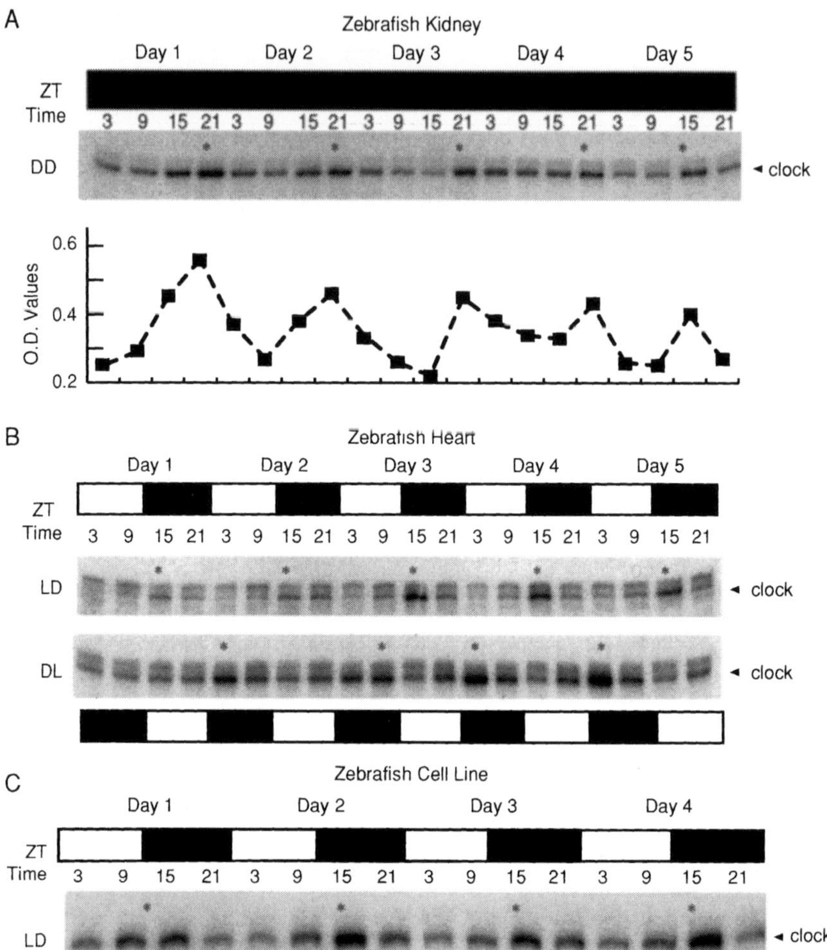

Fig. 6.1. Circadian oscillation in *Clock* gene transcript levels in a variety of tissues dissected from zebrafish maintained on 14:10 light-dark cycle (LD). (A) Clock oscillations in cultured zebrafish kidneys in constant darkness over a 5-day period. (B) Re-entrainment of cultured hearts placed on alternate light/dark cycles for 5 days. Hearts were dissected from an identical group of fish, and then placed onto out-of-phase light/dark cycles in "side-by-side" incubators, illuminated with a fiber-optic light source. Note that hearts in the DL reversed cycle re-entrain after only 1 day in the new lighting regime. (C) Clock oscillations in a zebrafish cell line when placed on a light-dark cycle. *Increased *Clock* expression. ZT time, "zeitgeber time," with ZT0 corresponding to lights on time; OD values, optical density measurements of *Clock* gene expression from (A). (Adapted from Tamai *et al.*, 2003.)

hormonal circadian signal to the entire organism, whereas the retina utilises its melatonin mostly for local, paracrine functions. In fish, this is further highlighted by the existence of two forms of the key enzyme for melatonin synthesis and differential expression of the genes encoding them in the pineal gland (AANAT 2) and retina (AANAT 1) (Benyassi *et al.*, 2000). This way, not only the physiological functions but also the timing of melatonin production by these organs may differ, as mentioned above.

In the retina, the circadian or light-dependent fluctuations in retinal dopamine and melatonin production define some of the critical light–dark adaptations in this major photoreceptive organ. For example, under a light–dark cycle or in constant darkness, dopamine release from isolated fish retina has a circadian pattern, being high during subjective day and low during subjective night (Li and Dowling, 2000). This daily fluctuation in retinal dopamine enables light adaptation in advance of the actual changes in illumination and can affect light sensitivity. Melatonin, acting via its specific receptors, defines this daily fluctuation in retinal dopamine levels, attenuating nighttime dopamine release (Figure 6.2), activating rod input and decreasing cone input through dopamine-mediated D2-like receptor activation (Ribelayga *et al.*, 2004).

The multiple physiological roles of daily variations in circulating melatonin and in the affinity of melatonin receptors from different tissues are only starting to emerge in all of their variety and complexity. Because, in the majority of fish species as well as other vertebrates, melatonin is secreted only at night and is suppressed by bright light, the duration of photoperiod is inversely related to the duration of melatonin production. While providing a circadian signal to multiple tissues and organs, melatonin affects such diverse processes as reproduction, locomotor activity, feeding, or sleep in fish.

Mating and spawning exhibit daily and seasonal variations, promoting offspring survival. Increased melatonin production, associated with short photoperiods, can significantly affect reproduction in fish (Zachmann *et al.*, 1992). In male sea bass, daily rhythms of pituitary LH positively correlate with the duration of melatonin production, and its daily rhythm of storage and release is altered by nighttime light exposure, which suppresses melatonin synthesis (Bayarri *et al.*, 2004). In masu salmon, a short photoperiod is known to stimulate gonadal maturation via activation of the brain-pituitary-gonadal axis. Melatonin treatment administered during long photoperiods, in such a way as to simulate melatonin profile during a short photoperiod, accelerates testicular development in this fish (Amano *et al.*, 2000). However, it is important to appreciate that these effects of melatonin depend on the

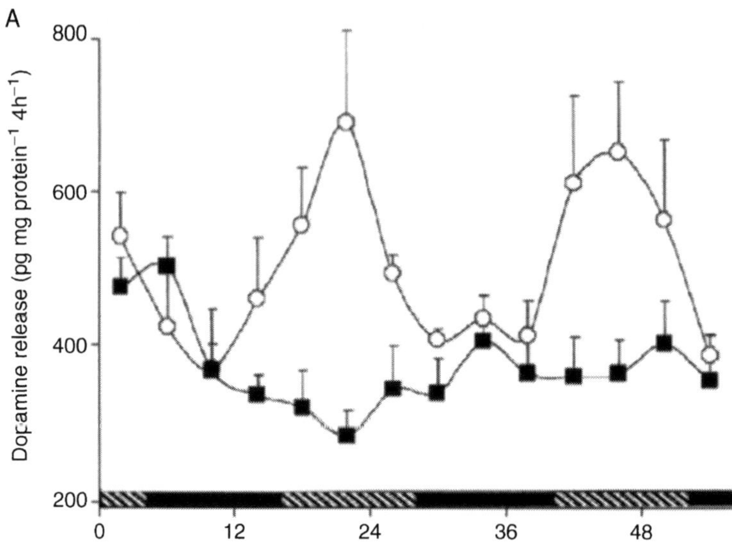

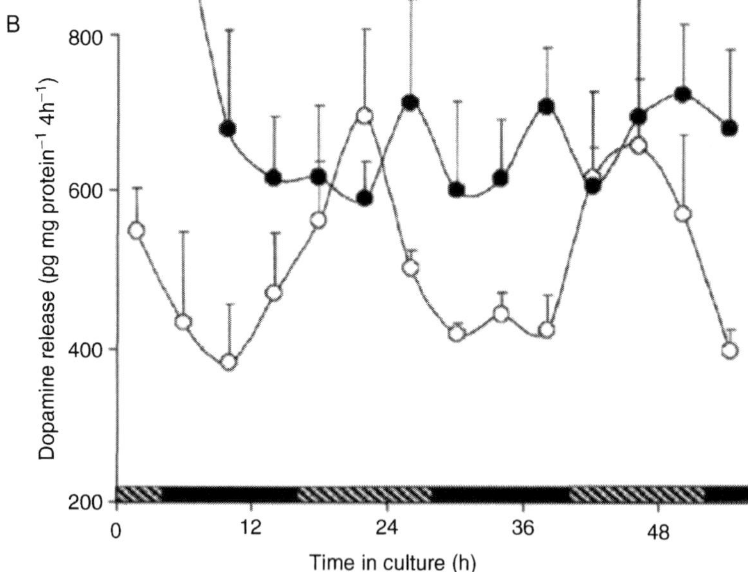

Fig. 6.2. The retinal clock uses melatonin to control the circadian release of dopamine. Isolated retinae were maintained for 56 hours in total darkness and constant temperature in a culture medium containing no drugs, melatonin (1 nM), or luzindole (1 μM). (A) Continuous application of melatonin (filled squares) abolished the rhythm of dopamine release by decreasing

maintenance of its normal circadian pattern. Increasing melatonin to supraphysiological levels and keeping them high throughout the day not only abolishes melatonin's promoting effect on reproduction but also lowers the gonadosomatic index and plasma testosterone levels in fish (Amano *et al.*, 2004).

The role of melatonin in fish locomotor activity and sleep has been studied in zebrafish (Zhdanova *et al.*, 2001). Similar to diurnal primates (Zhdanova *et al.*, 2002), melatonin treatment attenuates locomotor activity in this diurnal fish and induces a sleeplike state. This sleep-promoting effect of the hormone is mediated via specific melatonin receptors because melatonin receptor antagonists attenuate or block its effects (Figure 6.3). The location of the structures responsible for this effect of melatonin remains unknown.

Interestingly, melatonin administration has also been reported to influence food intake and diet selection in fish. In goldfish, intraperitoneal injection of melatonin inhibited food intake, whereas melatonin administration into the cerebral ventricles did not produce such effect (Pinillos *et al.*, 2001). Similarly, oral administration of melatonin inhibited total food intake and reduced carbohydrate intake in sea bass (Rubio *et al.*, 2004).

Knowledge of specific sites of melatonin actions in fish is critical for understanding its functions. Circadian patterns of melatonin production can be complemented by circadian variations in melatonin receptor density or affinity. In the goldfish brain, for example, a circadian variation in melatonin receptor (binding site) density has been documented under both entrained and constant conditions (i.e., under a light–dark cycle and in constant darkness), whereas the receptor affinity to its ligand remained stable (Iigo *et al.*, 2003b; Figure 6.4). It is not yet known whether other fish also exhibit daily or seasonal variations in melatonin receptor density or melatonin receptor affinity, and whether such variations might be coordinated in different tissues.

Further studies are needed to establish whether other circadian phenomena in fish—such as histamine levels in goldfish brain (Burns *et al.*, 2003), steroid levels in Japanese char (Yamada *et al.*, 2002), or electric discharges in gymnotid electric fish (Deng and Tseng, 2000)—are also under the control of melatonin production.

---

daytime levels to the nighttime values. (B) Continuous application of the selective melatonin receptor antagonist luzindole (filled circles) abolished the rhythm of dopamine release by increasing the nighttime values to the daytime levels. Each data point represents mean values ±S.E.M. for 5 independent retinae. Open circles represent a positive control performed at the same time, but with no test drugs added. Hatched and filled bars indicate the subjective day and night, respectively. (Adapted from Ribelayga *et al.*, 2004.)

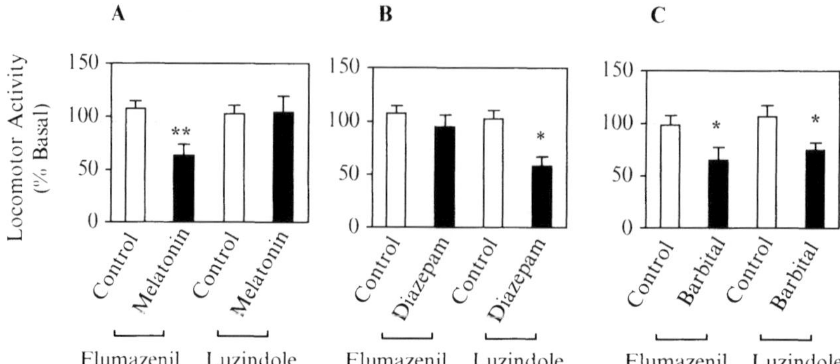

Fig. 6.3. Melatonin and diazepam affect locomotor activity in zebrafish via specific membrane receptors. Pretreatment with the specific antagonist for melatonin receptors, luzindole, blocked the decline in locomotor activity induced by (A) melatonin, but not by (B) diazepam or (C) pentobarbital. Pretreatment with specific benzodiazepine receptor antagonist, flumazenil, blocked reduction in locomotor activity following (B) diazepam, but not (A) melatonin or (C) pentobarbital treatment. Control solutions are vehicles for each treatment used. Data are expressed as mean ± S.E.M. group changes (%) in daytime locomotor activity, measured for 2 hours after treatment, relative to basal activity. $N = 30$ for each group. $^{**}P < 0.01$. (Adapted from Zhdanova *et al.*, 2001.)

4. ENTRAINMENT TO OTHER ENVIRONMENTAL CUES

In addition to the strong effect of light, other environmental factors such as water temperature and chemistry, food availability, social interaction, or predation risk could be considered as potential synchronisers. So far, there are only limited data available regarding these potentially entraining stimuli in fish. For example, daily variations in water chemistry (most notably oxygen content) have not been tried or reported as synchronisers of fish rhythms, at least to our knowledge.

4.1. Temperature

In homeothermic animals, temperature is considered to be a much weaker synchroniser than light. However, in poikilothermic vertebrates, ambient temperature can greatly influence metabolism and behaviour, including circadian periodicity (Rensing and Ruoff, 2002). Surprisingly, however, there seem to be no examples of entrainment by temperature cycles in fish. Only a few chronobiological studies in fish have mentioned temperature, the majority analysing it in conjunction with seasonal variations (e.g., Yokota and Oishi,

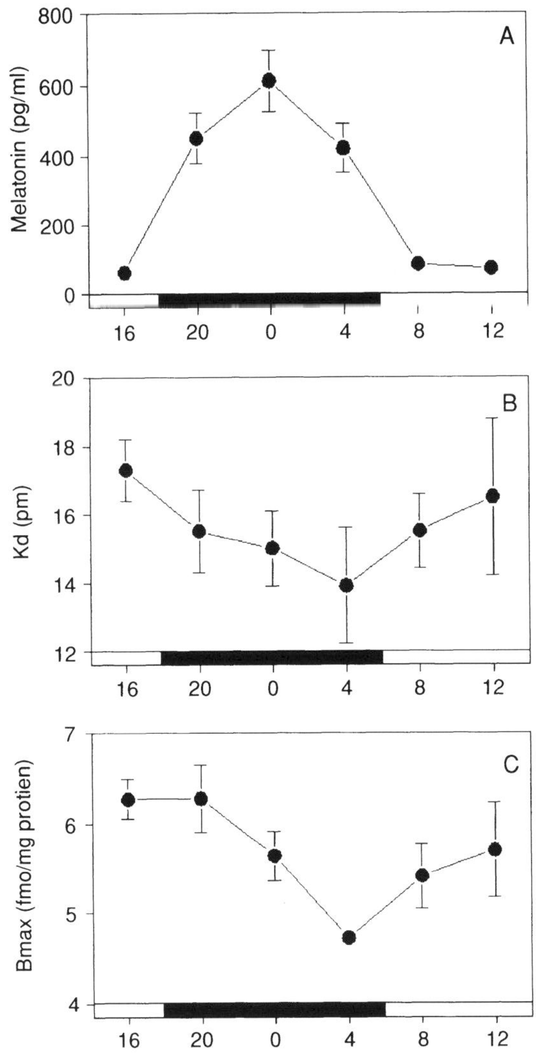

Fig. 6.4. Daily variations of melatonin levels in the plasma (A) and the K_d (B) and B_{max} (C) of melatonin-binding sites in the brain of the goldfish reared under LD 12:12. Solid and open bars along the X-axis represent the dark phase and the light phase, respectively. Plasma melatonin levels and the B_{max} of melatonin binding sites in the brain exhibited significant daily variations (ANOVA, $P < 0.05$) with no variation in the K_d. K_d, melatonin affinity for its receptor in radioligand binding experiments; B_{max}, receptor density. (Adapted from Iigo *et al.*, 2003b.)

1992; Aranda *et al.*, 1999a). For example, lake chubs captured in winter display longer free-running periods in constant darkness than individuals captured in summer (Kavaliers, 1978; Kavaliers and Ross, 1981). The reverse relationship is observed in burbot and longnose dace (Kavaliers, 1980e; 1981c). It is not clear whether this reflects an endogenous circannual rhythm of circadian periodicity or simply an aftereffect of temperature and/or photoperiod at the time of capture. Because both photoperiod and temperature change with seasons, it will be necessary to resort to an experimental approach to determine the relative influence of each in seasonal variation.

When studying the effects of temperature on circadian clocks, it is difficult to tell whether the effects represent true entrainment or simply a temperature-dependent change in metabolism that could affect clock outcomes, activity, or endocrine rhythms. For example, melatonin production can be modified by water temperature, producing a masking effect on its endogenous rhythm (Zachmann *et al.*, 1992; Max and Menaker, 1992; Iigo and Aida, 1995). Normal melatonin production by the pineal gland typically occurs during the night (i.e., during a colder phase of the 24-hour cycle). On the other hand, studies on isolated pineal glands from white sucker, lamprey, and sea bass show that low temperatures suppress melatonin production and attenuate its circadian rhythm (Zachmann *et al.*, 1992; Samejima *et al.*, 2000; Garcia-Allegue *et al.*, 2001; Masuda *et al.*, 2003). Similarly, changes in water temperature are associated with changes in the amplitude of fish plasma melatonin levels *in vivo* (Iigo and Aida, 1995; Garcia-Allegue *et al.*, 2001). Because melatonin provides a centralised circadian signal to all the cells of the organism, modification of this signal is one of the ways water temperature could affect the amplitude of circadian responses in fish. However, the extent to which such changes in melatonin secretion may affect the phase of circadian clocks under normal conditions is not yet clear.

4.2. Food Availability

When food delivery is restricted to the same time every day, fish, like other animals, display food-anticipatory activity (FAA) under a light–dark cycle (e.g., Spieler and Noeske, 1984; Laguë and Reebs, 2000a,b; Aranda *et al.*, 2001; Chen and Purser, 2001; Chen and Tabata, 2002; for a review, see Sánchez-Vázquez and Madrid, 2001). Even when maintained under constant lighting conditions, fish can rapidly synchronise their activity pattern to restricted food availability (e.g., Davis and Bardach, 1965; Gee *et al.*, 1994; Naruse and Oishi, 1994).

Studies in mammals suggest the existence of a separate food-dependent circadian oscillator(s) because SCN lesions do not abolish FAA. Fish may also have such a separate entraining pathway. For example, in medaka (Weber and Spieler, 1987) maintained under light–dark cycle and fed once a day, a daily rhythm of agonistic behaviour was entrained to the feeding schedule and persisted during a starvation period. In contrast, the daily rhythms of egg laying and courtship remained entrained to the light–dark cycle and were not affected by the time of food availability in this species. Similarly, trout held under constant light and a food restriction protocol displayed both a free-running rhythm, presumably defined by a light-dependent oscillator, and another one corresponding to the feeding schedule (Bolliet *et al.*, 2001). Goldfish have yielded data that suggest food could be as strong a synchroniser as light (Aranda *et al.*, 2001). Experiments with goldfish and sea bass have also provided data consistent with the existence of a food-entrainable oscillator (Sánchez-Vázquez *et al.*, 1995a,b; 1997).

Thus, light and food-dependent oscillators could coexist and interact with each other. Their relative power may vary between different fish species and this might depend on the habitual diurnal-nocturnal activity patterns and feeding habits. In a diurnally active trout, relying on its visual system to find food, light is a stronger zeitgeber; however, in a nocturnally active catfish, which uses its barbells to detect food, food availability has been shown to be a better entraining factor (Bolliet *et al.*, 2001).

It is also possible, however, to envision a circadian mechanism that would allow FAA based only on a light-entrainable oscillator. In such a model, the animal could store in memory a representation of the circadian phase at which meals are delivered. When clock time approaches the marked phase on a light-entrained oscillator, FAA would take place. Reebs and Laguë (2000) obtained some evidence for such a mechanism in golden shiners: FAA in trained fish disappeared when the light–dark cycle was removed, but persisted for a few days when the daily meals were withheld. Their results, however, could not eliminate the possibility of a food-entrainable oscillator linked to the light-entrainable one, disrupted by the damping out of the latter when the light–dark cycle was removed.

The mechanisms involved in FAA remain unknown, as well as the location(s) of any putative feeding-entrainable oscillator in fish or other animals. Perhaps multiple peripheral oscillators could be synchronised by food intake processes and/or energy availability, providing a powerful enough signal to entrain behavioural patterns. Interestingly, the properties of feeding-entrainable oscillators may depend more on meal size than on the amount of dietary energy supplied, suggesting that part of the FAA mechanism might involve gut distension (Sánchez-Vázquez *et al.*, 2001).

4.3. Social Environment

Many fish live in groups, ranging from small aggregates to shoals and schools (see Chapter 5). Social factors might affect the entraining properties of different synchronisers and define circadian patterns of activity and food intake.

When tested in laboratory conditions, groups of fish typically show more robust circadian rhythms than isolated animals, whether they are entrained by light or by a feeding schedule. This has been documented in killifish (Kavaliers, 1980d), white sucker (Kavaliers, 1980a), goldfish (Kavaliers, 1981a), and trout and catfish (Bolliet *et al.*, 2001). It is difficult to evaluate whether these instances reflect a direct effect of social factors on the circadian oscillator of fish, or more simply a social facilitation of movement that increases the chance of detection by the activity-recording devices.

In golden shiners, hungry individuals can discern the FAA of conspecifics and join them (Reebs and Gallant, 1997). They can also follow shoal leaders who know where and when food is available (Reebs, 2000). This attention to the activity of others raises the possibility that fish might display social entrainment of circadian activity and feeding rhythms, as has already been shown in various mammals (Bovet and Oertli, 1974; Crowley and Bovet, 1980; Marimuthu *et al.*, 1981). However, there have been no published attempts thus far to rigourously demonstrate social synchronisation of circadian activity rhythms in fish.

In an opposite vein, intraspecific competition for limited resources may also result in circadian rhythm fragmentation in fish groups. For example, trout were found to be almost exclusively diurnal while individually isolated, but they displayed both diurnal and nocturnal feeding behaviour soon after they were put in groups (Chen *et al.*, 2002a,b). The nocturnal feeding gradually decreased and disappeared after about 10 days of group housing. This phenomenon could be explained either by subordinate individuals initially, but not permanently, being forced to feed at an unfavorable time (night) or by an overall initial stress of resocialisation following isolation, which could temporarily disrupt the circadian rhythms in fish of different hierarchical ranks.

Thus, social environment per se may serve as an entraining signal but can also disrupt entrainment to other synchronisers. High plasticity of the fish circadian system, discussed below, may contribute to this phenomenon.

4.4. Predation Risk

Predation risk plays an important part in fishes' lives (Chapter 3; Smith, 1997). However, there is no evidence that predatory attacks occurring daily at the same clock time can entrain circadian rhythms. Time-place learning (the ability to associate specific locations with specific daily times, a process that is known to rely on circadian clocks for the estimation of time; see also Chapter 1) can develop in some fish when food is used as a reward (Reebs, 1996) but not when predatory attacks are simulated (Reebs, 1999). When threatened, fish seem to rely more on a general decrease in activity rather than shifting the timing of activity (e.g., Pettersson *et al.*, 2001). Given the abundance of fish predators in nature, probably active at all times of day and night as a whole, it may be unrealistic to expect that a mechanism could have evolved to entrain circadian oscillators to predation risk.

5. CIRCADIAN RHYTHMS AND DEVELOPMENT

External development in fish can be successful only if all the critical processes are provided by the egg (e.g., nutrition) or develop very early in ontogenesis. Based on this assumption, in fish the circadian system appears to be critical for the embryo's adaptation.

A study of the expression of one of the circadian clock genes, *Per3*, during zebrafish development showed a remarkable picture of an inherited functional circadian clock (Delaunay *et al.*, 2000). A circadian rhythm of *Per3* expression corresponding to that in the female was observed in unfertilised eggs, in the fertilised eggs prior to initiation of embryonic transcription, and in the developing embryos maintained under constant light or constant darkness. This rhythm can be entrained by light to either circadian or ultradian (e.g., light–dark 8:8 hours) periods. It should be noted, however, that the downstream effect of this clock (the expression of specific proteins under circadian regulation) seems to require embryonic development of the pineal gland and axonal connections. These data suggest that the products of the maternal clock gene play an important role in early fish development, along with the newly emerging circadian structures of the embryo itself.

Early onset of melatonin secretion by fish pineal glands, which occurs prior to hatching (Kazimi and Cahill, 1999; Roberts *et al.*, 2003; Danilova *et al.*, 2004), can provide a unifying neuroendocrine circadian signal, synchronising internal physiological processes with each other and with the periodically changing environmental conditions. This embryonic melatonin may both synchronise and potentiate development. In the zebrafish embryo, melatonin treatment can promote the S phase of the cell cycle and accelerate zebrafish growth and hatching via specific melatonin receptors, abundantly expressed starting 18 hours postfertilisation (Danilova *et al.*, 2004).

Circadian entrainment of young zebrafish larvae by light–dark cycle synchronises and enhances the S phase in the skin, heart, and gut, with peak levels observed around 3 hours before lights-on time (Dekens *et al.*, 2003).

These data further suggest that circadian rhythmicity plays an important role in early vertebrate development and may synchronise entry into the S phase and mitosis.

6. CIRCADIAN RHYTHMS AND PLASTICITY OF BEHAVIOURAL PATTERNS

The existence of a circadian clock within an animal imposes restrictions on the periodicity of this animal's activity, but not necessarily on the phasing of its activity relative to the light–dark cycle. Though this does not happen easily in humans, it is possible for some animals to live a diurnal life one day, and then switch to a nocturnal mode a few days later (or vice versa). Such cases are particularly numerous in fishes (for a review, see Reebs, 2002). Whether a fish is diurnal or nocturnal at any one time seems to depend mostly on food availability. For example, goldfish are diurnal when fed by day and nocturnal when fed at night (Spieler and Noeske, 1984; Gee *et al.*, 1994; Sánchez-Vázquez *et al.*, 1997) and can change the timing of their general activity when the timing of food availability suddenly changes (Aranda *et al.*, 2001; Sánchez-Vázquez *et al.*, 2001). Similarly, golden shiners can anticipate the arrival of food day or night, or during both day and night simultaneously (Laguë and Reebs, 2000b; for similar results in rainbow trout, see Chen and Tabata, 2002). Many species appear to be able to feed both visually and nonvisually (Diehl, 1988; Ehlinger, 1989; Collins and Hinch, 1993; Mussen and Peeke, 2001) and can vary their diet so that they are able to feed during the day or at night (Ebeling and Bray, 1976; Johnson and Dropkin, 1995; Pedersen, 2000) or forage at the time when food is more nutritious (Zoufal and Taborsky, 1991).

The choice between day- and night-living can also be influenced by competition, ontogeny, and light intensity, inasmuch as these factors all have a bearing on food availability. In addition to Chen *et al.* (2002a,b) already mentioned in Section 4.3. examples of intraspecific competition resulting in a temporal segregation of activity include Randolph and Clemens (1976) in channel catfish and Alanärä *et al.* (2001) in brown trout. Ontogenic shifts in activity patterns are widely documented (Reebs, 2002) and may reflect a strategy to alleviate competition between age classes, or a change in diet as the gape of a fish increases with growth; unfortunately, neither of these two hypotheses has been formally tested. The role of light intensity in helping or hindering food detection has been invoked to explain diurnal versus nocturnal activity in such studies as Løkkeborg (1998) and Thetmeyer (1997).

Predation risk may also determine whether a fish is diurnal or nocturnal, though the evidence here is weak, being inferential (there is no denying that one sensory modality is less available to predators during the night, and so it may pay for prey to be nocturnal) rather than experimental. One of the most interesting arguments involving predation risk to explain plastic activity rhythms in fish originated from a series of studies on Atlantic salmon. Fraser *et al.* (1993, 1995) experimentally confirmed that Atlantic salmon become proportionally more nocturnal when water temperatures are low, even when photoperiod is held constant. They argued that cold fish were more sluggish and felt more vulnerable to attacks by warm-blooded predators such as birds and mink (see also Chapters 3 and 5). Accordingly, such fish confined their activity to the relative safety of darkness at night. Warm fish could afford the risk of foraging by day and reap the benefit of better foraging due to good lighting conditions. Interestingly, in those individuals for whom foraging success was more important (e.g., salmon with low energy reserves or preparing to migrate at sea), the switch to nocturnalism in the cold was less pronounced, suggesting that the higher imperative for foraging tipped the balance towards increased diurnalism, despite the predation risk (Metcalfe *et al.*, 1998; see also Valdimarsson *et al.*, 1997; Metcalfe *et al.*, 1999; Johnston *et al.*, 2004).

Many other examples are known of switches between diurnal and nocturnal activity as summer makes way for winter (e.g., Andreasson, 1973; Müller, 1978; Cook and Bergersen, 1988; Clark and Green, 1990; Grant and Brown, 1998; Sánchez-Vázquez *et al.*, 1998; David and Closs, 2001, 2003). There is no solid evidence, however, to show whether these shifts are caused by changes in photoperiod, temperature, or food availability. One exception is research by Aranda *et al.* (1999a,b) on sea bass, in which experimental manipulations of photoperiod and temperature were tried. The results indicated that neither factor was at work, suggesting that the switch to nocturnal living shown by sea bass in winter is under the influence of a circannual clock.

It seems that freshwater species are more prone than marine ones to switch back and forth between diurnalism and nocturnalism (Reebs, 2002). Perhaps this reflects the fact that freshwater environments tend to be less stable than the sea, and therefore may have yielded fewer evolutionary cases of sensory specialisation for particular conditions, including daytime or nighttime. Being able to fare relatively well by day or by night, freshwater species can choose to be nocturnal as well as diurnal, depending on food availability and perhaps also temperature (Reebs, 2002).

The physiological mechanism that binds activity (permanently or not) to a particular phase of the circadian clock is unknown for any animal or plant taxon. Because of their flexibility in this regard, fish may represent a promising subject for future studies on this question.

7. CIRCADIAN RHYTHMS AND SLEEP

A rest–activity cycle, its regulation by the internal circadian clock, and its entrainment to environmental cues are evolutionary conserved phenomena. However, in higher vertebrates, rest is associated with sleep, a distinct behavioural and physiological state. Brain activity during sleep shows recognisable electrophysiological patterns significantly different from those observed during wakefulness. Moreover, sleep in mammals is not a uniform process but a series of alternating states, called "slow wave" (SWS) and "rapid eye movement" (REM) sleep. In addition to differences in brain activity during SWS and REM sleep, the peripheral organs and systems are in a different functional mode.

So far, the physiological function of sleep and its principal molecular mechanisms remain obscure. However, a substantial body of evidence on the regulation of sleep processes identifies two major forces driving sleep initiation and maintenance. They include a homeostatic control, defining an increase in sleep propensity proportional to the time spent awake, and a circadian control, determining a window of high sleep propensity or of increased wakefulness within each 24-hour period. The circadian mechanisms may synchronise sleep with daytime or nighttime, depending on whether the animal is nocturnal or diurnal.

Typical behavioural features of sleep include prolonged behavioural quietness, species-specific postures, elevated arousal threshold, and rapid reversibility from quietness to activity in response to moderately intense stimulation (Campbell and Tobler, 1984). These features are common to many fish species, with some completely ceasing their activity and others only slowing it down (Clark, 1973; Tauber, 1974; Shapiro and Hepburn, 1976; Karmanova, 1981; for reviews, see Reebs, 1992, 2002). The increase in arousal threshold associated with such sleeplike state can be substantial. For example, blueheads, Spanish hogfish, and several species of wrasses could be lifted by hand to the surface at night before "waking up" (Tauber, 1974).

Zebrafish larvae (7–14 days old) demonstrate sleeplike behaviour even at an early developmental stage (Zhdanova *et al.*, 2001). Long periods of immobility in zebrafish larvae are associated with either floating with the head down or staying in a horizontal position close to the bottom of a tank. A nocturnal decline in motor activity in these diurnal fish is accompanied by a significant increase in arousal threshold relative to daytime

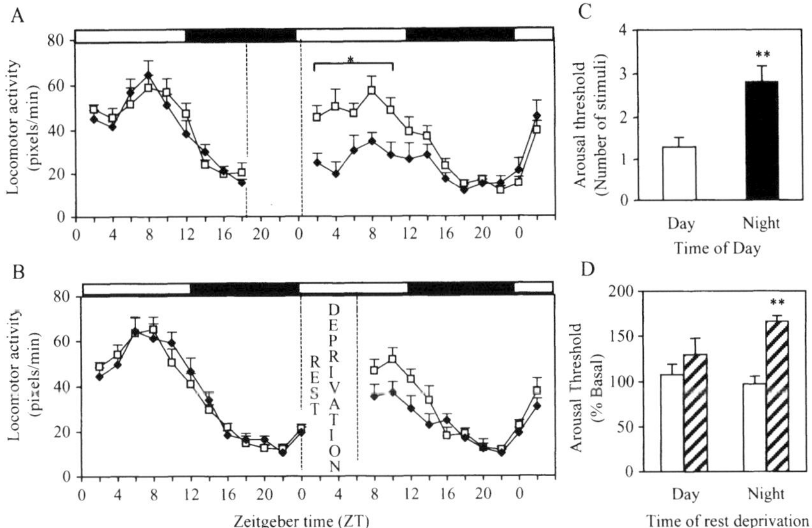

Fig. 6.5. Daily variation in locomotor activity and arousal threshold in larval zebrafish maintained in constant darkness and a compensatory reduction in locomotor activity and increase in arousal threshold following rest deprivation. (A, B) Zeitgeber time (ZT) and horizontal white/black bars indicate subjective day versus subjective night, according to 12:12 light–dark cycle prior to the beginning of recording, with ZT0 corresponding to lights on time. Each data point represents mean ± S.E.M. group locomotor activity for preceding 2 hours of recording (computerised image analysis data in pixels per minute). $N = 60$ for each group. The rest deprivation was scheduled either (A) during subjective night (ZT18–ZT24) or (B) during subjective day (ZT0–ZT6). Closed diamonds, rest deprivation group; open squares, control group. (C) Arousal threshold was measured in constant darkness during subjective day (ZT3–5) or subjective night (ZT15–17). $N = 20$ for each group. (D) Changes in daytime arousal threshold (% of basal) starting an hour after daytime or nighttime rest deprivation. White bars, control; striped bars, rest deprivation. $N = 20$ for each group. $*P < 0.05$; $**P < 0.01$. (Adapted from Zhdanova *et al.*, 2001.)

(Figure 6.5) and is maintained in constant darkness, confirming that sleeplike state in larval zebrafish is under the control of a circadian system.

As in mammals, preventing fish from resting or sleeping results in subsequent increases in sleeplike behaviour, as has been shown in cichlids (Tobler and Borbely, 1985) and zebrafish (Zhdanova *et al.*, 2001). Daytime rest deprivation in zebrafish larvae does not significantly affect rest behaviour thereafter, though some decrease in daytime locomotor activity and elevation in the arousal threshold have been noticed (Figure 6.5, B and D). In contrast, nighttime rest deprivation resulted in a significant decline in daytime locomotor activity and in a heightened arousal threshold, compared

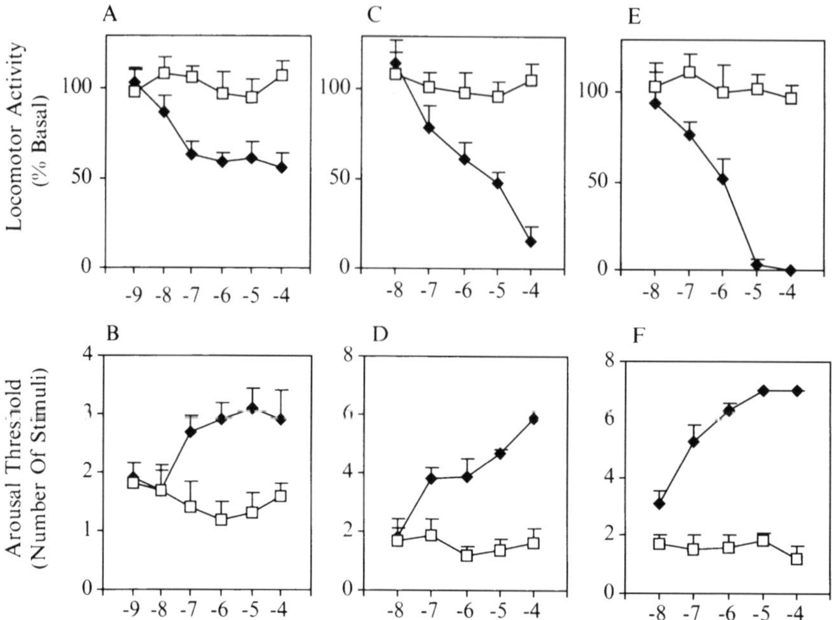

Fig. 6.6. Melatonin and conventional sedatives promote rest behaviour in larval zebrafish. Melatonin, diazepam, and sodium pentobarbital (barbital) significantly and dose-dependently reduced zebrafish locomotor activity (A, C, E) and increased arousal threshold (B, D, F). Each data point represents mean ± SEM group changes in a 2-hour locomotor activity relative to basal activity, measured in each treatment or control group for 2 hours prior to treatment administration. Arousal threshold data are expressed as the mean ± SEM group number of stimuli necessary to initiate locomotion in a resting fish. Closed diamond, treatment; open square, vehicle control. $N = 20$ for each group. (Adapted from Zhdanova *et al.*, 2001.)

to basal recordings. These data further confirm that sleeplike states in fish are under both circadian and homeostatic control.

Furthermore, a sleeplike state can be induced in fish with common hypnotic agents, such as benzodiazepines or barbiturates (Figure 6.6), suggesting that at least some of the mechanisms of sleep regulation must be similar in fish and mammals. Diurnal zebrafish are also sensitive to the sleep-promoting effects of melatonin (Figure 6.6), similar to diurnal monkeys and humans.

The lower vertebrates have obvious anatomical limitations for the expression of some of the electrographic patterns commonly found in mammalian sleep. For example, it is unlikely that animals such as fish with only a rudimentary neocortex, if any, would generate the brain waves characteristic of mammalian slow-wave sleep. However, the neurons with sleep-promoting properties are contained in structures within the isodendritic core of the brain, extending from the medulla through the brainstem, hypothalamus, and up into the basal forebrain (see Chapter 1 for details on brain anatomy; for a review, see Jones, 1993). Thus the lower vertebrates, with their "limbic brain," possess most of the brain structures that have been found to regulate the sleep process in mammals.

New methodological approaches, such as monitoring the expression of genes related to spontaneous sleep processes or inducing sleep by hypnotic agents, might help clarify the extent to which sleeplike states in fish are similar to those in mammals.

8. SUMMARY

Circadian rhythms play a critical role in fish development and daily activities. Although a major circadian "master" clock, like the SCN of mammals, has not yet been identified in fish, indirect evidence suggests that a light-entrainable oscillator is present in fish brain. Furthermore, the structural and functional design of fish circadian systems is remarkably complicated. Photosensitive CNS-related clock organs (the pineal gland and retina), peripheral photosensitive tissues with autonomic circadian clocks, and presumed food- or temperature-entrainable circadian oscillator(s) all make for complex circadian machinery that must remain well coordinated and still be able to ensure physiological adaptation to a periodically changing environment.

Such a multilevel structure of partially independent oscillators may explain the high interspecies variability observed in piscine circadian systems and substantial individual plasticity in fish behaviour and physiology. Studying these features will continue to contribute to a better understanding of the principal mechanisms involved in circadian clock functions.

Data accumulated so far show that rest in fish has fundamental similarities to the behavioural manifestations of sleep in higher vertebrates. Analogous to sleep in mammals, fish show a compensatory rest rebound, reducing locomotor activity and increasing arousal thresholds after a period of rest deprivation, suggesting that fish exert a homeostatic control on rest behaviour. Furthermore, rest in fish is regulated by the circadian system, because periodic reduction in locomotor activity and increase in arousal threshold are maintained in constant darkness and occur during the subjective night. These observations, together with the hypnotic effects of melatonin and sleep-inducing agents of the benzodiazepine and barbiturate families, indicate that rest behaviour in fish can be considered a sleeplike state. Studying sleep in fish may prove to be very productive in deciphering both the enigmatic function and the physiological mechanisms of sleep.

ACKNOWLEDGEMENTS

Some of the authors' own work was supported by the National Institutes of Health (I.Z.: MH 65528, DA 15418) and the Natural Sciences and Engineering Research Council of Canada (S.R.)

REFERENCES

Alanärä, A., Burns, M. D., and Metcalfe, N. B. (2001). Intraspecific resource partitioning in brown trout: The temporal distribution of foraging is determined by social rank. *J. Anim. Ecol.* **70,** 980–986.

Amano, M., Iigo, M., Ikuta, K., Kitamura, S., Okuzawa, K., Yamada, H., and Yamamori, K. (2000). Roles of melatonin in gonadal maturation of underyearling precocious male masu salmon. *Gen. Comp. Endocrinol.* **120,** 190–197.

Amano, M., Iigo, M., Ikuta, K., Kitamura, S., and Yamamori, K. (2004). Disturbance of plasma melatonin profile by high dose melatonin administration inhibits testicular maturation of precocious male masu salmon. *Zool. Sci.* **21,** 79–85.

Andreasson, S. (1973). Seasonal changes in diel activity of *Cottus poecilopus* and *C. gobio* (Pisces) at the Arctic Circle. *Oikos* **24,** 16–23.

Aranda, A., Sánchez-Vázquez, F. J., and Madrid, J. A. (1999a). Influence of water temperature on demand-feeding rhythms in sea bass. *J. Fish Biol.* **55,** 1029–1039.

Aranda, A., Madrid, J. A., Zamora, A., and Sánchez-Vázquez, F. J. (1999b). Synchronising effect of photoperiod on the dual phasing of demand-feeding rhythms in sea bass. *Biol. Rhythm Res.* **30,** 392–406.

Aranda, A., Madrid, J. A., and Sánchez-Vázquez, F. J. (2001). Influence of light on feeding anticipatory activity in goldfish. *J. Biol. Rhythms* **16,** 50–57.

Bayarri, M. J., Rol de Lama, M. A., Madrid, J. A., and Sánchez-Vázquez, F. J. (2003). Both pineal and lateral eyes are needed to sustain daily circulating melatonin rhythms in sea bass. *Brain Res.* **969,** 175–182.

Bayarri, M. J., Rodriguez, L., Zanuy, S., Madrid, J. A., Sánchez-Vázquez, F. J., Kagawa, H., Okuzawa, K., and Carrillo, M. (2004). Effect of photoperiod manipulation on the daily rhythms of melatonin and reproductive hormones in caged European sea bass (*Dicentrarchus labrax*). *Gen. Comp. Endocrinol.* **136,** 72–81.

Begay, V., Falcon, J., Cahill, G. M., Klein, D. C., and Coon, S. L. (1998). Transcripts encoding two melatonin synthesis enzymes in the teleost pineal organ: Circadian regulation in pike and zebrafish, but not in trout. *Endocrinology* **139,** 905–912.

Bellingham, J., Whitmore, D., Philp, A. R., Wells, D. J., and Foster, R. G. (2002). Zebrafish melanopsin: Isolation, tissue localisation and phylogenetic position. *Brain Res. Mol. Brain Res.* **107,** 128–136.

Benyassi, A., Schwartz, C., Coon, S. L., Klein, D. C., and Falcon, J. (2000). Melatonin synthesis: Arylalkylamine N-acetyltransferases in trout retina and pineal organ are different. *Neuroreport* **11,** 255–258.

Bolliet, V., Begay, V., Taragnat, C., Ravault, J. P., Collin, J.P, and Falcon, J. (1997). Photoreceptor cells of the pike pineal organ as cellular circadian oscillators. *Eur. J. Neurosci.* **9,** 643–653.

Bolliet, V., Aranda, A., and Boujard, T. (2001). Demand-feeding rhythm in rainbow trout and European catfish: Synchronisation by photoperiod and food availability. *Physiol. Behav.* **73,** 625–633.

Boujard, T., Gélineau, A., Corraze, G., Kaushik, S., Gasset, E., Coves, D., and Dutto, G. (2000). Effect of dietary lipid content on circadian rhythm of feeding activity in European sea bass. *Physiol. Behav.* **68,** 683–689.

Bovet, J., and Oertli, E. F. (1974). Free-running circadian activity rhythms in free-living beaver (Castor canadensis). *J. Comp. Physiol.* **92,** 1–10.

Burdeyron, H., and Buisson, B. (1982). On a Circadian Endogenous Locomotor Rhythm of Loaches (*Noemacheilus barbatulus* L., Pisces, Cobitidae). *Zool. Jb. Physiol.* **86,** 82–89.

Burns, T. A., Huston, J. P., and Spieler, R. E. (2003). Circadian variation of brain histamine in goldfish. *Brain Res. Bull.* **59,** 299–301.

Cahill, G. M. (2002). Clock mechanisms in zebrafish. *Cell Tissue Res.* **309,** 27–34.

Campbell, S. S., and Tobler, I. (1984). Animal sleep: A review of sleep duration across phylogeny. *Neurosci. Biobehav. Rev.* **8,** 269–300.

Cermakian, N., and Sassone-Corsi, P. (2002). Environmental stimulus perception and control of circadian clocks. *Curr. Opin. Neurobiol.* **12,** 359–365.

Cermakian, N., Whitmore, D., Foulkes, N. S., and Sassone-Corsi, P. (2000). Asynchronous oscillations of two zebrafish CLOCK partners reveal differential clock control and function. *Proc. Natl. Acad. Sci. USA* **97,** 4339–4344.

Chen, W. M., and Purser, G. J. (2001). The effect of feeding regime on growth, locomotor activity pattern and the development of food anticipatory activity in greenback flounder. *J. Fish Biol.* **58,** 177–187.

Chen, W. M., and Tabata, M. (2002). Individual rainbow trout *Oncorhynchus mykiss* Walbaum can learn and anticipate multiple daily feeding times. *J. Fish Biol.* **61,** 1410–1422.

Chen, W. M., Naruse, M., and Tabata, M. (2002a). The effect of social interactions on circadian self-feeding rhythms in rainbow trout *Oncorhynchus mykiss* Walbaum. *Physiol. Behav.* **76,** 281–287.

Chen, W., Naruse, M., and Tabata, M. (2002b). Circadian rhythms and individual variability of self-feeding activity in groups of rainbow trout *Oncorhynchus mykiss* (Walbaum). *Aquac. Res.* **33,** 491–500.

Chiu, J. F., Mack, A. F., and Fernald, R. D. (1995). Daily rhythm of cell proliferation in the teleost retina. *Brain Res.* **673,** 119–125.

Clark, D. S., and Green, J. M. (1990). Activity and movement patterns of juvenile Atlantic cod, *Gadus morhua*, in Conception Bay, Newfoundland, as determined by sonic telemetry. *Can. J. Zool.* **68,** 1434–1442.

Clark, E. (1973). "Sleeping" sharks in Mexico. *Underwat. Nat.* **8,** 4–7.

Colgan, P. (1975). Self-selection of photoperiod as a technique for studying endogenous rhythms in fish. *J. Interdiscipl. Cycle Res.* **6,** 203–211.

Collins, N. C., and Hinch, S. G. (1993). Diel and seasonal variation in foraging activities of pumpkinseeds in an Ontario pond. *Trans. Am. Fish. Soc.* **122,** 357–365.

Cook, M. F., and Bergersen, E. P. (1988). Movements, habitat selection, and activity periods of northern pike in Eleven Mile Reservoir, Colorado. *Trans. Am. Fish. Soc.* **117,** 495–502.

Coon, S. L., Begay, V., Deurloo, D., Falcon, J., and Klein, D. C. (1999). Two arylalkylamine N-acetyltransferase genes mediate melatonin synthesis in fish. *J. Biol. Chem.* **274,** 9076–9082.

Crowley, M., and Bovet, J. (1980). Social synchronisation of circadian rhythms in deer mice (*Peromyscus maniculatus*). *Behav. Ecol. Sociobiol.* **7,** 99–105.

Cummings, S. M., and Morgan, E. (2001). Time-keeping system of the eel pout, *Zoarces viviparus*. *Chronobiol. Int.* **18,** 27–46.

Danilova, N., Krupnik, V. E., Sugden, D., and Zhdanova, I. V. (2004). Melatonin stimulates cell proliferation in zebrafish embryo and accelerates its development. *FASEB J.* **18,** 751–753.

David, B. O., and Closs, G. P. (2001). Continuous remote monitoring of fish activity with restricted home ranges using radiotelemetry. *J. Fish Biol.* **59,** 705–715.

David, B. O., and Closs, G. P. (2003). Seasonal variation in diel activity and microhabitat use of an endemic New Zealand stream-dwelling galaxiid fish. *Freshwater Biol.* **48,** 1765–1781.

Davis, R. E. (1964). Daily "predawn" peak of locomotion in fish. *Anim. Behav.* **12,** 272–283.

Davis, R. E., and Bardach, J. E. (1965). Time-co-ordinated prefeeding activity in fish. *Anim. Behav.* **13,** 154–162.

Dekens, M. P., Santoriello, C., Vallone, D., Grassi, G., Whitmore, D., and Foulkes, N. S. (2003). Light regulates the cell cycle in zebrafish. *Curr. Biol.* **13,** 2051–2057.

Delaunay, F., Thisse, C., Marchand, O., Laudet, V., and Thisse, B. (2000). An inherited functional circadian clock in zebrafish embryos. *Science* **289,** 297–300.

Delaunay, F., Thisse, C., Marchand, O., Laudet, V., Thisse, B., Delaunay, F., Thisse, C., Thisse, B., and Laudet, V. (2003). Differential regulation of Period 2 and Period 3 expression during development of the zebrafish circadian clock. *Gene Expr. Patterns* **3,** 319–324.

Deng, T. S., and Tseng, T. C. (2000). Evidence of circadian rhythm of electric discharge in Eigenmannia virescens system. *Chronobiol. Int.* **17,** 43–48.

Diehl, S. (1988). Foraging efficiency of three freshwater fishes: Effects of structural complexity and light. *Oikos* **53,** 207–214.

Drivenes, O., Soviknes, A. M., Ebbesson, L. O., Fjose, A., Seo, H. C., and Helvik, J. V. (2003). Isolation and characterisation of two teleost melanopsin genes and their differential expression within the inner retina and brain. *J. Comp. Neurol.* **456,** 84–93.

Dunlap, J. C. (1999). Molecular bases for circadian clocks. *Cell* **96,** 271–290.

Ebeling, A. W., and Bray, R. N. (1976). Day versus night activity of reef fishes in a kelp forest off Santa Barbara, California. *Fish. Bull.* **74,** 703–717.

Ehlinger, T. J. (1989). Foraging mode switches in the golden shiner (*Notemigonus crysoleucas*). *Can. J. Fish. Aquat. Sci.* **46,** 1250–1254.

Erckens, W., and Martin, W. (1982a). Exogenous and endogenous control of swimming activity in *Astyanax mexicanus* (Characidae, Pisces) by direct light response and by a circadian oscillator. I. Analyses of the time-control systems of an epigean river population. *Z. Naturforsch.* **37,** 1253–1265.

Erckens, W., and Martin, W. (1982b). Exogenous and endogenous control of swimming activity in *Astyanax mexicanus* (Characidae, Pisces) by direct light response and by a circadian oscillator. II. Features of time-controlled behaviour of a cave population and their comparison to an epigean ancestral form. *Z. Naturforsch.* **37,** 1266–1273.

Erckens, W., and Weber, F. (1976). Rudiments of an ability for time measurement in the cavernicolous fish *Anoptichthys jordani* Hubbs and Innes (Pisces Characidae). *Experientia* **32,** 1297–1299.

Eriksson, L.-O., and Van Veen, T. (1980). Circadian rhythms in the brown bullhead, *Ictalurus nebulosus* (Teleostei): Evidence for an endogenous rhythm in feeding, locomotor, and reaction time behaviour. *Can. J. Zool.* **58,** 1899–1907.

Falcon, J. (1999). Cellular circadian clocks in the pineal. *Prog. Neurobiol.* **58,** 121–162.

Falcon, J., Gothilf, Y., Coon, S. L., Boeuf, G., and Klein, D. C. (2003). Genetic, temporal and developmental differences between melatonin rhythm generating systems in the teleost fish pineal organ and retina. *J. Neuroendocrinol.* **15,** 378–382.

Foster, R. G., Hankins, M., Lucas, R. J., Jenkins, A., Munoz, M., Thompson, S., Appleford, J. M., and Bellingham, J. (2003). Non-rod, non-cone photoreception in rodents and teleost fish. *Novartis Found. Symp.* **253,** 3–23.

Fraser, N. H. C., Metcalfe, N. B., and Thorpe, J. E. (1993). Temperature-dependent switch between diurnal and nocturnal foraging in salmon. *Proc. R. Soc. Lond. B* **252,** 135–139.

Fraser, N. H. C., Heggenes, J., Metcalfe, N. B., and Thorpe, J. E. (1995). Low summer temperatures cause juvenile Atlantic salmon to become nocturnal. *Can. J. Zool.* **73,** 446–451.

García-Allegue, R., Madrid, J. A., and Sánchez-Vázquez, F. J. (2001). Melatonin rhythms in European sea bass plasma and eye: Influence of seasonal photoperiod and water temperature. *J. Pineal Res.* **31,** 68–75.

Garg, S. K., and Sundararaj, B. I. (1986). Role of pineal in the regulation of some aspects of circadian rhythmicity in the catfish, *Heteropneustes fossilis* (Bloch). *Chronobiologia* **13,** 1–11.

Gee, P., Stephenson, D., and Wright, D. E. (1994). Temporal discrimination learning of operant feeding in goldfish (*Carassius auratus*). *J. Exp. Anal. Behav.* **62,** 1–13.

Gerkema, M. P., Videler, J. J., de Wiljes, J., van Lavieren, H., Gerritsen, H., and Karel, M. (2000). Photic entrainment of circadian activity patterns in the tropical labrid fish *Halichoeres chrysus. Chronobiology International* **17,** 613–622.

Gern, W. A., and Greenhouse, S. S. (1988). Examination of *in vitro* melatonin secretion from superfused trout (Salmo gairdneri) pineal organs maintained under diel illumination or continuous darkness. *Gen. Comp. Endocrinol.* **71,** 163–174.

Gibson, R. N. (1971). Factors affecting the rhythmic activity of *Blennius pholis* L. (Teleostei). *Anim. Behav.* **19,** 336–343.

Glossop, N. R., Lyons, L. C., and Hardin, P. E. (1999). Interlocked feedback loops within the Drosophila circadian oscillator. *Science* **286,** 766–768.

Godin, J.-G. J. (1981). Circadian rhythm of swimming activity in juvenile pink salmon (*Oncorhynchus gorbuscha*). *Mar. Biol.* **64,** 341–349.

Goudie, C. A., Davis, K. B., and Simco, B. A. (1983). Influence of the eyes and pineal gland on locomotor activity patterns of channel catfish *Ictalurus punctatus. Physiol. Zool.* **56,** 10–17.

Grant, S. M., and Brown, J. A. (1998). Diel foraging cycles and interactions among juvenile Atlantic cod (*Gadus morhua*) at a nearshore site in Newfoundland. *Can. J. Fish. Aquat. Sci.* **55,** 1307–1316.

Harden Jones, F. R. (1956). The behaviour of minnows in relation to light intensity. *J. Exp. Biol.* **33,** 271–281.

Herrero, M. J., Madrid, J. A., and Sánchez-Vázquez, F. J. (2003). Entrainment to light of circadian activity rhythms in tench (*Tinca tinca*). *Chronobiol. Int.* **20,** 1001–1017.

Hurd, M. W., and Cahill, G. M. (2002). Entraining signals initiate behavioural circadian rhythmicity in larval zebrafish. *J. Biol. Rhythms* **17,** 307–314.

Iigo, M., and Aida, K. (1995). Effects of season, temperature, and photoperiod on plasma melatonin rhythms in the goldfish, *Carassius auratus. J. Pineal Res.* **18,** 62–68.

Iigo, M., and Tabata, M. (1996). Circadian rhythms of locomotor activity in the goldfish *Carassius auratus. Physiol. Behav.* **60,** 775–781.

Iigo, M., Hara, M., Ohtani-Kaneko, R., Hirata, K., Tabata, M., and Aida, K. (1997a). Photic and circadian regulations of melatonin rhythms in fishes. *Biol. Signals* **6,** 225–232.

Iigo, M., Sanchez-Vazquez, F. J., Madrid, J. A., Zamora, S., and Tabata, M. (1997b). Unusual responses to light and darkness of ocular melatonin in European sea bass. *Neuroreport* **8,** 1631–1635.

Iigo, M., Mizusawa, K., Yokosuka, M., Hara, M., Ohtani-Kaneko, R., Tabata, M., Aida, K., and Hirata, K. (2003a). *In vitro* photic entrainment of the circadian rhythm in melatonin release from the pineal organ of a teleost, ayu (Plecoglossus altivelis) in flow-through culture. *Brain Res.* **932,** 131–135.

Iigo, M., Furukawa, K., Tabata, M., and Aida, K. (2003b). Circadian variations of melatonin binding sites in the goldfish brain. *Neurosci. Lett.* **347,** 49–52.

Iigo, M., Fujimoto, Y., Gunji-Suzuki, M., Yokosuka, M., Hara, M., Ohtani-Kaneko, R., Tabata, M., Aida, K., and Hirata, K. (2004). Circadian rhythm of melatonin release from the photoreceptive pineal organ of a teleost, ayu (*Plecoglossus altivelis*) in flow-thorough culture. *J. Neuroendocrinol.* **16,** 45–51.

Ishikawa, T., Hirayama, J., Kobayashi, Y., and Todo, T. (2002). Zebrafish CRY represses transcription mediated by CLOCK-BMAL heterodimer without inhibiting its binding to DNA. *Genes Cells* **7,** 1073–1086.

Jenkins, A., Munoz, M., Tarttelin, E. E., Bellingham, J., Foster, R. G., and Hankin, M. W. (2003). VA opsin, melanopsin, and an inherent light response within retinal interneurons. *Curr. Biol.* **13,** 1269–1278.

Johnson, J. H., and Dropkin, D. S. (1995). Diel feeding chronology of six fish species in the Juniata River, Pennsylvania. *J. Freshwat. Ecol.* **10,** 11–18.

Johnston, P., Bergeron, N. E., and Dodson, J. J. (2004). Diel activity patterns of juvenile Atlantic salmon in rivers with summer temperatures near the temperature-dependent suppression of diurnal activity. *J. Fish Biol.* **65,** 1305–1318.

Jones, B. E. (1993). The organisation of central cholinergic systems and their functional importance in sleep-waking states. *Prog. Brain Res.* **98,** 61–71.

Kabasawa, H., and Ooka-Souda, S. (1989). Circadian rhythms in locomotor activity of the hagfish, *Eptatretus burgeri* (IV). The effect of eye-ablation. *Zool. Sci.* **6,** 135–139.

Katz, H. M. (1978). Circadian rhythms in juvenile American shad, *Alosa sapidissima. J. Fish Biol.* **12,** 609–614.

Karmanova, I. G., Khomutetskaia, O. E., and Shilling, N. V. (1981). Comparative physiologic analysis of the evolutionary stages of sleep and its regulatory mechanisms. *Usp. Fiziol. Nauk.* **12,** 3–19.

Kavaliers, M. (1978). Seasonal changes in the circadian period of the lake chub, *Couesius plumbeus. Can. J. Zool.* **56,** 2591–2596.

Kavaliers, M. (1979). Pineal involvement in the control of circadian rhythmicity in the lake chub, *Couesius plumbeus. J. Exp. Zool.* **209,** 33–40.

Kavaliers, M. (1980a). Circadian activity of the white sucker, *Catostomus commersoni*: Comparison of individual and shoaling fish. *Can. J. Zool.* **58,** 1399–1403.

Kavaliers, M. (1980b). Pineal control of ultradian rhythms and short-term activity in a cyprinid fish, the lake chub, *Couesius plumbeus. Behav. Neur. Biol.* **29,** 224–235.

Kavaliers, M. (1980c). Retinal and extraretinal entrainment action spectra for the activity rhythms of the lake chub, *Couesius plumbeus. Behav. Neur. Biol.* **30,** 56–67.

Kavaliers, M. (1980d). Social groupings and circadian activity of the killifish, *Fundulus heteroclitus. Biol. Bull.* **158,** 69–76.

Kavaliers, M. (1980e). Circadian locomotor activity rhythms of the burbot, *Lota lota*: Seasonal differences in period length and the effect of pinealectomy. *J. Comp. Physiol.* **136,** 215–218.

Kavaliers, M. (1981a). Period lengthening and disruption of socially facilitated circadian activity rhythms of goldfish by lithium. *Physiol. Behav.* **27,** 625–628.

Kavaliers, M. (1981b). Circadian organisation in white suckers *Catostomus commersoni*: The role of the pineal organ. *Comp. Biochem. Physiol.* **68A,** 127–129.

Kavaliers, M. (1981c). Seasonal effects on the freerunning rhythm of circadian activity of longnose dace (*Rhinichthys cataractae*). *Env. Biol. Fish* **6,** 203–206.

Kavaliers, M. (1981d). Circadian rhythm of nonpineal extraretinal photosensitivity in a teleost fish, the lake chub, *Couesius plumbeus*. *J. Exp. Zool.* **216,** 7–11.

Kavaliers, M., and Abbott, F. S. (1977). Rhythmic colour change of the killifish, *Fundulus heteroclitus*. *Can. J. Zool.* **55,** 553–561.

Kavaliers, M., and Ross, D. M. (1981). Twilight and day length affects the seasonality of entrainment and endogenous circadian rhythms in a fish, *Couesius plumbeus*. *Can. J. Zool.* **59,** 1326–1334.

Kazimi, N., and Cahill, G. M. (1999). Development of a circadian melatonin rhythm in embryonic zebrafish. *Brain Res. Dev. Brain Res.* **117,** 47–52.

Kim, W. S., Kim, J. M., Yi, S. K., and Huh, H. T. (1997). Endogenous circadian rhythm in the river puffer fish *Takifugu obscurus*. *Mar. Ecol. Prog. Ser.* **153,** 293–298.

Kleerekoper, H., Taylor, G., and Wilton, R. (1961). Diurnal periodicity in the activity of *Petromyzon marinus* and the effects of chemical stimulation. *Trans. Amer. Fish Soc.* **90,** 73–78.

Kobayashi, Y., Ishikawa, T., Hirayama, J., Daiyasu, H., Kanai, S., Toh, H., Fukuda, I., Tsujimura, T., Terada, N., Kamei, Y., Yuba, S., Iwai, S., and Todo, T. (2000). Molecular analysis of zebrafish photolyase/cryptochrome family: Two types of cryptochromes present in zebrafish. *Genes Cells* **5,** 725–738.

Kojima, D., Mano, H., and Fukada, Y. (2000). Vertebrate ancient-long opsin: A green-sensitive photoreceptive molecule present in zebrafish deep brain and retinal horizontal cells. *J. Neurosci.* **20,** 2845–2851.

Laguë, M., and Reebs, S. G. (2000a). Phase-shifting the light-dark cycle influences food-anticipatory activity in golden shiners. *Physiol. Behav.* **70,** 55–59.

Laguë, M., and Reebs, S. G. (2000b). Food-anticipatory activity of groups of golden shiners during both day and night. *Can. J. Zool.* **78,** 886–889.

Le Comber, S. C., and Smith, C. (2004). Polyploidy in fishes: Patterns and processes. *Biol. J. Linnean Soc.* **82,** 431–442.

Leggatt, R. A., and Iwama, G. K. (2003). Occurrence of polyploidy in the fishes. *Rev. Fish Biol. Fish.* **13,** 237–246.

Li, L., and Dowling, J. E. (2000). Effects of dopamine depletion on visual sensitivity of zebrafish. *J. Neurosci.* **20,** 1893–1903.

Løkkeborg, S. (1998). Feeding behaviour of cod, *Gadus morhua*: Activity rhythm and chemically mediated food search. *Anim. Behav.* **56,** 371–378.

Marimuthu, G, Rajan, S., and Chandrashekaran, M. K. (1981). Social entrainment of the circadian rhythm in the flight activity of the microchiropteran bat *Hipposideros speoris*. *Behav. Ecol. Sociobiol.* **8,** 147–150.

Mashiko, K. (1979). The light intensity as a key factor controlling nocturnal action in the catfish, *Pseudobagrus aurantiacus*. *Jap. J. Ichthyol.* **25,** 251–258.

Masuda, T., Iigo, M., Mizusawa, K., Naruse, M., Oishi, T., Aida, K., and Tabata, M. (2003). Variations in plasma melatonin levels of the rainbow trout (*Oncorhynchus mykiss*) under various light and temperature conditions. *Zool. Sci.* **20,** 1011–1016.

Max, M., and Menaker, M. (1992). Regulation of melatonin production by light, darkness, and temperature in the trout pineal. *J. Comp. Physiol. A* **170,** 479–489.

Metcalfe, N. B., Fraser, N. H. C., and Burns, M. D. (1998). State-dependent shifts between \ nocturnal and diurnal activity in salmon. *Proc. R. Soc. Lond. B* **265,** 1503–1507.

Metcalfe, N. B., Fraser, N. H. C., and Burns, M. D. (1999). Food availability and the nocturnal vs. diurnal foraging trade-off in juvenile salmon. *J. Anim. Ecol.* **68,** 371–381.

Minamoto, T., and Shimizu, I. (2002). A novel isoform of vertebrate ancient opsin in a smelt fish, Plecoglossus altivelis. *Biochem. Biophys. Res. Commun.* **290,** 280–286.

Minh-Nyo, M., Tabata, M., and Oguri, M. (1991). Circadian locomotor activity in kawamutsu *Zacco temmincki. Nippon Suisan Gakkaishi* **57,** 771.

Mizusawa, K., Iigo, M., Suetake, H., Yoshiura, Y., Gen, K., Kikuchi, K., Okano, T., Fukada, Y., and Aida, K. (1998). Molecular cloning and characterisation of a cDNA encoding the retinal arylalkylamine N-acetyltransferase of the rainbow trout, *Oncorhynchus mykiss. Zool. Sci.* **15,** 345–351.

Mizusawa, K., Iigo, M., Masuda, T., and Aida, K. (2000). Photic regulation of arylalkylamine N-acetyltransferase1 mRNA in trout retina. *Neuroreport* **16,** 3473–3477.

Morita, Y., Tabata, M., Uchida, K., and Samejima, M. (1992). Pineal-dependent locomotor activity of lamprey, *Lampetra japonica,* measured in relation to LD cycle and circadian rhythmicity. *J. Comp. Physiol. A* **171,** 555–562.

Moutsaki, P., Bellingham, J., Soni, B. G., David-Gray, Z. K., and Foster, R. G. (2000). Sequence, genomic structure and tissue expression of carp (*Cyprinus carpio* L.) vertebrate ancient (VA) opsin. *FEBS Lett.* **473,** 316–322.

Moutsaki, P., Whitmore, D., Bellingham, J., Sakamoto, K, David-Gray, Z. K., and Foster, R. G. (2003). Teleost multiple tissue (tmt) opsin: A candidate photopigment regulating the peripheral clocks of zebrafish? *Brain Res. Mol. Brain Res.* **112,** 135–145.

Müller, K. (1978). The flexibility of the circadian system of fish at different latitudes. *In* "Rhythmic Activity of Fishes" (Thorpe, J. E., Ed.), pp. 91–104. Academic Press, New York.

Mussen, T. D., and Peeke, H. V. S. (2001). Nocturnal feeding in the marine threespine stickleback (*Gasterosteus aculeatus* L.): Modulation by chemical stimulation. *Behaviour* **138,** 857–871.

Naruse, M., and Oishi, T. (1994). Effects of light and food as zeitgebers on locomotor activity rhythms in the loach, *Misgurnus anguillicaudatus. Zool. Sci.* **11,** 113–119.

Nelson, D. R., and Johnson, R. H. (1970). Diel activity rhythms in the nocturnal, bottom-dwelling sharks *Heterodontus francisci* and *Cephaloscyllium ventriosum. Copeia* **1970,** 732–739.

Nishi, G. (1989). Locomotor activity rhythm in two wrasses, *Halichoeres tenuispinnis* and *Pteragogus flagellifera,* under various light conditions. *Jap. J. Ichthyol.* **36,** 350–356.

Nishi, G. (1990). Locomotor activity rhythm in four wrasse species under varying light conditions. *Jap. J. Ichthyol.* **37,** 170–181.

Nishi, G. (1991). The relationship between locomotor activity rhythm and burying behaviour in the wrasse, *Suezichthys gracilis. Jap. J. Ichthyol.* **37,** 402–409.

Nixon, A. J., and Gruber, S. H. (1988). Diel Metabolic and Activity Patterns of the Lemon Shark (*Negaprion brevirostris*). *J. Exp. Zool.* **248,** 1–6.

Oishi, K., Sakamoto, K., Okada, T., Nagase, T., and Ishida, N. (1998). Antiphase circadian expression between BMAL1 and period homologue mRNA in the suprachiasmatic nucleus and peripheral tissues of rats. *Biochem. Biophys. Res. Commun.* **253,** 199–203.

Okamura, H. (2003). Integration of molecular rhythms in the mammalian circadian system. *Novartis Found Symp.* **253,** 161–170.

Olla, B. L., and Studholme, A. L. (1972). Daily and seasonal rhythms of activity in the bluefish (*Pomatomus saltatrix*). *In* "Behaviour of Marine Animals" (Wina, H. E., and Olla, B. L., Eds.), Vol. 2, pp. 303–326. Plenum Press, New York.

Ooka-Souda, S., and Kabasawa, H. (1995). Circadian rhythms in locomotor activity of the hagfish, *Eptatretus burgeri* V. The effect of light pulses on the free-running rhythm. *Zool. Sci.* **12,** 337–342.

Ooka-Souda, S., Kabasawa, H., and Kinoshita, S. (1985). Circadian rhythms in locomotor activity of the hagfish, *Eptatretus burgeri*, and the effect of reversal of light-dark cycle. *Zool. Sci.* **2,** 749–754.

Pati, A. K. (2001). Temporal organisation in locomotor activity of the hypogean loach, *Nemacheilus evezardi*, and its epigean ancestor. *Env. Biol. Fishes* **62,** 119–129.

Pedersen, J. (2000). Food consumption and daily feeding periodicity: Comparison between pelagic and demersal whiting in the North Sea. *J. Fish Biol.* **57,** 402–416.

Pettersson, L. B., Andersson, K., and Nilsson, K. (2001). The diel activity of crucian carp, *Carassius carassius*, in relation to chemical cues from predators. *Environ. Biol. Fish* **61,** 341–345.

Pinillos, M. L., De Pedro, N., Alonso-Gomez, A. L., Alonso-Bedate, M., and Delgado, M. J. (2001). Food intake inhibition by melatonin in goldfish (*Carassius auratus*). *Physiol Behav.* **72,** 629–634.

Postlethwait, J. H., Yan, Y. L., Gates, M. A., Horne, S., Amores, A., Brownlie, A., Donovan, A., Egan, E. S., Force, A., Gong, Z., Goutel, C., Fritz, A., Kelsh, R., Knapik, E., Liao, E., Paw, B., Ransom, D., Singer, A., Thomson, M., Abduljabbar, T. S., Yelick, P., Beier, D., Joly, J. S., Larhammar, D., Rosa, F., Westerfield, M., Zon, L. I., Johnson, S. L., and Talbot, W. S. (1998). Vertebrate genome evolution and the zebrafish gene map. *Nat Genet.* **18,** 345–349.

Pradhan, R. K., Pati, A. K., and Agarwal, S. M. (1989). Meal scheduling modulation of circadian rhythm of phototactic behaviour in cave dwelling fish. *Chronobiol. Int.* **6,** 245–249.

Randolph, K. N., and Clemens, H. P. (1976). Some factors influencing the feeding behaviour of channel catfish in culture ponds. *Trans. Am. Fish. Soc.* **105,** 718–724.

Reebs, S. G. (1992). Sleep, inactivity and circadian rhythms in fish. *In* "Rhythms in Fishes" (Ali, M. A., Ed.), pp. 127–135. Plenum Press, New York.

Reebs, S. G. (1996). Time-place learning in golden shiners (Pisces, *Notemigonus crysoleucas*). *Behav. Proces.* **36,** 253–262.

Reebs, S. G. (1999). Time-place learning based on food but not on predation risk in a fish, the inanga (*Galaxias maculatus*). *Ethology* **105,** 361–371.

Reebs, S. G. (2000). Can a minority of informed leaders determine the foraging movements of a fish shoal? *Anim. Behav.* **59,** 403–409.

Reebs, S. G. (2002). Plasticity of diel and circadian activity rhythms in fishes. *Rev. Fish Biol. Fish* **12,** 349–371.

Reebs, S. G., and Colgan, P. W. (1991). Nocturnal care of eggs and circadian rhythms of fanning activity in two normally diurnal cichlid fishes, *Cichlasoma nigrofasciatum* and *Herotilapia multispinosa*. *Anim. Behav.* **41,** 303–311.

Reebs, S. G., and Gallant, B. (1997). Food-anticipatory activity as a cue for local enhancement in golden shiners (Pisces, Cyprinidae, *Notemigonus crysoleucas*). *Ethology* **103,** 1060–1069.

Reebs, S. G., and Laguë, M. (2000). Daily food-anticipatory activity in golden shiners: A test of endogenous timing mechanisms. *Physiol. Behav.* **70,** 35–43.

Rensing, L., and Ruoff, P. (2002). Temperature effect on entrainment, phase shifting, and amplitude of circadian clocks and its molecular bases. *Chronobiol. Int.* **19,** 807–864.

Reppert, S. M., Weaver, D. R., Cassone, V. M., Godson, C., and Kolakowski, L. F., Jr. (1995). Melatonin receptors are for the birds: Molecular analysis of two receptor subtypes differentially expressed in chick brain. *Neuron* **15,** 1003–1015.

Reynolds, W. W., and Casterlin, M. E. (1976a). Locomotor activity rhythms in the bluegill sunfish, *Lepomis macrochirus*. *Amer. Midl. Nat.* **96,** 221–225.

Reynolds, W. W., and Casterlin, M. E. (1976b). Activity rhythms and light intensity preferences of *Micropterus salmoides* and *M. dolomieui*. *Trans. Am. Fish. Soc.* **105,** 400–403.

Richardson, N. E., and McCleave, J. D. (1974). Locomotor activity rhythms of juvenile Atlantic salmon (*Salmo salar*) in various light conditions. *Biol. Bull.* **147,** 422–432.

Ribelayga, C., Wang, Y., and Mangel, S. C. (2004). A circadian clock in the fish retina regulates dopamine release via activation of melatonin receptors. *J. Physiol.* **554,** 467–482.

Roberts, D., Okimoto, D. K., Parsons, C., Straume, M., and Stetson, M. H. (2003). Development of rhythmic melatonin secretion from the pineal gland of embryonic mummichog (Fundulus heteroclitus). *J. Exp. Zool. Part A Comp. Exp. Biol.* **296,** 56–62.

Ross, L. G., and McKinney, R. W. (1988). Respiratory cycles in *Oreochromis niloticus* (L.), measured using a six-channel microcomputer-operated respirometer. *Comp. Biochem. Physiol.* **89A,** 637–643.

Rubio, V. C., Sanchez-Vazquez, F. J., and Madrid, J. A. (2004). Oral administration of melatonin reduces food intake and modifies macronutrient selection in European sea bass (*Dicentrarchus labrax, L.*). *J. Pineal Res.* **37,** 42–47.

Samejima, M., Shavali, S., Tamotsu, S., Uchida, K., Morita, Y., and Fukuda, A. (2000). Light- and temperature-dependence of the melatonin secretion rhythm in the pineal organ of the lamprey, *Lampetra japonica. Jap. J. Physiol.* **50,** 437–442.

Sánchez-Vázquez, F. J., and Madrid, J. A. (2001). Feeding anticipatory activity. *In* "Food Intake in Fish" (Houlihan, D., Boujard, T., and Jobling, M., Eds.), pp. 216–232. Blackwell Science, London.

Sánchez-Vázquez, F. J., and Tabata, M. (1998). Circadian rhythms of demand-feeding and locomotor activity in rainbow trout. *J. Fish Biol.* **52,** 255–267.

Sánchez-Vázquez, F. J., Madrid, J. A., and Zamora, S. (1995a). Circadian rhythms of feeding activity in sea bass, *Dicentrarchus labrax* L.: Dual phasing capacity of diel demand-feeding pattern. *J. Biol. Rhythms* **10,** 256–266.

Sánchez-Vázquez, F. J., Zamora, S., and Madrid, J. A. (1995b). Light-dark and food restriction cycles in sea bass: Effect of conflicting zeitgebers on demand-feeding rhythms. *Physiol. Behav.* **58,** 705–714.

Sánchez-Vázquez, F. J., Madrid, J. A., Zamora, S., Iigo, M., and Tabata, M. (1996). Demand feeding and locomotor circadian rhythms in the goldfish, *Carassius auratus*: Dual and independent phasing. *Physiol. Behav.* **60,** 665–674.

Sánchez-Vázquez, F. J., Madrid, J. A., Zamora, S., and Tabata, M. (1997). Feeding entrainment of locomotor activity rhythms in the goldfish is mediated by a feeding-entrainable circadian oscillator. *J. Comp. Physiol. A* **181,** 121–132.

Sánchez-Vázquez, F. J., Azzaydi, M., Martínez, F. J., Zamora, S., and Madrid, J. A. (1998). Annual rhythms of demand-feeding activity in sea bass: Evidence of a seasonal phase inversion of the diel feeding pattern. *Chronobiol. Int.* **15,** 607–622.

Sánchez-Vázquez, F. J., Iigo, M., Madrid, J. A., and Tabata, M. (2000). Pinealectomy does not affect the entrainment to light nor the generation of the circadian demand-feeding rhythms of rainbow trout. *Physiol. Behav.* **69,** 455–461.

Sánchez-Vázquez, F. J., Aranda, A., and Madrid, J. A. (2001). Differential effects of meal size and food energy density on feeding entrainment in goldfish. *J. Biol. Rhythms* **16,** 58–65.

Schwassmann, H. O. (1971). Biological rhythms. *In* "Fish Physiology" (Hoar, W. S., and Randall, D. J., Eds.), Vol. 6, pp. 371–428. Academic Press, New York.

Sevenster, P., Feuth-de Bruijn, E., and Huisman, J. J. (1995). Temporal structure in stickleback behaviour. *Behaviour* **132,** 1267–1284.

Shapiro, C. M., and Hepburn, H. R. (1976). Sleep in a schooling fish, *Tilapia mossambica. Physiol. Behav.* **16,** 613–615.

Shearman, L. P., Sriram, S., Weaver, D. R., Maywood, E. S., Chaves, I., Zheng, B., Kume, K., Lee, C. C., van der Horst, G. T., Hastings, M. H., and Reppert, S. M. (2000). Interacting molecular loops in the mammalian circadian clock. *Science* **288,** 1013–1019.

Smith, R. J. F. (1997). Avoiding and deterring predators. *In* "Behavioural Ecology of Teleost Fishes" (Godin, J.-G. J., Ed.), pp. 163–190. Oxford University Press, Oxford.

Soni, B. G., and Foster, R. G. (1997). A novel and ancient vertebrate opsin. *FEBS Lett.* **406,** 279–283.

Soni, B. G., Philp, A. R., Foster, R. G., and Knox, B. E. (1998). Novel retinal photoreceptors. *Nature* **394,** 27–28.

Spieler, R. E., and Clougherty, J. J. (1989). Free-running locomotor rhythms of feeding-entrained goldfish. *Zool. Sci.* **6,** 813–816.

Spieler, R. E., and Noeske, T. A. (1984). Effects of photoperiod and feeding schedule on diel variations of locomotor activity, cortisol, and thyroxine in goldfish. *Trans. Am. Fish. Soc.* **113,** 528–539.

Steele, C. W. (1984). Diel activity rhythms and orientation of sea catfish (*Arius felis*) under constant conditions of light and darkness. *Mar. Behav. Physiol.* **10,** 183–198.

Stickney, A. P. (1972). The locomotor activity of juvenile herring (*Clupea harengus harengus* L.) in response to changes in illumination. *Ecology* **53,** 438–445.

Tabata, M., Minh-Nyo, M., Niwa, H., and Oguri, M. (1989). Circadian rhythm of locomotor activity in a teleost, *Silurus asotus*. *Zool. Sci.* **6,** 367–375.

Tabata, M., Minh-Nyo, M., and Oguri, M. (1991). The role of the eyes and the pineal organ in the circadian rhythmicity in the catfish *Silurus asotus*. *Nippon Suisan Gakkaishi* **57,** 607–612.

Tamai, T. K., Vardhanabhuti, V., Arthur, S., Foulkes, N. S., and Whitmore, D. (2003). Flies and fish: Birds of a feather. *J. Neuroendocrinol.* **15,** 344–349.

Tauber, E. S. (1974). The phylogeny of sleep. *In* "Advances in Sleep Research" (Weitzman, E. D., Ed.), Vol. 1, pp. 133–172. Spectrum Public, New York.

Thetmeyer, H. (1997). Diel rhythms of swimming activity and oxygen consumption in *Gobiusculus flavescens* (Fabricius) and *Pomatoschistus minutus* (Pallas) (Teleostei, Gobiidae). *J. Exp. Mar. Biol. Ecol.* **218,** 187–198.

Thinès, G. (1967). Étude d'un dispositif actographique permettant de mesurer l'activité des poissons: Examen de quelques résultats obtenus sur *Carassius auratus*, *Pygocentrus piraya* et *Hyphessobricon rosaceus*. *In* "La Distribution Temporelle des Activités Animales et Humaines" (Médioni, J., Ed.), pp. 65–77. Masson, Paris.

Tobler, I., and Borbely, A. A. (1985). Effect of rest deprivation on motor activity of fish. *J. Comp. Physiol. A* **157,** 817–822.

Trajano, E., and Menna-Barreto, L. (2000). Locomotor activity rhythms in cave catfishes, genus Taunayia, from eastern Brazil (Teleostei: Siluriformes: Heptapterinae). *Biol. Rhythm Res.* **31,** 469–480.

Valdimarsson, S. K., Metcalfe, N. B., Thorpe, J. E., and Huntingford, F. A. (1997). Seasonal changes in sheltering: Effect of light and temperature on diel activity in juvenile salmon. *Anim. Behav.* **54,** 1405–1412.

Weber, D. N., and Spieler, R. E. (1987). Effects of the light-dark cycle and scheduled feeding on behavioural and reproductive rhythms of the cyprinodont fish, Medaka, *Oryzias latipes*. *Experientia* **43,** 621–624.

Whitmore, D., Foulkes, N. S., and Sassone-Corsi, P. (2000). Light acts directly on organs and cells in culture to set the vertebrate circadian clock. *Nature* **404,** 87–91.

Whitmore, D., Foulkes, N. S., Strahle, U., and Sassone-Corsi, P. (1998). Zebrafish Clock rhythmic expression reveals independent peripheral circadian oscillators. *Nat. Neurosci.* **1,** 701–707.

Yamada, H., Satoh, R., Ogoh, M., Takaji, K., Fujimoto, Y., Hakuba, T., Chiba, H., Kambegawa, A., and Iwata, M. (2002). Circadian changes in serum concentrations of steroids in Japanese char *Salvelinus leucomaenis* at the stage of final maturation. *Zool. Sci.* **19,** 891–898.

Yokota, T., and Oishi, T. (1992). Seasonal change in the locomotor activity rhythm of the medaka, *Oryzias latipes*. *Int. J. Biometeorol.* **36,** 39–44.

Zachmann, A., Falcon, J., Knijff, S. C., Bolliet, V., and Ali, M. A. (1992). Effects of photoperiod and temperature on rhythmic melatonin secretion from the pineal organ of the white sucker (*Catostomus commersoni*) *in vitro*. *Gen. Comp. Endocrinol.* **86,** 26–33.

Zhdanova, I. V., Geiger, D. A., Schwagerl, A. L., Leclair, O. U., Killiany, R., Taylor, J. A., Rosene, D. L., Moss, M. B., and Madras, B. K. (2002). Melatonin promotes sleep in three species of diurnal nonhuman primates. *Physiol. Behav.* **75,** 523–529.

Zhdanova, I. V., Wang, S. Y., Leclair, O. U., and Danilova, N. P. (2001). Melatonin promotes sleep-like state in zebrafish. *Brain Res.* **903,** 263–268.

Zhdanova, I. V., and Tucci, V. (2003). Melatonin, Circadian Rhythms, and Sleep. *Curr. Treat. Options Neurol.* **5,** 225–229.

Zoufal, R., and Taborsky, M. (1991). Fish foraging periodicity correlates with daily changes of diet quality. *Mar. Biol.* **108,** 193–196.

7

BEHAVIOURAL PHYSIOLOGY OF FISH MIGRATIONS: SALMON AS A MODEL APPROACH

SCOTT G. HINCH
STEVEN J. COOKE
MICHAEL C. HEALEY
A. P. (TONY) FARRELL

Behaviour and Physiology of fish: Volume 24
FISH PHYSIOLOGY

DOI: 10.1016/S1546-5098(05)24007-4

1. INTRODUCTION

The ecology, distribution, and behaviour of migrating fishes are areas of research that have received considerable attention with numerous historical reviews (e.g., Harden-Jones, 1968; Hasler, 1971; Leggett, 1977; McDowall, 1988) and several recent syntheses (e.g., Dodson, 1997; Lucas and Baras, 2001; Metcalfe *et al.*, 2002). The physiological basis for some aspects of migration have been recently reviewed (e.g., Ueda and Yamauchi, 1995; Hogasen, 1998); however, there have been no reviews that focus on the integration of physiology with behaviour and ecology. Here we explore the linkages between behaviour, environmental conditions, and physiological cues and constraints to understand the mechanisms that underlie and control migrations.

About 2.5% of all fish species undertake migrations (Riede, 2002, 2004), so why devote an entire chapter to understanding the physiology underlying behaviour of such a small group of species? First, migrations can be remarkable. They can involve synchronous movements of an entire population of thousands to millions of fish. Thus, the low number of migratory fish species belies the overall abundance of fish involved in migrations. Second, many migratory species figure prominently in large fisheries (e.g., Atlantic cod, *Gadus morhua*; bluefin tuna, *Thunnus thynnus*; Pacific and Atlantic salmon, *Oncorhynchus spp.*, *Salmo salar*; sturgeon, *Acipenser spp.*; herring, *Clupea spp.*; Pacific hake, *Merluccius productus*; and menhaden, *Brevoortia spp.*). Understanding their migratory behaviour is critical to their management. Third, migratory species seem to be at twice the risk of extinction than nonmigrating ones. About 4% of all fish species and 7.2% of all migratory fish species are listed as "threatened with extinction" (Riede, 2002, 2004). Successful conservation of migratory species will require an understanding of the mechanisms responsible for migratory behaviours.

2. MIGRATIONS: DEFINITION, TYPES, AND PREVALENCE

Over the past several decades, there have been many attempts to define or describe migration. Many have adopted an ecological or physiological basis for this endeavor. Others have focused on the magnitude and scope of movements. Unfortunately, this has limited the development of a broad intertaxa general definition (see review in Dingle, 1996). In his review of plant, animal, and fungi transport behaviours, Dingle (1996) attempted to generate a universal definition through a summary of key characteristics that identified migration as a distinct and specialised behaviour with evolutionary underpinnings. Migrations would involve most of the following:

1. Persistent movement of greater duration than occurs during station keeping or ranging;
2. Straightened-out movements which differ from those during station keeping or ranging;
3. Suppression or inhibition of responses to stimuli that normally would arrest movements (e.g., feeding);
4. Activity patterns particular to departure and arrival; and
5. Specific patterns of energy allocation to support movements.

Endler (1977) described migrations as long-distance movements made by large numbers of individuals in the same direction at the same time. Northcote (1978, 1984, 1997) used concepts consistent with several of those mentioned above to generate criteria to define general fish migrations:

1. Individuals must cyclically move among at least two, but maybe more, separated and distinctive habitats (e.g., between a spawning and a feeding area);
2. Movements must be directed and usually involve active swimming;
3. Movements must occur within a reasonably predictable time frame and sequence; and
4. A large segment of the population must be involved in the movements.

Fish migrations have been categorised and described based on origin, destination, function, and direction of movement. Migrations entirely within seawater, and the fish that make these migrations, are termed *oceanodromous*. Migrations between freshwater and marine environments are termed *diadromous*. There are three types of diadromy. Fish undertaking spawning migrations from seawater to freshwater are termed *anadromous*, whereas spawning migrations from freshwater to seawater are termed *catadromous*. Migrations between freshwater and seawater for nonreproductive purposes (e.g., feeding) are termed *amphidromous*. A *potamodromous* migration occurs solely within a stream or river (or between a lake and a river), whereas those that take place solely in a lake are termed *limnodromous* (McDowall, 1988). Of the approximately 35,000 species of fish, 874 species (about 2.5%) are known to undertake migrations (57% oceanodromous, 16% anadromous, 10% potamodromous, 8% amphidromous, 8% catadromous, 1% limnodromous; Riede, 2002, 2004).

3. COSTS AND BENEFITS OF MIGRATION

The evolutionary basis for migration is that the fitness benefits and costs associated with residing at a particular location change with the stages of a fish's lifecycle so that fitness advantages are gained by migrating from habitat to habitat. Each individual fish, within the context of its life history pattern, has the objective of surviving and maximising its genetic contribution to the next generation (i.e., lifetime fitness). Specific to a life stage, migration confers a number of benefits (Northcote, 1978) but these come at a cost. Each phase of migration force fish to deal with a number of ecological needs:

1. To accumulate surplus energy for growth and maturation;
2. To survive (e.g., avoid predators, harsh environmental conditions, starvation); and
3. To get positioned geographically to do well in the next life history stage or phase of migration.

These ecological needs are both interrelated and in conflict. Clearly, feeding can expose fish to predators or alter their geographic positioning. Therefore, avoiding predation and getting well positioned geographically for the next phase of migration may require ignoring feeding opportunities. To complicate these already demanding behavioural problems, fish are exposed to a dynamic range of environmental conditions and are faced with important time limitations on completing the different phases of migration (Healey, 2000). For example, there are times when migrations may be impossible (e.g., ice over lakes, water velocities that are too high, inappropriate tides and water temperatures, etc.). These challenges are intimately associated with physiological capabilities and processes, which can affect an individual's ability to meet specific life history stage objectives. This intersection of physiology, life history variables, and behaviours is increasingly being recognised as central to ecology (Ricklefs and Wikelski, 2002).

To address these needs and challenges, migratory fish must make three general types of decisions:

1. Should a migration be initiated?
2. If the decision is yes, then when should the migration start?
3. How will migrants swim and behave during the migration?

Decisions made by individual fish likely reflect an assessment of how well it is doing in its present location, what other conspecifics around it are doing, and how much time it has left to complete a particular migratory phase. This will lead to decisions reflecting a tradeoff between asset accumulation and asset protection, mediated by the fish's assessment of its current state relative to some state-dependent payoff function and time constraints (Houston *et al.*, 1988; Mangel and Clark, 1988). It is these considerations that provide the basis for the decisions used for migration within each stage.

4. SALMONIDS AS MODEL SPECIES

The study of behavioural physiology of fish migrations is a relatively new area of interdisciplinary research, and information for most migratory species is nonexistent. We must thus look to a few species and systems that can be used as models. The focus of this chapter will be predominantly on the salmonids because, with the exception of catadromous migrations, this group of fish exhibits all the migration types, and most of the migration studies that have linked behaviour to physiology have examined salmonids. The behavioural physiology of migration has been explored for some life history stages in a few nonsalmonids and, where applicable, the results of these studies will be incorporated into our review. Although only able to draw from a limited number of species (17 in total: 10 salmonidae, 3 scombridae, 1 cottidae, 1 gadidae, 1 clupeidae, 1 centrarchidae) and studies, it is hoped that this review will stimulate research in this area and in particular serve as a catalyst for the examination of other species and systems not reviewed herein.

Of all salmonids, sockeye salmon (*O. nerka*) is perhaps the single best model species because a considerable amount of research has recently focused on migration physiology of this species, and studies exploring linkages between behaviour and some aspects of physiology have been conducted on nearly every migratory phase. Thus, one can generate the most complete life-history perspective with work done on sockeye. However, even with a focus on sockeye, treatment of each migratory phase is uneven, reflecting the relative amount of behavioural physiology information available. Although other fish may not share all the same migratory phases, the migratory behaviour of most species is reflected in some subset of the migratory phases of sockeye.

5. SOCKEYE SALMON MIGRATORY PHASES

Here we briefly review eight general migratory phases for sockeye salmon, most of which pertain at least somewhat to all anadromous salmonid species. We must emphasise that defining the start and end of some phases is somewhat arbitrary but necessary to provide a framework for describing and analysing the interactions among behaviour, physiology, and environment. Sockeye eggs develop and hatch as alevins in gravel/cobble substrates, most commonly in rivers but sometimes in lakes. The first phase of migration is from the substrate to the water surface to fill the swim bladder. The next phase of migration is the movement of alevins from the spawning grounds to a nursery lake, where juvenile sockeye salmon (fry) grow typically for 1 or 2 years, unlike most other salmon species that either spend 1–2 years in a river or migrate directly to the sea. In the nursery lake, sockeye fry exhibit intralake movements, often including diel vertical migrations. After lake rearing, fry undergo the most studied phase of migration when they migrate downstream to the ocean and smoltify in preparation for entry into the seawater environment. In the ocean, juveniles engage in a directed coastal migration, and then migrate offshore to exploit rich feeding grounds and grow rapidly (up to 3 kg, though other salmon species may reach 20 kg) while traveling thousands of kilometers for a period of 2–3 years. When sexual maturation is initiated, sockeye salmon start a directed return migration towards the coast and their natal river. In the final phase of migration, sockeye re-enter freshwater and migrate upriver, in some cases covering more than 1000 km as they travel home to their natal stream.

Below we explore each phase of sockeye salmon migration, emphasising linkages between behavioural decisions, environmental variability, and physiological cues, capabilities, and constraints. If available, information from other salmonids and nonsalmonids are included. For each phase we start with a brief general description, and provide details on factors affecting its initiation, and the physiological, behavioural, and environmental changes associated with migration. Readers requiring detailed information on life history, ecology, distribution, and evolution of Pacific salmon in general or sockeye in particular should examine other sources (e.g., Groot and Margolis, 1991; Groot *et al.*, 1995; Hendry and Stearns, 2004).

6. PHASE 1: LARVAL DEVELOPMENTAL MIGRATION

Emergence of larval fish from eggs is a stressful and critical period for all fishes, but there is very little known about the interplay between physiology and behaviour. Fertilised sockeye eggs develop into alevins over several months, protected by the gravel cover of the redd and obtaining energy from endogenous yolk stores. When yolk stores are almost entirely depleted, alevins undertake their first migration to find food. Leaving the protection of the gravel substrate, alevins swim to the surface of the lake or river to fill the swim bladder with air, which is essential for regulating buoyancy (Saunders, 1965). In doing so, they often encounter moving water and predators for the first time. Until emergence, alevins are photonegative; even when they do emerge, it occurs at night, presumably to reduce predation threat (Bams, 1969).

6.1. Migration Initiation and Physiological Changes

The timing of emergence is critical. It coincides with the time when endogenous yolk reserves are dwindling, when swimming muscles are adequately developed, and when environmental conditions are optimal, but the specific environmental cues and physiological reactions responsible for initiating this migration are unclear. Emerging sockeye larvae are negatively buoyant and, at this stage, are physostomous: a duct connects the swim bladder to the pharynx, allowing gulps of air to pass into the swim bladder. Proper development of this duct and its control systems represents an essential physiological requirement. Nevertheless, the rate of alevin development is variable with respect to temperature. By being synchronised to the thermal regimes of the spawning stream, emergence can coincide with optimal food supplies (e.g., plankton blooms in nursery lakes; Brannon, 1987). A shift to positive phototaxis (e.g., orientation to light) may play a role in emergence. Photic sensitivity increases coincident with elevated thyroid hormones (thyroxine) in brook charr (*Salvelinus fontinalis*) (Meisenheimer, 1990), as well as with the development of paired and median fins in rainbow trout (*O. mykiss;* Carey and Noakes, 1981).

The common behaviour in sockeye is repeated attempts to visit the surface, using primarily pectoral fins (Harvey and Bothern, 1972) and punctuated with periods of rest until neutral buoyancy is achieved (Bams, 1969). Perhaps larvae have a limited development of the caudal musculature at this stage. Alevin benefit behaviourally from swim bladder buoyancy regulation because, when threatened, young sockeye expel air, sink to the bottom, and reinflate the bladder at the surface only when the perceived danger is past. The disadvantage of surfacing (announcing their whereabouts to predators) disappears later in ontogeny, when the swim bladder gas gland has developed and they vary buoyancy (albeit slower) through gas diffusion between the blood and swim bladder. Gas exchange is primarily a cutaneous route in these small fish and the primary initial role for the gills is for ionoregulation (Rombough and Ure, 1991). Dense capillary networks encompassing the yolk sac likely assist with cutaneous gas exchange.

7. PHASE 2: MIGRATIONS TO LARVAL FEEDING GROUNDS

After hatching, many species migrate to a nursery area by either passive drifting (as is common in many oceanodromous species) or volitional swimming. River spawning is by far the most common strategy among most anadromous and amphidromous fish and larvae often migrate downstream to other riverine, estuarine, or marine environments (McDowall, 1988). For sockeye, this migration typically involves fry moving from spawning areas to a nursery lake, unless they spawn in a lake. Fish spawned in lakes also make an initial migration, but this is characterised by dispersal along shorelines and eventual movement into deeper waters (Burgner, 1962).

7.1. Migration Initiation and Control

The mechanisms proposed for initiating and controlling fry migrations include: light, temperature and water flow relative to size, age, and swimming performance. Some evidence for sockeye suggests genetic control of migration timing and behaviour that includes racially specific responses to different water velocities (Brannon, 1972). Population-specific morphology influences swimming stamina in coho salmon (Taylor and McPhail, 1985). Nocturnal, negative phototaxis and rheotaxis (i.e., orientation to water currents) are also key behaviours. Fry are dark-adapted, emerge from gravel at night and usually migrate nocturnally to nursery lakes, presumably to reduce predation risk (Beauchamp, 1995). If fry fail to reach the nursery area in one night, they usually hold during the day, resuming migration at night. Sockeye fry even delay downstream migration if they encounter nocturnal light pollution from urban areas (Tabor *et al.*, 2004) but suffer increased predation (Tabor *et al.*, 2004), highlighting the importance of evolved phototactic responses to natural light regimes in controlling downstream migration and protecting fry from predation.

Overall, rheotaxis is likely the dominant orienting mechanism in the river. Outlet fry swim upstream to nursery areas (positive rheotaxis) and inlet fry swim downstream (positive rheotaxis). These two types of juveniles face differing constraints; thus, physiological and behavioural adaptations involved with these migrations also differ. Outlet fry must minimise drift and either actively swim, often at an angle to the current, to prevent being swept downstream, or exploit near bank regions of lower velocity (McCart, 1967). Thus, decisions of outlet fry regarding when and where to move will depend on locomotory capabilities and energy stores more so than inlet fry which, although they do actively swim downstream, can also rely on passive drift (Hartman *et al.*, 1962). Outlet fry experience increased risk of predation because, although they emerge from gravel at night, it is often daylight by the time they stop downstream drift and undertake upstream swimming (Brannon, 1972). Within the nursery lake, light and celestial cues apparently become the principal orientating mechanisms (Brannon, 1972), but magnetic compass orientation is possible during low-light conditions (Quinn, 1980; Quinn and Brannon, 1982).

Minimising energy use is essential when migrating to nursery habitats, particularly for fry that were spawned in outlets because yolk material is nearly depleted and feeding opportunities within the tributaries are generally limited (Brannon, 1972; for exceptions see McCart, 1967). The relationship between time and energy required to reach nursery areas is poorly understood. Some populations do not feed because they have sufficient yolk, whereas others feed even when yolk is not fully depleted (Brannon, 1972). The lack of feeding may reflect predator-avoidance taking precedence over energy needs, or simply a lack of available food en route. Initial migratory activity tends to be individualistic, but fish rapidly adopt schooling behaviours. The fact that fry swim as well as drift suggests the decision to migrate is made with ample energy stores. At the same time, fish spawned in tributaries switch to positive phototactic and rheotactic behaviours. Interestingly, outlet fry can emerge as positively rheotactic but switch during their migration to being negative rheotactic, such as Weaver Creek sockeye (Figure 7.1), which migrate

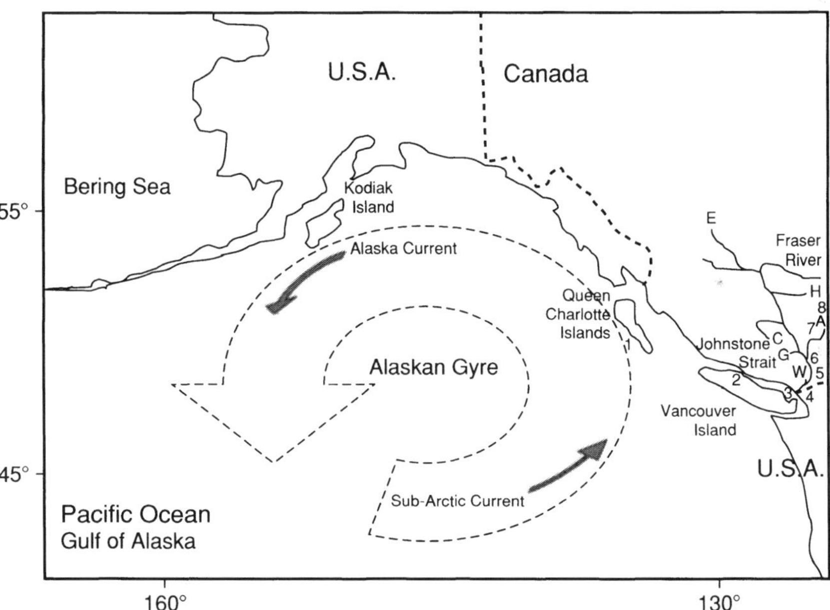

Fig. 7.1. Map of northwest region of North America and northeast Pacific Ocean showing locales, landmarks, and major current patterns that are associated with salmon migration phases. Within the Fraser River watershed, letters indicate location of spawning grounds for major sockeye populations that are discussed in Figure 7.6 and sections of the text: Weaver (W), Gates (G), Chilko (C), Horsefly (H), Adams (A), Early Stuart (E). Numbers indicate sites where adult sockeye were captured and biopsied during coastal and spawning migrations (see Figure 7.3).

downstream from their spawning stream but then migrate upstream to their rearing lake.

7.2. Physiological Changes

Using eggs and milt to create full-sib crosses from two reproductively isolated sockeye populations, one inlet and one outlet, Patterson *et al.* (2004a) discovered differences in enzymatic physiology for these groups. Outlet fry had higher mass specific activities for lactate dehydrogenase, citrate synthase, and cytochrome c oxidase, as well as higher protein specific activities of lactate dehydrogenase. The authors suggested that increased muscle development and anaerobic capacity were adaptations required by outlet fry for successful upstream migration to the nursery lake. Outlet fry also had larger eggs, which produced larger fry with higher absolute metabolic capacities and presumably more advanced muscle development. Thus, because selective pressures on this migratory phase differ among populations (or species) (Patterson *et al.*, 2004a), selection should be for high energy accrual into eggs to create fry that are strong swimmers, traits that were not necessarily present in or needed by the parents, which migrate only short distances from the ocean to spawning grounds (see Section 12). This would potentially add emergent fry migration to the list of possible factors shaping egg size in salmonids, which include difficulty of adult migration (Kinnison *et al.*, 2001), incubation environment (Quinn *et al.*, 1995), and resource availability (Hutchings, 1991).

8. PHASE 3: VERTICAL MIGRATIONS IN LAKES

Vertical migrations are the most widespread daily migration in the aquatic world, occurring throughout oceans and lakes by vertebrates and invertebrates, juveniles and adults alike.

8.1. Migration Initiation and Control

For the first weeks of lake residence, both migratory and nonmigratory sockeye fry concentrate in the littoral (near shore) area, noted for higher primary productivity and predator density compared with the pelagic (open water) zone. With growth, fry move to pelagic regions, where gradients (both seasonal and vertical) in prey abundance, light intensity, and temperature exist, leading to seasonal and diel patterns of vertical migration that vary according to a lake's physiochemical properties (Clark and Levy, 1988). In general, surface waters are favoured for feeding. Light intensity appears to be the primary orienting stimuli during diurnal vertical migrations, with temperature being the primary stimulus for nocturnal migration and distribution. Indeed, diel alterations in light intensity (Novales-Flamarique *et al.*, 1992) influence the timing of vertical migrations. However, feeding opportunities are limited to dawn and dusk, presumably reflecting the increased predation risk at the surface and the energy use of vertical migrations.

Vertical migrations also provide behavioural thermoregulation. Crepuscular feeding activity of sockeye fry in warm epilimnetic waters, interspersed with time in cold hypolimnetic waters, clearly suggests that fry are not selecting for a single optimal temperature (Brett, 1971). Brett suggested that this thermoregulatory behaviour maximises growth because growth conversion efficiency peaks at 11.5 °C, whereas growth optimum occurs at 15 °C (Brett, 1971). Because sockeye salmon rarely reach satiation under field conditions (Brett, 1971), bioenergetic constraints conserve energy when food is limited. Moreover, with reduced daily rations, maximum food conversion efficiency occurs at progressively lower temperatures (Brett *et al.*, 1969; Brett, 1971). Bear Lake sculpin (*Cottus extensus*) migrate to warmer water after feeding to promote digestion and enhance growth and so after feeding on benthos during the day, they ascend nocturnally to warmer surface water (Wurtsbaugh and Neverman, 1988). Thus, this daily vertical migration is an adaptation to exploit thermal gradients and maximise energy intake, which contrasts with that used by sockeye salmon where their thermoregulatory strategy is to lower metabolic rate, conserve energy when food is not abundant, and minimise predation risk.

Anadromous fry must either grow sufficiently to smolt in the subsequent spring (i.e., during one growing season) or delay smolting and spend a second growing season in freshwater (sockeye in northern lakes may spend 3 or more years in freshwater before migrating; Burgner, 1991). Bimodal growth and smoltification have been linked in a number of studies. However, how fish sense their size and suppress this cascade of physiological processes associated with smoltification is unknown. The benefit of a large smolt size is improved ocean survival. The tradeoffs of spending extra seasons to achieve this is greater predation risk (overwinter survival is size dependent), greater competition for food (intra- and interspecific), increased exposure to winter hypoxia, and more likely exposure to numerous parasites.

8.2. Physiological Changes

The visual ability of 3-cm sockeye fry is such that an average-sized prey item of 1.2 mm in length can be detected up to a distance of 7.8 cm (Novales-Flamarique and Hawryshyn, 1996). This distance is similar to the intermittent swimming distance of rainbow trout as it searches for prey (Browman *et al.*, 1994), highlighting the linkages between movement, vision, and prey detection. Under varied photic conditions, their retina with a square mosaic and predominant double cone surface area would maximise photon catch, increasing chances of prey detection and predator avoidance (Novales-Flamarique and Hawryshyn, 1996). In turbid lakes, vertical migration is less common, probably because visual acuity is reduced for both the fry and its predators.

Scheuerell and Schindler (2003) determined that juvenile sockeye salmon were not simply tracking their prey when they migrated vertically, but instead focused on maintaining a constant light environment, so that fish exploited an "antipredator window" where foraging potential and minimising predation risk were not mutually exclusive. In this regard, detection of light quality is also important. Prey detection and predator avoidance are enhanced by ultraviolet sensitivity, which develops in yearling sockeye (Novales-Flamarique and Hawryshyn, 1996). Due to scattering and attenuation, ultraviolet light only penetrates several centimeters below the water surface. Fry are known to feed in this surface zone (Novales-Flamarique *et al.*, 1992).

Tuna also migrate vertically. Fascinating discoveries have accrued from field telemetry (reviewed in Gunn and Block, 2001) that clearly show tuna regulate their red skeletal muscle temperature warmer (e.g., in Pacific bluefin tuna, 6.2–8.6 °C; Marcinek *et al.*, 2001) than surrounding waters. This endothermic adaptation enables foraging dives to deep, cool waters without losing locomotory capacity to capture prey (Marcinek *et al.*, 2001; Brill *et al.*, 2002). In fact, bluefin tuna have been observed to use the entire water column in some locations (Brill *et al.*, 2002). However, other tissues must operate at ambient temperature because of rapid heat exchange across the gills (Block, 1994). This creates a problem for the heart because it must work to supply blood to the endothermic red muscle; cold temperature decreases heart rates in fish, negatively affecting functioning of the sarcolemal L-type calcium channel, which delivers calcium to control the force of cardiac contraction (Shiels *et al.*, 2002). Consequently, the ability of tuna hearts to function at cold temperature may be critical to the animal's ability to forage at colder depths. The bluefin tuna heart continues to pump at 2 °C, unlike other tuna (Blank *et al.*, 2004), an adaptation that probably allows bluefin tuna to exploit colder, more northerly latitudes. The ability of salmon and tuna hearts to pump effectively at cold temperatures is in part related to adrenergic stimulation of the L-type calcium channel in salmonids (Shiels *et al.*, 2003) and enhanced ventricular sarcoplasmic reticular Ca^{2+} ATPase activity in bluefin tuna (Blank *et al.*, 2004). Some marine pelagic fish, including billfish, have evolved a heater organ that allows the optic muscles to be warmed above ambient, illustrating the additional importance of vision for these predatory fish (Block, 1994).

9. PHASE 4: SMOLTING AND MIGRATION TO SEA

All fish that leave freshwater to enter the marine environment must undergo physiological changes similar to those characterised by smolting in salmon. A suite of physiological, morphological, biochemical, and behavioural changes accompany smoltification, a process that has been extensively covered in entire chapters and books (e.g., Hogasen, 1998). Our discussion draws on smoltification only where it is relevant to the interaction of behaviour and physiology in downstream migration. There are several behavioural components to this migration, discussed separately, that involve decisions about migration initiation, smoltification, and movement patterns. This migratory phase is one of the best studied from physiological and movement perspectives; however, few studies have linked detailed behaviours or individual-specific behaviours with aspects of physiology. Though smolting does not play a role in downstream migrations of potamodromous fish, the transition from freshwater to seawater associated with amphidromous and catadromous fish may well reflect similar physiological and biochemical processes to those of anadromous fish, though there is scant research on this.

9.1. Migration Initiation

Salmon prepare for the downstream migration well before initiating downstream movements (Groot, 1982) and use seasonal cues to do so. Usually during their first summer growing season, fry must decide whether they are going to smolt or not the following spring. A fast growing "trajectory" can reflect either an inherent fast growth ability or an environmental opportunity for fast growth. Dominant salmon tend to grow faster and larger and so social dominance may play a role in segregating a population of fish into those that smolt and those that do not during the first summer (see Chapter 5, Section 4). Fast-growing fry enter the winter period (prior to downstream migration) larger and presumably with lower predation risk, which may partially offset the increased predation risk tradeoff associated with feeding more to grow faster.

The exact trigger to begin migration is unknown. Current evidence suggests that an innate, endogenous circannual rhythm is the primary controlling factor (Ueda and Yamauchi, 1995) that is triggered by the annual photoperiod cycle. The process of smoltification, however, is directly linked with initiating this migration. A short photoperiod (10 light:14 dark) acts to prime the fish and a long photoperiod (16 light:8 dark) induces smolting. Thorarensen *et al.* (1989) showed that brief exposure of coho fry to low-intensity nocturnal light slowed growth and inhibited smoltification. Later, Thorarensen and Clarke (1989) showed through the use of skeleton photoperiods that it was not the number of accumulated hours of exposure to light that initiated smolting, but rather the time during the day when light is experienced. Temperature, by regulating the rate of growth, is a modifying factor, just as is food availability (Clarke *et al.*, 1981; see also Chapter 6).

9.2. Smoltification and Other Physiological Changes

Smoltification is a complex physiological process involving changes to gill, kidney, and intestinal functions and especially ion and water exchange systems. Smoltification facilitates a directed migration by changing body shape and metabolism, and by reducing feeding (McCormick and Saunders, 1987). The transition between hypo-osmotic freshwater and hyper-osmotic marine conditions imposes the largest physiological challenge associated with this migration and, along with predation during the migration to sea, can result in high rates of mortality.

The endocrine control of the physiological changes associated with smoltification has been well described elsewhere (Iwata, 1995; Ueda and Yamauchi, 1995). The hypothalamus and possibly the pineal gland likely receive information about light, temperature, and food status and, through the pituitary gland, trigger the release of hormones that affect physiology, behaviour, and morphology. The interrenal gland, urophysis (caudal neurosecretory gland), thyroid gland, and corpuscules of Stannius all secrete hormones that typically target gill, kidney, and intestinal epithelia and influence osmoregulation, ionoregulation, growth, and presumably smoltification (see Ueda and Yamauchi, 1995). Strong evidence exists that the premigratory increase in Na^+/K^+-ATPase activity is predictive of future migratory behaviour. Using confamilial brown trout (*S. trutta*), Nielsen *et al.* (2004) measured ATPase activity in small gill biopsies and used the measurement to accurately predict those individuals that were going to migrate to the ocean. Fish preparing for migration had elevated levels of gill Na^+/K^+-ATPase activity 2–3 weeks before migrating, highlighting the level of preparedness and the fact that decisions regarding migration are made much earlier than when actual migration commences. A recent multivariate study of the physiology of smolting chinook salmon (*O. tshawytscha*) revealed that combined indicators of nutritional status, tissue damage, lipid metabolism, and stress were all important (Wagner and Congleton, 2004). Interestingly, smoltification-related indicators did not contribute to the observed variation. This study is one of the first multivariate studies of migration (and in animal physiology) and reveals the power of analyses that look at composite variables rather than individual ones.

Premigration, chum salmon (*O. keta*) fry are negatively rheotatic, but treatment with the thyroid hormone makes them positively rheotactic (Iwata, 1995). Thus, the natural surge of thyroxine that occurs 2 weeks before downstream migration in underyearling sockeye salmon is likely an important trigger for downstream migration. Similarly, plasma thyroxine levels peak in masu salmon (*O. masou*) during the new moon and it is during or immediately after this period that peaks in downstream migration have been observed (Yamauchi *et al.*, 1984; see also Chapter 1, Section 5.4.1. of this volume). Thyroxine may also be associated with the disappearance of ultraviolet sensitive cones in the eyes of sockeye salmon experiencing smoltification, (Beaudet *et al.*, 1997) but the functional and behavioural significance of alterations in ultraviolet photosensitivity are currently unknown (Browman and Hawryshyn, 1994). In addition to thyroxine, growth hormone, insulinlike growth factor, and cortisol all increase during smolting and contribute to development of gill chloride cells, which are the primary site for ion regulation in seawater fish (McCormick and Saunders, 1987).

Some migrating sockeye salmon fail to adapt to seawater, resulting in mortality. Although it is unclear what percentage of fish enter the ocean volitionally only to die because of osmoregulatory incompetence, it is clear that stress is a contributing factor. Franklin *et al.* (1992) compared the physiological status of successfully and unsuccessfully adapted sockeye. Successful adaptation was characterised by a brief but large fluctuation in cortisol, and plasma ionic concentrations rose initially but were regulated after 24–48 hours. Salmon that failed to adapt to seawater exhibited elevated cortisol and haematocrit, unregulated plasma ionic concentrations, severe dehydration, and ultimately death. Recently, Price and Schreck (2003) evaluated the effects of different stressors on the saltwater tolerance in juvenile chinook salmon. Stress reduced the saltwater preference of smolts that would normally select for full-strength seawater. Moreover, the stress of rearing coho smolts in a hatchery has been shown to elevate plasma cortisol, downregulate gill cortisol receptors, reduce saltwater tolerance, reduce relative swimming speed, and inhibit rapid recovery from exhaustion relative to wild coho (Brauner *et al.*, 1994). Thus, even though smoltification prepares fish for seawater residency by elevating cortisol, high cortisol levels are indicative of, and cause, physiological stress by decreasing corticosteroid receptors (Shrimpton and Randall, 1994). Therefore, migrants are in a conflicting physiological state between factors promoting smolting and those inhibiting it.

9.3. Swimming, Behaviour, and Energy Use Patterns

Migration usually begins within days of ice breakup (most sockeye nursery lakes are covered by ice), which varies considerably among stocks and species. It is believed that sockeye have the most well developed direction finding abilities of all salmonids (Hoar, 1976) and that interindividual and population-specific differences seem to exist in the direction-finding strategies used to locate lake outlets. Migration out of lakes is mainly during dawn and dusk, likely using an innate compass that directs movement. Groot (1972) experimentally confirmed that sockeye can use a sun-compass and polarised light cues. During periods of low light or overcast conditions, orientation can occur with use of the Earth's magnetic field (Quinn, 1980).

The downstream migration requires modifications to swimming abilities, energy storage, and energy use tactics (Groot, 1982). Fish eat little during this migration and appear energy deficient (Stefansson *et al.*, 2003). Reduced condition in Atlantic salmon is associated with increased catabolism of carbohydrates, lipids, and proteins (Sheridan *et al.*, 1985; Sheridan and Bern, 1986), which is accompanied by declining values of plasma nutritional indices (triglycerides, cholesterol, and total protein). Atlantic salmon increase white muscle phosphofructokinase several-fold and heart size 37–69% during smoltification (Leonard and McCormick, 2001), which appears adaptive for downstream and possibly ocean migration by enhancing metabolic capacity.

Downstream migrating smolts orient with the flow and, by actively swimming downstream, generate ground speeds that often exceed that of the water flow in hydraulically simple reaches (Smith *et al.*, 2002). Active downstream migration may reduce predation risk and ensure prompt arrival at the ocean. In reaches with flow turbulence or obstructions, sockeye may orient themselves upstream and pass through tail first (Hartman *et al.*, 1967). Turbulence resulted in 1.3- to 1.6-fold increases in swimming costs of juvenile Atlantic salmon relative to nonturbulent areas (Enders *et al.*, 2003). Thus, energy requirements are strongly influenced by environmental and geomorphological conditions affecting hydrology. Changing river environments can have a large effect on ocean arrival times. For example, in the Columbia River system, downstream migrations are prolonged by several weeks as a result of dams (Giorgi *et al.*, 1997). The prolonged migration, as well as the energetic costs and physiological disturbances associated with dam passage, can deplete energy and markedly affect smolt and marine survival. This problem is exacerbated in years with low flows or poor near-shore ocean productivity.

10. PHASE 5: DIRECTED COASTAL MIGRATION

The directed coastal migrations of juvenile salmon, and most other species, are poorly studied in terms of linkages between behaviour and physiology. The behavioural information that exists for salmonids is largely anecdotal from observations of distribution surveys coupled with growth, energy, and movement modeling exercises.

10.1. Behaviour and Energy Use Patterns

Some of the most detailed observations of juvenile behaviour during coastal migration arise from research conducted on Fraser River salmon (Figure 7.1). Groot and Cooke (1987) found that the direction and ground speed of smolts were consistent with surface water circulation patterns along the coast. However, Peterman *et al.* (1994), who modeled juvenile movements, discovered that coastal circulation patterns alone were insufficient to explain the migration patterns. Instead, if fish chose to swim at an average speed of about 4 cm/s to the northwest, then there was a good match between observed and modeled movements, which suggests young sockeye actively swim and are assisted by the wind-driven surface currents. In contrast to the constant speed of migration in this model, visual observations revealed sporadic movements, with fish holding and actively feeding in areas before moving to other locations (Healey, 1967). In particular, juveniles congregated in sheltered bays, taking advantage of food concentrating hydraulic features like eddies and tidal shear zones, which allow them to minimise energetic costs of foraging while maximising energy intake. In view of this intermittent active migration behaviour, it must involve directed swimming speeds considerably greater than 4 cm/s. Intermittent migration may allow the fish to take maximum advantage of favourable tidal surface currents (e.g., Levy and Cadenhead, 1994; Forward and Tankersley, 2001).

Fish attain efficiency in terms of energy expended per distance traveled by swimming at speeds with a low cost of transport (COT; Brett, 1995). Figure 7.2 illustrates our calculation of COT using Brett's respirometry data for 18-cm sockeye and clearly shows three important points. Foremost, COT increases with temperature. Second, optimum swimming speed increases with increasing temperature. Third, at all temperatures COT is essentially independent of swimming speeds close to the optimum; that is, it is relatively insensitive to a wide range of swimming speeds, such that COT remains within 10% of optimum between 0.8 and 2.0 body lengths (BL)/s (14–36 cm/s) at 10 °C. Thus, physiological plasticity ensures a broad range of migratory speeds without compromising energy expenditure unduly.

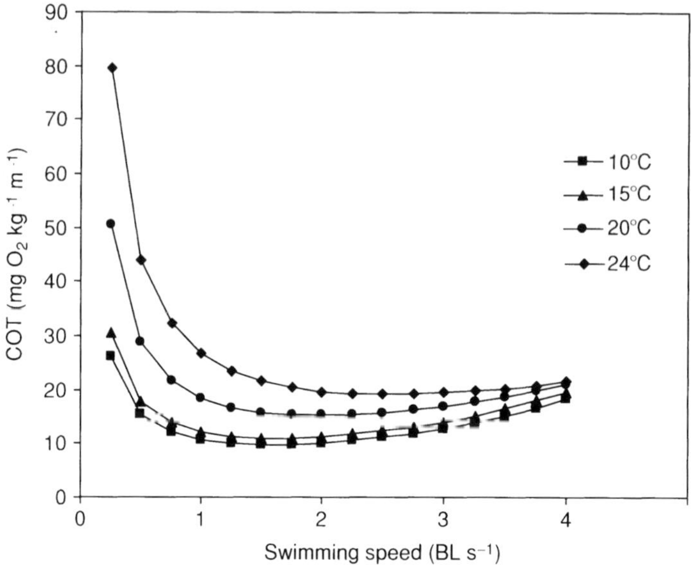

Fig. 7.2. Relationships between cost of transport (COT) and swimming velocity for juvenile sockeye at different water temperatures. Optimal swimming speeds are those which minimise COT. Across all temperatures, large ranges of swimming speeds are very close to the optimal speed. (Adapted from Brett, 1995.)

Later during coastal migration, sockeye move further from shore (Healey, 1991; Jaenicke and Celewycz, 1994), which is possibly a dispersal behaviour, tuned to finding foraging patches in response to lower abundance of large zooplankton during the mid to late summer (Healey, 1991). Despite low average prey abundances in these locales, their growth rates are rapid and so sockeye must then be capable of cueing in on small-scale patchiness in their zooplankton prey. Mortality is high during this phase of migration and rapid growth has been postulated to be an effective means of reducing predation mortality as larger fish are less vulnerable to predation (Wood *et al.*, 1993; see also Chapter 3). Beamish *et al.* (2004) have shown that slow growth during the first marine year of juvenile coho salmon in coastal British Columbia is associated with high mortality. They suggest that critical weights and energy reserves must be attained before autumn if fish are to survive poor winter feeding opportunities. Though speculative, periods of active feeding, punctuated by periods of active migration, may be the best tactic to address the tradeoffs between energy accumulation, survival, and positioning.

Little is known about the physiology of salmonids after they have initiated directed coastal migration. During the early phases of coastal migration, Atlantic salmon postsmolts mobilise energy, reducing lipid and glycogen stores, allocate protein to somatic growth, and ultimately decrease plasma levels of growth hormone (Stefansson *et al.*, 2003). Thus, juvenile salmon in coastal waters employ a strategy of rapid anabolic growth at the expense of energy reserves presumably to increase immediate survival potential (MacFarlane and Norton, 2002).

11. PHASE 6: OPEN OCEAN MIGRATION

Unlike the coastal migration phase (Section 10), open ocean migrations involve fast growth during early stages, and rapid accumulation of energy reserves during later stages. This phase is the most poorly studied for all anadromous fish. Behavioural patterns have largely been inferred from distributional data obtained either from sampling fish in different locations or from recapturing marked fish. Energetic information has been generated from knowledge of growth, food availability, and physical oceanographic features. There is virtually no data on the physiological aspects of this migration. Thus, our discussion will focus on linkages that have been investigated between behavioural and energetic patterns.

11.1. Behaviour and Energy Use Patterns

After about 1 year as ocean residents, sockeye move off the continental shelf into the open ocean. For Fraser River sockeye, whose ocean distribution have been well examined, a transition into the Gulf of Alaska (Figure 7.1) occurs that seems to be dependent on current patterns and temperature (French *et al.*, 1976; Walter *et al.*, 1997). Subsequently, in open ocean their migration is dominated by the Alaskan Gyre (Healey, 2000). Speeds of the sub-Arctic current (Figure 7.1) are about 10–25% of the normal cruising speed of the fish (Thomson *et al.*, 1992) and might be expected to have a significant influence on oceanic migrations. Walter *et al.* (1997) argued that swimming speeds required to override the influence of ocean currents would be prohibitively costly (i.e., a nonoptimal COT) and that little directional preference should be exhibited. Regardless of whether these fish make one (Walter *et al.*, 1997) or two (French *et al.*, 1976) seasonal "loops" of the Alaskan Gyre for each year at sea, it is clear that currents play a dominant role in determining the migration pattern of sockeye salmon in the Gulf of Alaska.

In the Gulf of Alaska, sockeye salmon must locate regions favourable to growth, as demonstrated by an order of magnitude change in salmon growth potential over a narrow range of sea surface temperatures (ca. 6–7 °C; Rand, 2002). Even so, the Alaskan Gyre has a complex thermal structure, with temperatures ranging from <1 °C to >15° C, depending on season and year. The search for zones of positive growth potential may lead fish to avoid areas of high temperature, particularly when food is scarce, because of the energetic cost of routine metabolic rates at high temperature. Moreover, the southern distributions of sockeye and other salmon in the Gulf of Alaska have sharp thermal limits, which are higher in summer than in winter, implying that fish migrate further north in winter than in summer and are behaviourally adapting to colder water to conserve energy when winter food supplies are low (Welch *et al.*, 1998). These behaviours may be analogous to those reported for Albacore tuna (*Thunnus alalunga*) in relation to thermal fronts (Laurs *et al.*, 1977).

Bioenergetic modelling of sockeye ocean migration suggests that finding food in the most energetically efficient manner is the most important activity enabling successful completion of subsequent phases of migration and spawning. To achieve their typical final ocean weights, sockeye must obtain near maximum rations virtually continuously during their ocean life (Hinch *et al.*, 1995; Davis *et al.*, 1998). Analysis of the distribution of opportunities for positive growth based on temperature and prey (i.e., zooplankton) patterns indicate that most of the Gulf of Alaska is not a good growing environment (Rand *et al.*, 1997). Zooplankton distribution is heterogeneous and exhibits patch structure on scales of tens to several hundreds of kilometers (Rand and Hinch, 1998a). Thus, sockeye must have finely tuned physiological systems enabling detection and evaluation of subtle differences in oceanic thermal and growth potential conditions.

Ogura and Ishida (1992) followed open ocean movements for about a week by attaching depth-sensing acoustic transmitters to four coho salmon in the central North Pacific Ocean. All fish tended to swim within a few meters of the surface and ground speeds (~10–20 km/day) were less than half of the typical speeds exhibited by maturing salmon during their directed migrations towards the coast (see Ogura and Ishida, 1995). Slower ground speeds can be accounted by two behaviours: occasional depth forays (generally <50 m) for periods of a few minutes to a few hours that seemed more prevalent at night; and horizontal meandering, particularly at night, occasionally backtracking (Ogura and Ishida, 1995). These movement, speed, and diel patterns fit with the conclusion of bioenergetic models that fish are expending most of their energy searching for patches of food (i.e., not swimming fast, making depth forays), and when they find food, they attempt to remain near it (i.e., backtracking).

Open ocean distribution and potential migration patterns have been studied in other pelagic fish such as tuna and billfish using satellite "pop-up" transmitter telemetry (see Akesson, 2002; Gunn and Block, 2001), but there has been little work on their energetics or physiology at sea, with the exception of research on Atlantic cod that examined the role of thyroid hormones on migration (Comeau *et al.*, 2001). In that study, the researchers determined that levels of T3 and thyroxine both increase at the start of the autumn migration. They suggest that thyroid hormones may facilitate the onset of the autumn migration by enhancing metabolism, sensory physiology, and swimming capacity (Woodhead, 1970). In particular, T3 may have had an excitatory effect on metabolism. The authors presume that the external cue for thyroid hormone expression and thus migration initiation is alteration in photoperiod.

12. PHASE 7: DIRECTED MIGRATION TO THE COAST AND NATAL RIVER

Foraging remains an important activity for the fish during this phase. Salmon gain about half of their mature mass during their last 6 months at sea, so they must forage continuously en route to the coast (Hinch *et al.*, 1995). This is the period when most of the energy reserves needed for upriver migrations and spawning are accumulated. However, how they balance their time budget for foraging and directed migration is unknown. Considerable behavioural and bioenergetic modelling and direct observations of movements using telemetry have helped understand growth patterns of Pacific salmon and mechanisms for locating natal rivers. Assessments of reproductive states and experiments evaluating initiation factors also exist. However, few studies have truly integrated the physiology and behaviour behind decisions by migrants regarding timing and swimming activity.

12.1. Migration Initiation

The cues and physiological mechanisms responsible for the decision to depart the open ocean environment are poorly understood, but the decision seems to be closely related with the initiation of gonadal maturation, which is regulated mainly by the hypothalamo-pituitary-gonadal axis (Ueda *et al.*, 2000b). Specifically, changing levels of reproductive hormones seem to be the key. Gonadotropin-releasing hormone (GnRH) triggers the initiation of gonad growth in salmon (Ueda and Yamauchi, 1995) and the onset of gonadal growth is associated with the shift from foraging to homing migration. Sockeye salmon in Lake Toya, Japan that were implanted with artificial GnRH ceased feeding in open water and began homeward migration prematurely (Kitahashi *et al.*, 1998). Artificial release of GnRH in sockeye salmon in Lake Shikotsu, Japan, caused increased levels of serum steroid hormones and increased development rate of gonads (Sato *et al.*, 1997). GnRH appears to accelerate gonadal maturation specifically by elevating levels of serum DHP (Fukaya *et al.*, 1998). Japanese chum salmon showed a distinct pattern of increasing testosterone and estradiol concentrations as fish migrated from the open ocean to spawning areas (Ueda *et al.*, 2000b). Injection of testosterone in experimentally castrated masu salmon stimulated migratory behaviours (Munakata *et al.*, 2001), further supporting the notion that gonadal maturation is a key factor responsible for migration initiation in this species. The trigger for the initial release of GnRH is not clear, but some have suggested that changing photoperiod plays a role. However, body size and energy levels must also have a role because Pacific salmon that do not reach a critical size early in the year are likely to delay spawning and remain in the ocean for an additional year (Groot and Margolis, 1991). Possibly, photoperiod triggers GnRH release only if a certain growth or energy accrual threshold is realised.

12.2. Direction and Speed of Migrations

The arrival time of salmon populations to the Pacific coast seldom varies by more than a week on either side of the long-term average. Modelling studies have explored how ocean circulation patterns influence homeward migration behaviour in Pacific salmon in terms of direction and timing (Thomson *et al.*, 1992; Healey *et al.*, 2000). For some species and stocks, the choice of migration route has important implications for energy and time conservation. For example, Fraser sockeye travelling with a strong compared with a weak sub-Arctic current (Figure 7.1) would reach the coast much farther north (Thomson *et al.*, 1994), causing them to reach the Fraser River later unless they swam faster. On the other hand, ocean currents can facilitate salmon migrations, as shown by Healey *et al.* (2000) who modeled the effects of the 300 km in diameter Sitka Eddy, located northwest of the Queen Charlotte Islands (Figure 7.1). The rotation of the eddy had a negligible effect on the latitude at which salmon reached the coast, but slightly boosted migration speed and conferred a modest energy benefit. As mesoscale eddies are a common feature of surface circulation in the Alaskan Gyre (Mysak, 1985; Thomson *et al.*, 1990), other eddies might influence homeward migration.

Tagging studies on Fraser River sockeye suggest that movements from the open ocean to the natal river are well directed and rapid, averaging 20–30 km/day but as fast as 60–70 km/day (Groot and Quinn, 1987).

Maturing sockeye, chum, and pink (*O. gorbuscha*) salmon in the Bering Sea (Figure 7.1), tagged with acoustic telemetry transmitters, traveled at similar speeds (~30–50 km/day; Ogura and Ishida, 1995), whereas nonmature chinook salmon moved much slower (<20 km/day). Similarly, sockeye returning to the natal rivers in British Columbia (Quinn and terHart, 1987; Quinn *et al.*, 1989) traveled 20–25 km/day, with periods of strongly directed swimming interspersed with wandering behaviours. The wandering could be related to either currents in coastal areas being stronger, more variable, and less predictable than that in the open ocean, or responses to incorrect route selections into deep inlets or fjords, or some element of foraging. Tagged salmon did not seem to be timing their coastal migrations to take advantage of tidal flows, though they did appear to use selective tidal transport in estuaries (Quinn *et al.*, 1989). Average swim speeds of acoustically tagged sockeye in coastal areas near natal rivers were close to the energetically optimal speed (Quinn *et al.*, 1989).

12.3. Physiological Changes

During homeward migration, salmon are physiologically preparing for the next and final migration phase, which involves freshwater entry, high activity exercise, and eventual reproduction. Even swimming patterns may change to save energy as fish approach the coast. Sockeye slow their migration and drift with tidal currents as they near natal estuaries (Levy and Cadenhead, 1994), a behaviour not observed in nonnatal estuaries.

Salmon must also position themselves at the mouth of the river at a time that coincides with some level of osmoregulatory preparedness for freshwater and when river conditions (e.g., temperature and discharge) are appropriate for enabling the upriver migration. Returning adults should avoid low-salinity plumes in the same way that outmigrating smolts do, but only if that water does not contain odours of their natal river. Encountering low-salinity water that smells like "home" could not only direct fish towards natal rivers but also lead to the initiation of the upriver migration (Døving and Stabell, 2003).

Based on samples taken from maturing sockeye intercepted 700 km and 250 km away from the mouth of the Fraser River (Figures 7.1 and 7.3), it is clear that fish are preparing for the transition to freshwater well before they enter freshwater. For example, gill Na^+/K^+ ATPase activity is lower at 250 km than the levels found ~700 km away, yet both locales are full-strength seawater (Figure 7.1). Also, the reproductive hormone titers in plasma suggest that the reproductive machinery is already in full swing at both locations. For example, estradiol increased from 2.5 to 5 ng/ml between the Queen Charlotte Islands and Johnstone Strait and remained at this level when the fish reached the river (Figure 7.4). Gonadosomatic index is also

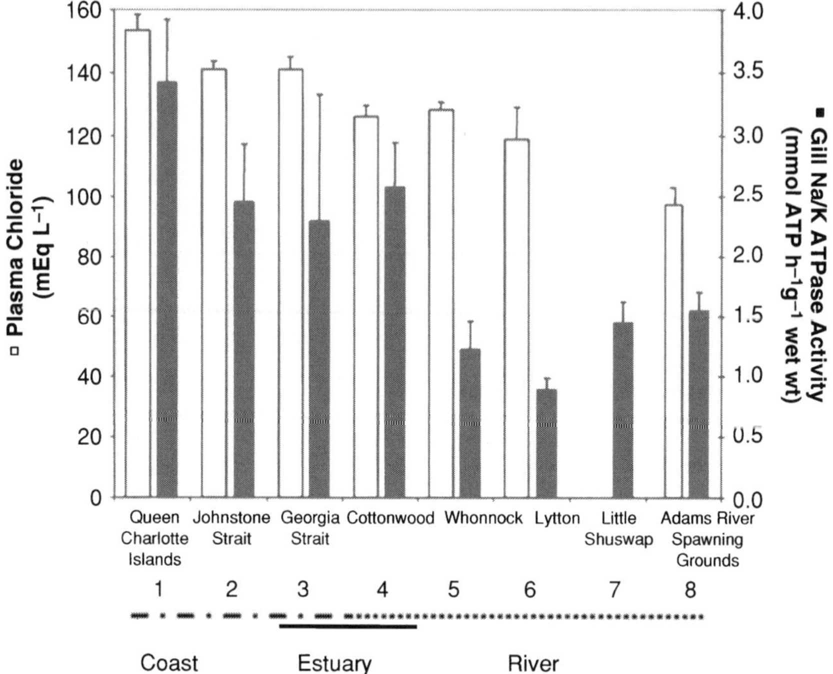

Fig. 7.3. Mean (+SE) plasma chloride concentrations (open bars) and gill Na^+/K^+ ATPase activity (solid bars) based on biopsy samples from adult sockeye (Adams River population) collected at eight sites that span the physical extent of the directed coastal and upriver spawning migrations. Sites are numbered according to migration sequence and correspond to locales in Figure 7.1. Sites are classified based on general habitat types and their salinity transitions (ocean, estuary, river). Plasma chloride data are missing for site 7. The sampling procedure may have caused some stress to fish so comparisons are best made with the plasma ion values for postexercise rather than the routine resting state values in Table 7.1. (Adapted from A. P. Farrell, unpublished data.)

increasing, suggesting lipid deposition in eggs is active and vitellogenin levels are elevated. This suggests relocation of lipid stores within the fish because they have ceased feeding. Understanding the endocrine control of this process has eluded research to date. As with smoltification, physiological changes are being made several weeks before the transition from seawater to freshwater and several months before spawning.

12.4. How Do Ocean Migrants Find Their Natal Rivers?

Much debate centers on the sensory cues and physiological systems used by salmon to direct homeward migration, owing in part to disparities among studies and species. Indeed, the controversy regarding salmonid orientation

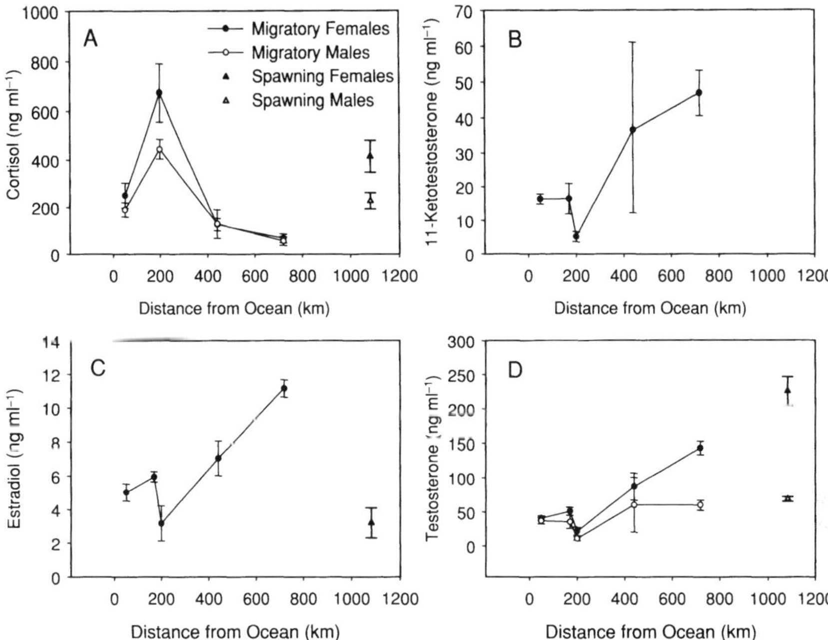

Fig. 7.4. Plasma levels of selected hormones (mean ±1 SE, A-cortisol, B-11 ketotestosterone, C-estradiol, D-testosterone) from Early Stuart sockeye salmon sampled during upstream spawning migration in 2001 (0 km = mouth of Fraser River). At 200 km, fish negotiate a very difficult hydraulic barrier (Hell's Gate) where plasma cortisol levels are significantly elevated. This stress is also associated with depression of all three sex hormones. (Adapted from A. P. Farrell, unpublished data.)

is long-standing (e.g., Nordeng, 1971). Døving and Stabell (2003) advanced a generalised scenario to describe homing orientation. In the open ocean, salmon should swim with small scale vertical and horizontal oscillations, which could allow them to sample odours from different water layers. Salmon recognise four major aquatic odourants (amino acids, bile salts, steroid hormones, and prostaglandins) at very low concentrations (Døving *et al.*, 1980; see also Chapter 9). Freshwater layers containing characteristic odours of the natal stream, collectively termed the *home stream olfactory bouquet* (Carruth *et al.*, 2002), would play a dominant role in the final stages of locating the natal river (Hasler and Scholz, 1983; see also Chapter 1, Section 5.4.1 of this volume). The attraction of adults to freshwater is the converse of the behaviour of outmigrating juvenile salmon.

Døving and Stabell (2003) further postulate that all phases of migration discussed in this chapter may be based on rheotactic responses to conspecific odours. Smoltification and downstream migration involves a shift from positive to negative rheotaxis, which could be triggered by the presence of chemical signals from conspecifics in its population. During sexual maturation, salmon are proposed to switch from negative to positive rheotaxis in response to the same set of chemical signals. Døving and Stabell (2003) further suggest that odour trails may be followed to and from feeding areas in the open ocean during a full migration cycle.

This orientation model assumes that ocean dwelling salmonids detect natal odours; however, other sensory systems are needed to bring fish to the proximity of these odours. Acoustic tracking of sockeye in Lake Toya, Japan has been revealing in this regard (Ueda *et al.*, 1998, 2000a). Fish carrying magnets found spawning areas just as quickly as control fish, suggesting that geomagnetic cues did not play a strong role in homing behaviour. In contrast, blinded fish released in the middle of the lake had difficulty locating their spawning area only 7 km away. Vision was clearly of paramount importance for homing behaviour, and fish may be using a sun compass, polarised lighting patterns, or, as they approach land, memory of landmarks (Ueda *et al.*, 1998). Blinded fish did locate spawning areas after days of random swimming, presumably by using olfactory cues, which they eventually encountered.

Olfactory nerve studies on chum and masu salmon have revealed that amino acids elicit olfactory responses and that salmonids can discriminate among different intensities of these chemicals (Sato *et al.*, 2000). Ocean migrating adults might be able to cue in on amino acids, pheromones, or population-specific odours released by ocean-bound juvenile migrants (Groot *et al.*, 1986). This latter possibility lends theoretical support to the Døving and Stabell (2003) model. However, Quinn *et al.* (1989) present arguments against this odour-based model. As sockeye approach the mouth of the Fraser River, vertical questing movements begin (Quinn and terHart, 1987) and are particularly frequent at inlets that do not contain natal rivers.

Other environmental variables could play roles as orientation cues, particularly as fish approach natal rivers. For instance, coastal temperature, which can be affected by riverine input, could be used, as could presence of conspecifics or predators, which may be higher near riverine areas. Migrants often begin to school in conspecific groups close to river mouths but whether this is simply a consequence of similar timing of arrival and limited space, a recognition of conspecifics, or an antipredator mechanism is unknown.

Though different species of salmon may enter the same river at similar times, their behaviours at the river mouth can be quite distinctive. Sockeye salmon entering Alberni Inlet (west coast of Vancouver Island; Figure 7.1; Spohn *et al.*, 1996) and near the Fraser River mouth (Rick Thomson, unpublished data, 2003) show a clear preference for the deep, cold, halocline rather than warmer, less saline, surface waters. Ironically, this exposes sockeye to very low oxygen concentrations (~4 mg O_2/L). This behavioural strategy brings into question the dogma that salmon are generally hypoxia sensitive. It may be that behavioural selection of cold temperature moderates the hypoxia sensitivity, as has been shown for isolated rainbow trout hearts (Overgaard *et al.*, 2004), as well as slowing routine metabolic rate. In contrast to sockeye, pink salmon entering the Fraser River occupy much shallower and less saline parts of the water column prior to river entry (A. P. Farrell, unpublished data, 2003).

13. PHASE 8: ADULT UPRIVER MIGRATION TO NATAL STREAM

Migration upriver for purposes of reproduction is common among potamodromous fishes, ranging from small cyprinds to large ictalurids (see Lucas and Baras, 2001), as well as being essential for anadromous fishes. Considerable effort has been placed on assessing physiological and energetic status of migrants and patterns of movements from many taxa sampled at different times or locales during migrations. Research on this migration phase thus provides the best and most detailed examples of linkages between behaviour and physiology. This knowledge is greatest for salmon, particularly for sockeye salmon.

13.1. Migration Initiation and Control

The cues and physiological mechanisms responsible for the decision to depart the coastal environment and head upriver are not well studied, though they must involve extremely precise links between behaviour and physiology. For example, in Fraser River sockeye, freshwater entry dates for a given population rarely deviate more than 1 week (~3%) from historical averages (Woodey, 1987). Migration windows vary widely, however, among populations of sockeye (Hodgson and Quinn, 2002). Therefore, different cues must exist among many salmon stocks entering the same river system. In general, increasing flows and decreasing temperatures stimulate upriver migrations in streams and small rivers, but this may not apply to large rivers (Quinn *et al.*, 1997). Initiation times in large rivers are related to spawning ground location, migration distance (Gilhousen, 1990), and river temperature (Quinn and Adams, 1996). For example, Fraser River sockeye populations that travel the longest distances also enter the river first in early summer, just as the spring freshet is beginning to subside but before water temperatures become high. Migration is therefore timed to miss the worst flow conditions and yet provide sufficient incubation time for fertilised eggs to hatch before streams freeze in winter. Conversely, sockeye populations that migrate only short distances to coastal spawning grounds do so in late summer and fall, in association with lower river flows and often cooler temperatures en route and at the spawning bed. They thus avoid peak water temperatures (>18 °C) that might promote lethal stress or disease in spawners.

It is thought that changes to osmoregulatory and ionoregulatory systems could play key physiological roles in initiating migrations. As salmon move from seawater to freshwater, they reorganise gill and gut transport mechanisms (e.g., ion channels, exchangers, ATPase pumps, paracellular permeabilities) and upregulate the activity of their kidneys (Wood and Shuttleworth, 1995) and plasma ion concentrations change (Table 7.1). The trigger for the osmoregulatory switch is not clear, but changing photoperiod probably plays a role and has been suggested as a likely cue affecting the migratory timing in sockeye (Quinn and Adams, 1996). Comparatively little is known about osmoregulation associated with upriver migrating salmon, in contrast with what is known about downstream migrating smolts (see Section 9).

There are additional physiological and environmental factors that may influence the timing of freshwater entry but may not serve as primary triggers of migration. Associated with the onset of salmon spawning migrations is a cessation of feeding (see discussion below) and fish remain in a true catabolic state for weeks to months, as hormonal systems are diverting energy and activities away from digestive functions towards gonad development. Even if fish had the capabilities to feed, opportunities for foraging would be curtailed because times available for spawning are often limited so migrations are fairly directed. Because adequate energy reserves are critical for successful migration and spawning, levels of energy could play a role. We have found with Fraser River sockeye that abnormally early migrants have lower reserves and are more advanced reproductively than normal timed migrants (S. Cooke and S. Hinch, unpublishd data, 2003). Initiation times can be affected by marine parasites. In Ireland, sea lice infestations were associated with premature river migrations of sea trout (*Salmo trutta*; Tully *et al.*, 1993). Tidal cycles mediate migration initiation. Fraser sockeye position themselves close to the sea and river mouth beds and initiate upriver swimming during high tides rather than low tides, presumably to receive forward assists from the inflowing tide (Levy and Cadenhead, 1994). However, lengthy delays in estuaries can increase the risk of predation by marine mammals and humans, and expose fish to hypoxic, highly saline areas, which if occurring for more than 2 days results in high hematocrit values (Korstrom *et al.*, 1996), suggesting elevated stress levels and possibly dehydration. Recent swim tunnel respirometry on maturing Fraser sockeye collected in the ocean and held in either sea or freshwater revealed that the

Table 7.1

Plasma Ion Concentrations in Adult Sockeye Salmon, Coho Salmon, and Atlantic Salmon

Plasma variable	Salinity	Routine values			Postexercise values		
		Sockeye	Atlantic	Coho	Sockeye	Atlantic	Coho
Sodium (mmol/L)	FW	154 ± 2	143	—	159 ± 3	151	—
	SW	185 ± 3*	149	159 ± 11	213 ± 7*,†	178	186 ± 5†
Chloride (mmol/L)	FW	122 ± 7	134	—	121 ± 3	134	—
	SW	164 ± 2*	143	141 ± 10	173 ± 4*	160	168 ± 7†
Potassium (mmol/L)	FW	2.7 ± 0.9	3.8	—	2.0 ± 0.4	4.4	—
	SW	1.5 ± 0.3	3.7	4.3 ± 0.2	6.3 ± 0.8*,†	5.4	5. 2 ± 0.1
Osmolality (mOsm/kg)	FW	283 ± 2	307	—	334 ± 17	314	—
	SW	334 ± 17	314	305 ± 15	329 ± 24	362	377 ± 11†
Glucose (mmol/L)	FW	6.4 ± 0.7	—	—	7.1 ± 0.4	—	—
	SW	5.9 ± 0.6	—	—	8.6 ± 1.0	—	—
Lactate (mmol/L)	FW	0.8 ± 0.1	—	—	5.0 ± 1.5†	—	—
	SW	1.3 ± 0.3	—	1.0 ± 0.1	5.5 ± 1.8†	—	12.9 ± 2.3†
Cortisol (ng/ml)	FW	112 ± 14	—	—	229 ± 82	—	—
	SW	354 ± 95	—	71 ± 16	365 ± 96	—	194 ± 36†

*Significant difference ($P < 0.05$) between freshwater and seawater.

†Significant change postexercise.

Sources: G. Wagner and T. Farrell, unpublished data (adult sockeye salmon); Cech *et al.*, 2004 (coho salmon); and Byrne *et al.*, 1972 (Atlantic salmon).

Data are means ±1 SE. Samples were held in either saltwater or freshwater and were taken either under routine resting state or 30 minutes postexercise. No SE or statistics were available for Atlantic salmon.

costs of routine metabolism and swimming were 30% higher in seawater (Figure 7.5), a difference that could reflect ionoregulatory costs as well as restlessness (see Table 7.1 for elevated routine plasma cortisol and lactate levels). Therefore, energetic and survival advantages may accrue by not delaying the initiation of migration into freshwater.

A group of sockeye populations, termed "late-run" Fraser River stocks, which exhibit unique migration behaviour and a recent migration problem, exemplify the linkage between physiological and environmental factors in affecting migration initiation. Late-run populations, like "summer-run" populations, arrive at the river mouth in August, but unlike other sockeye population, late-runs remain near the river mouth milling in the ocean for 4–6 weeks without feeding before initiating river migration. The milling behaviour allows reproductive development to continue while limiting the migrant's contact with a naturally occurring freshwater kidney parasite and relatively warm river migration and spawning ground temperatures, which would escalate the parasite's disease development (Cooke *et al.*, 2004b). Prolonged exposure to this parasite affects swimming performance and osmoregulatory abilities (Wagner *et al.*, 2005a). Since 1994, these fish have begun upriver migrations earlier than normal, reducing or eliminating this milling behaviour. The consequences are severe for these early migrants, as they suffer very high mortality before spawning (up to 95% in some years; Cooke et *al.*, 2004b). The precise causes of this curious and maladaptive changed behaviour are not completely understood, though early migrants have been found to have lower energy reserves and advanced reproductive states (S. Cooke and S. Hinch, unpublished data, 2003). Preliminary evidence has associated this changed behaviour with reductions in coastal zone salinity levels at depth (Rick Thomson, unpublished data, 2004). The premature initiation of migration may be caused by fish reacting to premature exposure to low salinity in coastal migration areas. These findings lend further support to the general role that osmoregulatory systems have in affecting migratory behaviour, and that decisions to initiate river migrations may be made considerable distances away from natal freshwater.

13.2. Energy Use

There is considerable anecdotal evidence from commercial and recreational anglers that most species of migratory salmonids stop eating shortly before reaching freshwater and do not eat during upriver migration. The latter is a common observation in studies of Pacific salmon migration behaviour and energetics (see Margolis and Groot, 1991; Brett, 1995). In our work on Fraser River sockeye and pink salmon energetics (e.g., Crossin *et al.*, 2003, 2004), we have never encountered food in the stomachs of river

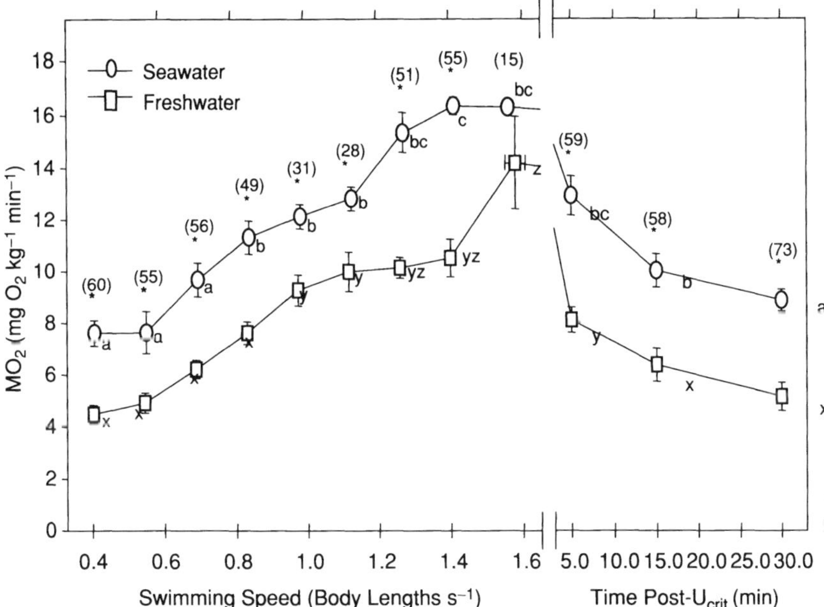

Fig. 7.5. Oxygen uptake of adult sockeye salmon in saltwater (circles) and freshwater (squares) during critical swimming speed (U_{crit}) tests (left side of figure). At all levels of activity except at the maximum U_{crit} oxygen uptake was lower ($*P < 0.05$) in freshwater than in seawater (from Wagner *et al.*, 2005b). This indicates that in terms of oxygen costs, life in seawater would be more expensive than in freshwater. Note that the cost of swimming theoretically increases exponentially with swimming speed based on the cost of drag through water and so an exponential relationship should exist between MO_2 and swimming speed. This relationship does not always occur. In these data, oxygen is consumed exponentially at the lower speeds but this relationship changes at faster speeds. The main reason this phenomenon occurs is that as salmon approach their critical swimming speed, they power swimming anaerobically through glycolytic activity in white muscle, and associated with this are changes in swimming gait. The true cost of this swimming is deferred to the recovery period. This phenomenon, known as excess postexercise oxygen consumption (EPOC), has been modeled for salmon to increase MO_2 in an exponential fashion with increasing swim speeds beyond 60% of U_{crit} (from Lee *et al.*, 2003a). Applying these findings to this figure, EPOC would become a factor at about one body length per second, thereby accounting in part for the nonexponential nature of these data. It is also important to note that the percentage difference (in brackets) between MO_2 in saltwater and freshwater can be as small as 15%. The difference in oxygen cost can be largely deferred, but this is temporary because as soon as the recovery starts, EPOC is 58–73% higher in saltwater than in freshwater (see Wagner *et al.*, 2005b for details on EPOC). Oxygen recovery measurements (right side of figure) were made at a swimming speed of 0.45 m/s, which allowed fish to rest on the bottom of the swim chamber and orientate into the water current. The vertical and horizontal bars around the mean values indicate the SEM for MO_2 and swimming speed, respectively.

migrating fish (S. Hinch, personal observation, 2005). However, there is less quantitative information on where feeding actually stops in the marine environment, probably because the traditional approaches (i.e., capturing fish at different locales and different times, and looking at stomach contents) have not generally confirmed stock origin using genetics, and have not been able to identify if specific fish are truly in migration mode or not. For example, chinook salmon caught off the west coast of Vancouver Island (Figure 7.1) in September had little food in their stomachs, and none in October when captured in the lower Fraser River (Healey, 1991). This may indicate feeding largely stopped somewhere in coastal waters while migrating the 200–300 km between those two locales. However, whether the Vancouver Island samples were Fraser River fish, or in true migration mode, is not certain. More definitive results, however, were recently provided by G. Crossin's work on salmon bioenergetics (unpublished data, 2003). On July 29–31, 2003, 30 adult pink and 30 sockeye salmon (DNA analyses confirmed all fish were from the Fraser River) were captured in the ocean west of the Queen Charlotte Islands while migrating southward (Figure 7.1, site 1). Every fish's stomach was full to capacity of prey, primarily euphausid shrimps. Two weeks later, over 50 pink and 50 sockeye salmon were captured in Johnstone Strait while migrating southward (Figure 7.1, site 2) and DNA analyses again confirmed fish were Fraser origin and were in fact from the same sockeye stock groups as those sampled at the Queen Charlotte Islands. Most of the stomachs of these fish were empty, though a few contained small amounts of residual digested matter. What is clear from these results is that feeding was maximal at 800 km, but ceased before 250 km, from natal freshwater. Depending on the stock, sockeye and pink salmon will take 2–4 weeks to travel from Johnstone Strait into the Fraser River, and are thus largely in a catabolic state during this time.

Because most migrants do not feed during river migrations and may well not feed for weeks prior to initiating migrations, appropriate allocation of declining energy reserves is essential for long distance migrations, reproductive development, and spawning behaviours. Body constituent analyses for fish sampled before and after upriver migrations reveal that swimming activity of salmonids typically deplete 75–95% of body fat, which is about half of the salmon's total energy reserves (reviewed in Brett, 1995). Roughly 55% of a female sockeye salmon's available energy is used for swimming activity, whereas 23% is used to ripen gonads and reproduce (Brett, 1983). American shad (*Alosa sapidissima*), which spawn in southern regions with relatively stable and warm spawning areas, are semelparous (i.e., spawn once and die) and use as much as 70–80% of endogenous energy stores (Leggett and Carscadden, 1978). They only use 35–65% (Leonard and McCormick, 1999) in more northerly systems where water is cooler, spawning areas are less stable, and populations tend to be iteroparous (i.e., can repeat spawn).

Fish size also directly affects energy use, with large Atlantic salmon (Jonsson *et al.*, 1997) and American shad (Leonard and McCormick, 1999) using relatively more energy, after correcting for body size, than small fish to complete the migration.

Populations of sockeye bound for distant inland spawning locales start their migration with nearly twice the somatic energy concentration as that of populations bound for nearby coastal watersheds (Figure 7.6) because of the long distances and steep grades that are encountered by the former (Figure 7.1). Short distance coastal populations also initiate upriver migrations with more mature gonads (Crossin *et al.*, 2004), thus distant inland migrations require a greater amount of energy for gonad ripening, as well as locomotion (Figure 7.6). Average egg size for coastal spawning populations is larger than that for distant inland populations (Crossin *et al.*, 2004). Similar patterns of premigration energy storage and egg size in relation to migratory distance are evident in other Pacific salmon species and regions (reviewed in Brett, 1995).

Hendry and Berg (1999) related reproductive development and specific energy stores for Alaskan sockeye salmon. Mass-specific somatic energy declined from freshwater entry (6.7% lipid, 20.6% protein, 6.6 MJ/kg) to the start of spawning (1.6% lipid, 18.0% protein, 4.5 MJ/kg). Stored lipid appeared to be used primarily for upriver migration and egg production, whereas stored protein seemed to be used primarily for the development of secondary sexual characters and metabolism during spawning. Most development of secondary sexual characters occurred late in maturation, perhaps to delay deterioration of muscle tissue. Similar results exist for populations of Fraser sockeye (Figure 7.6) and bioenergetic modeling indicates that swimming metabolism is the primary consumer of energy, with aerobic swimming activity dominating and consuming 10–25 times more of the energy reserves than standard metabolism (Figure 7.7). In terms of energy management, it would seem that minor delays in migration could occur with little energetic consequence provided the fish did not have to swim excessively (i.e., encounter hydraulic barriers requiring anaerobic metabolism). Assuming energy management issues are minimal, reproductive timing is probably the primary driving force for prompt river passage.

Lipid constituents associated with upriver migration have been explored in detail. Fatty acids are the main energy source for swimming during the migration. Total plasma nonesterified free fatty acid (NEFA) concentrations declined in both males and females to 60% and 40% of their respective initial levels during upstream migration of Fraser sockeye salmon (Ballantyne *et al.*, 1996). Nevertheless, palmitic (10:0), oleic (18:1), docosahexaenoic (22:6 n-3), and eicosapentaenoic (20:5 n-3) acids remained at 66% to 77% of the total plasma NEFAs. Some lipid constituents are also important for gonadal development (Ballantyne *et al.*, 1996). Interestingly, a difference

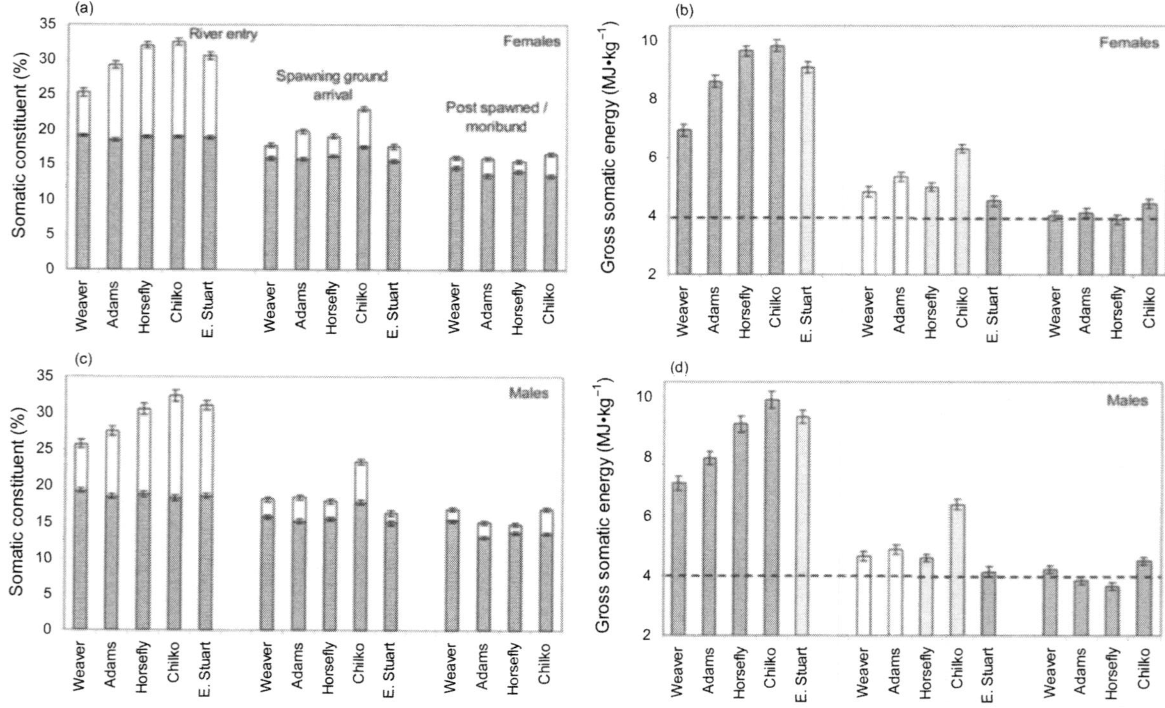

Fig. 7.6. Mean (±1 SE) protein and lipid percentages (a and c) and gross energy concentrations (b and d) in the soma of female and male Fraser River sockeye collected in 1999. Collection locations are indicated in (a), and are the same in subsequent panels. Dark bars in panels (a) and (c) are for protein, and light bars are for lipid. N for each bar in each panel is ≥10. The dashed line in panels (b) and (d) is speculation at the approximate energetic threshold to sustain life. See Figure 7.1 for location of spawning grounds. (Adapted from Crossin *et al.*, 2004.)

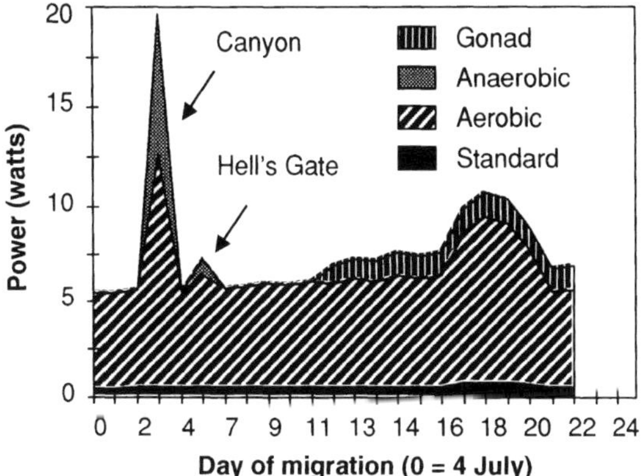

Fig. 7.7. Time series of simulated power consumption during the spawning migration, from the mouth of the Fraser River to spawning grounds, for an average female Early Stuart sockeye. Power has been partitioned among swimming costs (aerobic and anaerobic), standard metabolism, and gonad development. Metabolic costs were estimated using EMG telemetry and bioenergetics modeling. Gonad development was invoked on day 10; however, recent work (Patterson *et al.*, 2004b) has revealed that gonads are developing from day 1. Swimming costs are elevated during passage through hydraulic barriers (Canyon and Hell's Gate) and through very warm tributaries encountered on day 16. (Adapted from Rand and Hinch, 1998b.)

in monoene levels between sexes suggested that females utilised these, particularly oleic acid, for yolk production.

Lipids are not the sole source of energy for upriver migration. Proteins in white muscle can be metabolised during migration. Mommsen *et al.* (1980) determined that soluble and insoluble protein from white muscle decreased by 70% during migration. Many of the major muscle enzymes also decreased; however, the activities of cathepsin D and carboxypeptidase A increased considerably. Protein levels and enzyme activities of the red muscle and heart changed little, with the exception of glucose-6-phosphate dehydrogenase, which increased threefold to sixfold. In liver, the activities of metabolic enzymes and the levels of soluble protein decreased, whereas proteolytic enzyme activities increased slightly during migration.

13.3. Physiological Changes

During upriver migration, changing levels of reproductive hormones under mainly brain-pituitary-gonadal axis controls, regulate the development of secondary sexual characteristics, gonads, and other maturation indices (Ueda *et al.*, 2000b; Leonard *et al.*, 2001). Figure 7.4 illustrates an example of the changes in plasma sex hormones during the migration of the Early

Stuart sockeye. Salmon gonadotropin-releasing hormone (sGnRH), produced from the ventrocaudal telencephalon and the preoptic area, gonadotropins (GtH I and II) from pituitary gland, and gonadal steroid hormones all seem to influence both the homing migration and gonadal maturation in salmonids (Ueda *et al.*, 2000b). Upriver migrating female sockeye typically maintain high levels of estradiol-17 beta until immediately before reaching the spawning grounds, at which time it decreases to a minimal level (Truscott *et al.*, 1986). Free testosterone remains extremely high throughout the migration, whereas free and conjugated 17 alpha, 20 beta-dihydroxy-4-pregnen-3-one (17 alpha, 20 beta-P) peaks immediately before reaching spawning grounds. Males maintain high serum levels of 11-ketotestosterone throughout all phases of upriver migration (Truscott *et al.*, 1986).

Interrenal cell hypertrophy leads to increased (near maximal) plasma concentrations of cortisol in sockeye salmon (Schmidt and Idler, 1962), which are elevated to levels that would be observed following exposure to chronic stress in nonmigrating fish (Carruth *et al.*, 2002; Figure 7.4; Table 7.1). These elevated levels of plasma cortisol are likely to be adaptive and maladaptive in cortisol's three principle roles: a stress hormone, a mineral hormone, and a substrate for the production of sex hormones. High water temperatures or high levels of swimming activity can further elevate cortisol (Macdonald *et al.*, 2000) and in extreme conditions the elevation can be associated with a reduction in sex hormone titers (Figure 7.4). Elevated plasma cortisol is associated with the somatic and neural degeneration observed during the upriver migration (Carruth *et al.*, 2002). Excessively high cortisol levels prior to spawning reduce sperm quality (Parenskiy and Podlesnykh, 1995). Nevertheless, high cortisol titers are likely to be important in a high turnover of plasma sex hormones, but their role in osmoregulatory challenges is unclear. Plasma cortisol levels can peak on the spawning ground and it is unclear whether this is a reflection of the stresses of spawning behaviours or a trigger for them, especially because cortisol is so implicated in dominant-subordinate behaviours (see Chapter 5). Carruth *et al.* (2002) propose that increased plasma cortisol enhances the ability of migratory fish to recall the imprinted memory of home-stream olfactory odours. The authors argue that elevated plasma cortisol primes the hippocampus or other olfactory regions of the brain to recall the home-stream olfactory memory, thus aiding in the homing to natal streams. Thus, there may be a fine line between cortisol helping and hindering migration (see also Section 9.2.).

13.4. Swimming Speeds

The pattern of swimming speeds largely dictates how energy gets used during the migration. There are several lines of evidence indicating that behaviours and physiological systems that can facilitate energy conservation are critically important for successful migration. First, sockeye populations with difficult river migrations depart the ocean with high reserve energy (Figure 7.6) and morphologies that favour energy conservation (i.e., fusiform, with short and round bodies) relative to populations with less difficult migrations (Crossin *et al.*, 2004). Second, body constituent analyses on two species (sockeye and pink salmon) sampled in different years and from disparate regions have revealed that in general, spawning ground salmon die when energy levels decline to about 4 MJ/kg (Hendry and Berg, 1999; Crossin *et al.*, 2003; Figure 7.6). This suggests an energetic threshold is responsible in part for mortality in salmonids. Third, high energy use is associated with mortality during river migrations. In the Fraser River when discharge is unusually high, migration rates of some stocks are slowed, extending migration duration by several weeks. This can deplete energy reserves to levels below critical thresholds (Rand and Hinch, 1998b). In years of extremely high discharge, hundreds of thousands of Fraser sockeye have died during their migration and energy exhaustion is thought to be partly responsible (Macdonald and Williams, 1998).

Few studies have explored swimming speeds and their patterns of adult salmon in natural riverine habitats, although behaviours and swimming speeds in fishways and culverts have been studied (reviewed in Cooke et *al.*, 2004a). Electromyogram (EMG) telemetry (Cooke *et al.*, 2004a) with adult sockeye and pink salmon migrating through challenging but typical reaches in the Fraser River revealed that fish exhibited sustained speeds 76% of the time, prolonged speeds 18% of the time, and burst speeds 6% of the time (Hinch *et al.*, 2002). Neither sex nor species differed in these proportions, and mean swimming speeds did not differ between the two species. However, sockeye were twice as variable in their speeds because they alternated between burst and sustained (occasionally intermixed with coasting; Figure 7.8). In contrast, pink alternated between prolonged and sustained (occasionally intermixed with bursts; Hinch *et al.*, 2002). Pauses in active movement (i.e., coasting), as well as alterations in the duration and speed of movement, form a dynamic system of intermittent locomotion in which animals adjust their locomotory behaviour to changing environmental conditions (Kramer and McLaughlin, 2001). EMG telemetry revealed that adult chinook salmon migrating up the Klickitat River, Washington used burst speeds 2% of the time (Brown and Geist, 2002). By being 50% longer and 250% heavier than the Fraser sockeye studied above, it is possible that either the larger chinook with their higher aerobic speeds do not need to invoke anaerobiosis as often, or alternatively the hydraulics of the two river systems were different. Although burst swimming behaviours are critical to negotiate hydraulic challenges, they are used sparingly, likely because it is energetically costly.

The metabolic cost of burst (or sprint) swimming has not been accurately estimated for adult salmon. At speeds approaching 60–70% of the critical

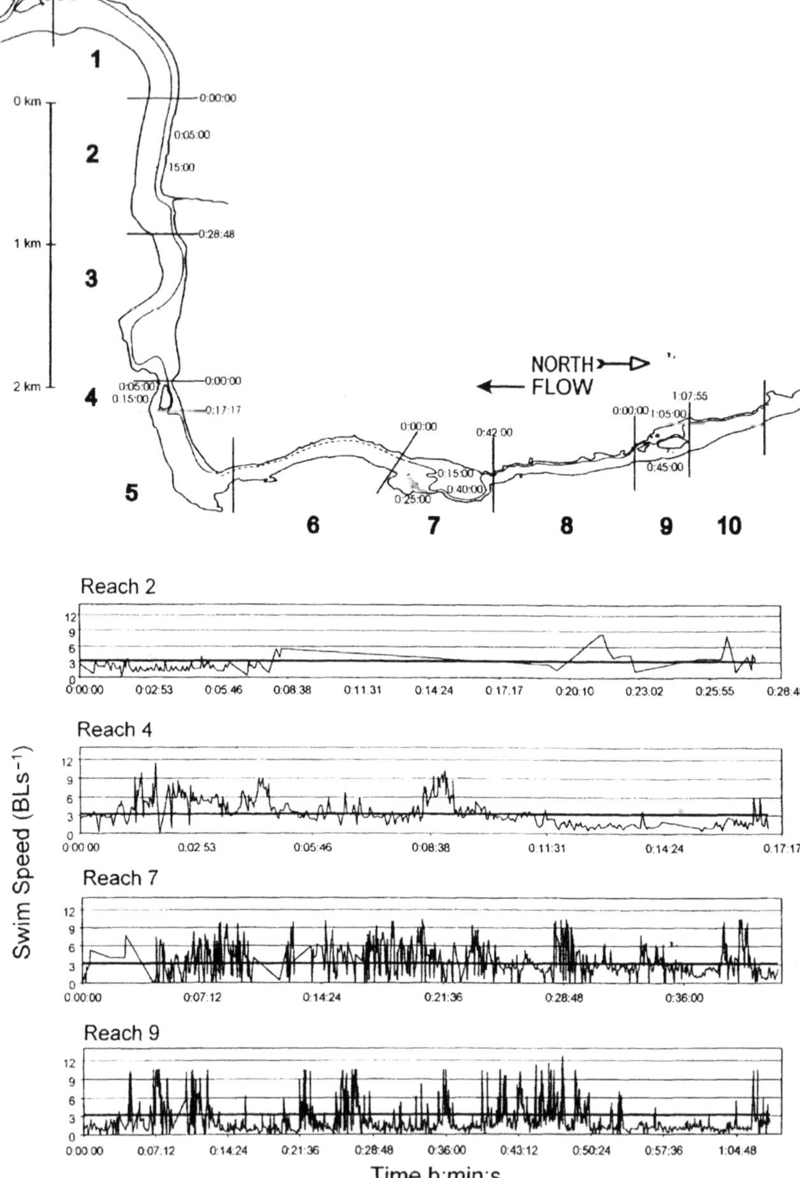

Fig. 7.8. Upper panel is a map of a 7-km section of the Fraser River indicating 10 reaches through which individual adult sockeye were tracked with EMG telemetry. The meandering thin, solid line between the two outer lines (the river banks) indicates the migratory path of a single fish. The fish migrated east then north upstream through the study area. Times of entry,

swimming speed (U_{crit}), Pacific salmon and rainbow trout begin to utilise anaerobic metabolism (Burgetz *et al.*, 1998; Lee *et al.*, 2003b). By measuring excess postexercise oxygen consumption following U_{crit}, the nonaerobic costs in adult sockeye and coho were estimated to add an additional 20–50% to the oxygen consumption measured at U_{crit} (Lee *et al.*, 2003b). Anaerobic costs can equal one-third of total daily energy use for sockeye migrating through rapids in the Fraser River canyon (Figure 7.7). Furthermore, extended bouts of bursting lead to exhaustion and the accumulation of wastes such as lactic acid, which in both laboratory and field situations can cause metabolic acidosis and delayed mortality (Black, 1958; Wood *et al.*, 1983; Farrell *et al.*, 2001). Indeed, hyperactive swimming by Fraser sockeye, recorded by EMG telemetry at Hell's Gate fishway, was associated with passage failure and inriver mortality (Hinch and Bratty, 2000). In fact, when passing through specific rapids or waterfalls, both Fraser sockeye and Klickitat chinook spent considerable time (50% and 60%, respectively) burst swimming. In adult salmonids and other species, not surprisingly, the white muscle group used for burst swimming comprises 60–70% of total muscle mass (Bone, 1978), underscoring its importance for fish survival by serving as an "insurance policy" despite high costs for development and maintenance.

Females allocate significantly more stored energy to gonad development during migration than do males (14% vs. 2%, respectively; reviewed in Brett, 1995). To compensate, it is logical that energy saving swimming patterns should be more strongly selected for, and exist more prominently, in females, which are also physically smaller than males. EMG studies in Fraser River salmon support the notion that females have different swim patterns and are more efficient in their use of energy. At reaches with river bank constrictions and complex river hydraulics where passage is presumably most difficult, female sockeye and pink salmon exhibited disproportionately high levels of sustained and prolonged swimming speeds compared with males, which exhibited disproportionately high burst rates (Hinch *et al.*, 2002). As a result, the energetic costs for male sockeye were an order of magnitude higher than that estimated for females at those reaches (Standen *et al.*, 2002). However,

exit, and intermediary points are indicated for four specific reaches. These four reaches span the range of difficulty of migratory habitats in terms of variability in flow dynamics. Reach 2 is a typical migratory reach with parallel banks and no unusual flow patterns. Reaches 4, 7, and 9 contain islands or peninsular outcroppings which create multidirectional and fast flow patterns. The dotted sections of the migratory path are estimated locations when telemetry data were not available. The bottom panel shows time-series plots of instantaneous swimming speeds for the sockeye tracked in the upper panel from time of entry to exit for each of the four reaches. The solid line represents U_{crit}. (Adapted from Hinch *et al.*, 2002.)

there were no differences between sexes (or species) in energetic costs to migrate through reaches with straight banks and relatively simple river hydraulic flows—conditions that represent their typical migratory reaches. Males do, however, invest more energy into distinctive morphologies (a large dorsal hump and extended kype), which likely increase their hydrodynamic drag (Videler, 1993) and potentially force them to increase their swimming speed and energy expenditure to reach the spawning grounds at the same time as females. A difference between sexes in hydrodynamic drag may be most evident at reaches with fast and complex flows. Cardiac and metabolic processes also differ between males and females as they near reproductive maturity. Male salmonids typically have a larger ventricular mass (Farrell *et al.*, 1988). During respirometer trials with sockeye, males swam at consistently higher VO_2 max (maximum volume of oxygen consumed by a swimming fish) than females of the same size swimming against the same water speeds, indicating that females are more efficient swimmers (MacNutt, 2003).

Aside from the influence of sex or river reach, there is evidence that fish swim at individual-specific swimming speeds (Hinch *et al.*, 2002). Size may play a role. Small male sockeye have faster absolute swim speeds than large males (Hinch and Rand, 1998) though the reasons for this are not clear. Size did not explain the variation in swim speeds and energy use of pink salmon among reaches (Standen *et al.*, 2002). Neither fish size nor condition was correlated with successful passage of chinook through waterfalls on the Klickitat River (Brown and Geist, 2002). It is possible there is a genetic propensity for certain swimming behaviours in some individuals. Efforts evaluating to what degree body size (according to different scaling relationships) affects migratory costs and puts constraints on migratory behaviour are a common theme throughout migration biology research on many animal taxa (Alerstam *et al.*, 2003).

Anadromous and potamodromous upriver migrants should be naturally selected to be efficient in their use of energy and attempt to minimise costs of swimming wherever possible. Transport costs can be reduced by swimming at hydrodynamic or metabolic optimal speeds that minimise total energy expended in moving unit mass through unit distance. Bernatchez and Dodson (1987) compared energetic costs of migration for short- and long-distance upriver-migrating anadromous fishes and concluded that long distance migrants swam at optimal speeds. Their approach involved estimating swimming speeds in the field from laboratory respirometry trials based on total energy expenditure, migration duration, and mean river current speeds. They assumed fish experienced mean current speeds, swam at steady speeds, and incurred no additional costs from anaerobic metabolism (i.e., no burst swimming). However, these assumptions have not been realised.

Long-distance migrating sockeye select low current paths where possible (Standen *et al.*, 2004), exhibit nonsteady sustained and prolonged speeds, and occasionally exhibit bursts thus invoking anerobiosis (e.g., Figure 7.8). It is unclear how these assumption violations would bias conclusions.

Hinch and Rand (2000) used underwater stereo videography to directly measure encountered water flows, and swim and ground speeds of several populations of sockeye at different sites along migratory routes. They found that salmon adjusted their swimming speeds, based on encountered currents, and swam at optimal speeds when currents were slow (e.g., <20 cm/s). They concluded that this energy conservation behaviour may be more broadly used than was originally surmised by Bernatchez and Dodson (1987). Using swim tunnel respirometry, Lee *et al.* (2003a) demonstrated three significant findings about optimal swim speeds in mature adult Pacific salmon. First, optimal swim speeds differed between species with similar migratory routes and spawning locations (Figure 7.9). Second, optimal swim speeds differed among stocks within a species, and third, they differed among levels of maturation within a given stock.

Trump and Leggett (1980), in theoretically analysing the optimal swimming speeds for fish in currents, suggested that minimum energy expenditure in the absence of water currents occurs near to the optimal cruising speed (i.e., minimum COT) but in the nonoscillating currents such as rivers, fish should realise an average ground speed of about one body length per second. Although this theoretical calculation is consistent with empirical estimates of ground speed from travel times of salmon and other fishes, Trump and Leggett (1980) noted that no data supported the hypothesis that fish can anticipate and integrate periodic changes in current velocity. Recent work with smallmouth bass (*Micropterus dolomieu*) voluntarily ascending a 50-m raceway against different water currents has clearly revealed that migrants can maintain constant ground speed independent of water velocity until they reach swimming speeds that require mixed gait swimming (steady aerobic swimming and burst-and-coast swimming gaits; Peake and Farrell, 2004). Once this threshold is reached, ground speed increases and passage time decreases. Using underwater stereovideography to monitor migrations of sockeye swimming up tributaries of the Fraser River, a similar phenomenon has been observed (Hinch and Rand, 2000). Sockeye maintained swimming speeds around 100 cm/s under low to moderate encountered water speeds (up to 20 cm/s) but accelerated their swimming speeds (>160 cm/s) when currents exceeded 20 cm/s.

Although intermittent swimming behaviour would be expected to increase energetic costs due to additional expenditure for acceleration and deceleration, there are also a number of energetic benefits that can be derived if forward movement continues even when active movement has stopped

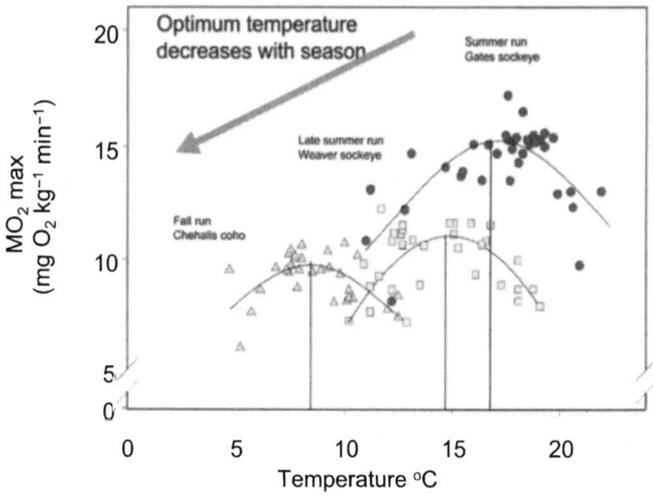

Fig. 7.9. Population-specific relationships between oxygen consumption at U_{crit} (MO_{2max}) and water temperature. The within- and among-population relationships between aerobic scope-for-activity and temperature are not shown but are consistent with those reported in this figure for U_{crit}. Each datum represents a different fish. Fish were swum at either their ambient capture temperature or at some adjusted temperature. Most fish were tested in the field. The vertical lines indicate the average ambient water temperature normally encountered by upriver migrants within each population, which corresponded closely with population-specific peak MO_{2max}. Note that as the migration season progresses from early summer to late fall, the optimum temperature becomes cooler. Chehalis coho (triangles) and Weaver sockeye (squares) spawn in locales near each other but migrate and spawn at different times of the year (winter and fall respectively) thus have different ambient migration temperatures. Gates sockeye (circles) migrate in mid-summer, travelling 500 km to spawning areas inland, and must ascend through fast flows and rapids en route (see Figure 7.1). Chehalis and Weaver populations travel very short distances to their coastal spawning areas, encountering relatively low flow en route conditions. Selection for superior aerobic swim performance (i.e., the relatively high U_{crit} values for Gates sockeye) may thus not be strong for Chehalis and Weaver populations. The specific numerical and statistical models for these relationships are found in Lee *et al.* (2003a). (Adapted from Lee *et al.*, 2003a.)

(Kramer and McLaughlin, 2001). Indeed, swimming in a "burst-then-coast" fashion has been suggested to reduce energetic costs of transport in nonsalmonids by 60% (Weihs, 1974). There is some evidence from EMG telemetry research that sockeye (Hinch *et al.*, 2002; Figure 7.8) and chinook (Brown and Geist, 2002) may occasionally invoke this pattern when ascending fast flowing reaches.

River temperature influences several aspects of swimming performance (e.g., MO_2 max, U_{crit}, scope for activity), all of which have optimal values at specific temperatures (reviewed in Brett, 1995). Using field respirometry,

Farrell *et al.* (2003) demonstrated that some species, and populations within species, respond to temperature in ways that reflect their migratory and natal stream environments. For instance, adult Gates Creek and Early Stuart sockeye, which migrate to distant inland lakes (Figure 7.1) in midsummer have maximum scope for activity and maximum MO_2 at 17 °C, whereas the optima have a maximum at 15 °C for Weaver Creek sockeye migrating a short distance to a coastal lake in early fall (Figure 7.9). Nevertheless, the temperature optima were broad and fish could reach 90% of peak maximum MO_2 over a 5 °C temperature range (Lee *et al.*, 2003a). Moreover, other Pacific salmonid species such as Fraser River pink salmon (MacNutt, 2003) exhibit weak thermal optima for U_{crit}, perhaps reflecting a broader range of encountered natal river temperatures. Swimming performance experiments with cutthroat trout (*Oncorhynchus clarki clarki*) have shown that fish can thermally adjust to temperature changes of up to 5 °C either side of the acclimation temperature at 1 °C per day (MacNutt *et al.*, 2004).

13.5. Selection of Migratory Paths

The selection of specific travel paths is influenced by several environmental and physiological factors. Although migrants rely largely on olfactory cues to select and perhaps navigate towards spawning areas, direction of flow is the primary guide. Migrants are negatively rheotactic. By orientating into the current, salmon derive an energetic benefit from ram gill ventilation, which is estimated to save up to 10% of metabolic rate (Farrell and Steffensen, 1987). Nevertheless, EMG studies have shown that sockeye can override negative rheotaxis for short periods when they encounter flows that are heading in the upstream direction as a result of flow direction changes associated with river bank peninsulas (Hinch *et al.*, 1996, 2002; Hinch and Rand, 1998). However, complex flow patterns arising from bank constrictions, islands, and river bends delay upstream progress for hours to days, suggesting a difficulty in locating clear flow cues (Hinch and Rand, 1998; Brown and Geist, 2002). Fish may then select the near shore to take advantage of boundary layer conditions, even though this increases the risk of exposure to mammalian predators. Even if appropriate directional flow cues can be located, migrants may not be able to pass through some areas if flows are extremely fast and surpass burst swimming capabilities. This phenomenon was observed in the Fraser River in 1997 at Hells Gate where high water speeds prevented about half of the Early Stuart stock (~500,000 sockeye) from passing; this ultimately resulted in en route mortality of these fish (Macdonald and Williams, 1998).

Under slow and moderate flow conditions, path selection can play a significant role in reducing energetic costs of transport. Underwater stereovideography has revealed that adult migrants select low speed paths. Current speeds encountered during the migration of adult Gates Creek Fraser sockeye were lower than mean current speeds measured in the immediate vicinity of the migrating fish at those locales (Standen *et al.*, 2004). EMG telemetry and videography studies have shown that sockeye move faster through water than their tails are propelling them, thus are receiving forward-assists during their migrations (Hinch and Rand, 1998, 2000). Migrants are likely locating and exploiting very small-scale reverse-flow vortices created by rough substrates or banks (Vogel, 1994). When exposed to flow vortices under laboratory conditions, rainbow trout adopt a locomotory behaviour akin to salmon in between vortices by activating only their anterior axial muscles. Energetic costs of forward motion are thus reduced by using vortices in this manner (Liao *et al.*, 2003).

Path selection varies among species, which may be related to differences in their size. Generally, migration paths for larger-bodied fish are situated further from river banks than for smaller-bodied fish. In the Fraser River, pink salmon migrate very close to shore, swim in tight aggregations, and rarely make sojourns away from the bank to cross the river (Hinch *et al.*, 2002). Sockeye tend to migrate farther from shore in deeper water, in much looser aggregations (Xie *et al.*, 1997) and cross the thalwag (i.e., the deepest portion of the river) repeatedly during migrations (Figure 7.8). Sockeye seem to utilise more diverse migration paths than pinks. One reason for this is that sockeye are physically able to roam more freely as they are larger than pink salmon (58% heavier, 13% longer; Hinch *et al.*, 2002), so they should be able to generate more power for a given tailbeat and thus could make forward progress against faster and more diverse currents than pink salmon. An alternative reason for this lateral segregation is wave drag, which is the resistance associated with the generation of surface waves when swimming close to the surface (Hughes, 2004). Large fish may swim further from the bank to avoid wave drag, which scales according the ratio of maximum body diameter to submergence depth, so bigger fish need to swim deeper to escape its effects. In the Nushagak River, Alaska, the wave drag model accurately predicted that the migration corridor for the large-bodied chinook salmon should be well away from the banks compared to that of the relatively small-bodied sockeye salmon, which are found nearer the banks (Hughes, 2004) and can take advantage of the boundary layer.

Migrants are faced with an energetic dilemma. They need to conserve energy during migrations to ensure that they can reach spawning areas and still have enough energy to mature gonads and successfully spawn. However, they have a limited amount of time available to reach spawning areas—significant migration delays could have a negative impact on fitness. It is thus not surprising that different tactics are used when faced with different flow conditions. When currents are slow to moderate, energy conserving behaviours like swimming at optimal speeds, selection of low water speed paths, or the location of areas offering forward-assisted propulsion can be adopted. Under high flow conditions, sockeye migrants do not select low-speed paths (Standen *et al.*, 2004). This may occur because it is either difficult to locate or energetically costly to relocate to microsite flow patterns when surrounding flows are fast, and/or when the risk of experiencing delays is large. Thus, migrants may use paths where they have little choice but to invoke fast swimming speeds (Standen *et al.*, 2002, 2004). Minimising travel time could become more important than minimising energy cost.

Path selection can also be influenced by water temperature. In particular, rapid temperature changes encountered, for instance, when entering tributaries that are either extremely warm or cold can halt anadromous migrations (Bjornn and Reiser, 1991). Columbia River sockeye migrating towards Okanagan Lake ceased migrations when river temperatures exceeded 21 °C, but resume upriver swimming when temperatures fall below this level (Major and Mighell, 1966). Though the physiological reasons behind such delays are not fully understood, effects of oxygen deficiencies, stress, and thermal acclimation are probably important. Levels of stress hormones are known to be elevated in Fraser sockeye that are experiencing above normal temperatures (Macdonald *et al.*, 2000) and this could affect swimming abilities. Although the optimal temperature for U_{crit} and maximum MO_2 appear to be well matched to normally encountered river temperatures (Lee *et al.*, 2003a; Figure 7.9), it is important to stress that these temperatures may be precariously close to the upper lethal temperature limit. As salmon approach this temperature limit, there appears to be cardiorespiratory collapse (Farrell, 1997).

13.6. Upriver Migration: Not Just a Means to an End?

Although the benefits of a successfully completed upriver migration are obvious (i.e., the opportunity to spawn), it has recently been shown that the migration act is also important for optimising reproductive maturation. Adult sockeye salmon captured at the mouth of the Fraser River at the start of their 1100-km migration were held in large tanks and allowed to mature for 3 weeks (a typical migration time) in either a moderate-flow (captive exercise group) or no-flow (captive nonexercise group) environment (Patterson *et al.*, 2004b). The expectation was that restricting exercise would result in a reallocation of energy to reproduction (i.e., increased fecundity, egg size, and gonad composition). However, this was not the case as these reproductive parameters were unchanged among the groups, even though at the completion of the trials, energy density for nonexercised fish was

50% higher than that of natal fish and 20% higher than that for exercised fish. Instead, nonexercised females had delayed maturity, lower egg deposition rates, and were more likely to die prior to egg ovulation than exercised females and natal spawners. Also, eggs from captive exercise adult females were more likely to survive to the eyed stage than eggs from captive non-exercise females. The mechanisms by which exercise improved reproductive development remain unclear, especially since plasma levels of glucose, lactate, cortisol, and reproductive hormones did not differ among the groups. No studies have investigated the role of exercise in mobilising lipids and proteins as a means to alter to egg depositional rates. It is well recognised that exercise training does improve the catabolism of fatty acids in feeding fishes (Johnston and Moon, 1980). Exercise also increases the deposition of lipids to white muscles during growth (East and Magnan, 1987). It is also possible that exercise stimulates vitellogenesis and final egg maturation. Results from Patterson *et al.* (2004b) support the idea that the normal physiological changes associated with exercise during adult freshwater migrations may play a key role in regulating the timing and quality of sexual maturation.

14. CONCLUSIONS

Our overview of the behavioural physiology of fish migrations has revealed that, for most phases of migration, we know very little. The noted exceptions are for the river migration phases and, even then, our knowledge comes from just a few species and studies that link behaviour with energetics and endocrinology. It is understandable that the least studied phases of migration are those in the ocean because of the logistic challenges in studying fish in such vast environments. However, we contend that the lack of behaviour-physiology linkages in most studies of migratory fish is not due to logistic constraints, but rather dogmatic ones. Reductionist approaches used in the study of both fish behaviour and physiology have led to isolated work focused on very specific questions. Necessary are research teams that transcend reductionism and create true interdisciplinarity, which is required to study migration biology of fish. Similar calls for more interdisciplinary research have been recently proposed for migration biology in general (Alerstam *et al.*, 2003).

Interdisciplinary research would facilitate needed integration between physiological laboratory experiments and empirical field observations. Behavioural ecology has been focused on field environments for some time, but it is only recently that field research has become a focus of physiologists (Costa and Sinervo, 2004). This likely reflects the growing recognition of the interplay between physiology and life history (Ricklefs and Wikelski, 2002), and the development of a more mechanistic field of ecology. Instead of just understanding patterns of distribution and behaviour, there is increased emphasis on understanding how fish achieve a specific behaviour (Altmann and Altmann, 2003). Efforts to assess the fitness implications of different behaviours and physiologies represent a frontier of modern biology (Irschick and Garland, 2001).

ACKNOWLEDGEMENTS

We thank Glenn Wagner and Glenn Crossin for providing unpublished data and assisting with formatting and editing, and Jeff Young for assisting with literature reviewing. We also acknowledge our collaborators who have helped to develop the ideas presented in this paper and have provided unpublished data. Funding for our salmon research has come from the Natural Sciences and Engineering Research Council of Canada Discovery and Strategic Research Programs. We thank Dave Patterson and the Environmental Watch Program of the Canadian Department of Fisheries and Oceans for research funding and logistic support. Several anonymous referees provided detailed comments that improved the manuscript.

REFERENCES

Akesson, S. (2002). Tracking fish movements in the ocean. *Trends Ecol. Evol.* **17,** 56–57.

Alerstam, T., Hedenström, A., and Åkesson, S. (2003). Long-distance migration: Evolution and determinants. *Oikos* **103,** 247–260.

Altmann, S. A., and Altmann, J. (2003). The transformation of behaviour field studies. *Anim. Behav.* **65,** 413–423.

Ballantyne, J. S., Mercure, F., Gerrits, M. F., Van Der Kraak, G., McKinley, S., Martens, D. W., Hinch, S. G., and Diewert, R. E. (1996). Plamsa nonesterified fatty acid profiles in male and female sockeye salmon, *Oncorhynchus nerka*, during the spawning migration. *Can. J. Fish. Aquat. Sci.* **53,** 1418–1426.

Bams, R. A. (1969). Adaptations of sockeye salmon associated with incubation in stream gravels. *In* "Symposium on Salmon and Trout in Streams. H.R. MacMillan Lectures in Fisheries" (Northcote, T. G., Ed.), pp. 71–87. Institute of Fisheries, University of British Columbia, Vancouver.

Beamish, R. J., Mahnken, C., and Neville, C. M. (2004). Evidence that reduced early marine growth is associated with lower marine survival of Coho salmon. *Trans. Am. Fish. Soc.* **133,** 26–33.

Beauchamp, D. A. (1995). Riverine predation on sockeye salmon fry migrating to Lake Washington. *N. Am. J. Fish. Mgmt.* **15,** 358–365.

Beaudet, L., Novales Flamarique, I., and Hawryshyn, C. W. (1997). Cone photoreceptor topography in the retina of sexually mature Pacific salmonid fishes. *J. Comp. Neurol.* **383,** 49–59.

Bernatchez, L., and Dodson, J. J. (1987). Relationship between bioenergetics and behaviour in anadromous fish migration. *Can. J. Fish. Aquat. Sci.* **44,** 399–407.

Bjornn, T. C., and Reiser, D. W. (1991). Habitat requirements of salmonids in streams. *In* "Influences of Forest and Rangeland Management on Salmonid Fishes and Their Habitats" (Meehan, W. R., Ed.), Vol. 19, pp. 83–138. American Fisheries Society.

Black, E. C. (1958). Hyperactivity as a lethal factor in fish. *J. Fish. Res. Bd. Can.* **15,** 573–584.

Blank, J. M., Morrissette, J. M., Landeira-Fernandez, A. M., Blackwell, S. B., Williams, T. D., and Block, B. A. (2004). *In situ* cardiac performance of Pacific bluefin tuna hearts in response to acute temperature change. *J. Exp. Biol.* **207,** 881–890.

Block, B. A. (1994). Thermogenesis in Muscle. *Annu. Rev. Physiol.* **56,** 535–577.

Bone, Q. (1978). Locomotor muscle. *In* "Fish Physiology" (Hoar, W. S., and Randall, D. J., Eds.), Vol. VII, pp. 361–424. Academic Press, New York.

Brannon, E. L. (1972). Mechanisms controlling migration of sockeye salmon fry. *Inter. Pac. Salm. Fish. Comm. Bull.* **21,** 1–86.

Brannon, E. L. (1987). Mechanisms stabilising salmonid fry emergence timing. *Can. Spec. Pub. Fish. Aquat. Sci.* **96,** 29–63.

Brauner, C. J., Iwama, G. K., and Randall, D. J. (1994). The effect of short-duration seawater exposure on the swimming performance of wild and hatchery-reared juvenile coho salmon (*Oncorhynchus kisutch*) during smoltification. *Can. J. Fish. Aquat. Sci.* **51,** 2188–2194.

Brett, J. R. (1971). Energetic responses of salmon to temperature: A study of some thermal relations in the physiology and freshwater ecology of sockeye salmon (*Oncorhynchus nerka*). *Am. Zool.* **11,** 99–113.

Brett, J. R. (1983). Life energetics of sockeye salmon, Oncorhynchus nerka. *In* "Behavioural Energetics: The Cost of Survival in Vertebrates" (Aspey, W. P., and Lustick, S. I., Eds.), pp. 29–63. Ohio State University Press, Columbus, Ohio.

Brett, R. (1995). Energetics. *In* "Physiological Ecology of Pacific Salmon" (Groot, C., Margolis, L., and Clarke, W. C., Eds.), pp. 1–68. UBC Press, Vancouver.

Brett, R. J., Shelbourne, J. E., and Shoop, C. T. (1969). Growth rate and body composition of fingerling sockeye salmon, *Oncorhynchus nerka*, in relation to temperature and ration size. *J. Fish. Res. Board Can.* **26,** 2363–2394.

Brill, R., Lutcavage, M., Metzger, G., Bushnell, P., Arendt, M., Lucy, J., Watson, C., and Foley, D. (2002). Horizontal and vertical movements of juvenile bluefin tuna (*Thunnus thynnus*) in relation to oceanographic conditions of the western North Atlantic determined with ultrasonic telemetry. *Fish. Bull.* **100,** 155–167.

Browman, H. I., and Hawryshyn, C. W. (1994). Retinoic acid modulates retinal development in the juvenile of a teleost fish. *J. Exp. Biol.* **193,** 191–207.

Browman, H. I., Novales-Flamarique, I., and Hawryshyn, C. W. (1994). Ultraviolet photoreception contributes to prey search behaviour in 2 species of zooplanktivorous fishes. *J. Exp. Biol.* **186,** 187–198.

Brown, R. S., and Geist, D. R. (2002). Determination of swimming speeds and energetic demands of upriver migrating fall chinook salmon (*Oncorhynchus tshawytscha*) in the Klickitat River, Washington. Project 22063, Contract 42663A. Pacific Northwest Laboratory, Bonneville Power Administration, Richland, Washington.

Burgetz, I. J., Rojas-Vargas, A., Hinch, S. G., and Randall, D. J. (1998). Initial recruitment of anaerobic metabolism during submaximal swimming in rainbow trout (*Oncorhynchus mykiss*). *J. Exp. Biol.* **201,** 2711–2721.

Burgner, R. L. (1962). Sampling red salmon fry by lake trap in the Wood River Lakes, Bristol Bay, Alaska. PhD Thesis, University of Washington.

Burgner, R. L. (1991). Life history of sockeye salmon (*Oncorhynchus nerka*). *In* "Pacific Salmon Life Histories" (Groot, C., and Margolis, L., Eds.), pp. 3–117. UBC Press, Vancouver.

Byrne, J. M., Beamish, F. W. H., and Saunders, R. L. (1972). Influence of salinity, temperature, and exercise on plasma osmolality and ionic concentration in Atlantic salmon (*Salmo salar*). *J. Fish. Res. Bd. Can.* **29,** 1217–1220.

Carey, W. E., and Noakes, D. L. G. (1981). Development of photobehavioural responses in young rainbow trout, *Salmo gairdneri* Richardson. *J. Fish Biol.* **19,** 285–296.

Carruth, L. L., Jones, R. E., and Norris, D. O. (2002). Stress and Pacific salmon: A new look at the role of cortisol in olfaction and home-stream migration. *Int. Comp. Biol.* **42,** 574–581.

Cech, J. J., McEnroe, M, and Randall, D. J. (2004). Coho salmon haematological, metabolic and acid-base changes during exercise and recovery in sea water. *J. Fish Biol.* **65,** 1223–1232.

Clark, C. W., and Levy, D. A. (1988). Diel vertical migrations by juvenile sockeye salmon and the antipredation window. *Am. Nat.* **131,** 271–290.

Clarke, W. C., Shelbourn, J. E., and Brett, R. J. (1981). Effect of artificial photoperiod cycles, temperature, and salinity on growth and smolting in underyearling coho (*Oncorhynchus kisutch*), chinook, (*O. tshawytscha*) and sockeye (*O. nerka*). *Aquaculture* **22,** 105–116.

Comeau, L. A., Campana, S. E., Chouinard, G. A., and Hanson, J. M. (2001). Timing of Atlantic cod *Gadus morhua* seasonal migrations in relation to serum levels of gonadal and thyroidal hormones. *Mar. Ecol. Prog. Ser.* **221,** 245–253.

Cooke, S. J., Thorstad, E. B., and Hinch, S. G. (2004a). Activity and energetics of free-swimming fish: Insights from electromyogram telemetry. *Fish and Fisheries* **5,** 21–52.

Cooke, S. J., Hinch, S. G., Farrell, A. P., Lapointe, M., Healey, M. C., Patterson, D. A., Macdonald, J. S., Jones, S., and Van Der Kraak, G. (2004b). Early-migration and abnormal mortality of late-run sockeye salmon in the Fraser River, British Columbia. *Fisheries* **29,** 22–33.

Costa, D. P., and Sinervo, B. (2004). Field physiology: Physiological insights from animals in nature. *Ann. Rev. Physiol.* **66,** 209–238.

Crossin, G. T., Hinch, S. G., Farrell, A. P., Whelly, M. P., and Healey, M. C. (2003). Pink salmon (*Oncorhynchus gorbuscha*) migratory energetics: Response to migratory difficulty and comparisons with sockeye salmon (*Oncorhynchus nerka*). *Can. J. Zool.* **81,** 1986–1995.

Crossin, G. T., Hinch, S. G., Farrell, A. P., Higgs, D. A., Lotto, A. G., Oakes, J. D., and Healey, M. C. (2004). Energetics and morphology of sockeye salmon: Effects of upriver migratory distance and elevation. *J. Fish Biol.* **65,** 788–810.

Davis, N. D., Myers, K. W., and Ishida, Y. (1998). Caloric value of high-seas salmon prey organisms and simulated salmon ocean growth and prey consumption. NPAFC Bulletin Number 1: Assessment and status of Pacific Rim salmonid stocks. no. 1, pp. 146–162. North Pacific Anadromous Fish Commission, Vancouver.

Dingle, H. (1996). Migration: The biology of life on the move Oxford University Press, Oxford.

Dodson, J. J. (1997). Fish migration: An evolutionary perspective. *In* "Behavioural Ecology of Teleost Fishes" (Godin, J. G., Ed.), pp. 10–36. Oxford University Press, Oxford.

Døving, K. B., and Stabell, O. B. (2003). Trails in open water: Sensory cues in salmon migration. *In* "Sensory Processing in Aquatic Environments" (Collin, S. P., and Marshall, N. J., Eds.), pp. 39–52. Springer-Verlag, New York.

Døving, K. B., Selset, R., and Thommesen, G. (1980). Olfactory sensitivity to bile acids in salmonid fishes. *Acta Physiol. Scand.* **108,** 123–131.

East, P., and Magnan, P. (1987). The effect of locomotor activity on the growth of brook charr, *Salvelinus fontinalis* Mitchell. *Can. J. Zool.* **65,** 843–846.

Enders, E. C., Boisclair, D., and Roy, A. G. (2003). The effect of turbulence on the cost of swimming for juvenile Atlantic salmon (*Salmo salar*). *Can. J. Fish. Aquat. Sci.* **60,** 1149–1160.

Endler, J. A. (1977). Geographical variation, speciation, and clines Princeton University Press, Princeton, NJ.

Farrell, A. P. (1997). Effects of temperature on cardiovascular performance. *In* "Global Warming Implications for Freshwater and Marine Fish" (Wood, C. M., and McDonald, D. G., Eds.), pp. 135–158. Cambridge University Press, Cambridge.

Farrell, A. P., and Steffensen, J. F. (1987). An analysis of the energetic cost of the branchial and cardiac pumps during sustained swimming in trout. *Fish Physiol. Biochem.* **4,** 73–79.

Farrell, A. P., Hammons, A. M., Graham, M. S., and Tibbits, G. F. (1988). Cardiac growth in rainbow trout, *Salmo gairdneri*. *Can. J. Zool.* **66,** 2368–2373.

Farrell, A. P., Gallaugher, P. E., and Routledge, R. (2001). Rapid recovery of exhausted adult coho salmon after commercial capture by troll fishing. *Can. J. Fish. Aquat. Sci.* **58,** 2319–2324.

Farrell, A. P., Lee, C. G., Tierney, K., Hodaly, A., Clutterham, S., Healey, M. C., Hinch, S. G., and Lotto, A. G. (2003). Field-based measurements of oxygen uptake and swimming performance with adult Pacific salmon (*Oncorhynchus* sp.) using a mobile respirometer swim tunnel. *J. Fish Biol.* **62,** 64–84.

Forward, R. B., and Tankersley, R. A. (2001). Selective tidal transport of marine animals. *Oceanogr. Mar. Biol.* **39,** 305–353.

Franklin, C. E., Forster, M. E., and Davison, W. (1992). Plasma cortisol and osmoregulatory changes in sockeye salmon transferred to sea water: Comparison between successful and unsuccessful adaptation. *J. Fish. Biol.* **41,** 113–122.

French, R., Bilton, H., Osako, M., and Hartt, A. (1976). Distribution and origin of sockeye in offshore waters of the North Pacific Ocean. *Int. N. Pac. Fish. Comm. Bull.* **34,** 1–113.

Fukaya, M., Ueda, H., Sato, A., Kaeriyama, M., Ando, H., Zohar, Y., Urano, A., and Yamauchi, K. (1998). Acceleration of gonadal maturation in anadromous maturing sockeye salmon by gonadotropic-releasing hormone analog implantation. *Fish. Sci.* **64,** 948–951.

Gilhousen, P. (1990). Prespawning mortalities of sockeye salmon in the Fraser River system and possible causal factors. *Int. Pac. Salm. Fish. Comm. Bull.* **26,** 1–58.

Giorgi, A. E., Hillman, T. W., Stevenson, J. R., Hays, S. G., and Peven, C. M. (1997). Factors that influence the downstream migration rates of juvenile salmon and steelhead through the hydroelectric system in the mid-Columbia River basin. *N. Am. J. Fish. Mgmt.* **17,** 268–282.

Groot, C. (1972). Migration of yearling sockeye salmon (*Oncorhynchus nerka*) as determined by time-lapse photography of sonar observations. *J. Fish. Res. Board Can.* **29,** 1431–1444.

Groot, C. (1982). Modifications on a theme: A perspective on migratory behaviour of Pacific salmon. *In* "Proceedings of the Salmon and Trout Migratory Behaviour" (Brannon, E. L., and Salo, E. O., Eds.), pp. 1–21. Fisheries Department, University of Washington, Seattle.

Groot, C., and Cooke, K. (1987). Are the migrations of juvenile and adult Fraser River sockeye salmon (*Oncorhynchus nerka*) in nearshore waters related? *Can. Spec. Pub. Fish. Aquat. Sci.* **96,** 53–60.

Groot, C., and Margolis, L. (1991). "Pacific Salmon Life Histories." UBC Press, Vancouver.

Groot, C., and Quinn, T. (1987). Homing migration of sockeye salmon, *Oncorhynchus nerka*, to the Fraser River. *Fish. Bull. US Fish Wildlife Ser.* **85,** 455–469.

Groot, C., Quinn, T. P., and Hara, T. J. (1986). Responses of migrating adult sockeye salmon (*Oncorhynchus nerka*) to population-specific odors. *Can. J. Zool.* **44,** 926–932.

Groot, C., Margolis, L., and Clarke, W. C. (1995). "Physiological Ecology of Pacific Salmon." University of British Columbia Press, Vancouver.

Gunn, J., and Block, B. (2001). Advances in acoustic, archival and satellite tagging of tunas. *In* "Tunas: Physiology, Ecology and Evolution" (Block, B., and Stevens, E., Eds.), pp. 167–224. Academic Press, San Diego.

Harden-Jones, F. R. (1968). Fish Migration. St. Martins Press, New York.

Hartman, W. L., Heard, W. R., and Drucker, B. (1967). Migratory behaviour of sockeye salmon fry and smolts. *J. Fish. Res. Board Can.* **24,** 2069–2099.

Hartman, W. L., Strickland, C. W., and Hoopes, D. T. (1962). Survival and behaviour of sockeye salmon fry migrating into Brooks Lake, Alaska. *Trans. Am. Fish. Soc.* **91,** 133–139.

Harvey, H. H., and Bothern, C. R. (1972). Compensatory swimming in the kokanee and sockeye salmon *Onchorhyncus nerka* (Walbaum). *J. Fish Biol.* **4,** 237–247.

Hasler, A. D. (1971). Orientation and fish migration. *In* "Fish Physiology" (Hoar, W. S., and Randall, D. J., Eds.), Vol. VI, pp. 429–510. Academic Press, New York.

Hasler, A. D., and Scholz, A. T. (1983). "Olfactory Imprinting and Homing in Salmon." Springer Verlag, New York.

Healey, M. C. (1967). Orientation of pink salmon (*Oncorhynchus gorbuscha*) during early marine migration from Bella Coola River system. *J. Fish. Res. Board Can.* **24,** 2321–2338.

Healey, M. C. (1991). Diets and feeding rates of juvenile pink, churn and sockeye salmon in Hecate Strait, British Columbia. *Trans. Amer. Fish. Soc.* **120,** 303–318.

Healey, M. C. (2000). Pacific salmon migrations in a dynamic ocean. *In* "Fisheries Oceanography: An Integrative Approach to Fisheries Ecology and Management" (Harrison, P. J., and Parsons, T. R., Eds.), pp. 29–54. Blackwell Science, Oxford.

Healey, M. C., Thomson, K., Le Blond, P., Huato, L., Hinch, S., and Walters, C. (2000). Computer simulations of the effects of the Sitka eddy on the migration of sockeye salmon returning to British Columbia. *Fish. Oceanogr.* **9,** 271–281.

Hendry, A. P., and Berg, O. K. (1999). Secondary sexual characters, energy use, senescence, and the cost of reproduction in sockeye salmon. *Can. J. Zool.* **77,** 1663–1675.

Hendry, A. P., and Stearns, S. C. (2004). "Evolution Illuminated – Salmon and their Relatives." Oxford University Press, Oxford.

Hinch, S. G., and Bratty, J. (2000). Effects of swim speed and activity pattern on success of adult sockeye salmon migration through an area of difficult passage. *Trans. Am. Fish. Soc.* **129,** 598–606.

Hinch, S. G., and Rand, P. S. (1998). Swim speeds and energy use of upriver–migrating sockeye salmon (*Oncorhynchus nerka*): Role of local environment and fish characteristics. *Can. J. Fish. Aquat. Sci.* **55,** 1821–1831.

Hinch, S. G., and Rand, P. S. (2000). Optimal swimming speeds and forward-assisted propulsion: Energy-conserving behaviours of upriver-migrating adult salmon. *Can. J. Fish. Aquat. Sci.* **57,** 2470–2478.

Hinch, S. G., Diewert, R. E., Lissimore, T. J., Prince, A. M. J., Healey, M. C., and Henderson, M. A. (1996). Use of electromyogram telemetry to assess difficult passage areas for river-migrating adult sockeye salmon. *Trans. Am. Fish. Soc.* **125,** 253–260.

Hinch, S. G., Healey, M. C., Diewert, R. E., Thomson, K. A., Hourston, R., Henderson, M. A., and Juanes, F. (1995). Potential effects of climate change on marine growth and survival of Fraser River sockeye salmon. *Can. J. Fish. Aquat. Sci.* **52,** 2651–2659.

Hinch, S. G., Standen, E. M., Healey, M. C., and Farrell, A. P. (2002). Swimming patterns and behaviour of upriver migrating adult pink (*Oncorhynchus gorbuscha*) and sockeye (*O. nerka*) salmon as assessed by EMG telemetry in the Fraser River, British Columbia. *Hydrobiologia* **483,** 147–160.

Hoar, W. S. (1976). Smolt transformation: Evolution, behaviour, and physiology. *J. Fish. Res. Board Can.* **33,** 1234–1252.

Hodgson, S., and Quinn, T. P. (2002). The timing of adult sockeye salmon migration into fresh water: Adaptations by populations to prevailing thermal regimes. *Can. J. Zool.* **80,** 542–555.

Hogasen, H. R. (1998). Physiological changes associated with the diadromous migration of salmonids. *Can. Spec. Pub. Fish. Aquat. Sci.* **127,** 1–128.

Houston, A., Clark, C., McNamara, I., and Mangel, M. (1988). Dynamic models in behavioural and evolutionary ecology. *Nature* **332,** 29–34.

Hughes, N. F. (2004). The wave-drag hypothesis: An explanation for size-based lateral segregation during the upstream migration of salmonids. *Can. J. Fish. Aquat. Sci.* **61,** 103–109.

Hutchings, J. A. (1991). Fitness consequences of variation in egg size and food abundance in brook trout *Salvelinus fontinalis*. *Evolution* **45,** 1162–1168.

Irschick, D. J., and Garland, T. (2001). Integrating function and ecology in studies of adaptation: Investigations of locomotor capacity as a model system. *Ann. Rev. Ecol. Syst.* **32,** 367–396.

Iwata, M. (1995). Downstream migratory behaviour of salmonids and its relationship with cortisol and thyroid hormones: A review. *Aquaculture* **135,** 131–139.

Jaenicke, H. W., and Celewycz, A. G. (1994). Marine distribution and size of juvenile Pacific salmon in Southeast Alaska and northern British Columbia. *Fish. Bull.* **92,** 79–90.

Johnston, I. A., and Moon, T. W. (1980). Exercise training in skeletal muscle of brook trout (*Salvelinus fontalis*). *J. Exp. Biol.* **87,** 177–194.

Jonsson, N., Jonsson, B., and Hansen, L. P. (1997). Changes in proximate composition and estimates of energetic costs during upstream migration and spawning in Atlantic salmon *Salmo salar*. *J. Anim. Ecol.* **66,** 425–436.

Kinnison, M. T., Unwin, M. J., Hendry, A. P., and Quinn, T. P. (2001). Migratory costs and the evolution of egg size and number in introduced and indigenous salmon populations. *Evolution* **55,** 1656–1667.

Kitahashi, T., Ando, H., Ban, M., Ueda, H., and Urano, A. (1998). Changes in the levels of gonadotropin subunit mRNAs in the pituitary of pre-spawning chum salmon. *Zool. Sci.* **15,** 753–760.

Korstrom, J. S., Birtwell, I. K., Piercey, G. E., Spohn, S., Langton, C. M., and Kruzynski, G. M. (1996). Effect of hypoxia, fresh water, anaesthesia and sampling technique on the hematocrit values of adult sockeye salmon (*Oncorhynchus nerka*). *Can. Tech. Rep. Fish. Aquat. Sci.* **2101,** 1–29.

Kramer, D. L., and McLaughlin, R. L. (2001). The behavioural ecology of intermittent locomotion. *Amer. Zool.* **41,** 137–153.

Laurs, R. M., Yuen, H. S. H., and Johnson, J. H. (1977). Small-scale movements of Albacore, *Thunnus alalunga*, in relation to ocean features as indicated by ultrasonic tracking and oceanographic sampling. *Fish. Bull.* **75,** 347–355.

Lee, C. G., Farrell, A. P., Lotto, A. G., Hinch, S. G., and Healey, M. C. (2003a). Excess post-exercise oxygen consumption in adult sockeye (*Oncorhynchus nerka*) and coho (*O. kisutch*) salmon following critical speed swimming. *J. Exp. Biol.* **206,** 3253–3260.

Lee, C. G., Farrell, A. P., Lotto, A. G., MacNutt, M. J., Hinch, S. G., and Healey, M. C. (2003b). Effects of temperature on swimming performance and oxygen consumption in adult sockeye (*Oncorhynchus nerka*) and coho (*O. kisutch*) salmon stocks. *J. Exp. Biol.* **206,** 3239–3251.

Leggett, W. C. (1977). The ecology of fish migrations. *Ann. Rev. Ecol. Sys.* **8,** 285–308.

Leggett, W. C., and Carscadden, J. E. (1978). Latitudinal variation in reproductive characteristics of American shad (*Alosa sapidissima*): Evidence for population specific life history strategies in fish. *J. Fish. Res. Board. Can.* **35,** 1469–1478.

Leonard, J. B. K., and McCormick, S. D. (1999). The effect of migration distance and timing on metabolic enzyme activity in an anadromous clupeid, the American shad (*Alosa sapidissima*). *Fish Physiol. Biochem.* **20,** 163–179.

Leonard, J. B. K., and McCormick, S. D. (2001). Metabolic enzyme activity during smolting in stream- and hatchery-reared Atlantic salmon (*Salmo salar*). *Can. J. Fish. Aquat. Sci.* **58,** 1585–1593.

Leonard, J. B. K., Iwata, M., and Ueda, H. (2001). Seasonal changes of hormones and muscle enzymes in adult lacustrine masu (*Oncorhynchus masou*) and sockeye salmon (*O. nerka*). *Fish Physiol. Biochem.* **25,** 153–163.

Levy, D. A., and Cadenhead, A. D. (1994). Selective tidal stream transport of adult sockeye salmon (*Oncorhynchus nerka*) in the Fraser River estuary. *Can. J. Fish. Aquat. Sci.* **52,** 1–12.

Liao, J. C., Beal, D. N., Lauder, G. V., and Triantafyllou, M. S. (2003). Fish exploiting vortices decrease muscle activity. *Science* **302,** 1566–1569.

Lucas, M. C., and Baras, E. (2001). "Migration of Freshwater Fishes." Blackwell Science Ltd, Oxford.

Macdonald, J. S., and Williams, I. V. (1998). Effects of environmental conditions on salmon stocks: The 1997 run of Early Stuart sockeye salmon. *In* "Speaking for the Salmon" (Gallaugher, P., and Wood, L., Eds.), pp. 46–51. Simon Fraser University Press, Vancouver.

Macdonald, J. S., Foreman, M. G. G., Farrell, T., Williams, I. V., Grout, J., Cass, A., Woodey, J. C., Enzenhofer, H., Clarke, W. C., Houtman, R., Donaldson, E. M., and Barnes, D. (2000). The influence of extreme water temperature on migrating Fraser River sockeye salmon (*Oncorhynchus nerka*) during the 1998 spawning season. *Can. Tech. Rep. Fish. Aquat. Sci.* **2326,** 1–117.

MacFarlane, R. B., and Norton, E. C. (2002). Physiological ecology of juvenile chinook salmon (*Oncorhynchus tshawytscha*) at the southern end of their distribution, the San Francisco Estuary and Gulf of the Farallones, California. *Fish. Bull.* **100,** 244–257.

MacNutt, M. J., Hinch, S. G., Farrell, A. P., and Topp, S. (2004). The effect of temperature and acclimation period on repeat swimming performance in cutthroat trout (*Oncorhynchus clarki clarki*). *J. Fish. Biol.* **65,** 324–335.

MacNutt, M. J. (2003). Effects of temperature on swimming performance, energetics and aerobic capacities of adult migrating pink salmon (*Oncorhynchus gorbuscha*) with a comparison to sockeye salmon (*O. nerka*). MSc Thesis, University of British, Columbia.

Major, R. L., and Mighell, J. L. (1966). Influence of Rocky Reach Dam and the temperature of the Okanogan river on the upstream migration of sockeye salmon. *Fish. Bull.* **66,** 131–147.

Mangel, M., and Clark, C. (1988). "Dynamic Optimisation in Behavioural Ecology." Princeton University Press, Princeton.

Marcinek, D., Blackwell, S., Dewar, H., Freund, E., Farwell, C., Dau, D., and Seitz, A. (2001). Depth and muscle temperature of Pacific bluefin tuna examined with acoustic and pop-up satellite archival tags. *Mar. Biol.* **138,** 869–885.

McCart, P. (1967). Behaviour and ecology of sockeye salmon fry in the Babine River. *J. Fish. Res. Board Can.* **24,** 375–428.

McCormick, S. D., and Saunders, R. L. (1987). Preparatory physiological adaptations for marine life in salmonids: Osmoregulation, growth and metabolism. *Am. Fish. Soc. Symp.* **1,** 211–229.

McDowall, R. M. (1988). "Diadromy in Fishes: Migrations Between Freshwater and Marine Environments." Croom-Helm, London.

Meisenheimer, P. (1990). Condition, thyroid hormones, and emergence behaviour of brook (*Salvelinus fontinalis*) and lake (*Salvelinus namaycush*) charr. MSc thesis, University of Guelph.

Metcalfe, J., Arnold, G., and McDowall, R. M. (2002). Migration. *In* "Handbook of Fish and Fisheries" (Hart, P. J. B., and Reynolds, J. D., Eds.), pp. 175–199. Blackwell Scientific, Oxford.

Mommsen, T. P., French, C. J., and Hochachka, P. W. (1980). Sites and patterns of protein and amino acid utilisation during the spawning migration of salmon. *Can. J. Zool.* **58,** 1785–1799.

Munakata, A., Amano, M., Ikuta, K., Kitamura, S., and Aida, K. (2001). The effects of testosterone on upstream migratory behaviour in masu salmon, *Oncorhynchus masou*. *Gen. Comp. Endocrin.* **122,** 329–340.

Mysak, L. (1985). On the interannual variability of eddies in the Northeast Pacific Ocean. *In* "El Nino North: Nino Effects in the Eastern Subarctic Pacific Ocean" (Wooster, W. S., and Fluharty, D. L., Eds.), pp. 97–106. Washington Sea Grant Publication, University of Washington, Seattle.

Nielsen, C., Aarestrup, K., Norum, U., and Madsen, S. S. (2004). Future migratory behaviour predicted from premigratory levels of gill $Na^{+}/K(^{+}$-)ATPase activity in individual wild brown trout (*Salmo trutta*). *J. Exp. Biol.* **207,** 527–533.

Nordeng, H. (1971). Is the local orientation of anadramous fishes determined by pheromones? *Nature* **233,** 411–413.

Northcote, T. G. (1978). Migratory strategies and production in freshwater fishes. *In* "Ecology of freshwater fish production" (Gerking, S. D., Ed.), pp. 326–359. Wiley, New York.

Northcote, T. G. (1984). Mechanisms of fish migration in rivers. *In* "Mechanisms of Migration in Fishes" (McCleave, J. D., Dodson, J. J., and Neill, W. H., Eds.), pp. 317–355. Plenum, New York.

Northcote, T. G. (1997). Potamodromy in salmonidae: Living and moving in the fast lane. *N. Am. J. Fish Mgmt.* **17,** 1029–1045.

Novales-Flamarique, I., and Hawryshyn, C. W. (1996). Retinal development and visual sensitivity of young Pacific sockeye salmon (*Oncorhynchus nerka*). *J. Exp. Biol.* **199,** 869–882.

Novales-Flamarique, I., Hendry, A. P., and Hawryshyn, C. W. (1992). The photic environment of a salmonid nursery lake. *J. Exp. Biol.* **169,** 121–141.

Ogura, M., and Ishida, Y. (1992). Swimming behaviour of coho salmon, *Oncorhynchus kisutch*, in the open sea as determined by ultrasonic telemetry. *Can. J. Fish. Aquat. Sci.* **49,** 453–457.

Ogura, M., and Ishida, Y. (1995). Homing behaviour and vertical movements of four species of Pacific salmon (*Oncorhynchus* spp.) in the central Bering Sea. *Can. J. Fish. Aquat. Sci.* **52,** 532–540.

Overgaard, J., Stecyk, J. A. W., Gesser, H., Wang, T., and Farrell, A. P. (2004). Effects of temperature and anoxia upon the performance of *in situ* perfused trout hearts. *J. Exp. Biol.* **207,** 655–665.

Parenskiy, V. A., and Podlesnykh, A. V. (1995). Sperm quality in sockeye males in connection with their morpho-physiological characteristics. *Russ. J. Mar. Biol.* **20,** 148–153.

Patterson, D.A, Guderley, H., Bouchard, P., Macdonald, J. S., and Farrell, A. P. (2004a). Maternal influence and population differences in activities of mitochondrial and gylcolytic enzymes in emergent sockeye salmon (*Oncorhynchus nerka*) fry. *Can. J. Fish. Aquat. Sci.* **61,** 1225–1234.

Patterson, D. A., Macdonald, J. S., Hinch, S. G., Healey, M. C., and Farrell, A. P. (2004b). The effect of exercise and captivity on energy partitioning, reproductive maturation, and fertilisation success in adult sockeye salmon. *J. Fish Biol.* **64,** 1039–1059.

Peake, S. J., and Farrell, A. P. (2004). Locomotory behaviour and post-exercise physiology in relation to swimming speed, gait transition and metabolism in free-swimming smallmouth bass (*Micropterus dolomieu*). *J. Exp. Biol.* **207,** 1563–1575.

Peterman, R. M., Marinone, S. G., Thomson, K. A., Jaradine, I. D., Crittenden, R. N., Le Blond, P. H., and Walters, C. J. (1994). Simulation of juvenile sockeye salmon (*Oncorhynchus nerka*) migrations in the Strait of Georgia, British Columbia. *Fish. Oceanogr.* **3,** 221–235.

Price, C. S., and Schreck, C. B. (2003). Stress and saltwater-entry behaviour of juvenile chinook salmon (*Oncorhynchus tshawytscha*): Conflicts in physiological motivation. *Can. J. Fish. Aquat. Sci.* **60,** 910–918.

Quinn, T. (1980). Evidence for celestial and magnetic compass orientation in lake migrating sockeye salmon fry. *J. Comp. Physiol.* **137,** 243–248.

Quinn, T. P., and Adams, D. J. (1996). Environmental changes affecting the migratory timing of American shad and sockeye salmon. *Ecology* **77,** 1151–1162.

Quinn, T. P., and Brannon, E. L. (1982). The use of celestial and magnetic cues by orienting sockeye salmon smolts. *J. Comp. Physiol.* **147,** 547–552.

Quinn, T. P., and terHart, B. A. (1987). Movements of adult sockeye salmon (*Oncorhynchus nerka*) in British Columbia Coastal Waters in relation to temperature and salinity stratification: Ultrasonic telemetry results. *Can. Spec. Pub. Fish. Aquat. Sci.* **96,** 61–77.

Quinn, T. P., Hendry, A. P., and Wetzel, L. A. (1995). The influence of life history trade-offs and the size of incubation gravels on egg size variation in sockeye *salmon (Oncorhynchus nerka*). *Oikos* **74,** 425–438.

Quinn, T. P., Hodgson, S., and Peven, C. (1997). Temperature, flow, and the migration of adult sockeye salmon (*Oncorhynchus nerka*) in the Columbia River. *Can. J. Fish. Aquat. Sci.* **54,** 1349–1360.

Quinn, T. P., terHart, B. A., and Groot, C. (1989). Migratory orientation and vertical movements of homing adult sockeye salmon *Oncorhynchus nerka* in coastal waters. *Anim. Behav.* **37,** 587–599.

Rand, P. S. (2002). Modeling feeding and growth in Gulf of Alaska sockeye salmon: Implications for high-seas distribution and migration. *Mar. Ecol. Prog. Ser.* **234,** 265–280.

Rand, P. S., and Hinch, S. G. (1998a). Spatial patterns of zooplankton biomass in the northeast Pacific Ocean. *Mar. Ecol. Prog. Ser.* **171,** 181–186.

Rand, P. S., and Hinch, S. G. (1998b). Swim speeds and energy use of upriver-migrating sockeye salmon (*Oncorhynchus nerka*): Simulating metabolic power and assessing risk of energy depletion. *Can. J. Fish. Aquat. Sci.* **5,** 1832–1841.

Rand, P., Scandol, J., and Walter, E. (1997). Nerkasim: A research and educational tool to simulate the marine life history of Pacific salmon in a dynamic environment. *Fisheries* **22,** 6–13.

Ricklefs, R. E., and Wikelski, M. (2002). The physiology-life history nexus. *Trends Ecol. Evol.* **17,** 462–468.

Riede, K. (2002). Global register of migratory species. German Federal Agency for Nature Conservation, Project 808 05 081.

Riede, K. (2004). Global register of migratory species – from global to regional scales. German Federal Agency for Nature Conservation, Project 808 05 081. Final Report.

Rombough, P. J., and Ure, D. (1991). Partitioning of oxygen uptake between cutaneous and branchial surfaces in larval and joung juvenile chinook salmon (*Oncorhynchus tshawytscha*). *Physiol. Zool.* **64,** 717–727.

Sato, A., Shoji, T., and Ueda, H. (2000). Olfactory discriminating ability of lacustrine sockeye and masu salmon in various freshwaters. *Zool. Sci.* **17,** 313–317.

Sato, A., Ueda, H., Kaeriyama, M., Zohar, Y., Urano, A., and Yamauchi, K. (1997). Sexual differences in homing profiles and shortening of homing duration by gonadotropin-releasing hormone analog implantation in lacustrine sockeye salmon (*Oncorhynchus nerka*) in lake Shikotsu. *Zool. Sci.* **14,** 1009–1014.

Saunders, R. L. (1965). Adjustments of bouyancy in young Atlantic salmon and brook trout by changes in swim bladder volume. *J. Fish. Res. Board Can.* **22,** 335–352.

Scheuerell, M. D., and Schindler, D. E. (2003). Diel vertical migration by juvenile sockeye salmon: Empirical evidence for the antipredation window. *Ecology* **84,** 1713–1720.

Schmidt, P. J., and Idler, D. R. (1962). Steroid hormones in the plasma of salmon at various states of maturation. *Gen. Comp. Endocrinol.* **2,** 204–214.

Sheridan, M. A., and Bern, H. A. (1986). Both somatostatin and the caudal neuropeptide, urotensin II, stimulate lipid mobilisation from coho salmon liver incubated *in vitro*. *Regul. Peptides* **14,** 333–344.

Sheridan, M. A., Allen, W. V., and Kerstetter, T. H. (1985). Changes in the fatty acid composition of steelhead trout, *Salmo gairdnerii* Richardson, associated with parr-smolt transformation. *Comp. Biochem. Physiol.* **80B,** 671–676.

Shiels, H. A., Vornanen, M., and Farrell, A. P. (2002). Force-frequency relationships in fish cardiac muscle. *Comp. Biochem. Physiol.* **132A,** 811–826.

Shiels, H. A., Vornanen, M., and Farrell, A. P. (2003). Acute temperature change modulates the response of I_{Ca} to adrenergic stimulation in fish cardiomyocytes. *Physiol. Biochem. Zool.* **76,** 816–824.

Shrimpton, J. M., and Randall, D. J. (1994). Downregulation of corticosteroid receptors in the gills of coho salmon due to stress and cortisol treatment. *Am. J. Physiol.* **267,** R432–R438.

Smith, S. G., Muir, W. D., Williams, J. G., and Skalski, J. R. (2002). Factors associated with travel time and survival of migrant yearling chinook salmon and steelhead in the lower Snake River. *N. Am. J. Fish. Mgmt.* **22,** 385–405.

Spohn, S., Birtwell, I. K., Hohndorf, H., Korstrom, J. S., Langton, C. M., and Piercey, G. E. (1996). Preliminary studies on the movement of adult sockeye salmon (*Oncorhynchus nerka*) in Alberni Inlet, British Columbia, using ultrasonic telemetry. *Can. Manuscript Rep. Fish. Aquat. Sci.* **2355,** 1–54.

Standen, E. M., Hinch, S. G., Healey, M. C., and Farrell, A. P. (2002). Energetic costs of migration through the Fraser River Canyon, British Columbia, in adult pink (*Oncorhynchus gorbuscha*) and sockeye (*Oncorhynchus nerka*) salmon as assessed by EMG telemetry. *Can. J. Fish. Aquat. Sci.* **59,** 1809–1818.

Standen, E. M., Hinch, S. G., and Rand, P. S. (2004). Influence of river speed on path selection by migrating adult sockeye salmon. *Can. J. Fish. Aquat. Sci.* **61,** 905–912.

Stefansson, S. O., Bjornsson, B. T., Sundell, K., Nyhammer, G., and McCormick, S. D. (2003). Physiological characteristics of wild Atlantic salmon post-smolts during estuarine and coastal migration. *J. Fish Biol.* **63,** 942–955.

Tabor, R. A., Brown, G. S., and Luiting, V. T. (2004). The effect of light intensity on sockeye salmon fry migratory behaviour and predation by cottids in the Cedar River, Washington. *N. Am. J. Fish. Mgmt.* **24,** 128–145.

Taylor, E. B., and McPhail, J. D. (1985). Variation in burst and prolonged swimming performance among British Columbia populations of coho salmon, *Oncorhynchus kisutch. Can. J. Fish. Aquat. Sci.* **42,** 2029–2033.

Thomson, K. A., Ingraham, W. J., Healey, M. C., Le Blond, P. H., Groot, C., and Healey, C. (1994). Computer simulations of the influence of ocean currents on Fraser River sockeye salmon (*Oncorhynchus nerka*) return times. *Can. J. Fish. Aquat. Sci.* **51,** 441–449.

Thomson, K. A., Ingraham, W. J., Healey, M. C., Leblond, P. H., Groot, C., and Healey, C. G. (1992). The influence of ocean currents on latitude of landfall and migration speed of sockeye salmon returning to the Fraser River. *Fish. Oceanogr.* **1,** 163–179.

Thomson, R. E., Le Blond, P. H., and Emery, W. J. (1990). Analysis of deep-drogued satellite-tracked drifter measurements in the Northeast Pacific. *Atmosphere Ocean* **28,** 409–443.

Thorarensen, H., and Clarke, W. C. (1989). Smoltification induced by a "skeleton" photoperiod in underyearling coho salmon (*Oncorhynchus kisutch*). *Fish Physiol. Biochem.* **6,** 11–18.

Thorarensen, H., Clarke, W. C., and Farrell, A. P. (1989). Effects of photoperiods and various intensities of night illumination on growth and seawater adaptability of juvenile coho salmon (*Oncorhynchus kisutch*). *Aquaculture* **82,** 39–49.

Trump, C. L., and Leggett, W. C. (1980). Optimum swimming speeds in fish: The problem of currents. *Can. J. Fish. Aquat. Sci.* **37,** 1086–1092.

Truscott, B., Idler, D. R., So, Y. P., and Walsh, J. M. (1986). Maturational sterioids and gonadotropin in upstream migratory sockeye salmon. *Gen. Comp. Endocrin.* **62,** 99–110.

Tully, O., Poole, W. R., Whelan, K. F., and Merigoux, S. (1993). Parameters and possible causes of epizootics of *Lepeophtheirus salmonis* (Krøyer) infesting sea trout (*Salmo trutta* L.) off the west coast of Ireland. *In* "Pathogens of Wild and Farmed Fish: Sea Lice" (Boxshall, G. A., and Defaye, D., Eds.), pp. 202–218. Ellis Horwood, London.

Ueda, H., and Yamauchi, K. (1995). Biochemistry of fish migration. *In* "Environmental and Ecological Biochemistry" (Hochachka, P. W., and Mommsen, T. P., Eds.), pp. 265–279. Elsevier, Amsterdam.

Ueda, H., Kaeriyama, M., Mukasa, K., Urano, A., Kudo, H., Shoji, T., Tokumitsu, Y., Yamauchi, K., and Kurihara, K. (1998). Lacustrine sockeye salmon return straight to their natal area from open water using both visual and olfactory cues. *Chem. Senses* **23,** 207–212.

Ueda, H., Leonard, J. B. K., and Naito, Y. (2000a). Physiological biotelemetry research on the homing migration of salmonid fishes. *In* "Proceedings of the Third Conference on Fish Telemetry in Europe: Advances in Fish Telemetry" (Moore, A., and Russell, I., Eds.), pp. 89–97. Centre for Environment Fisheries and Aquaculture Science, Lowestoft, U.K.

Ueda, H., Urano, A., Zohar, Y., and Yamauchi, K. (2000b). *In* Hormonal control of homing migration in salmonid fishes "Proceedings of the 6th International Symposium on the Reproductive Physiology of Fish," pp. 95–98. Department of Fisheries and Marine Biology, University of Bergen, Bergen, Norway.

Videler, J. (1993). "Fish Swimming." Chapman and Hall, London.

Vogel, S. (1994). Life in Moving Fluids. *In* "The Physical Biology of Flow." Princeton University Press, Princeton.

Wagner, T., and Congleton, J. L. (2004). Blood chemistry correlates of nutritional condition, tissue damage, and stress in migrating juvenile chinook salmon (*Oncorhynchus tshawytcha*). *Can. J. Fish. Aquat. Sci.* **61,** 1066–1074.

Wagner, G. N., Hinch, S. G., Kuchel, L. J., Lotto, A., Jones, S. R. M., Patterson, D. A., Macdonald, J. S., Van Der Kraak, G., Shrimpton, M., English, K. K., Larsson, S., Cooke, S. J., Healey, M. C., and Farrell, A. P. (2005a). Metabolic rates and swimming performance of adult Fraser River sockeye salmon (*Oncorhynchus nerka*) after a controlled infection with *Parvicapsula minibicornis*. *Can. J. Fish. Aquat. Sci.* **62,** 2124–2133.

Wagner, G. N., Kuchel, L., Lotto, A., Patterson, D. A., Shrimpton, M., Hinch, S. G., and Farrell, A. P. (2005b). Routine and active metabolic rates of migrating, adult sockeye salmon (*Oncorhynchus nerka* Walbaum) in seawater and freshwater. *Physiol. Biochem. Zool.* in press.

Walter, E. E., Scandol, I. P., and Healey, M. C. (1997). A reappraisal of the ocean migration patterns of Fraser River sockeye salmon (*Oncorhynchus nerka*) by individual-based modeling. *Can. J. Fish. Aquat. Sci.* **54,** 847–858.

Weihs, D. (1974). Energetic advantages of burst swimming of fish. *J. Theor. Biol.* **48,** 215–229.

Welch, D. W., Ishida, Y., and Nagasawa, K. (1998). Thermal limits and ocean migrations of sockeye salmon (*Oncorhynchus nerka*): long term consequences of global warming. *Can. J. Fish. Aquat. Sci.* **55,** 937–948.

Wood, C. M., Turner, J. D., and Graham, M. S. (1983). Why do fish die after severe exercise? *J. Fish Bio.* **22,** 189–201.

Wood, C., Hargreaves, N. B., Rutherford, D., and Emmett, B. (1993). Downstream and early marine migratory behaviour of sockeye salmon (*Oncorhynchus nerka*) smolts entering Barkley Sound, Vancouver Island. *Can. J. Fish. Aquat. Sci.* **50,** 1329–1337.

Wood, C. M., and Shuttleworth, T. J. (1995). "Cellular and Molecular Approaches to Fish Ionic Regulation." Academic Press, San Diego.

Woodey, J. C. (1987). In season management of Fraser River sockeye salmon (*Oncorhynchus nerka*): Meeting multiple objectives. *Can. Spec. Pub. Fish. Aquat. Sci.* **96,** 367–374.

Woodhead, P. M. J. (1970). An effect of thyroxine upon the swimming of cod. *J. Fish. Res. Bd. Canada* **27,** 2337–2338.

Wurtsbaugh, W. A., and Neverman, D. (1988). Post-feeding thermotaxis and daily vertical migration in a larval fish. *Nature* **333,** 846–848.

Xie, Y., Cronkite, G., and Mulligan, T. J. (1997). A split beam echosounder perspective on migratory salmon in the Fraser River: A progress report on the split-beam experiment at Mission, B.C., in 1995. *Pac. Salm. Comm. Tech. Rep* **8,** 1–32.

Yamauchi, K., Kagawa, H., Ban, M., Kasahara, N., and Nagahama, Y. (1984). Changes in plasma estradiol-17 beta and 17 alpha , 20 beta-dihydroxy-4-pregnen-3-one levels during final oocyte maturation of the masu salmon *Oncorhynchus masou*. *Bull. Jap. Soc. Sci. Fish. (Nissuishi)* **50,** 21–37.

8

NEUROENDOCRINE MECHANISMS OF ALTERNATIVE REPRODUCTIVE TACTICS IN FISH

RUI F. OLIVEIRA

1. INTRODUCTION

With over 20,000 species described to date, teleost fish are the most diverse taxon of living vertebrates, representing a very successful lineage of recently evolved organisms (Nelson, 1994). Teleosts also exhibit the widest range of reproductive modes and mating systems across the vertebrates (Breder and Rosen, 1966; Thresher, 1984). The diversity in reproductive patterns ranges from gonochoristic species, to sequential hermaphroditic species (both protogynous and protandrous), to serial sex-changing species (e.g., goby *Trimma okinawae*; Sunobe and Nakazono, 1993) to simultaneous hermaphrodites (see Demski, 1987 for a review). Asexual reproduction also

DOI: 10.1016/S1546-5098(05)24008-6

occurs in some parthenogenic species (e.g., the amazon molly *Poecilia formosa*; Schartl *et al.*, 1995). Although most species are external fertilisers, live-bearers also occur in phylogenetic independent lines (Goodwin *et al.*, 2002) and the mating system can vary from monogamous to polygamous to promiscuous (Turner, 1993). Also the patterns of parental care are the most diverse among vertebrates, with most species showing no care, to species with biparental, paternal, or maternal care (Sargent and Gross, 1993). This rich variation in reproductive modes is not only present at the interspecific level but also within species, with population differences in mating systems (e.g., sex-role reversal in courtship behaviour in lagunar populations of the peacock blenny, *Salaria pavo*, Ruchon *et al.*, 1995; Gonçalves *et al.*, 1996), to highly flexible breeding systems within the same population (e.g., monogamy vs. polygamy and biparental *vs* maternal care or paternal care in the St. Peter's fish, *Sarotherodon galilaeus*; Balshine Earn and Earn, 1998; Fishelson and Hilzerman, 2002) to the occurrence of alternative sexual phenotypes within the same sex (e.g., Taborsky, 1994). This wide variation in reproductive patterns makes teleost fish the group of election for the study of the proximate causes of sexual plasticity in vertebrates, namely for alternative reproductive tactics.

2. ALTERNATIVE REPRODUCTIVE STRATEGIES AND TACTICS

2.1. Patterns of Variation

Intrasexual variation in reproductive behaviour is widespread among animals, occurring in many invertebrate taxa such as insects (Forsyth and Alcock, 1990), crustaceans (Shuster, 1992), cephalopods (Norman *et al.*, 1999), and in all vertebrate classes such as fish (Taborsky, 1994, 2001), amphibians (Perril *et al.*, 1978), reptiles (Sinervo and Lively, 1996), birds (Lank *et al.*, 1995), and mammals (Hogg, 1984; for general reviews see Arak, 1984 and Austad, 1984). In vertebrates, the intrasexual variation in reproductive behaviour has been mainly documented for males and it may be continuous or discrete (Rhen and Crews, 2002). Discrete polymodal behavioural phenotypes within adults of the same sex in a given species have been called *alternative reproductive tactics* (ART) (Brockmann, 2001). For the sake of brevity, this review will limit the focus to species that display ARTs. Continuous variation in sexual behaviour is beyond the scope of this review and has been published elsewhere (e.g., Crews, 1998; Oliveira *et al.*, 2005).

From an evolutionary point of view, a classic distinction is made between alternative strategies and tactics. According to Gross (1996), a strategy is a genetically based program, whereas a tactic is a phenotype that results from a strategy. Two main types of strategies are possible: alternative and conditional strategies (Gross, 1996; Brockmann, 2001). Alternative strategies are based on genetic polymorphism, are under frequency-dependent selection, and thus the alternative phenotypes have equal fitness. On the other hand, in conditional strategies the expression of alternative phenotypes is based on an individual's assessment of status-dependent cues, not genetic differences (i.e., status-dependent selection; Gross, 1996). In this case, fitness is not expected to be equal among alternative morphs. Theoretically, a third strategy is possible (a "mixed strategy") if frequency dependent selection can result in equal fitness outcomes for the alternative phenotypes. However, an empirical example of such a case has never been reported in the literature (Gross, 1996). According to this view, alternative phenotypes may represent either alternative strategies or alternative tactics, depending on the relative contribution of genetic and environmental factors for their evolution.

Based on the published empirical studies, conditional strategies (i.e., alternative tactics) appear to be more common than alternative strategies (Gross, 1996). However, this apparent prevalence of conditional strategies may result from the fact that this model seems to explain most observed cases of ART and genetic studies to unravel underlying genetic polymorphisms are only rarely conducted (but see Zimmerer and Kallman, 1989 and Ryan *et al.*, 1992). Moreover, the assumption of lack of heritability in the determination of the tactic switch-point (at which it pays off for individuals to switch from one tactic to the other) in conditional strategies has also been recently challenged by Shuster and Wade (2003), who have proposed that most cases of conditional strategies must have an underlying genetic polymorphism.

The new genetic approaches to the description of fish mating systems have also provided some interesting new data with major implications for the interpretation of the evolution and maintenance of alternative reproductive phenotypes. It has been shown that the alternative male morph may have higher fitness than the conventional tactic as a result of both an unsuspected higher reproductive success or by producing more viable offspring. In the bluegill sunfish, it has been shown that bourgeois males achieve lower egg fertilisation rates than parasitic males under sperm competition (Fu *et al.*, 2001). It has also been shown, both in bluegill sunfish and in Atlantic salmon, that the offspring of parasitic males grow faster and to a larger size than the offspring of bourgeois males (Garant *et al.*, 2002; Neff, 2004). Moreover, in field conditions bluegill sunfish fry from nests with a proportionally higher incidence of cuckoldry, are larger, and show a threefold increase in survival when faced with their major predator (Neff, 2004). Thus, the computation of relative fitness between alternative phenotypes, needed for the clarification of the evolutionary mechanisms underlying the

ART, should include not only the (lifetime) reproductive success of both male types but also the potential differential survivorship of their offspring.

Alternative reproductive phenotypes can also be classified based on the descriptive patterns of the observed behaviour (Brockmann, 2001). According to this classification scheme, ART can be categorised as fixed or plastic (Moore, 1991). In fixed alternative phenotypes, the individuals adopt one of the tactics for their entire lifetime. In plastic (or flexible) alternative phenotypes, individuals may change tactic during their lifetime. Within plastic ART, two subcategories can be further distinguished: irreversible sequential patterns, when individuals switch from one tactic to another at a particular moment in their lifetime (developmental switches), and reversible patterns, when individuals can change back and forth between patterns (Moore, 1991, Moore *et al.*, 1998; Brockmann, 2001; Figure 8.1).

Therefore, this chapter adopts a classification of ART based on observed patterns as recommended by Brockmann (2001) (i.e., fixed vs. sequential vs. reversible tactics; Figure 8.1).

2.2. Terminology of Alternative Reproductive Phenotypes

A vast number of terms have been used to describe alternative reproductive phenotypes (Table 8.1). Some of these terms fail to reveal the functional role of each tactic (e.g., type I vs. type II males in plainfin midshipman,

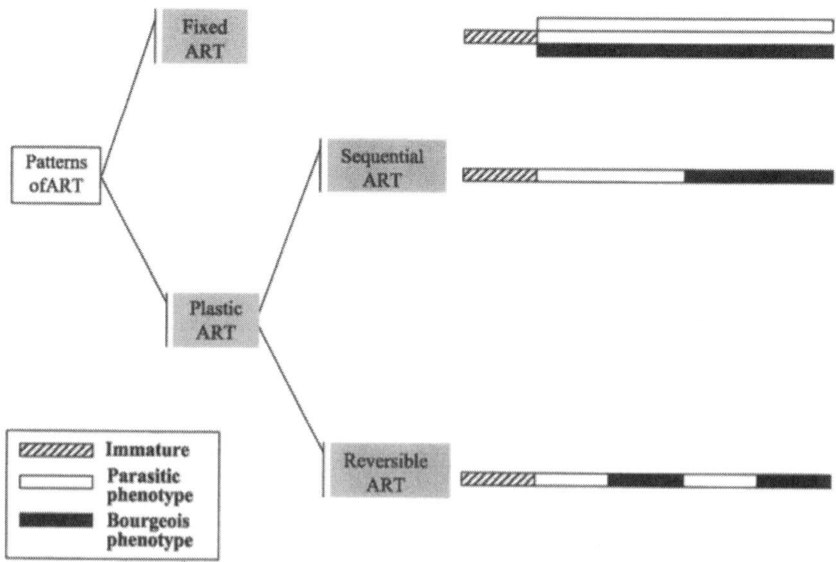

Fig. 8.1. Classification of alternative reproductive tactics (ART) based on observed patterns.

Table 8.1
Examples of the Diversity of Terms Used to Describe Alternative Reproductive Phenotypes in Male Fish

Conventional male type	Alternative male type
Territorial	Sneaker (most widely used term to name the alternative phenotype to designate a male that surreptitiously mates with a female within the territory of the conventional male; e.g. Gross, 1982; Taborsky, 1994)
Nest-holder	
Parental	
Cuckold	
Type I (term used in Batrachoids, e.g. Bass, 1993; Modesto and Canário, 2003a)	Female-mimic/ Pseudo-female (e.g. alternative males mimic females to achieve parasitic fertilisations, Taborsky, 1994, 1997)
Terminal phase (term used in Wrasses, e.g. Warner and Robertson, 1978)	Streaker (e.g. alternative males in pelagic spawning species that rush-in to join the mating pair, Warner and Robertson, 1978)
Courting (unifying term proposed by Brantley *et al.*, 1993b)	Satellite (e.g. alternative morphs associated with particular territories or conventional males, which usually tolerate their presence, Gross, 1982; Taborsky, 1994; Oliveira *et al.*, 2002b)
Bourgeois (functional unifying term proposed by Taborsky, 1997)	Opportunistic
	Cuckolder
	Type II (term used in Batrachoids, e.g. Bass, 1993; Modesto and Canário, 2003a)
	Initial phase (term used in Wrasses, e.g. Warner and Robertson, 1978)
	Non-courting (unifying term proposed by Brantley et al., 1993b)
	Parasitic (functional unifying term proposed by Taborsky, 1997)

Porichthys notatus; Bass, 1993), whereas others represent particular aspects of the expression of the tactic in a given species (initial *vs* terminal phase males in wrasses; Warner and Robertson, 1978). Therefore, the use of unified terms is highly recommendable to allow a comparative approach in the study of ART. However, the use of the pair of terms *territorial* versus *sneaker* that is present in several species (see Taborsky, 1994 for examples) is not appropriate because conventional male phenotypes do not need to necessarily be territorial (e.g., peacock blenny; Almada *et al.*, 1994). Brantley *et al.* (1993b) used the terms *courting* versus *noncourting* male morphotypes, but once again courtship behaviour is not present in all conventional reproductive phenotypes (e.g., peacock blenny; Almada *et al.*, 1995; Gonçalves *et al.*, 1996). Finally, Taborsky (1997) proposed the use of the terms *bourgeois* versus *parasitic* as the most adequate to describe the functional asymmetry in investment to gain access to a mate between the conventional and the alternative phenotype. Bourgeois males actively compete among themselves for the access to females, whereas parasitic males exploit the investment of bourgeois males to fertilise eggs (Taborsky, 1997). Thus, bourgeois males invest in mate attraction traits such as morphological ornaments (e.g., extension of the tail in swordtails, *Xiphophorus spp.*; Basolo, 1990), mating vocalisations (e.g., humming calls in plainfin midshipman, *Porichthys notatus*; Brantley and Bass, 1994; see also Chapter 2), the release of sex-pheromones (e.g., sex-pheromone producing anal gland in blennies; Laumen *et al.*, 1974; see also Chapter 9), or the elaboration and ornamentation of nests (e.g., Mediterranean wrasses; Lejeune, 1985). Conversely, parasitic males exploit the investment of bourgeois males in various possible ways: (1) they may try to approach the spawning site without being noticed, possibly even using female mimicry (e.g., adoption of female colours and courtship behaviours by sneaker males in the peacock blenny; Gonçalves *et al.*, 1996); (2) darting to the mating pair and releasing their sperm before the bourgeois male can react to its presence (e.g., streaking males in wrasses; Warner, 1984); or (3) cooperating with bourgeois males so that they tolerate their presence in the breeding grounds (e.g., satellite behaviour in rock-pool blenny, *Parablennius parvicornis*; Oliveira *et al.*, 2002b). Thus, without the initial investment by the bourgeois male, the alternative phenotype would not be functional.

It should be stressed here that the term *parasitic male* in this context merely indicates that these males exploit the investment of the bourgeois males in mate attraction, and it does not refer to the relationship between the alternative phenotypes, which can range from almost true mutualism when both types of males appear to benefit, to true parasitic where only the parasitic males appear to benefit (Taborsky, 1999, 2001).

2.3. Why Are Alternative Reproductive Phenotypes So Common in (Male) Fish?

Although ART are present in all vertebrate taxa (for a review see Caro and Bateson, 1986), teleosts are by far the taxon with the highest incidence of species with alternative reproductive phenotypes (Taborsky, 1994, 1999, 2001). The last published count of ART in fish identified 140 species of 28 different families (Taborsky, 1998; Figure 8.2), and new examples of species with ART are described annually, even among common temperate species that are studied on a regular basis such as Mediterranean parrotfish, *Sparisoma cretense* (de Girolamo *et al.*, 1999), grass goby, *Zosterisessor ophiocephalus* (Mazzoldi *et al.*, 2000), two species of triple fin blennies (Neat,

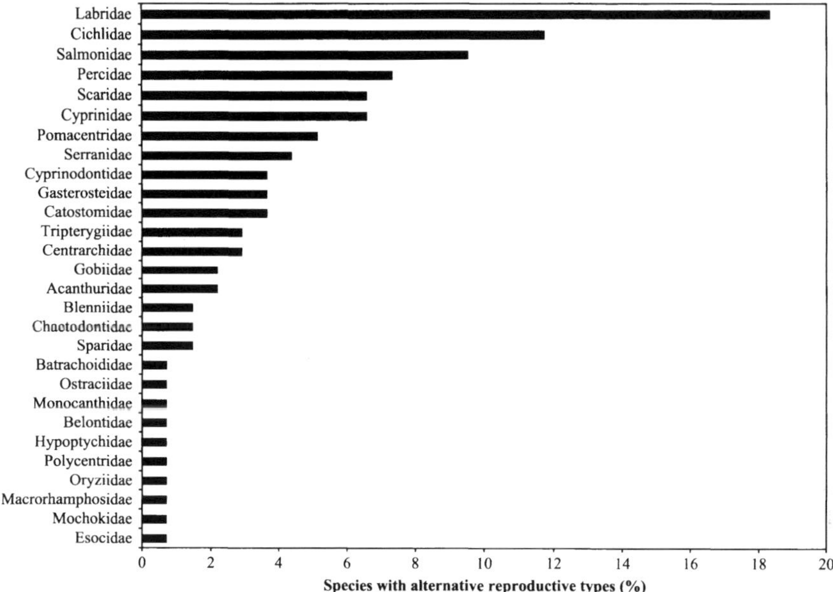

Fig. 8.2. Relative abundance (%) of alternative reproductive tactics in teleosts described in the literature up to 1998 distributed by taxonomic family. (Adapted from Taborsky, 1998.)

2001), black goby, *Gobius niger* (Mazzoldi and Rasotto, 2002), and dusky frillgoby, *Bathygobius fuscus* (Taru *et al.*, 2002). Thus, the numbers given by Taborsky (1998) should be regarded as an underestimate of the prevalence of ART among fish.

Three main factors have been identified that may predispose male fish to ART:

1. Indeterminate growth resulting in large size differences among sexually mature males;
2. External fertilisation permitting simultaneous parasitic spawning; and
3. Male parental care which provides parasitic males with a high payoff or benefit from breeding without incurring the high costs (in terms of time and energy allocation) associated with parental care (Taborsky, 1999).

Indeterminate growth can create a significant asymmetry among the sex competing for mates (usually the males). Taborsky (1999) estimated that the average difference between the largest and the smallest reproductive males for species for which data was available, and concluded that on average the largest breeding males were 18 times larger than their smallest counterparts.

With such extreme size differences among breeding competitors, smaller males may be more successful if they adopt a parasitic tactic than if they act as bourgeois males. However, this does not mean that parasitic males should always be the smallest individuals among their age class. The particular aspects of the parasitic tactic being adopted may determine the optimal relative body size for that specific tactic. For example, satellite males in the Azorean rock-pool blenny help the territorial nest-holder males to defend their breeding territory and thus are the largest of their age class (Oliveira *et al.*, 2002b). On the other hand, in the peacock blenny, sneaker males mimic females to get access to nests during spawning episodes, with larger sneakers being more easily detected by nest-holder males than smaller ones (Gonçalves *et al.*, 2005). Thus, in the peacock blenny, sneaker males are the smallest of their age class (Oliveira *et al.*, 2001f; Figure 8.3). Indeterminate growth also creates the need for the central nervous system to continue growing in adulthood (Zupanc, 2001). The consequent evolution of adult neurogenesis in teleosts in theory would have facilitated the plasticity of the neural circuits underlying reproductive behaviour during the lifetime of an individual and thus created the opportunity for the emergence of ART.

A second factor that may favor the occurrence of ART among fish is the prevalence of external fertilisation, which allows simultaneous parasitic spawning, not possible in species with internal fertilisation (Taborsky, 1999). Nevertheless, ARTs are also present in teleosts with internal fertilisation. This paradox may be explained by the fact that the evolution of ART in live-bearing teleosts is accompanied by the evolution of specialisations in the parasitic morph that increase the probability of fertilisation. For example, in poeciliids—in which males may inseminate females after courting them (bourgeois tactic) or by using forced copulations (parasitic tactic; for a review of alternative reproductive behaviour in poeciliid fish, see Farr, 1989)—an association between the frequency at which each behaviour is used and male morphology is present both at the intra- and interspecific levels. In guppies (*Poecilia reticulata*), males with longer gonopodia engaged significantly more often in forced copulations than males with shorter gonopodia (Reynolds *et al.*, 1993); in the sailfin molly (*Poecilia latipinna*), small males, specialised in gonopodial thrusts, have higher relative gonopodium lengths than large males (R. F. Oliveira, D. M. Gonçalves, and I. Schlupp, unpublished data, 2000). Also, there is an interspecific association among poeciliids between gonopodium length and the use of gonopodium thrusts rather than courtship displays (Rosen and Tucker, 1961).

Finally, because paternal care is widespread among teleosts (Sargent and Gross, 1993), in species with male parental care, parasitic males would benefit from a high payoff by being able to breed without incurring the high costs (in terms of time and energy allocation) associated with parental care.

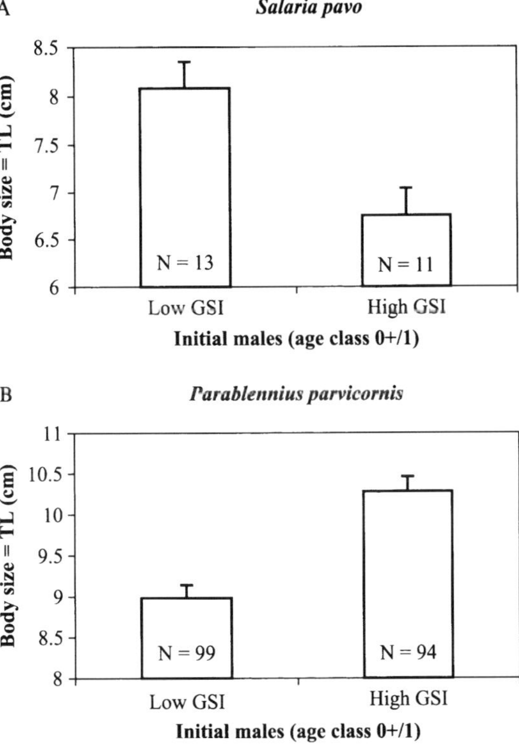

Fig. 8.3. Investment in gonadal tissue by the young-of-the-year (age class 0+ and 1) in two closely related blenniid species differs according to different types of alternative reproductive tactics. (A) In the peacock blenny, *Salaria pavo*, the parasitic tactic consists of female-mimicry and the smaller individuals among the 0+/1 year-old are the ones that display the tactic. (B) In the Azorean rock-pool blenny, *Parablennius parvicornis*, the parasitic tactic consists of behaving as territory satellites of nest-holders helping them in territorial defense and the larger individuals of the 0+/1 year-olds are the ones that exhibit the tactic. (Adpated from Oliveira *et al.*, 2001f.)

However, in both wrasses and cichlids, the two fish families with the highest prevalence of ART (see Figure 8.2), there was no association between the occurrence of ART and the mode of parental care (based on data compiled by Taborsky, 1994, 1999).

In summary, among the three factors usually used to explain the high incidence of ART in fish, high intrasexual size differences among reproductive competitors seems to be the most powerful factor. This may result from the fact that ART may have evolved in response to scenarios of high levels of intrasexual selection (but see Jones *et al.*, 2001). Apart from these ultimate factors there are also proximate factors that might help to explain the high numbers of fish species with ART, namely the high variability and lability of sex determination and differentiation mechanism and the brain bipotentiality of fish. These two factors will be discussed below (see Sections 3 and 4, respectively).

The occurrence of ART described above considers male alternative phenotypes. Reports of female ART are less common, which can be due to two factors: (1) most of the studies so far have focused on alternative male behaviours, possibly because in most fish species female sexual behaviour is less conspicuous than male behaviour (Breder and Rosen, 1966; Thresher, 1984); and/or (2) because intrasexual selection is usually higher among males than among females, indeterminate growth and its associated pattern of high intrasexual size differences would mainly have consequences for the competing sex, that is the males, because females seldom compete for breeding resources or for males, except in species with sex-role reversal (e.g., some Syngnathidae; Berglund and Rosenqvist, 1993). Nevertheless, female alternative reproductive behaviours have also been documented (Henson and Warner, 1997). For example, in the blueheaded wrasse (*Thalassoma bifasciatum*), pair-spawning sites and group-spawning sites exist, and females have a consistent pattern of their preferred mode of mating: some prefer group-spawnings, whereas others prefer pair-spawnings (Warner, 1987, 1990). Interestingly, almost all large-size females prefer pair-spawning irrespective of their previous preference at smaller size (Warner, 1985). This data suggests a mixed situation with fixed ART among smaller females followed by sequential ART when they grow older.

Despite the interest of female ART, this chapter will concentrate on male ART in the following sections and will address the proximate mechanism underlying the expression of discrete alternative phenotypes.

3. SEX DETERMINATION, SEXUAL DIFFERENTIATION, AND ALTERNATIVE REPRODUCTIVE PHENOTYPES

Historically, the conceptual framework of sex determination and sexual differentiation in vertebrates has been greatly influenced by the mammalian model of genotypic sex determination (GSD). In mammals, the sex-determining gene (*Sry*) in the Y chromosome induces the differentiation of a testis from the primordial bipotential gonad, and the differentiated testis then produces hormones that will start a cascade of events, which will simultaneously promote the masculinisation of some traits and the defeminisation of others, leading to sexual differentiation (Zarkower, 2001). In the differentiated gonad, Sertoli cells produce Müllerian inhibiting substance

(MIS) that promotes the regression of the Müllerian ducts, whereas Leydig cells produce testosterone (T) that differentiates a set of male traits: (1) the Wolffian ducts differentiate into male accessory reproductive organs (e.g., vas deferens, seminiferous tubules); (2) the differentiation of the male genitalia from the genital bud induced by dihydrotestosterone (DHT), which results from the local metabolisation of T by the enzyme 5-α-reductase; and (3) the organisation of the brain (hypothalamus/preoptic area) and the pituitary gland in a male fashion, that translates into a tonic release of gonadotropins from the pituitary and in the expression of male sexual behaviour (mounting and intromission) later in adulthood. In XX embryos, the lack of *Sry* allows ovary differentiation, and in the absence of testicular androgens and MIS, female traits differentiate: (1) the lack of the repressor effect of MIS allows the Müllerian ducts to develop into the female reproductive accessory organs (i.e., uterus and fallopian tubes), and the lack of T leads to the regression of the Wolffian ducts; (2) in the absence of DHT, a female genitalia develops; and (3) in the absence of T, a female brain organisation emerges that results in a female pattern of cyclic gonadotropin release from the pituitary, and in female typical behaviour (sexual receptivity; e.g., lordosis in rodents). Thus, in mammals the cascade of events that leads to morphological and behavioural sex differentiation is mainly orchestrated by gonadal sex steroids.

In fish, sex determination is more labile than in mammals. Cases of genotypic sex determination (GSD) and environmental sex determination (ESD) have been described in fish, and GSD mechanisms are more flexible and open to environmental influences in fish than in mammals (Devlin and Nagahama, 2002). GSD mechanisms in fish are very diverse and range from polygenic systems to systems with dominant sex-determining factors mixed with autosomal controls, to sex chromosomes with either heterogametic males (XY) or females (ZW) (Devlin and Nagahama, 2002). Cytogenetical data revealed that sex chromosomes are present in approximately half the species in which they were searched for (Devlin and Nagahama, 2002). However, this number may be underestimated because many related species have been investigated thus promoting a phylogenetic bias in the outcome (Devlin and Nagahama, 2002). Nevertheless, it is interesting to note that the number of species that display male heterogamety is twice the number of those with female heterogamety (Devlin and Nagahama, 2002). Although a sex-determining gene has not yet been found in fish, a strong candidate has recently been identified on the Y chromosome of medaka. Like the mammalian *Sry* gene, this gene (*DMY*, belonging to the DM-domain gene family, named after the *Dsx* and *mab-3* male specific regulatory genes in *Drosophila* and *Caenorhabditis elegans*, respectively) is expressed only in somatic cells of the male gonad at the time of sex determination (Matsuda *et al.*, 2002).

There are two main classes of ESD in fish: one is where sex is regulated by temperature (i.e., temperature sex determination [TSD]), with high temperatures promoting male differentiation and more females being produced at low temperatures; and the second is where sex is regulated by social factors (behavioural sex determination [BSD]). Both of these may occur before (primary ESD or BSD) or after (secondary ESD or BSD) sexual maturation (Nakamura *et al.*, 1998; Strüssman and Nakamura, 2002; Godwin *et al.*, 2003). Primary BSD has been documented in the Midas cichlid, *Cichlasoma citrinellum*, where the larger brood mates differentiate as males and the smaller as females, an effect that was demonstrated to be due to relative size by experimentally reversing the within-broods relative body sizes (Francis, 1990, 1992). Also, in the paradise fish, *Macropodus opercularis*, some evidence for prematurational BSD is present. In an artificial selection experiment for divergent lines of social dominance, Francis (1984) found that the high-dominance line consisted almost entirely of males and the low-dominance line of females. Thus, social dominance seems to be affecting sexual differentiation in this species. Postmaturational BSD has been more widely documented and comprises cases of functional sex-change in adulthood (Francis, 1992; Godwin *et al.*, 2003).

Sex steroids have been implicated in all three mechanisms of fish sex determination (i.e., GSD, TSD, and BSD; Devlin and Nagahama, 2002; Godwin *et al.*, 2003). In primary sex differentiation of species with GSD, endogenous estrogen acts as an ovarian inducer (Nakamura *et al.*, 2003), with aromatase playing a key role in all fish species investigated so far such as rainbow trout, *Oncorhynchus mykiss* (Guiguen *et al.*, 1999), Nile tilapia, *Oreochromis niloticus* (Guiguen *et al.*, 1999; Kown *et al.*, 2001), Mozambique tilapia, *Oreochromis mossambicus* (Tsai *et al.*, 2000), Japanese flounder, *Paralychthys olivaceus* (Kitano *et al.*, 1999; Trant *et al.*, 2001), and zebrafish, *Danio rerio* (Trant *et al.*, 2001). The involvement of androgens as testicular inducers is less clear and seems to vary among species (e.g., Govoroun *et al.*, 2001; Liu *et al.*, 2000; Nakamura *et al.*, 2003). Apart from steroids, other factors have been involved in fish sexual determination and differentiation including genes already known to be involved in sex differentiation pathways in other vertebrate and invertebrate taxa (e.g., *DMRT1*, *Sox9*, *DAX1*; see Baron and Guiguen, 2003 and references therein). Thus, a complete synthesis of fish sex determination and differentiation pathways is still difficult to achieve, which in part may result from the high diversity of sex-determining mechanisms found among teleosts.

Sex-changing species with BSD after sexual maturation are relevant for the understanding of the differentiation of alternative phenotypes within the same sex (more than any species with other ESD mechanisms) because they also exhibit a phenotypic reorganisation in adulthood. In these species, an association between changes in circulating sex steroid levels and sex change is present. In protogynous species (i.e., sequential hermaphrodites with female-to-male sex change), the levels (plasma concentrations or synthesis rates of *in vitro* incubations of gonadal tissue) of 11-ketotestosterone (KT), the most potent androgen in fish (Borg, 1994), increase but estradiol concentrations decline during female-to-male sex change, such as in saddleback wrasse, *Thalassoma duperrey* (Hourigan *et al.*, 1991) and stoplight parrotfish, *Sparisoma viride* (Cardwell and Liley, 1991). However, in protandrous fish (male-to-female sex change), the opposite changes occur, such as in anemonefishes, *Amphiprion melaropus, A. frenatus* (Godwin and Thomas, 1993; Nakamura *et al.*, 1994), and black porgy, *Acanthopagrus schlegeli* (Lee *et al.*, 2001). Similarly, KT levels are generally higher in the male phase and E2 levels in the female phase in both protogynous species such as saddleback wrasse, *Thalassoma duperrey* (Hourigan *et al.*, 1991), bambooleaf wrasse, *Pseudolabrus japonicus* (Morita *et al.*, 1997), blackeye goby, *Coryphopterus nicholsii* (Kroon and Liley, 2000), red grouper, *Epinephelus morio* (Johnson *et al.*, 1998), Hong Kong grouper, *Epinephelus akaara* (Tanaka *et al.*, 1990), and black seabass, *Centropristis striatus* (Cochran and Grier, 1991), and in protandrous species such as black porgy, *Acanthopagrus schlegeli* (Chang *et al.*, 1994), goldlined seabream, *Rhabdosargus sarba* (Yeung and Chan, 1987), and seabass, *Lates calcarifer* (Guiguen *et al.*, 1993).

In general, T levels are not reliable indicators of the sex phase in sequential hermaphroditic fish (Devlin and Nagahama, 2002). Moreover, in simultaneous hermaphrodites, KT levels tend to be positively correlated with the adoption of a male role by the individual (Cheek *et al.*, 2000). However, socially induced behavioural sex change can apparently occur in gonadectomised individuals in the bluehead wrasse, *Thalassoma bifasciatum* (Godwin *et al.*, 1996), which indicates that although sex steroids might be playing a major role in morphological sex differentiation in sequential hermaphroditic fish, they are not needed for behavioural sex change. The occurrence of BSD in sequential hermaphrodites also stresses the fact that the brain must have primacy over the gonads in the decision-making process of sexual differentiation. As Francis (1992) phrased it: "the only way the behaviour can affect the gonads is through the brain." Thus, one fundamental difference between teleost and mammalian sex differentiation is the fact that, although in mammals the differentiation of the gonad will lead the whole process of sexual differentiation including the sexual differentiation of brain and behaviour, mainly through organisational actions of sex steroids (but see Carruth *et al.*, 2002), in fish the brain seems to be the initial site of sex differentiation and the pattern of brain sex differentiation will then determine gonadal sex differentiation (Francis, 1992; Grober, 1998; Reavis and Grober, 1999). This crucial difference may explain the extreme sexual plasticity found among fish and the environmental and social influences on teleost sexuality that are present even after sexual maturation both in sex-changing and gonochoristic species.

How can the understanding of the sex differentiation mechanism in fish contribute to the understanding of the differentiation of alternative reproductive phenotypes? First, GSD mechanisms might be seen as homologous to the mechanisms underlying fixed ART, whereas ESD can be regarded as homologous to the differentiation of plastic ART, because the former impose lower degrees of freedom in the making of the sexual phenotype (i.e., there is a genetic constraint in the development of the alternative phenotypes) than the latter. Second, the sequential (irreversible) versus reversible patterns within plastic ART can be seen as a parallel to primary versus secondary patterns of ESD, respectively. Thus, the study of sex determination and differentiation mechanisms may provide important hints for the investigation of the proximate mechanisms of alternative reproductive tactics. To investigate the potential relationship between the types of ART displayed and the modes of sex determination in fish, data were compiled on these two parameters from the literature, which is summarised in Table 8.2. The following conclusions can be drawn from an analysis of Table 8.2:

1. Fixed ART are equally distributed among GSD and ESD, but most species listed with fixed ART and GSD (i.e., 60%) are salmonids, whereas species with fixed ART and ESD are centrarchids (also 60% of the species). These results suggest that fixed tactics do not have necessarily to be based on genetic mechanisms and can rely both on genetic and on environmental mechanisms, depending on the historical evolutionary pathway of the ART on a specific phylogenetic group. This implies that the same type of ART might have evolved separately in different teleost taxa using different mechanisms, which were probably constrained by the existing sex-determining mechanism of each species.
2. All but one of the species displaying plastic irreversible ART have ESD, which suggests that individuals from species with ESD are more prone to respond to environmental cues during their adult lives and thus maintain their phenotypic plasticity during their lifetime. This allows them to reorganise (or redifferentiate) their phenotyes at a postmaturational life stage.
3. Reversible ARTs are also equally distributed among GSD and ESD, but again a phylogenetic bias is present (i.e., 50% of species with reversible ART and GSD are cichlids and 50% of the species with reversible ART and ESD are sticklebacks). This result might sound

Table 8.2

Relationship Between Sex Determination Mechanisms and Alternative Reproductive Phenotypes in Fish

		Sex determining mechanism	
		GSD	ESD
Alternative reproductive tactic or strategy	Fixed	Coho salmon, *Oncorhynchus kisutch*	Bluegill sunfish, *Lepomis macrochirus*
		Pacific salmon, *Oncorhynchus tshawytscha*	Longear sunfish, *Lepomis megalotis*
		Atlantic salmon, *Salmo salar*	Spotted sunfish, *Lepomis punctatus*
		Guppy, *Poecilia reticulata*	Plainfin midshipman, *Porichthys notatus*
		Swordtail, *Xiphophorus nigrensis*	Corkwing wrasse, *Symphodus melops*
	Plastic irreversible (Sequential)	Rainbow wrasse, *Coris julis*	Rock-pool blenny, *Parablennius parvicornis*
			Peacock blenny, *Salaria pavo*
			Grass goby, *Zosterisessor ophiocephalus*
			Black goby, *Gobius niger*
			Common goby, *Pomatoschistus microps*
			Sand goby, *Pomatoschistus minutus*
			Dusky frillgoby, *Bathygobius fuscus*
			Stoplight parrotfish, *Sparisoma viride*
			Blueheaded wrasse, *Thalassoma bifasciatum*
			Mediterranean wrasse, *Symphodus ocellatus*
	Plastic reversible	Sailfin molly, *Poecilia velifera*	Three-spined stickleback, *Gasterosteus aculeatus*
		Mozambique tilapia, *Oreochromis mossambicus*	Fifteen-spined stickleback, *Pungitus pungitus*
		St. Peter's fish, *Sarotherodon galilaeus*	Damselfish, *Chromis chromis*
		Sergeant major, *Abudefduf saxatilis*	Dusky farmerfish, *Stegastes nigricans*

Sources: sex-determining mechanisms based on fish cytogenetic data presented in Devlin, R. H., and Nagahama, Y. (2002) and in Froese and Pauly (2004); type of ART based on Taborsky (1994, 1999, 2001) and references therein.

odd because reversible ART represents the extreme expression of phenotypic plasticity, and thus a prevalence of ESD could be expected. However, it should be stressed that in all cases listed in Table 8.2 in this category (reversible ART), the differences between alternative phenotypes are mostly behavioural with little or no morphological differentiation between the two male types (e.g., three-spined stickleback, *Gasterosteus aculeatus*; Jamieson and Colgan, 1992). In fact, in most cases listed, the same individual can adopt one of the alternative tactics depending on context, such as nest density in breeding colonies of pomacentrids (Tyler, 1989; Picciulin *et al.*, 2004) or internest distance in sticklebacks (Goldschmidt *et al.*, 1992). Thus, high levels of phenotypic plasticity, which result in facultative reversible tactics, might require rapid and transient changes in neural activity (Zupanc and Lamprecht, 2000; Hofmann, 2003) independent of hormone-induced changes in gene expression and usually underlie ARTs that are stable over longer time periods (i.e., fixed and sequential ART). Hence, reversible ART may have become independent from a sex-differentiation mechanism ruled by sex hormones.

Interestingly, using examples from all vertebrate groups, Crews (1998) came to a different conclusion: he argued that among vertebrates all species with fixed tactics displayed GSD, but species with plastic tactics might display either GSD or ESD. The discrepancy between Crews' results and those presented here might reflect the fact that all the nonpiscine species (i.e., two amphibia, five reptiles, one bird, and three mammals) included in his analysis had GSD (see Table 1 in Crews, 1998). Furthermore in that study, a uniform taxon level of analysis was not used (i.e., in most cases the species level was used but in some other cases families were entered as the taxonomic unit, such as wrasse, angelfish, and parrotfish; see Table 1 in Crews, 1998). This is understandable because the author was providing examples rather than making a comprehensive coverage of all published studies. Unfortunately, these might have confounding effects for a subsequent quantitative analysis.

The analogy between processes of sexual differentiation (i.e., males *vs* females) and the differentiation of discrete alternative reproductive phenotypes within the same sex made it also plausible to consider a role for sex steroids in the differentiation of intrasexual alternative phenotypes. The view of sex steroids as candidates for the proximate control of ART becomes even stronger if one considers the fact that they are one of the pillars of fish reproduction because they are involved in the control of sexual maturation, in the development of secondary sex characters, and in the expression of reproductive behaviour (Borg, 1994).

4. SEX STEROIDS AND ALTERNATIVE REPRODUCTIVE PHENOTYPES

4.1. Patterns of Sex Steroids in Species with Alternative Reproductive Tactics

Sex steroids are good candidates to play a key role in the expression of male ART in fish considering their involvement in teleost sex differentiation described above. Also, sex steroids can be considered as intercellular honest signals (*sensu* animal communication theory; Bradbury and Vehrencamp, 1998) sent by the gonads as they mature to the rest of the body, to coordinate the expression of sexual behaviour and the differentiation of sexual characters with the availability of mature gametes to be released (Oliveira and Almada, 1999). Among the sex steroids, the androgens are a first choice because they are deeply involved in male reproductive physiology (e.g., Borg, 1994).

4.1.1. Androgens and ART

A first review of the available data on androgen levels in species with ART was given in Brantley *et al.* (1993b). These authors found that alternative phenotypes differ in their circulating 11-ketotestosterone (KT) levels, with bourgeois males having significantly higher levels of circulating KT than parasitic males, and that there was no clear pattern regarding testosterone (T): it was higher in bourgeois males in some species, higher in parasitic males in others, and there was also a species in which no differences were found between the two morphs. The six species surveyed were not closely related to each other, and thus phylogenetical bias is probably not present in their qualitative analysis. The author has updated the database used by Brantley *et al.* (1993b) with the data that has been published since then. Currently data on 16 species from nine different teleost families is available (Table 8.3). A reanalysis of the data yields similar results to those obtained by Brantley *et al.* (1993b): there is a strong association between the ART and KT, but not with T. In 13 out of the 16 species (81.3%) with ART for which sex steroid levels (or synthesis rates of *in vitro* incubations of gonadal tissue) are available, KT is higher in the bourgeois than in the parasitic males (Table 8.3).

The three exceptions to this rule are two species with reversible ART (monogamous vs. sequential polygamous males in the St. Peter's fish, *Sarotherodon galilaeus* and large courting males vs. small courting males in the sailfin molly, *Poecilia velifera*) and one species with sequential ART (breeders vs. sexually active helpers in the cooperative brooder cichlid *Neolamprologus pulcher*). Two kinds of factors that may help to explain these exceptions will be discussed below.

Table 8.3
Comparison of Sex Steroid Levels Between Bourgeois and Parasitic Males, in Relation to the Type of Alternative Tactic and the Presence/Absence of Intrasexual Dimorphism

FAMILY/ Species	Alternative phenotypes	Intrasexual dimorphism	KT	T	E2	DHP
BATRACHOIDIDAE						
Lusitanean toadfish, *Halobatrachus didactylus*[1]	Fixed?	+	$B > P$	$B = P$	$B = P$	$B = P$
Plainfin midshipman, *Porichtys notatus*[2,3]	Fixed: Type I calling vs. Type II non-calling males	+	$B > P$	$B < P$	$B = P$	—
BLENNIIDAE						
Peacock blenny, *Salaria pavo*[4,5]	Sequential: Nest-holders vs. female-mimic sneakers	+	$B > P^a$	$B > P^a$	—	—
Rock-pool blenny, *Parablennius parvicornis*[6]	Sequential: Nest-holders vs. satellites	+	$B > P$	$B = P$	—	—
CENTRARCHIDAE						
Bluegill sunfish, *Lepomis macrochirus*[7,8]	Fixed: Parentals vs.sneakers and satellites	+	$B > P$	$B = P$	—	—
Longear sunfish, *Lepomis megalotis*[9]	Fixed?	+	$B > P^b$	$B = P^b$	—	—
CICHLIDAE						
Princess of Burundi, *Neolamprologus pulcher*[10,*]	Sequential: breeders vs. helpers	—	$B = P^c$	$B = P^c$	—	—
Mozambique tilapia, *Oreochromis mossambicus*[*, 11,12]	Reversible: territorial courting vs. non-territorial female-mimics	+	$B > P^d$	$B > P^d$	—	$B > P^d$
St. Peter's fish, *Sarotherodon galilaeus*[13,*]	Reversible: monogamous vs. polygynous males	—	$B = P$	$B = P$	—	$B = P$
LABRIDAE						

Corkwing wrasse, *Symphodus melops*[14,15]	Fixed: Territorial vs. female-mimics	+	B > P	B < P	B<P	—
Rainbow wrasse, *Coris julis*[16]	Sequential: Initial phase vs. terminal phase males	+	B > P[f]	—	—	—
Saddleback wrasse, *Thalassoma duperrey*[17]	Sequential: Initial phase vs. terminal phase males	+	B > P	B = P	B = P	B = P
POECILIIDAE						
Sailfin molly, *Poecilia velifera*[18,*]	Reversible: large courting vs. small non-courting males	—	B = P[c]	B = P[c]	—	—
SALMONIDAE						
Atlantic salmon, *Salmo salar*[19]	Fixed: mature parr vs. anadromous males	+	B > P[e]	B<P[e]	—	B = P[e]
SCARIDAE						
Stoplight parrotfish, *Sparisoma viride*[20]	Sequential: Initial phase vs. terminal phase males	+	B > P	B > P	B<P	—
SERRANIDAE						
Belted sandfish, *Serranus subligarius*[21]	Reversible: streakers vs. pair spawners in a simultaneous hermaphrodite	—	B > P	—	—	B > P

KT, 11-ketotestosterone; T, testosterone; E2, estradiol; DHP, 17α, 20β-dihydroxy–4-pregen–3-one; B, bourgeois; P, parasitic.

[a]Testicular androgen levels (ng steroid/g of tissue).

[b]Values extrapolated from graph in Knapp, 2004, Figure 3.

[c]Steroid levels in fish holding water (ng steroid/h/g body mass).

[d]Urinary sex steroid levels (ng steroid/ml urine).

[e]Values for late summer, when GSI values peak.

[f]In vitro gonadal production from [^{14}C] T incubation.

[1]Modesto and Canário, 2003a; [2]Brantley *et al.*, 1993b; [3]Bass, 1992; [4]Oliveira *et al.*, 2001b; [5]Gonçalves *et al.*, 1996; [6]Oliveira *et al.*, 2001c; [7]Kindler *et al.*, 1989; [8]Neff *et al.*, 2003; [9]Knapp, 2004; [10]Oliveira *et al.*, 2003; [11]Oliveira and Almada, 1998a; [12]Oliveira *et al.*, 1996; [13]Ros *et al.*, 2003; [14]Uglem *et al.*, 2000; [15]Uglem *et al.*, 2002; [16]Reinboth and Becker, 1984; [17]Hourigan *et al.*, 1991; [18]R. F. Oliveira, D. M. Gonçalves and I. Schlupp, unpublished data, 2000; [19]Mayer *et al.*, 1990; [20]Cardwell and Liley, 1991; [21]Cheek *et al.*, 2000. *Data based upon lab/pond studies that remain to be verified in the field.

First, differences in androgen (KT) levels may be less important in species with ART that lack major tactic-specific morphological specialisations (e.g., the expression of male secondary sex characters in bourgeois males) because KT has been demonstrated to be the most potent androgen in the induction of secondary sex characters in male teleosts (Borg, 1994). Indeed, in all three species, the main differences between the alternative male types are behavioural and of relative body size in the sailfin molly and in *N. pulcher*. No major morphological differences are present between these alternative morphs. Thus, differential KT levels are not necessary to induce or to maintain tactic-specific morphological characters. Therefore, if this hypothesis is correct, an association between the degree of dimorphism among the alternative male phenotypes and the differential variation in KT levels among alternative morphs is to be expected. In fact, from all listed species for which the ART involves a morphological intrasexual dimorphism (apart from differences in body size), the KT levels are higher in the bourgeois than in the parasitic male, and irrespective of the ART they display (Table 8.3). This suggests that androgens may play a major role in morphological differentiation between the alternative phenotypes, but will not be essential for the occurrence of behavioural sexual plasticity among fish. Two cases listed in Table 8.3 are worth mentioning in this context: the Mozambique tilapia, *Oreochromis mossambicus*, that exhibits a reversible ART and displays differential KT levels between the two male types, and the simultaneous hermaphrodite, *Serranus subligarius*, that exhibits differences in KT levels depending on the mating behaviour displayed (pair-spawning *vs* streaking) in the absence of sexual polymorphism (Cheek *et al.*, 2000). In the Mozambique tilapia, although the tactic is reversible, it is more constant in time and a morphological differentiation between territorial courting males and subordinate female-mimic males is present (Oliveira and Almada, 1998b). In *S. subligarius*, no polymorphism is present but in this simultaneous hermaphrodite, pair-spawning behaviour is associated with a larger body size, although no association was found between KT levels and testis mass (Cheek *et al.*, 2000).

Second, in species with a high degree of behavioural phenotypic plasticity, such as the case of species of reversible ART in which individuals can very rapidly switch reproductive tactics according to local or temporal conditions, differences in androgen levels underlying these rapid changes in behaviour are not necessarily expected, with changes on the activity of neural pathways being a more parsimonious explanation (Hofmann, 2003; but see Oliveira *et al.*, 2001g; Remage-Healey and Bass, 2004). This could explain the lack of differences in KT levels between alternative male types both in the St. Peter's fish and in the sailfin molly. In the St. Peter's fish, Fishelson and Hilzerman (2002) found that in captivity the mating system is very flexible and males varied their reproductive behaviour very quickly according to local conditions. For example, in environments with female-biased operational sex ratios (OSR), the expression of polygyny is promoted and conversely male-biased OSR facilitates the expression of monogamous behaviour in *S. galilaeus* males (Ros *et al.*, 2003). Also, in the sailfin molly, although there is a trend for a specialisation of large males in courtship and of small males in gonopodial thrusting, both small and large males can use both behaviours facultatively (Schlupp *et al.*, 2001). Finally, in the case of *N. pulcher*, male helpers may share the paternity of the brood (Dierkes *et al.*, 1999) and queue for vacated territories to become breeders (Balshine-Earn *et al.*, 1998). Thus, they are sexually active and must be ready for rapid behavioural changes once the opportunity for a nest takeover occurs.

An inspection of Table 8.3 also shows that in all species with fixed ART, KT levels are always higher in the bourgeois male morph. In fact, the incidence of the differences in KT levels among alternative phenotypes does not seem to be independent of the type of ART present in the species (Table 8.3). This fact prompted the author to analyze the dataset on androgens and ART in a quantitative way and compute, for each species with an available direct or an indirect measurement of androgen circulating levels (this criteria excluded *C. julis* and *S. subligarius* from this analysis), the ratio between the androgen levels in the bourgeois and in the parasitic male (e.g., bourgeois average KT level/parasitic average KT level). Data was collected from the references listed in Table 8.3, and in cases in which the original data was not in the text or in tables, the published figures were scanned to extrapolate the data from the graphs using the UnGraph software package (Biosoft, 1998). If one plots these ART relative androgen levels for each species against its type of ART, there is a trend for a positive association between the degree of phenotypic plasticity (with fixed ART being considered high, sequential ART intermediate, and reversible ART low in phenotypic plasticity) and the magnitude of the difference in KT levels between the male morphs. That is, species with fixed ART tend to display larger differences in KT levels between their alternative male types than do species with plastic ART. For T such a trend is not observed. These results suggest differential roles for the two androgens in the expression of phenotypic plasticity in fish.

Depending on each species, the alternative male types may differ from each other in a set of phenotypic traits, namely in their reproductive behaviour, in the differentiation of secondary sex characters, in the relative investment in gonadal development, and in the differentiation of accessory reproductive organs (e.g., testicular glands in blenniids; Patzner and Lahnsteiner, 1999). The available evidence shows that among androgens KT is more potent in eliciting all characters of this constellation of tactic-specific traits:

1. Androgens and reproductive behaviour: Although there is a large body of literature on the role of androgens in the expression of reproductive behaviours (for reviews see Liley and Stacey, 1983; Borg, 1994), there are few studies that have tested the specific role of KT. In species with male parental care, circulating levels of KT are higher in the courting phase than in the parental phase (Borg, 1994; Oliveira *et al.*, 2002a), and in castration and hormone replacement experiments KT is more effective than T in restoring the different aspects of male reproductive behaviour including territoriality, nest construction, and courtship, such as in the three-spined stickleback, *Gasterosteus aculeatus* (Borg, 1987) and bluegill sunfish, *Lepomis macrochirus* (Kindler *et al.*, 1991). In addition, KT treatment in field conditions is also effective in promoting male "bourgeois" behaviour in the rock-pool blenny (*Parablennius parvicornis*, Ros *et al.*, 2004b). These examples suggest that indeed KT plays a role in the expression of reproductive behaviours, namely in the establishment of a reproductive territory and/or in courtship behaviours.
2. Androgens and secondary sex characters: There is a large body of literature that shows that most of the morphological traits exhibited by bourgeois males are androgen-dependent, and that KT is the most potent androgen in inducing the differentiation of these characters, such as the sonic motor system in Type I male batrachoids (Brantley *et al.*, 1993a; but see Modesto and Canário, 2003b) and sex-pheromone producing anal gland in blenniids (Oliveira *et al.*, 2001b,d,e). Another aspect that supports this view is the fact that in gonochoristic species without sexual dimorphism males present very low levels of KT, such as the *Sardinops melanosticus* (Matsuyama *et al.*, 1991) and *Syngnathus typhle* (Mayer *et al.*, 1993).
3. Androgens and gonadal allocation: A trait that differs considerably among alternative male types is the relative size of the testis. In most species studied, parasitic males have higher gonadosomatic indices (GSI; gonad weight/body weight $\times$ 100) than bourgeois males, indicating a higher relative gonadal investment by the former male type (Taborsky, 1994, 1998; but see Tomkins and Simmons, 2002 for a critical assessment of the use of GSI values to measure relative gonadal investment in species with ART). Besides gonadotrophins, androgens also participate in the regulation of spermatogenesis. However, KT and T seem to play different roles: KT stimulates later stages of this process, whereas T is involved in the negative-feedback mechanisms needed to control KT-dependent spermatogenesis (Schulz and Miura, 2002). Consequently, a balance between T and KT is needed for the control of spermatogenesis (Schulz and Miura, 2002).

This suggests that the ratio between KT and T should be more informative than absolute androgen levels for the understanding of differential gonadal allocation between alternative male types. Data was gathered from the literature on GSI values for 9 out of the 14 species for which we have both KT and T levels, and thus a KT:T ratio can be computed. For these nine species (*P. notatus, H. didactylus, L. macrochirus, S. melops, S. pavo, P. parvicornis, T. duperrey, O. mossambicus* and *S. galilaeus*) there was a nonsignificant trend for the magnitude of the difference in the KT:T to be negatively correlated with the magnitude of the difference in GSI values between alternative male types ($Rs = -0.63$, $P = 0.067$). That is, in species for which the magnitude of the difference in KT levels between bourgeois and parasitic is larger, there is a smaller difference in relative gonad size. This also means that among parasitic males a higher GSI is associated with a lower KT:T ratio, which probably allows them to have larger testis without a linked expression of the secondary sex characters and of bourgeois male behaviour.

The results reported above (Figure 8.4) also suggest that in fixed ART a higher magnitude of differences in KT levels between phenotypes might be the result of early (i.e., prematurational), long-lasting, and irreversible actions of KT in the differentiation of the ART, resembling the organisational

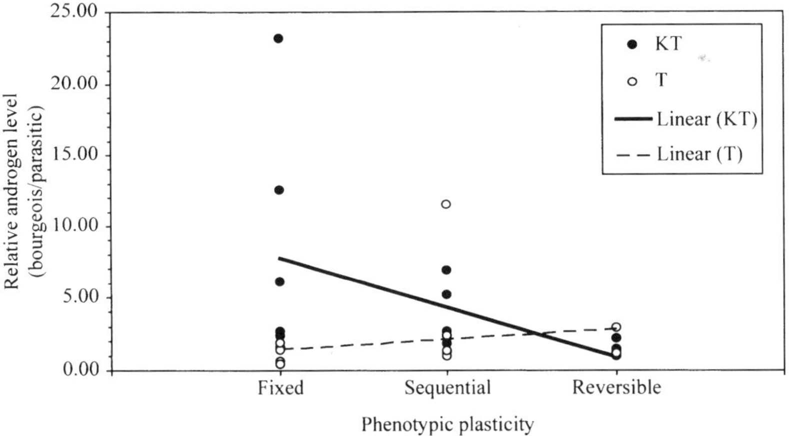

Fig. 8.4. Relative androgen levels between bourgeois and parasitic males (i.e., ratio between the androgen concentrations in the bourgeois male over the androgen concentration in the parasitic male) in teleost species with different types of alternative reproductive tactics. KT, 11-ketotestosterone; T, testosterone; solid line, linear fit curve for KT values; dashed line, linear fit curve for T values.

effects of sex steroids in mammalian sex differentiation. In the other extreme, the low magnitude in KT ratios between morphs in species with reversible ART would reflect the transient nature of this type of tactic, which parallels the activational effects of steroids on behaviour (see Section 4.2 for a further discussion of this hypothesis).

4.1.2. Estrogens and ART

Data on estrogen levels in species with ART is scarce (Table 8.3). This paucity of data probably reflects the focus of research on male ART. Only in five studies has estradiol (E2) been measured. Among the Batrachoididae (i.e., the plainfin midshipman and the Lusitanian toadfish), E2 levels do not differ between males of both phenotypes, with the majority of males having undetectable levels (Brantley *et al.*, 1993b; Modesto and Canário, 2003a). However, in the midshipman, a large proportion of type I males (ca. 50%) display low E2 levels during the breeding season (Sisneros *et al.*, 2004). The functional significance of this finding is unclear. Recent work found that aromatase levels in the testis of midshipman is very low or even undetectable, contrasting to high concentrations found in the brain (Forlano *et al.*, 2001), which suggests the brain as one potential source for these circulating low levels of E2 found in type I males during the breeding season.

In the saddleback wrasse, E2 levels were detectable at similar low levels both in initial and terminal phase males, and no seasonal variation was detected (Hourigan *et al.*, 1991). Finally, both in the stoplight parrotfish and in the corkwing wrasse the parasitic males have significantly higher E2 levels than the bourgeois males (Cardwell and Liley, 1991; Uglem *et al.*, 2002). Because in both species parasitic males mimic females, in particular their colouration pattern, it has been proposed that the higher levels of E2 found in parasitic males may have a feminising function (Uglem *et al.*, 2002). This hypothesis is supported by the fact that suppression of E2 synthesis is needed for sex and colour change in protogynous hermaphroditic fish (Nakamura *et al.*, 2003).

4.1.3. Progestogens and ART

Progestogens are very diverse among teleost species and have been associated with the acquisition of sperm motility and spermiation in male salmonids and cyprinids (Scott and Baynes, 1982; Asahina *et al.*, 1990, 1993). A pheromonal role has also been described for these compounds (e.g., Dulka *et al.*, 1987; Chapter 9). The progestin 17,20β-dihydroxy-4-pregen-3-one (17,20βP) has been reported to rise to high levels at times when males are expressing courtship behaviour such as in rainbow trout, *Oncorhynchus mykiss* (Liley *et al.*, 1986), Atlantic salmon (Mayer *et al.*, 1990), demoiselles, *Chromis dispilus* (Barnett and Pankhurst, 1994), brown trout,

Salmo trutta (Olsén *et al.*, 1998), and the exogenous administration of 17,20βP to castrates restored courtship behaviour in male rainbow trout, a species in which androgen replacement therapy is inefficient (Mayer *et al.*, 1994).

Data on progestogens is available for six species with ART (Table 8.3). There is no clear pattern with progestogens being higher in the bourgeois males than the parasitic males in two species (Mozambique tilapia and belted sandfish, Table 8.3) and no differences being present in the remaining four species (Lusitanian toadfish, saddleback wrasse, St. Peter's fish and Atlantic salmon; Table 8.3). However, the progestogen(s) measured varied from species to species. In the Lusitanian toadfish, although type I and type II males showed no differences in the levels of progestogens, 17,20β,21-trihydroxy-4-pregen-3-one (17,20β21P) peaked in the breeding season, whereas 17,20α-dihydroxy-4-pregen-3-one (17,20αP) levels remained stable over the year, and 17,20βP was undetectable in the plasma (Modesto and Canário, 2003a). In the saddleback wrasse, initial and terminal phase males showed similar levels of 17,20βP that peak in the breeding season (Hourigan *et al.*, 1991), whereas in the St. Peter's fish there was no difference between monogamous and polygynous males in 17,20βP levels (Ros *et al.*, 2003). Finally, in the Atlantic salmon, 17,20βP levels did not differ between male morphs, but a peak of this progestin has been observed in large males during the spawning season (Mayer *et al.*, 1990). Thus, 17,20β21P in the toadfish, 17,20βP in the saddleback wrasse, and 17,20βP in the Atlantic salmon may play a role in male reproductive function, probably in spermiation. The lack of differences in 17,20βP between alternative male morphs in the Atlantic salmon, where levels have been associated with the expression of courtship behaviour, might be explained by physiological constraints imposed by 17,20βP effects on spermiation in both morphs.

In the Mozambique tilapia, territorial males have higher levels of both 17,20αP and 17,20βP than nonterritorial female mimicking males, but only 17,20αP plasma concentrations increase in the presence of females, when courtship behaviour is expressed by the males (Oliveira *et al.*, 1996). This result suggests that in male tilapia 17,20αP may play a major role in spawning behaviour and/or spermiation (potentially induced by the presence of the females).

Finally, in the belted sandfish, a simultaneous hermaphrodite fish, 17,20β21P levels increased with increasing body size, whereas 17,20βP concentrations did not vary with size. Because larger individuals preferentially play the male role during spawning and get involved in pair spawning rather than in streaking attempts, 17,20β21P rather than 17,20βP seems to be associated with male reproductive behaviour (Cheek *et al.*, 2000).

In summary, different progestogens appear associated with male reproductive function in the different species analysed. However, for most species it is difficult to disentangle potential effects of progestins on male courtship behaviour from effects on spermiation, which may co-occur in time. Therefore, only by carefully planned experiments can these two effects be isolated in future studies.

4.2. Organisational and Activational Effects of Sex Steroids in the Differentiation of Alternative Reproductive Phenotypes: The Relative Plasticity Hypothesis

Moore (1991) proposed a conceptual framework for the hormonal basis of alternative reproductive tactics that has been named the *relative plasticity hypothesis*. The rationale behind this hypothesis is that the effects of hormones in the differentiation of ART are equivalent to their effects in primary sex differentiation, and thus the hormonal basis of ART would have a parallel in the activational-organisational effect of hormones (Arnold and Breedlove, 1985) depending on the plasticity of the tactics. The relative plasticity hypothesis predicts an organisational role of hormones in the expression of fixed alternative phenotypes and an activational role in the case of plastic alternative phenotypes (Moore, 1991; Moore *et al.*, 1998). Two predictions can then be extracted from this hypothesis:

1. In species with plastic ART, hormone levels should differ between adult alternative morphs, whereas in species with fixed ART adult hormone profiles should be similar among alternative morphs, except when morphs are affected by different social experiences (Moore, 1991).
2. In species with plastic ART, the effect of hormone manipulations on phenotypic differences in behaviour and morphology should be effective in adults but not during early development (i.e., activational effect), whereas in fixed ART hormone manipulations should be effective during early development but not in adults (i.e., organisational effect) (Moore, 1991).

The first prediction of the relative plasticity hypothesis is flawed in a Popperian sense as it cannot be disproved. In fact, Moore (1991) explicitly mentioned that in fixed ART adult hormone profiles should be similar among alternative male phenotypes, except if the two male types experience different social environments (which is expected in individuals opting for alternative modes of reproduction). Thus, both positive and negative associations, and even the lack of a relationship between sex steroid levels and the ART type, all find support from it. Therefore, a test of the first prediction is inconclusive. Nevertheless, it may be mentioned that this first prediction of Moore's hypothesis is not confirmed for KT by the dataset presented here, because differences in KT levels between male types are present in most species irrespective of whether their tactics are fixed or plastic (i.e., among species with plastic ART, this prediction would be apparently confirmed in *S. pavo*, *P. parvicornis*, *S. viride, T. duperrey, C. julis*, and *O. mossambicus*, but not in *N. pulcher*, *S. galilaeus*, and *P. velifera*; in species with fixed ART, the prediction would only be confirmed in *S. subligarius* and rejected in *P. notatus*, *L. macrochirus*, *S. melops*, and *S. salar*; see Section 4.1. and Table 8.3 for details).

In contrast, the second prediction does not suffer from epistemological flaws and hence a test of the allowable. Unfortunately, few hormone manipulation experiments have been conducted in teleost species with ART. To my knowledge, only in four species have the effects of androgens on parasitic males been tested so far. These consisted of three species with plastic ART: the peacock blenny (Oliveira *et al.*, 2001d), the rock-pool blenny (Oliveira *et al.*, 2001e), and the sailfin molly (R. F. Oliveira, I. Schlupp and D. M. Gonçalves, unpublished data, 2000); and of one species with fixed ART, the plainfin midshipman (Lee and Bass, 2004). In the three species with plastic ART, the exogeneous administration of androgens to parasitic males did not induce a clear behavioural switch towards a bourgeois tactic. A 7-day androgen implant treatment failed to induce bourgeois behaviours but failed to promote male typical behaviour (e.g., establishing a nest and attracting females to it) in parasitic (i.e., sneaker/satellite) males of both blenniid species (Oliveira *et al.*, 2001d,e). However, in the peacock blenny, the KT treatment resulted in the inhibition of sneaking behaviour and female nuptial colouration in sneaker males (Oliveira *et al.*, 2001d). In the sailfin molly, an immersion treatment with methyl-testosterone did not induce small males to display courtship behaviour (R. F. Oliveira, I. Schlupp and D. M. Gonçalves, unpublished data, 2000). On the other hand, the same treatments were efficient in promoting the expression of "bourgeois" morphological traits. In both blennies, the treatment with KT induced the differentiation of a sex-pheromone producing anal gland, the differentiation of a gonadal accessory gland (i.e., testicular gland) involved in sperm maturation and steroidogenesis, and the development of a male genital papillae (Oliveira *et al.*, 2001d,e). Thus, the second prediction of the relative plasticity hypothesis is partially confirmed for species with plastic ART, in relation to the differentiation of morphological characters typical of bourgeois males. In the plainfin midshipman, type II males that received an intraperitoneal silastic implant of KT neither show an increase in the size of the brain sonic motor nucleus, which controls the courtship calls typical of type I males, nor expressed the type I male territorial or courtship behaviours. Nevertheless, this treatment increased the sonic muscle mass and the frequency of sneaking behaviour in type II males (Lee and Bass, 2004). Thus, the prediction of the relative plasticity hypothesis is again partially fulfilled (the lack of response to hormone manipulations in adulthood in a species with fixed ART) but KT still induces the development of the sonic muscle. Taken together, this set of results suggests that irrespective of the ART type KT is more efficient in differentiating morphological than behavioural alternative reproductive traits.

A second generation of the relative plasticity hypothesis was subsequently proposed to accommodate the diversity of mechanisms that occur in ART (Moore *et al.*, 1998). In particular, the revised version of the relative plasticity hypothesis places emphasis on the distinction between reversible and irreversible phenotypes among plastic tactics and between conditional and unconditional fixed tactics. According to the new version of this hypothesis, the plastic reversible tactics would be the true equivalents to activational effects of hormones and thus the original predictions of the relative plasticity hypothesis would only apply to this type of alternative tactic. Conversely, plastic irreversible ART would represent an intermediate situation between organisational and activational effects; that is, a postmaturational reorganisation effect, in which the phenotypical outcome is immediately produced (Moore *et al.*, 1998). Thus, hormone differences needed to differentiate the two alternative phenotypes need not be permanent and may only be present during the transitional phase. Among the fixed ART, the distinction between conditional and unconditional fixed tactics has no consequences for the predictions concerning the endocrine mechanisms of ART, with organisational actions being predicted in both cases (Moore *et al.*, 1998).

The predictions of the second generation of the relative plasticity hypothesis remain untested for most species with ART, except for the main species studied by Moore and collaborators, the tree lizard (*Urosaurus ornatus*; Moore *et al.*, 1998; Knapp *et al.*, 2003; Jennings *et al.*, 2004). The available data on fish androgens does not support the new predictions generated by the second generation of the relative plasticity hypothesis, because KT differences are still present in species with plastic irreversible ART (Table 8.3), where differences would only be expected during the transitional period between the alternative phenotypes. Moreover, the magnitude of these differences is even larger in species with plastic irreversible ARTs than in species with plastic reversible ART (Figure 8.4), where pure activational effects (and thus clear differences is hormone profiles) have been predicted.

In conclusion, although the relative plasticity hypothesis represents a major effort to develop a conceptual framework for the study of the endocrine mechanisms underlying ART, it does not seem to be generalised across vertebrate taxa. At least in teleost fish, a taxon that contains most described cases of ART, the relative plasticity hypothesis does not fit the available data. One of the major reasons for this mismatch may reside in the fact that this hypothesis is deeply based in the mammalian paradigm of sex differentiation, which is not applicable to all other vertebrate taxa, in particular those with nongenetic sex determination mechanisms.

4.3. Androgens and Alternative Reproductive Tactics: Cause or Consequence?

So far, this Chapter has interpreted the association between KT and ART as a proof for the involvement of KT in the differentiation of ART. However, androgen levels not only influence vertebrate reproductive physiology and behaviour but can also be influenced by the social environment in which the animal lives (Wingfield *et al.*, 1990; Oliveira *et al.*, 2002a).

4.3.1. Social Interactions Feed Back to Influence Hormone Levels

Several studies have shown the effects of social interactions on the short-term modulation of androgen circulating levels (for a review see Oliveira, 2004). In teleost fish, endocrine responses to social and sexual stimuli have been documented. For example, the presence of ovulated females induces a rise in sex steroid and gonadotropin levels and an increase in milt production in male salmonids (Liley *et al.*, 1986, 1993; Rouger and Liley, 1993). Anosmic salmonid males (i.e., rainbow trout, *Onchorhynchus mykiss*, and kokanee salmon, *Onchorhynchus nerka*) in the presence of sexually active females have lower levels of sex steroids and a lower sperm production than males with intact olfactory epithelia, which suggests that chemical signals may play an important role in this social modulation of hormone levels (Liley *et al.*, 1986, 1993; Rouger and Liley, 1993). In the Mozambique tilapia, males are sensitive to the sexual maturity stage of females courting more intensively ovulated females (Silverman, 1978), an effect that also seems to be mediated by chemical signals emitted by receptive females. It has also been shown that male Mozambique tilapia experience a rise in KT following courtship (Borges *et al.*, 1998). Not only sexual stimuli but also male–male competition may induce an endocrine response in the participating individuals, a response especially sensitive in the case of androgens. In different vertebrate groups, including humans, short-term fluctuations in androgen concentrations related to social interactions have been demonstrated (Oliveira, 2004). However, it seems that the association between androgen levels and agonistic behaviour is stronger in periods of social instability (challenge), as is the case of the establishment of dominance hierarchies, the foundation of a new territory, the response to territorial intrusions, or the active competition with other males for access to females (Oliveira *et al.*, 2002a). During periods of social inertia, the levels of aggression drop to a breeding baseline and a decoupling between androgens and aggression may occur.

These results have been interpreted as an adaptation for the individuals to adjust their agonistic motivation to changes in the social environment in which they are living. Thus, male–male interactions would stimulate the production of androgens and the levels of androgens would be a function of the stability of the social environment in which the animal is placed. This hypothesis was first proposed by Wingfield and collaborators and is currently known as the *challenge hypothesis* (Wingfield *et al.*, 1990).

4.3.2. The Challenge Hypothesis

Wingfield and coauthors proposed that the variation in androgen levels could be more closely associated with temporal patterns of aggressive behaviour than with changes in reproductive physiology (challenge hypothesis; Wingfield *et al.*, 1990). According to this hypothesis, when the breeding season starts, androgen levels rise from a nonbreeding baseline to a higher breeding baseline which is sufficient for normal reproduction, that is for gametogenesis, the expression of secondary sexual characters, and the performance of reproductive behaviour. Androgen levels can further increase until they reach a maximum physiological level in response to such environmental stimuli as male–male interactions and/or the presence of receptive females (Figure 8.5A). However, above this breeding baseline any increase in androgen levels does not influence the development of secondary sex characters nor the expression of reproductive behaviour (Figure 8.5B). Thus, androgen patterns during the breeding season are predicted to vary between species according to the amount of social interactions to which the individuals are exposed. In monogamous species with high levels of parental care, androgen levels should increase above the breeding baseline only when males are challenged by other males or by mating. At other times, androgens should remain at the breeding baseline so that they do not interfere with paternal care. Conversely, androgen levels in polygynous males should be near physiological maximum throughout the breeding season due to high levels of male–male competition in this type of breeding system. Wingfield *et al.* (1990) reviewed the available literature on testosterone and aggression in free-living birds and the results supported the above predictions of the challenge hypothesis, which led the authors to suggest that this hypothesis may be valid not only for birds but for vertebrates in general. We have recently confirmed this hypothesis for teleost fish (Oliveira *et al.*, 2002a; Hirschenhauser *et al.*, 2004).

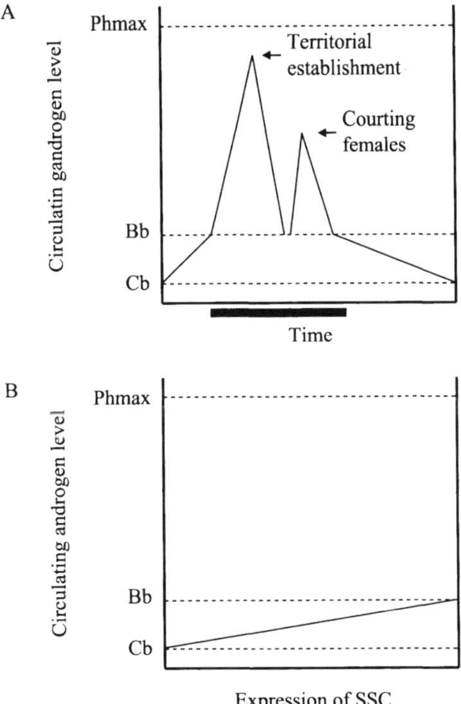

Fig. 8.5. Social influences on androgen levels. (A) Androgen levels rise from a constitutive baseline (Cb) to a breeding baseline (Bb) needed for the expression of reproductive behaviour, the differentiation of secondary sex characters (SSC) and spermatogenesis in breeding individuals; during the breeding season (marked by the black bar at the bottom of the graph) transitory androgen surges above the breeding baseline up to the physiological maximum (Phm) may be induced by social challenges posed by social interactions, either with competing males or with potential mates. (B) Theoretical variation in the expression of reproductive traits according to the challenge hypothesis (Wingfield *et al.*, 1990); note that above the breeding baseline further increases in androgen levels will have no effect in a subsequent increase in the expression of reproductive traits.

4.3.3. Social Modulation of Androgen Levels in Alternative Male Phenotypes

Considering the potential social influences on circulating androgen levels in fish, the differences in KT levels between alternative reproductive male types reported above (Section 4.1.1) might not reflect different hormone profiles due to an activational effect of KT on the expression of the bourgeois tactic, but rather merely reflect the responsiveness of this hormone to the expression of the tactic itself (i.e., they are a consequence and not a cause of the expression of alternative mating tactics). This can be the case if the alternative phenotypes experience different social environments, which is expectable since by definition bourgeois males defend resources to get access to mates and thus are expected to face higher levels of social challenges than parasitic males. There are few data available to test this hypothesis. However, three examples will illustrate how the social modulation of androgens may confound the interpretation of the data on KT levels on species with ART.

In the two closely related blenniid species with ART that the author and his collaborators have been studying, the peacock blenny and the rock pool blenny, there is a fundamental difference in the nature of the tactic used by parasitic males to achieve fertilisations. In the peacock blenny, parasitic males mimic female nuptial colouration and female courtship behaviour in order to get access to the nests when a spawning is taking place (Gonçalves *et al.*, 1996). In contrast, in the rock-pool blenny, parasitic males act as satellites of nest-holders taking a share in the defense of the breeding territory, engaging in a high number of agonistic interactions (Santos, 1985; Oliveira *et al.*, 2002b). Thus, the social challenge regime seems to be more asymmetric between male types in the peacock blenny than in the rock-pool blenny, where both male types participate in territorial defense. Thus, we predict that the magnitude of the difference in KT levels between bourgeois and parasitic males should be higher in the peacock blenny than in the rock-pool blenny. The available data is in accordance with this prediction because the difference in KT levels is fivefold higher in nest-holder males than in sneakers in the peacock blenny and approximately twofold higher in nest-holders than in satellites in the rock-pool blenny. Another piece of evidence comes from the study on the peacock blenny, in which KT levels in bourgeois males were shown to vary with the temporal variation in the rate of sneaking attempts (social challenges) by parasitic males (Oliveira *et al.*, 2001a).

In the St. Peter's fish, with facultative monogamous/polygynous males (Fishelson and Hilzerman, 2002), we tested the relationship between pair-bond intensity (i.e., quantitative measure of the partner preference when given a choice to interact with female partner vs. novel female) and KT levels, and found a significant negative correlation between KT levels and partner preference, suggesting that the more polygynous males had higher KT levels (Oliveira *et al.*, 2001h). In a second experiment, the effects of exogenous administration of androgens on pair-bonding were assessed by treating males with silastic implants of either T or vehicle only. Because in the first experiment a relationship between higher KT levels and a higher propensity to become polygynous emerged, it was predicted that androgen-treated males would display a lower partner preference than controls. Contrary to expectations, the androgen treatment had no effect on partner preference (Oliveira *et al.*, 2001h). The results from these two experiments suggest that the association between partner preference and androgen levels in St. Peter's fish is not due to a causal effect of androgens on partner preference, but they probably reflect variation in partner-preference behaviour observed among males (i.e., a raise in KT levels induced by the amount of interaction with the novel female). This interpretation is supported by data collected in Lake Kinneret (Israel), where KT levels of polygynous males did not differ from those of monogamous males (Ros *et al.*, 2003).

These three studies stress the fact that the differences in KT levels found between bourgeois and parasitic males may reflect, at least partially, differences in social experiences. This does not mean that these socially-induced androgen variations will not reflect in the behaviour subsequently adopted by the males, thus promoting the divergence between the two male types.

4.4. Steroid-Binding Globulins, Steroidogenic Enzymes, and Steroid Receptors

The previous sections have concentrated on differences in circulating levels of steroids. However, circulating hormones are just the "most visible" part of the endocrine system with the techniques that have been used more readily (i.e., hormone assays such as radioimmunoassay or enzyme-immunoassays). The activity of the steroidogenic pathways in the endocrine cells, the affinity to binding globulins in the plasma and local conditions in the target tissue, will all influence the action of the hormone (Knapp, 2004). Thus differences between morphs in total levels of circulating hormones are difficult to interpret per se and these other factors should also be considered and investigated.

4.4.1. Steroid-Binding Globulins

Binding globulins can regulate the availability of circulating steroids to target tissues, because only the free (unbound) fraction is biologically active. Despite their potential role in the modulation of endocrine mechanisms mediating the expression of reproductive traits (e.g., reproductive behaviour, morphological sex characters), to the author's knowledge, there is only one published study for all vertebrate taxa that documents differences in binding capacity of a steroid-binding globulin among alternative morphs (tree lizards; Jennings *et al.*, 2000). In teleost fish, sex hormone-binding globulins have been described in a number of species, and their affinity for different sex steroids varies across species (Pasmanik and Callard, 1986; Laidley and Thomas, 1994; Hobby *et al.*, 2000b) and with the reproductive cycle (Hobby *et al.*, 2000a). In most species studied, steroid hormone-binding globulins bind E2 with higher affinity than androgens, and among androgens T usually shows the highest relative affinity, whereas KT has a low affinity. For example, in zebrafish the relative binding affinity of KT is 10 times lower than that of T (Miguel-Queralt *et al.*, 2004), which indicates that KT is potentially more readily available to target tissues than T.

4.4.2. Steroidogenic Enzymes

Because in most teleosts studied to date, T is also found in females during the breeding season at higher or similar levels compared to males but KT has mainly been detected in males (Borg, 1994), some authors proposed that testosterone should be seen as a prohormone that can be metabolised into other biologically active steroids: KT in males and E2 in females (e.g., Bogart, 1987; Oliveira and Canário, 2001). The conversion of T into KT involves two steps catalysed by the enzymes 11-β-hydroxylase (11-β-H) and 11β-hydroxysteroid dehydrogenase (11-β-HSD). Testosterone, but not KT, can also be converted into E2 by a sequence of reactions in which the C19-methyl group is removed and the steroid A ring is aromatised. This cascade of reactions is catalysed by the cytochrome P450 aromatase complex (aromatase). Because both pathways have a common substrate and may be present in males in the same tissues, they may regulate each other's activity. For instance, one potential role for the high levels of aromatase found in the teleost brain (100- to 1000-fold higher than in the brain of mammals; Callard *et al.*, 1990) could be a downregulation of the KT biosynthetic pathway. By varying the local activity of these steroidogenic enzymes, an organism could dramatically change the availability of the biological active steroid in different target tissues. The occurrence or magnitude of the response of the target cell to the hormone depends on hormone levels in close proximity to targets, which can thus be different from circulating levels. 11β-H mRNA expression has been detected in the brain of terminal phase males of the saddleback wrasse after sex change, suggesting that levels of KT can indeed be increased at the target tissue (Morrey and Nagahama, 2000). Additionally, there is a shift in the expression of aromatase versus 11β-H in the gonad during sex change, with a downregulation of the former and an upregulation of the latter. The fact that in this species sex change is controlled by social factors, suggests that changes in steroidogenic enzyme activity may be influenced by social factors. Studies on steroid metabolism in the Siamese fighting fish (*Betta splendens*) further support this view, where subordinate individuals have lower expression of secondary sex characters and male display behaviour, both aggressive and sexual, and the activity of 11β-H in the testis is decreased (Leitz, 1987). This is consistent with different roles for the two androgens, with testosterone acting as a pool of androgens, which will allow the individual to shift behaviour according to the social context by a differential activation of aromatase and 11β-H.

In species with ART, a difference is thus expected in the activity of both enzymes between the two male types in tactic-sensitive tissues: bourgeois males are expected to upregulate 11β-H and downregulate aromatase and the opposite action is expected in parasitic males. Unfortunately, there is no data available on brain 11β-H expression in alternative reproductive phenotypes. In contrast, there is some evidence on brain aromatase that in general corroborates this hypothesis.

In the plainfin midshipman, aromatase activity in the hindbrain was three- to five-fold higher in type II males than in type I males (Schlinger *et al.*, 1999). Because only type I males produce a courting "humming" call that is controlled by a vocal motor circuit located in the hindbrain (Bass and Baker, 1990), these differences in aromatase activity may explain the differentiation of alternative male morphs regarding calling behaviour in this species. In the same species it has been found, using both immunocytochemistry and *in situ* hybridisation with specific probes for midshipman, that aromatase levels were high in the hindbrain sonic motor nucleus, in the periaqueductal gray area in the midbrain, and in the hypothalamus and telencephalon. More interestingly, this study shows that aromatase expression is almost exclusively located in glial cells and not in neurons (Forlano *et al.*, 2001). Thus, a role for glia in modulating brain steroid levels is plausible, and the ventricular location of most aromatase-immunoreactive cells would facilitate the bathing of the brain in E2 converted from T, and concomitantly would decrease the local availability of KT. Moreover, high levels of estradiol may be needed in the adult teleost brain because neurogenesis is present during adulthood (Zupanc, 2001) and estrogens are known to act as neurotrophic factors (e.g., Contreras and Wade, 1999; Dittrich *et al.*, 1999). In species with plastic ART, brain aromatase might also be important due to its role on synaptic plasticity that may underlie the reorganisation of neural circuits during the differentiation of alternative brain morphs.

The regulation of steroidogenesis in secreting tissues will also determine the amount of the steroid to be released in the plasma, and ultimately its availability to target tissues. Thus, as a crude exercise to assess the hypothesis raised above, the author has used the ratio KT:T (already used in Section 4.1.1.) as an indicator of the activity of the 11β-H + 11β-HSD pathway, and the E2:T ratio as an indicator of the activity of aromatase. The magnitude of the difference between bourgeois and parasitic male KT:T ratios is higher in species with fixed phenotypes, and lower in species with plastic reversible phenotypes. This suggests a higher upregulation of the 11-oxygenated androgens biosynthetic pathway in fixed ART species when compared to plastic ART species. These differences may be explained by the fact that the differentiation of divergent phenotypes in fixed species is not constrained by the subsequent phenotypic reorganisation that occurs in plastic species. Therefore, bourgeois males of fixed species might differentiate into more divergent phenotypes and 11-oxygenated steroids play a critical role in this process. For the ratio E2:T data is only available on five species and thus a clear pattern is not detectable. Nevertheless, if one compares the magnitude of the differences between tactics for both ratios, in all five species (*P. notatus, H. didactylus, S. viride, T.duperrey* and *S. melops*), the KT:T ratio is higher than the E2:T ratio, indicating a dominance of the 11-oxygenated-androgen pathway over the aromatisation pathway in the bourgeois tactic.

Recently, a model for the differentiation of alternative phenotypes based on glucocorticoid-androgen interactions has been independently proposed by Knapp *et al.* (2002), Knapp (2004) and Perry and Grober (2003). This model is based on the fact that the enzymes that participate in KT biosynthesis are also involved in the synthesis (11β-H) and inactivation (11β-HSD) of glucocorticoids. The functional significance of having the same enzymes catalysing parallel pathways is that they can be regulated by reciprocal competitive inhibition. In species with plastic ART or with socially controlled sex change, this commonality creates the possibility that the activity of these enzymes may mediate the transduction of social signals into endocrine ones that will modulate the adoption of a certain ART or sex change (Knapp *et al.*, 2002; Perry and Grober, 2003; Knapp, 2004). It is easy to conceive that parasitic males should have higher cortisol levels than bourgeois males, due to aggressive interactions among the two morphs, which would explain the lower levels of KT observed in parasitic males due to competitive inhibition of the KT-biosynthetic pathway. However, because inhibition of KT production does not affect T, a decoupling of the expression of male traits that are KT dependent from those that can be elicited by T is predicted. Moreover, inactivation of the KT pathway in parasitic males would also result in an accumulation of T, and thus in an increased availability of the substrate for the enzyme aromatase. Therefore, this would also predict a higher aromatase activity in parasitic males, which has been observed in midshipman type II males (Schlinger *et al.*, 1999). Data on cortisol levels in species with ART is very scarce. In the longear sunfish, parasitic males have both higher levels of cortisol and lower KT levels than bourgeois males, suggesting that parasitic males may have a lower activity of 11β-HSD both in the interrenal glands and in the testes relative to bourgeois males (Knapp, 2004). The role of glucocorticoids in phenotypic plasticity is further supported by the fact that in the bidirectional sex-changing goby, *Gobiodon histrio*, small males when stressed by competition with dominant males can change sex back to females (Munday and Jones, 1998). In accordance with this finding, a glucocorticoid responsive element has been identified as a putative promoter regulatory factor of the aromatase gene CYP19A1 (gonadal isoform) that promotes transcription of this gene, resulting in the expression of aromatase which subsequently enhances E2 synthesis involved in male-to-female sex change (Gardner *et al.*, 2003).

4.4.3. Steroid Receptors

The classic mechanism of steroid action involves the diffusion of the free (unbound) fraction of the circulating steroid into the target cell where it binds to intracellular receptors. The activated receptor is then translocated to the nucleus where it binds to specific genes acting as transcription factors. Thus, for steroids to exert their effects, their specific receptors have to be present in the target tissue.

In the brain, where the action of androgens is expected to modulate reproductive behaviour, unusually high levels of androgen receptors (AR) have been detected in goldfish, *Carassius auratus* (Pasmanik and Callard, 1988). The immunolocalisation of these AR, using an antibody to human/rat AR, identified numerous cells in the preoptic area and in the periventricular nuclei of the hypothalamus (Gelinas and Callard, 1997), which are brain key areas in the control of reproduction, namely activity of the hypothalamic-pituitary-gonadal (HPG) axis. A seasonal variation in AR levels was also found with a fivefold peak during the breeding season (Pasmanik and Callard, 1988). This suggests that brain AR expression can be regulated by sex steroids. Recently, a similar seasonal variation in a nuclear AR during the reproductive cycle in the brain of the Atlantic croaker (*Micropogonias undulatus*) was shown, which has been confirmed to be regulated by sex steroids (Larsson *et al.*, 2002).

To date, two different types of androgen receptors (AR1 and AR2; Sperry and Thomas, 1999, 2000; Ikeuchi *et al.*, 2001) and three forms of the estrogen receptor (ERα, ERβ, and ERγ; Hawkins *et al.*, 2001) have been identified both in the brain and in the gonads of teleost fish. All these receptors are ligand-activated transcription factors and thus mediate steroid genomic effects. Intriguingly, KT which has been shown to be the most potent androgen in fish in controlling reproductive function (see Section 3) is a poor ligand of known AR receptors (Pasmanik and Callard, 1988; Sperry and Thomas, 1999; Thomas, 2000). Two possible explanations to solve this paradox are the following:

1. Because KT has also a poor binding to steroid hormone-binding globulin, it is possible that the available fraction of KT to the target cell is relatively higher than that of its competitors that are better

ligands of the binding-globulin and thus are less available for the target cell. Therefore, the local kinetics of the receptor influenced by the relative availability of the ligands near the target may compensate KT for its low affinity to AR.

2. The KT actions on behaviour might be mediated by nongenomic mechanisms of steroid action that are mediated by a membrane steroid receptor. Evidence has accumulated over the years for rapid effects of steroids that are too fast to be mediated by the classic genomic mechanism that acts over a wider time scale (i.e., hours to days), suggesting the existence of specific receptors on the surface of target cells (Thomas, 2003). Recently, a membrane progestin receptor was identified that has characteristics of G protein-coupled receptors (Zhu *et al.*, 2003).

Therefore, it is possible that behavioural actions of KT on bourgeois males are being mediated by one of these two mechanisms. To the author's knowledge, the relative levels of brain steroid receptors between alternative reproductive phenotypes have only been documented for the protogynous wrasse, *Halichoeres trimaculatus* (Kim *et al.*, 2002). It was found using competitive reverse-transcription polymerase chain reaction that the levels of AR transcripts were significantly higher in the brain of terminal phase males than in initial phase males. No other significant differences in gene expression were observed either for AR in the gonads or for ER both in the brain and in the gonads. Thus, by varying the expression of AR in specific tissues (e.g., brain *vs* gonad), bourgeois males can both increase their sensitivity to circulating androgen levels in specific targets (the brain), and at the same time allow a compartmentalisation of the effects of androgens. In other words, deleterious effects can be avoided by varying AR densities in different tissues (Ketterson and Nolan, 1994). This mechanism hypothetically makes it possible to activate the expression of an androgen-dependent reproductive behaviour in bourgeois males without having the side effect of increasing spermatogenesis or the expression of a sex character, because the androgen action can be independently modulated at each compartment (brain *vs* gonad *vs* morphological secondary sex character).

5. NEURAL MECHANISMS OF ALTERNATIVE REPRODUCTIVE BEHAVIOURS

Historically, sex hormones were seen as causal agents of behaviour, acting directly on the display of a given behaviour. This classic view was the result of early studies of castration and hormone-replacement therapy that showed that some behaviours were abolished by castration and restored by exogenous administration of androgens (Nelson, 2000). Currently, this view has shifted towards a probabilistic approach and hormones are seen as facilitators of behavioural expression and not as deterministic factors (Simon, 2002). According to this paradigm, hormones may increase the probability of the expression of a given behaviour by acting as modulators of the neural pathways underlying that behavioural pattern. For example, the effects of androgens on the expression of aggressive behaviour in mammals are mediated by modulatory effects on central serotonergic and vasopressin pathways (Simon, 2002).

The phenotypic divergence between alternative male morphs is present at two main levels: sexual behaviour and morphological differentiation. Two main neurochemical systems have been studied that may help to understand the role of androgens on the differentiation of the ART at each of these two levels: gonadotropin-releasing hormone (GnRH), and arginine vasotocin (AVT) respectively. Both systems have been reviewed recently in the light of vertebrate sexual plasticity elsewhere (Foran and Bass, 1999; Bass and Grober, 2001). Thus, this section will only provide a brief update of these reviews, including data published more recently.

5.1. GnRH and the Differentiation of Alternative Reproductive Morphs

Across all vertebrates, GnRH plays a central role in the control of reproduction by orchestrating the functioning of the HPG axis. Nine different forms of GnRH have been identified to date in vertebrates (Parhar, 2002). Of these, three have been found in the brains of different fish species: salmon-GnRH (sGnRH or GnRH3), chicken-GnRH (cGnRH-II or GnRH2) and seabream-GnRH (sbGnRH or GnRH1). In some species, the three forms are present (e.g., the African cichlid *Astatotilapia burtoni*; White *et al.*, 1995), whereas only two forms (sGnRH and cGnRH-II) have been found in other species (e.g., goldfish; Lin and Peter, 1996). The neural distribution of these three forms suggests multiple GnRH systems in the fish brain. GnRH neurons have been identified in the terminal nerve (TN), in the preoptic area (POA), and in the midbrain, and the three forms of GnRH show a differential localisation across these three systems. In teleosts with the three forms present, GnRH1 is located in the POA, GnRH2 in the midbrain region, and GnRH3 in TN area (White *et al.*, 1995). In species with only two forms present, GnRH2 still occurs in the midbrain and the second form, which can vary with species, is present in the POA and olfactory-forebrain areas. Thus some variation occurs in the form of GnRH present in the POA across fish species (Yu *et al.*, 1997). However, in more advanced teleosts (e.g., cichlids) where three GnRH forms are present, each GnRH variant might have their specific receptor (Parhar, 2003).

Unlike other vertebrates in which the GnRH-POA neurons communicate with the pituitary via the blood portal system, in teleosts a direct innervation of the pituitary by GnRH-POA fibers occurs (Peter *et al.*, 1990; Yu *et al.*, 1997). This anatomical peculiarity allows for a faster action of the cascade of events present in the reproductive axis; that is, GnRH secretion will be more rapidly translated in peripheral effects, such as sex-steroid secretion in the gonads and the consequent differentiation of reproductive traits (Foran and Bass, 1999).

Only four studies on GnRH systems addressing the occurrence of morph differences have been conducted in species with ART. These studies revealed differences either in the number or in the size of GnRH-immunoreactive (GnRH-ir) cells between alternative male phenotypes. For example, in the bluehead wrasse, terminal-phase males have larger numbers of GnRH-ir cells than the initial phase males or the females (Grober and Bass, 1991; Table 8.4). A change from initial male into a terminal-phase male is followed by an increase in the number of GnRH-ir cells in the POA (Grober *et al.*, 1991). On the other hand, in the plainfin midshipman (*Porichthys notatus*), there are no observable differences in cell number (except if corrected for body mass; see Foran and Bass, 1999), but type I males have larger cells than type II males (Grober *et al.*, 1994; Table 8.4). Because otolith readings indicate that type II males are younger than type I (Bass *et al.*, 1996), the difference in this species is that the neural system is activated early in life in type I males. In the platyfish, *Xiphophorus maculatus*, small noncourting males have larger numbers of cells than large courting males and no differences are found in cell size between the two morphs (Halpern-Sebold *et al.*, 1986; Table 8.4). In the grass goby, a species with sequential ART, both the number and the size of GnRH-ir neurons is higher/larger in nest-holders than in sneaker males (Scaggiante *et al.*, 2004; Table 8.4). In this species, the experimentally induced transition from sneaker to nest-holder is accompanied by an increase in the number of GnRH-ir cells in the preoptic area, whereas no changes are detectable in their size (Scaggiante *et al.*, 2004). Terminal nerve GnRH-ir neurons seem not to vary in respect to ART (Foran and Bass, 1999; Scaggiante *et al.*, 2004).

Thus, in the two species with fixed ART (i.e., midshipman and platyfish), the early sexual maturation of the parasitic males is matched with the maturation of the GnRH system as revealed by the fact that parasitic males have similar numbers of GnRH-ir cells to bourgeois males but at a smaller body size (Table 8.4). Similarly, in the two species with sequential ART (i.e., grass goby and bluehead wrasse), the number of GnRH-ir neurons is higher in bourgeois males than in parasitic males (Table 8.4) and there is

Table 8.4
Differences in GnRH and AVT Neurons Between Alternative Reproductive Types

Species	ART	Neuropeptide	Cell size	Cell number	mRNA expression/cell
Plainfin midshipman, *Porichthys notatus*	Fixed	GnRH[1, 2]	B > P	B = P*	—
		AVT[3]	B > P	B = P*	—
Platyfish, *Xiphophorus maculatus*	Fixed	GnRH[4]	B = P	B < P	—
Grass goby, *Zosterisessor ophiocephalus*	Sequential	GnRH[5]	B > P	B > P	—
Bluehead wrasse, *Thalassoma bifasciatum*	Sequential	GnRH[6]	B = P	B > P	—
		AVT[7]	—	B > P	B > P
Saddleback wrasse, *Thalassoma duperrey*	Sequential	AVT[8]	B > P	B > P	—
Peacock blenny, *Salaria pavo*	Sequential	AVT[9]	B = P*	B = P*	B < P
Rock-pool blenny, *Parablennius parvicornis*	Sequential	AVT[10]	B = P*	B = P*	—

*If corrected for body mass then B < P.

[1]Grober *et al.* (1994); [2]Foran and Bass (1999); [3]Foran and Bass (1998); [4]Halpern-Sebold *et al.* (1986); [5]Scaggiante *et al.* (2004); [6]Grober and Bass (1991); [7]Godwin *et al.* (2000); [8]Grober (1998); [9]Grober *et al.*, 2002;[10]Miranda *et al.*, 2003.

a significant increase in GnRh-ir cells during the transition from parasitic to bourgeois male (Grober *et al.*, 1991; Scaggiante *et al.*, 2004). In summary, for the four species studied, changes in the GnRH-POA neural system seem to play a role in the sexual maturation among juveniles of species with fixed ART or in the reorganisation of the reproductive phenotype in plastic species.

5.2. AVT and the Differentiation of Alternative Reproductive Behaviours

The hypothalamic neuropeptide arginine-vasopressin (AVP), or its homologue arginine-vasotocin (AVT) in nonmammalian vertebrates, influences the expression of social behaviours in a wide range of vertebrates, from teleosts to mammals (Goodson and Bass, 2001). Exogenous administration of these neuropeptides induces the expression of different types of reproductive behaviours across all vertebrate taxa (Goodson and Bass, 2001), including the male spawning reflex and courtship behaviour in fish (Pickford and Strecker, 1977; Bastian *et al.*, 2001; Semsar *et al.*, 2001; Salek *et al.*, 2002; Carneiro *et al.*, 2003).

Species with ART offer the possibility of investigating how AVT may be involved in the differential expression of reproductive behaviour between male types. In all species studied so far, an association has been found between the expression of alternative tactics and forebrain AVT (i.e., soma size, number of AVT-ir neurons or AVT mRNA expression per cell; see Table 8.4). However, this association is not linear. In the midshipman and in the two wrasses, bourgeois males exhibit courtship behaviour and concomitantly have AVT-POA neurons with larger soma sizes (Foran and Bass, 1998; Grober, 1998) or with a higher mRNA expression on a per-cell basis (Godwin *et al.*, 2000) than parasitic males. In the peacock blenny, in which sex-role reversal occurs in courtship behaviour (i.e., females and sneaker males are the major courting phenotypes; Almada *et al.*, 1995) and thus the expression of courtship behaviour is decoupled from the bourgeois tactic, AVT mRNA expression is higher in both females and sneaker males than in nest-holders, suggesting that mRNA levels are correlated with the expression of courtship behaviour rather than with the alternative morph (Grober *et al.*, 2002). Finally, no differences were found between alternative male morphs in the rock-pool blenny both in the absolute size and number of AVT-ir cells in the POA (Miranda *et al.*, 2003). Interestingly, in this species the parasitic morph consists of satellite males that help the nest-holder males in the defense of the breeding territory and to some extent in female attraction (Santos, 1985; Oliveira *et al.*, 2002b), and thus the behavioural differentiation between morphs is smaller in this species than in the other studied species. Therefore, the reported differences in the AVT-POA system between alternative sexual morphs are clearly associated with the expression of "morph" typical behaviour.

Fewer studies have manipulated AVT by exogenous administration, either intraventricular or systemic, in fish species with ART. Using this approach, three species with alternative morphs have been studied to date: the midshipman, the bluehead wrasse, and the peacock blenny (Goodson and Bass, 2000; Carneiro *et al.*, 2003; Semsar and Godwin, 2003). These studies will be briefly reviewed below.

In the midshipman, administration of AVT directly in the anterior region of the hypothalamus has different effects in eliciting vocal activity in the two male morphs (Goodson and Bass, 2000). Both male morphs and females produce grunt vocalisations in an agonistic context (not to be confounded with the courtship humming calls emitted exclusively by type I males). However, type I males emit long trains of grunts associated with the defense of their nests, whereas both type II males and females produce short grunts in nonreproductive contexts. The central administration of AVT inhibits the production of electrically stimulated grunts in type I males, but not in type II males and females. Interestingly, isotocin (the teleost homologue of oxytocin) had the opposite effect, inhibiting vocal behaviour in type II males and females but not in type I males (Goodson and Bass, 2000).

In the bluehead wrase, AVT systemic injections increase courtship behaviour in both territorial and nonterritorial terminal-phase males but only increase aggressive behaviour and territoriality in nonterritorial TP males in field conditions (Semsar *et al.*, 2001). Moreover, the administration of an AVP V1 antagonist both to nonterritorial terminal-phase males and to females prevented them establishing breeding territories in vacated areas (Semsar and Godwin, 2003). However, exogenous AVT treatment did not increase the expression of display behaviours, typical of terminal-phase males, in an inhibitory social context, that is, in the presence of terminal-phase males in the field (Semsar and Godwin, 2003). In fact, AVT-treated initial-phase males continued to participate in group spawns in the reef. Interestingly, treating females with KT did not increase their responsiveness to AVT but induced the expression of terminal-phase male colouration and courtship behaviour, suggesting that courtship behaviour in females is KT-dependent (Semsar and Godwin, 2003). Thus, the behavioural effects of AVT in the bluehead wrasse seem to be dependent on the sexual phenotype.

In the peacock blenny, AVT treatment of sneaker males increased the time spent in female nuptial colouration and the frequency of the femalelike courtship behaviour displayed towards nest-holder males. In contrast, sneakers treated with AVT and presented with a gravid female failed to express male courtship behaviour (Carneiro *et al.*, 2003). Accordingly, AVT induced the expression of both nuptial colouration and courtship behaviour in females, but failed to promote any expression of male courtship behaviour in nest-holders (Carneiro *et al.*, 2003). As already mentioned above, KT treatment of sneaker males in this species inhibited the expression of female-courtship behaviour in sneaker males, but had no effect on the number or on the soma size of AVT-POA neurons (Oliveira *et al.*, 2001d), suggesting that the inhibitory effect of KT on AVT-dependent femalelike courtship behaviour is probably regulated by changes in mRNA expression. This hypothesis remains to be tested, but is supported by studies on other vertebrates species that present sex differences and/or seasonal variation in AVP brain elements. These results suggest that plasticity in AVP-POA neurons is regulated by T or by one of its metabolites (E2, DHT) acting on AVP mRNA expression and not on AVP-ir (Goodson and Bass, 2001 and references therein).

Thus, both in the midshipman and in the peacock blenny, AVT seems to activate neural mechanisms underlying behavioural patterns shared by parasitic males and females, whereas in the bluehead wrasse AVT is associated with the expression of courtship behaviour typical of bourgeois males. These within- and between-species variations in the effects of AVT in courtship behaviours indicate that a comparative approach using closely related species that vary in mating systems can be a very rewarding approach to this field in the future.

5.3. The Sexually Bipotential Brain of Teleost Fish

One interesting point related to the neural mechanisms underlying behavioural and morphological phenotypic plasticity is the question of whether alternative morphs possess separate brain mechanisms controlling their divergent behaviour or if both mechanisms are present and can be differentially activated depending on the tactic that is being expressed by the animal. The former situation would be expected in fixed and sequential ART and the latter in reversible ART.

As seen above in teleosts, the display of male sexual behaviour has been associated with KT and AVT, whereas female sexual behaviour is stimulated by prostaglandins, in particular by prostaglandin-$F_{2\alpha}$ ($PGF_{2\alpha}$) (Stacey, 1987). In teleosts, there is ample evidence that brain sexual bipotentiality is maintained through adulthood. A series of studies performed in one gonochoristic (goldfish) and in one gynogenetic (crucian carp, *Carassius auratus langsdorfii*) species have elegantly elucidated this point. In the goldfish, sexually mature males treated with $PGF_{2\alpha}$ display female sexual behaviour, including a pseudospawning (Stacey and Kyle, 1983). In both species, KT-treated females display male sexual behaviour when paired with conspecific ovulated females (Stacey and Kobayashi, 1996; Kobayashi and Nakanishi, 1999). More interestingly, KT-implanted females also showed female spawning behaviour when injected with $PGF_{2\alpha}$ (Kobayashi and Nakanishi, 1999). These results indicate that induced behavioural masculinisation of females and feminisation of males is reversible. Thus, sexual plasticity is present even in species that do not display it in natural situations, as is the case of sex-changing species. Finally, it should be mentioned that in the parthenogenic Amazon molly (*Poecilia formosa*) spontaneous masculinisation occurs naturally with some individuals exhibiting pseudomale behaviour and a differentiation of a rudimentary gonopodium (Schlupp *et al.*, 1992).

These results are particularly interesting in the context of ART because they suggest that alternative male morphs would have the potential to express both alternative reproductive behaviours. However, exogenous administration of KT to parasitic males fails to induce the display of bourgeois male behaviour in species with either fixed or sequential ART (i.e., midshipman, peacock blenny, rock-pool blenny, see Section 2.1.). Therefore, ART species may have tactic-specific constraints that prevent them expressing bipotential behaviours even when manipulated experimentally.

6. INTEGRATING PROXIMATE AND ULTIMATE QUESTIONS IN THE STUDY OF ALTERNATIVE REPRODUCTIVE TACTICS

Negative genetic correlations and life-history tradeoffs can be better understood if their proximate mechanisms are known (Sinervo and Svensson, 1998). Hormones are good candidates to play a major role as physiological mediators of life-history tradeoffs because they may have opposite effects on two or more traits (Ketterson and Nolan, 1992; Sinervo and Svensson, 1998). Phenotypic plasticity is a life-history trait that might have evolved to allow animals to shift resources from one life-history stage to another, (e.g., from reproduction into growth or vice versa, in a condition- or frequency-dependent fashion; West-Eberhard, 1989). These shifts between life-history stages may be controlled by endocrine mechanisms, and it is expected that in some cases the life-history tradeoffs and the associated phenotypic plasticity have the same underlying physiological mechanism (Sinervo and Svensson, 1998). Because androgens are both involved in the animal's investment in current reproduction and have multiple effects on different phenotypic traits, they are excellent candidates to orchestrate transitions between life-history stages. Thus, it is not surprising that an association between KT and the expression of ART has been found.

The costs associated with specific tactics imposed by their underlying physiological mechanisms are another potential constraint for their evolution.

Little research has been conducted on a cost/benefit analysis of ART in teleosts. Bourgeois males display a set of androgen-dependent behavioural traits that help them in competition with other males for resources or for the access to females (e.g., territoriality). Thus, costs associated with keeping androgen levels elevated should be associated with the bourgeois tactic. These include increased energy consumption, interference with immunocompetence, increased risk of predation, and a higher incidence of injuries from agonistic interactions (for a review on costs of elevated T levels see Wingfield *et al.*, 1999, 2001). For example, in the rock-pool blenny, a field study recently demonstrated that lymphocyte counts and the antibody response to a nonpathogenic antigen is lower in nest-holder males than in satellite males, and both immune parameters correlated negatively with KT levels (Ros *et al.*, 2004a).

Because hormones act on different target tissues, many traits may have a common underlying physiological mechanism and thus can be phenotypically linked (Ketterson and Nolan, 1999; Ketterson *et al.*, 2001), as is the case with androgen-dependent traits such as the expression of male morphological characters, muscle hypertrophy, and the expression of reproductive behaviour. Thus, it is likely that selection acting on any one of these traits will affect the others, so that beneficial traits may evolve indirectly as exaptations (Gould and Vrba, 1982). According to Ketterson and Nolan (1999), one way to distinguish between adaptations and exaptations in hormone-dependent traits would be to assess whether these traits arose either in response to selection on circulating hormone levels, or whether they arose in response to variation in the responsiveness of the target tissues to invariant hormone levels (Figure 8.6). In the former case, selection probably did not act on all correlated traits and thus the ones that subsequently conferred an advantage to its carriers should be viewed as exaptations (Figure 8.6). In the latter case, selection probably acted independently on target-tissue sensitivity to constant hormone levels (e.g., by varying density of receptors or the expression of enzymes for particular biosynthetic pathways; Figure 8.6). Although a mixed scenario may occur, in which both circulating levels and target-tissue sensitivity are under selection, this dichotomy provides us with a framework to address the issue of endocrine-mediated adaptive traits. In this respect, alternative reproductive tactics that involve the differential development of androgen-dependent traits within the same phenotype, such as the differentiation of larger testis in parasitic males without displaying secondary sex characters, suggests a compartimentalisation of androgen effects on different target tissues that can be achieved by varying the densities of androgen receptors (AR) in different targets (e.g., gonadal *vs* secondary sex characters). Therefore, ARTs that involve compartimentalisation effects evolved most probably as adaptations,

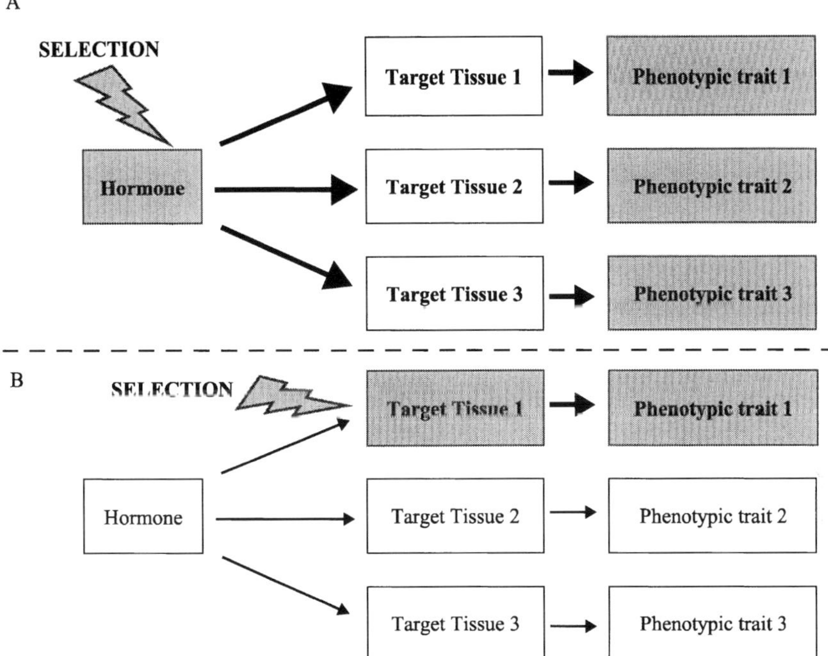

Fig. 8.6. Evolution of endocrine-mediated adaptive traits. (A) Exaptation: trait 1 is being selected by increasing the circulating level of the hormone that controls its expression. If this trait is linked to other traits (traits 2 and 3) because their expression is controlled by the same hormone, then traits 2 and 3 will also be selected, due to the selected increase in hormone levels. Traits 2 and 3 are exaptations. (B) Adaptation: trait 1 is being selected by increasing the expression of receptors to the hormone that controls its expression in a specific target tissue. Although trait 1 is linked to other traits (traits 2 and 3) because their expression is controlled by the same hormone, traits 2 and 3 will not be selected because hormone levels will not be subject to selection.

whereas ARTs that involve no compartimentalisation effects (e.g., some reversible tactics, such as the facultative use of sneaking behaviour by nest-holder males in sticklebacks) represent exaptations. Thus, this question stresses the importance that studies of proximate mechanisms may have to help to understand the evolution of alternative reproductive phenotypes.

ACKNOWLEDGEMENTS

I would like to thank all the collaborators with whom I have been working over the last years in the fascinating world of alternative reproductive tactics in fish, namely (by alphabetical order) Vitor Almada, Eduardo Barata, Adelino Canário, Luis Carneiro, Teresa Fagundes,

David Gonçalves, Emanuel Gonçalves, Matthew Grober, Albert Ros, João Saraiva, and Mariana Simões. They have not only contributed with their efforts on the original research reported in this review, but also with fruitful discussions on the topic that certainly influenced my way of seeing this subject. I thank also Sigal Balshine, Kath Sloman, Rod Wilson, and two anonymous reviewers for very thoughtful comments on earlier versions of this manuscript. I would like to dedicate this chapter to Matthew Grober, a long-term collaborator and friend whose enthusiasm on sexual plasticity in vertebrates has had a major impact on the research on fish ART conducted in our lab. The research from our lab described in this review was supported by a series of grants from the Fundação para a Ciência e a Tecnologia (FCT). The writing of this chapter was directly funded both by the Pluriannual Program of FCT (UI&D 331/2001) and by the FCT research grant POCTI/BSE/38395/2001. Finally, I would like to express my gratitude to my family (Xana, Catarina, and João) for tolerating my continuous unbalanced academic life.

REFERENCES

Almada, V. C., Gonçalves, E. J., Santos, A. J., and Baptista, C. (1994). Breeding ecology and nest aggregations in a population of *Salaria pavo* (Pisces: Blenniidae) in an area where nest sites are very scarce. *J. Fish Biol.* **45,** 819–830.

Almada, V. C., Gonçalves, E. J., Oliveira, R. F., and Santos, A. J. (1995). Courting females: Ecological constraints affect sex roles in a natural population of the blenniid fish, *Salaria pavo*. *Anim. Behav.* **49,** 1125–1127.

Arak, A. (1984). Sneaky breeders. *In* "Producers and Scroungers" (Barnard, C. J., Ed.), pp. 154–194. Croom Helm, London.

Arnold, A. B., and Breedlove, S. M. (1985). Organisational and activational effects of sex steroids on brain and behaviour: A reanalysis. *Horm. Behav.* **19,** 469–498.

Asahina, K., Aida, K., and Higashi, T. (1993). Biosynthesis of 17α,20α-dihydroxy-4-pregnen-3-one from 17α-hydroxyprogesterone by goldfish (*Carassius auratus*) spermatozoa. *Zool. Sci.* **10,** 381–383.

Asahina, K., Barry, T. P., Aida, K., Fusetani, N., and Hanyu, I. (1990). Biosynthesis of 17α, 20α-dihydroxy-4-pregnen-3-one from 17α-hydroxyprogesterone by spermatozoa of the common carp, *Cyprinus carpio*. *J. Exp. Zool.* **255,** 244–249.

Austad, S. N. (1984). A classification of alternative reproductive behaviours, and methods for field testing ESS models. *Am. Zool.* **24,** 309–320.

Balshine-Earn, S., and Earn, D. J. D. (1998). On the evolutionary pathway of parental care in mouthbrooding cichlid fishes. *Proc. Roy. Soc. London B* **265,** 2217–2222.

Balshine– Earn, S., Neat, F., Reif, H., and Taborsky, M. (1998). Paying to stay or paying to breed? Field evidence for direct benefits of helping behaviour in a cooperatively breeding fish. *Behav. Ecol.* **9,** 432–438.

Barnett, C. W., and Pankhurst, N. W. (1994). Changes in plasma concentrations of gonadal steroids and gonad morphology during the spawning cycle of male and female demoiselles *Chromis dispilus* (Pisces: Pomacentridae). *Gen. Comp. Endocrinol.* **93,** 260–274.

Baron, D., and Guiguen, Y. (2003). Gene expression during gonadal sex differentiation in rainbow trout (*Oncorhynchus mykiss*): From candidate genes studies to high throughout genomic approach. *Fish Physiol. Biochem.* **28,** 119–123.

Basolo, A. (1990). Female preference for male sword length in the green swortail, *Xiphophorus helleri* (Pisces: Poeciliidae). *Anim. Behav.* **40,** 332–338.

Bass, A. H. (1992). Dimorphic male brains and alternative reproductive tactics in a vocalising fish. *TINS* **15,** 139–145.

Bass, A. H. (1993). From brains to behaviour: Hormonal cascades and alternative mating tactics in teleost fishes. *Rev. Fish Biol. Fisher.* **3,** 181–186.

Bass, A. H., and Baker, R. (1990). Sexual dimorphisms in the vocal control system of a teleost fish: Morphology of physiologically identified neurons. *J. Neurobiol.* **21,** 1155–1168.

Bass, A. H., and Grober, M. S. (2001). Social and neural modulation of sexual plasticity in teleost fish. *Brain Behav. Evol.* **57,** 293–300.

Bass, A. H., Horvath, B. J., and Brothers, E. B. (1996). Nonsequential developmental trajectories lead to dimorphic vocal circuitry for males with alternative reproductive tactics. *J. Neurobiol.* **30,** 493–504.

Bastian, J., Schniederjan, S., and Nguyenkim, J. (2001). Arginine vasotocin modulates a sexually dimorphic communication behaviour in the weakly electric fish *Apteronotus leptorhynchus. J. Exp. Biol.* **204,** 1909–1923.

Berglund, A., and Rosenqvist, G. (1993). Selective males and ardent females in pipefishes. Behav. *Ecol. Sociobiol.* **32,** 331–336.

Bogart, M. H. (1987). Sex determination: A hypothesis based on steroid ratios. *J. Theor. Biol.* **128,** 349–357.

Borg, B. (1987). Stimulation of reproductive behaviour by aromatizable and non-aromatizable androgens in the male three-spined stickleback, *Gasterosteus aculeatus. In* "Proceedings of the 5th Congress of European Ichthyologists" (Kullander, S. O. K., and Fernholm, B., Eds.), pp. 269–271. Swedish Museum of Natural History, Stockholm.

Borg, B. (1994). Androgens in teleost fishes. *Comp. Biochem. Physiol. C* **109,** 219–245.

Borges, R. A., Oliveira, R. F., Almada, V. C., and Canário, A. V. M. (1998). Short-term social modulation of 11-Ketotestosterone urinary levels in males of the cichlid fish *Oreochromis mossambicus* during male-female interaction. *Acta Ethol.* **1,** 43–48.

Bradbury, J. W., and Vehrencamp, S. L. (1998). "Principles of Animal Communication." Sinauer, Sunderland, MA.

Brantley, R. K., and Bass, A. H. (1994). Alternative male spawning tactics and acoustic signals in the plainfin midshipman fish, *Porichthys notatus* (Teleostei, Batrachoididae). *Ethology* **96,** 213–232.

Brantley, R. K., Marchaterre, M. A., and Bass, A. H. (1993a). Androgen effects on vocal muscle structure in a teleost fish with inter- and intra-sexual dimorphisms. *J. Morphol.* **216,** 305–318.

Brantley, R. K., Wingfield, J. C., and Bass, A. H. (1993b). Sex steroid levels in *Porichthys notatus*, a fish with alternative reproductive tactics, and a review of the hormonal bases for male dimorphism among teleost fishes. *Horm. Behav.* **27,** 332–347.

Breder, C. M., and Rosen, D. E. (1966). "Modes of Reproduction in Fishes." Natural History Press, New York.

Brockmann, H. J. (2001). The evolution of alternative strategies and tactics. *Adv. Study Behav.* **30,** 1–51.

Callard, G. V., Schlinger, B. A., and Pasmanik, M. (1990). Nonmammalian vertebrate models in studies of brain-steroid interactions. *J. Exp. Zool.* **4** (Suppl.), 6–16.

Cardwell, J. R., and Liley, N. R. (1991). Hormonal control of sex and color change in the stoplight parrotfish, *Sparisoma viride. Gen. Comp. Endocrinol.* **81,** 7–20.

Carneiro, L. A., Oliveira, R. F., Canário, A. V. M., and Grober, M. S. (2003). The effect of arginine vasotocin on courtship behaviour in a blenniid fish with alternative reproductive tactics. *Fish Physiol. Biochem.* **28,** 241–243.

Caro, T. M., and Bateson, P. (1986). Organisation and ontogeny of alternative tactics. *Anim. Beh.* **34,** 1483–1499.

Carruth, L. L., Reisert, I., and Arnold, A. P. (2002). Sex chromosome genes directly affect brain sexual differentiation. *Nature Neurosci.* **5,** 933–934.

Chang, C. F., Lee, M. F., and Chen, M. R. (1994). Estradiol-17-β associated with the sex reversal in protandrous black porgy, *Acanthopagrus schlegeli*. *J. Exp. Zool.* **268,** 53–58.

Cheek, A. O., Thomas, P., and Sullivan, C. V. (2000). Sex steroids relative to alternative mating behaviours in the simultaneous hermaphrodite *Serranus subligarius* (Perciformes: Serranidae). *Horm. Behav.* **37,** 198–211.

Cochran, R. C., and Grier, H. J. (1991). Regulation of sexual succession in the protogynous black sea bass *Centropristis striatus*, Osteichthyes Serranidae. *Gen. Comp. Endocrinol.* **82,** 69–77.

Contreras, M. L., and Wade, J. (1999). Interactions between nerve growth factor binding and estradiol in early development of the zebra finch telencephalon. *J. Neurobiol.* **40,** 149–157.

Crews, D. (1998). On the organisation of individual differences in sexual behaviour. *Am. Zool.* **38,** 118–132.

de Girolamo, M., Scaggiante, M., and Rasotto, M. B. (1999). Social organisation and sexal pattern in the Mediteranean parrotfish *Sparisoma cretense* (Teleostei: Scaridae). *Mar. Biol.* **135,** 353–360.

Demski, L. (1987). Diversity in reproductive patterns and behaviour in fishes. *In* "Psychobiology of Reproductive Behaviour: An Evolutionary Perspective" (Crews, D., Ed.), pp. 1–27. Prentice Hall, Englewood Cliffs, N.J.

Devlin, R. H., and Nagahama, Y. (2002). Sex determination and sex differentiation in fish: An overview of genetic, physiological, and environmental influences. *Aquaculture* **208,** 191–364.

Dierkes, P., Taborsky, M., and Kohler, U. (1999). Reproductive parasitism of broodcare helpers in a cooperatively breeding fish. *Behav. Ecol.* **10,** 510–515.

Dittrich, F., Feng, Y., Metzdorf, R., and Gahr, M. (1999). Estrogen-inducible sex-specific expression of brain-derived neurotrophic factor mRNA in a forebrain song control nucleus of the juvenile zebra finch. *Proc. Natl. Acad. Sci. USA* **96,** 8241–8246.

Dulka, J. G., Stacey, N. E., Sorensen, P. W., and van der Kraak, G. J. (1987). A sex steroid pheromone synchronises male-female spawning readiness in goldfish. *Nature* **325,** 251–253.

Farr, J. A. (1989). Sexual selection and secondary sexual differentiation in poeciliids: Determinants of male mating success and the evolution of female choice. *In* "Ecology and Evolution of Livebearing Fishes (Poeciliidae)" (Meffe, G. K., and Snelson, F. J., Jr., Eds.), pp. 91–123. Prentice Hall, Englewood Cliffs, NJ.

Fishelson, L., and Hilzerman, F. (2002). Flexibility in reproductive styles of male St. Peter's tilapia, *Sarotherodon galilaeus* (Cichlidae). *Env. Biol. Fish* **63,** 173–182.

Foran, C. M., and Bass, A. H. (1998). Preoptic AVT immunoreactive neurons of a teleost fish with alternative reproductive tactics. *Gen. Comp. Endocrinol.* **111,** 271–282.

Foran, C. M., and Bass, A. H. (1999). Preoptic GnRH and AVT: Axes for sexual plasticity in teleost fish. *Gen. Comp. Endocrinol.* **116,** 141–152.

Forlano, P. M., Deitcher, D. L., Myers, D. A., and Bass, A. H. (2001). Anatomical distribution and cellular basis for high levels of aromatase activity in the brain of teleost fish: Aromatase enzyme and mRNA expression identify glia as a source. *J. Neurosci.* **21,** 8943–8955.

Forsyth, A., and Alcock, J. (1990). Female mimicry and resource defense polygyny by males of a tropical rove beetle, *Leistotrophus versicolor* (Coleoptera: Staphylinidae). *Behav. Ecol. Sociobiol.* **26,** 325–330.

Francis, R. C. (1984). The effects of bidirectional selection for social dominance on agonistic behaviour and sex ratios in the paradise fish (*Macropodus opercularis*). *Behaviour* **90,** 25–45.

Francis, R. C. (1990). Temperament in a fish: A longitudinal study of the development of individual differences in aggression and social rank in the Midas cichlid. *Ethology* **86,** 311–325.

Francis, R. C. (1992). Sexual lability in teleosts: Developmental factors. *Quart. Rev. Biol.* **67,** 1–18.

Froese, R., and Pauly, D. (Eds.) (2004). FishBase. http://www.fishbase.org. September 7, 2004.

Fu, P., Neff, B. D., and Gross, M. R. (2001). Tactic-specific success in sperm competition. *Proc. R. Soc. Lond. B* **268,** 1105–1112.

Garant, D., Fontaine, P.-M., Good, S. P., Dodson, J. J., and Bernatchez, L. (2002). The influence of male parental identity on growth and survival of offspring in Atlantic salmon (*Salmo salar*). *Evol. Ecol. Res.* **4,** 537–549.

Gardner, L., Anderson, T., Place, A. R., and Elizur, A. (2003). Sex change strategy and the aromatase genes. *Fish Physiol. Biochem.* **28,** 147–148.

Gelinas, D., and Callard, G. V. (1997). Immunolocalisation of aromatase- and androgen receptor-positive neurons in the goldfish brain. *Gen. Comp. Endocrinol.* **106,** 155–168.

Godwin, J. R., and Thomas, P. (1993). Sex-change and steroid profiles in the protandrous anemonefish *Amphiprion melanopus* (Pomacentridae, Teleostei). *Gen. Comp. Endocrinol.* **91,** 144–157.

Godwin, J., Crews, D., and Warner, R. R. (1996). Behavioural sex change in the absence of gonads in a coral reef fish. *Proc. R. Soc. Lond. B* **263,** 1683–1688.

Godwin, J., Luckenbach, J. A., and Borski, R. J. (2003). Ecology meets endocrinology: Environmental sex determination in fishes. *Evol. Dev.* **5,** 40–49.

Godwin, J., Sawby, R., Warner, R. R., Crews, D., and Grober, M. S. (2000). Hypothalamic arginine vasotocin mRNA abundance variation across sexes and with sex change in a coral reef fish. *Brain Behav. Evol.* **55,** 77–84.

Goldschmidt, T., Foster, S. A., and Sevenster, P. (1992). Inter-nest distance and sneaking in the three–spined stickleback. *Anim. Behav.* **44,** 793–795.

Gonçalves, D. M., Matos, R., Fagundes, T., and Oliveira, R. F. (2005). Do bourgeois males of the peacock blenny, *Salaria pavo*, discriminate females from female-mimicking sneaker males? *Ethology* **111,** 559–772.

Gonçalves, E. J., Almada, V. C., Oliveira, R. F., and Santos, A. J. (1996). Female mimicry as a mating tactic in males of the blenniid fish *Salaria pavo*. *J. Mar. Biol. Ass. UK* **76,** 529–538.

Goodson, J., and Bass, A. H. (2000). Forebrain peptides modulate sexually polymorphic vocal circuitry. *Nature* **403,** 769–772.

Goodson, J., and Bass, A. H. (2001). Social behaviour functions and related anatomical characteristics of vasotocin/ vasopressin systems in vertebrates. *Brain Res. Rev.* **35,** 246–265.

Goodwin, N. B., Dulvy, N. K., and Reynolds, J. D. (2002). Life history correlates in the evolution of live-bearing in fishes. *Phil. Trans. R. Soc. B* **357,** 259–267.

Gould, S. J., and Vrba, E. S. (1982). Exaptation: A missing term in the science of form. *Paleobiology* **8,** 4–15.

Govoroun, M., McMeel, O. M., D'Cotta, H., Ricordel, M. J., Smith, T., Fostier, A., and Guiguen, Y. (2001). Steroid enzyme gene expression during natural and androgen-induced gonadal differentiation in the rainbow trout, *Oncorhynchus mykiss*. *J. Exp. Zool.* **290,** 558–566.

Grober, M. S. (1998). Socially controlled sex change: Integrating ultimate and proximate levels of analysis. *Acta Ethol.* **1,** 3–17.

Grober, M. S., and Bass, A. H. (1991). Neuronal correlates of sex/role change in labrid fishes: LHRH-like immunoreactivity. *Brain Behav. Evol.* **38,** 302–312.

Grober, M. S., George, A. A., Watkins, K. K., Carneiro, L. A., and Oliveira, R. F. (2002). Forebrain AVT and courtship in a fish with alternative reproductive tactics. *Brain Res. Bull.* **57,** 23–25.

Grober, M. S., Jackson, I. M. D., and Bass, A. H. (1991). Gonadal steroids affect LHRH preoptic cell number in a sex/role reversing fish. *J. Neurobiol.* **2,** 734–741.

Grober, M. S., Laughlin, C., Fox, S., and Bass, A. (1994). GnRH cell size and number in a teleost fish with two male reproductive morphs: Sexual maturation, final sexual status and body size allometry. *Brain Behav. Evol.* **43,** 61–78.

Gross, M. R. (1982). Sneakers, Satellites and Parentals: Polymorphic mating strategies in North American sunfishes. *Z. Tierpsychol.* **60,** 1–26.

Gross, M. R. (1996). Alternative reproductive strategies and tactics: Diversity within sexes. *TREE* **11,** 92–98.

Guiguen, Y., Jalabert, B., Thouard, E., and Fostier, A. (1993). Changes in plasma and gonadal steroid hormones in relation to the reproductive cycle and the sex inversion process in the protandrous seabass, *Lates calacarifer*. *Gen. Comp. Endocrinol.* **92,** 327–338.

Guiguen, Y., Baroiller, J. F., Ricordel, M.-J., Iseki, K., McMeel, O. M., Martin, S. A. M., and Fostier, A. (1999). Involvement of estrogens in the process of sex differentiation in two fish species: the rainbow trout (*Oncorhynchus mykiss*) and a tilapia (*Oreochromis niloticus*). *Mol. Reprod. Dev.* **54,** 154–162.

Halpern-Sebold, L., Schreibman, M. P., and Margolis-Nunno, H. (1986). Differences between early- and late-maturing genotypes of the platyfish (*Xiphophorus maculatus*) in the morphometry of their immunoreactive luteinising hormone releasing hormone containing cells: A developmental study. *J. Exp. Zool.* **240,** 245–257.

Hawkins, M. B., Thornton, J. W., Crews, D., Skipper, J. K., Dotte, A., and Thomas, P. (2001). Identification of a third distinct estrogen receptor and reclassification of estrogen receptors in teleosts. *Proc. Natl. Acad. Sci. USA* **97,** 10751–10756.

Henson, S. A., and Warner, R. R. (1997). Male and female alternative reproductive behaviours in fishes: A new approach using intersexual dynamics. *Annu. Rev. Ecol. Syst.* **28,** 571–592.

Hirschenhauser, K., Taborsky, M., Oliveira, T., Canario, A. V. M., and Oliveira, R. F. (2004). A test of the 'challenge hypothesis' in cichlid fish: Simulated partner and territory intruder experiments. *Anim. Behav.* **68,** 741–750.

Hobby, A. C., Geraghty, D. P., and Pankhurst, N. W. (2000a). Differences in binding characteristics of sex steroid binding protein in reproductive and non reproductive female rainbow trout (*Oncorhynchus mykiss*), black bream (*Acanthopagrus butcheri*) and greenback flounder (*Rhombosolea tapirina*). *Gen. Comp. Endocrinol.* **120,** 249–259.

Hobby, A. C., Pankhurst, N. W., and Geraghty, D. P. (2000b). A comparison of sex steroid binding protein (SBP) in four species of teleost fish. *Fish Physiol. Biochem.* **23,** 245–256.

Hofmann, H. A. (2003). Functional genomics of neural and behavioural plasticity. *J. Neurobiol.* **54,** 272–282.

Hogg, J. T. (1984). Mating in bighorn sheep: Multiple creative male strategies. *Science* **225,** 526–529.

Hourigan, T. F., Nakamura, N., Nagahama, Y., Yamauchi, K., and Grau, E. G. (1991). Histology, ultrastructure, and *in vitro* steroidogenesis of the testes of two male phenotypes of the protogynous fish, *Thalassoma duperrey* (Labridae). *Gen. Comp. Endocrinol.* **83,** 193–217.

Ikeuchi, T., Todo, T., Kobayashi, T., and Nagahama, Y. (2001). Two subtypes of androgen and progestogen receptors in fish testes. *Comp. Biochem. Physiol. B* **129,** 449–455.

Jamieson, I. G., and Colgan, P. W. (1992). Sneak spawning and egg stealing by male three-spine sticklebacks. *Can. J. Zool.* **70,** 963–967.

Jennings, D. H., Moore, M. C., Knapp, R., Matthews, L., and Orchinik, M. (2000). Plasma steroid-binding globulin mediation of differences in stress reactivity in alternative male phenotypes in tree lizards, *Urosaurus ornatus*. *Gen. Comp. Endocrinol.* **120,** 289–299.

Jennings, D. H., Painter, D. L., and Moore, M. C. (2004). Role of the adrenal gland in early post-hatching differentiation of alternative male phenotypes in the tree-lizard (*Urosaurus ornatus*). *Gen. Comp. Endocrinol.* **135,** 81–89.

Johnson, A. K., Thomas, P., and Wilson, R. R., Jr. (1998). Seasonal cycles of gonadal development and plasma sex steroid levels in *Epinephelus morio*, a protogynous grouper in the eastern Gulf of Mexico. *J. Fish Biol.* **52,** 502–518.

Jones, A. G., Walter, D., Kvarnemo, C., Lindström, K., and Avise, J. C. (2001). How cuckoldry can decrease the opportunity for sexual selection: Data and theory from a genetic parentage análisis of the sand goby, *Pomatoschistus minutus*. *Proc. Natl. Acad. Sci. USA* **98,** 9151–9156.

Ketterson, E. D., and Nolan, V., Jr. (1992). Hormones and life histories: An integrative approach. *Am. Nat.* **140,** S33–S62.

Ketterson, E. D., and Nolan, V., Jr. (1994). Hormones and life histories: An integrative approach. *In* "Behavioural Mechanisms in Evolutionary Ecology" (Real, L. A., Ed.), pp. 327–353. University of Chicago Press, Chicago.

Ketterson, E. D., and Nolan, V., Jr. (1999). Adaptation, exaptation, and constraint: A hormonal perspective. *Am. Nat.* **154,** S4–S25.

Ketterson, E. D., Nolan, V., Jr., Casto, J. M., Buerkle, C. A., Clotfelter, E., Grindstaff, J. L., Jones, K. J., Lipar, J. L., McNabb, F. M. A., Neudorf, D. L., Parker-Renga, I., Schoech, S. J., and Snajdr, E. (2001). Testosterone, phenotype and fitness: A research program in evolutionary behavioural endocrinology. *In* "Avian Endocrinology" (Dawson, A., and Chaturvedi, C. M., Eds.), pp. 19–40. Narosa Publishing House, New Delhi.

Kim, S. J., Ogasawara, K., Park, J. G., Takemura, A., and Nakamura, M. (2002). Sequence and expression of androgen receptor and estrogen receptor gene in the sex types of protogynous wrasse, *Heliochoeres trimaculatus*. *Gen. Comp. Endocrinol.* **127,** 165–173.

Kindler, P. M., Philipp, D. P., Gross, M. R., and Bahr, J. M. (1989). Serum 11-ketotestosterone and testosterone concentrations associated with reproduction in male bluegill (*Lepomis macrochirus*: Centrarchidae). *Gen. Comp. Endocrinol.* **75,** 446–453.

Kindler, P. M., Bahr, J. M., and Philipp, D. P. (1991). The effects of exogenous 11-ketotestosterone, testosterone, and cyproterone acetate on prespawning and parental care behaviours of male bluegill. *Horm. Behav.* **25,** 410–423.

Kitano, T., Takamune, K., Kobayashi, T., Nagahama, Y., and Abe, S. I. (1999). Supression of P450 aromatase gene expression in sex-reversed males produced by rearing genetically female larvae at a high water temperature during a period of sex differentiation in the Japanese flounder (*Paralichthys olivaceus*). *J. Mol. Endocrinol.* **23,** 167–176.

Knapp, R., Carlisle, S. L., and Jessop, T. S. (2002). A model for androgen-glucocorticoid interactions in male alternative reproductive tactics: Potential roles for steroidogenic enzymes. *Horm. Behav.* **41,** 475.

Knapp, R., Hews, D., Thompson, C. W., Ray, L. E., and Moore, M. C. (2003). Environmental and endocrine correlates of tactic switching by non–territorial male tree lizards (*Urosaurus ornatus*). *Horm. Behav.* **43,** 83–92.

Knapp, R. (2004). Endocrine mediation of vertebrate male alternative reproductive tactics: The next generation of studies. *Integr. Com. Biol.* **43,** 658–668.

Kobayashi, M., and Nakanishi, T. (1999). 11-ketotestosterone induces male-type sexual behaviour and gonadotropin secretion in gynogenetic crucian carp, *Carassius auratus langsdorfii*. *Gen. Comp. Endocrinol.* **115,** 178–187.

Kown, J. Y., McAndrew, B. J., and Penman, D. J. (2001). Cloning of brain aromatase gene and expression of brain and ovarian aromtase genes during sexual differentiation in genetic male and female Nile tilapia, *Oreochromis niloticus*. *Mol. Reprod. Dev.* **59,** 359–370.

Kroon, F. J., and Liley, N. R. (2000). The role of steroid hormones in protogynous sex change in the blackeye goby, *Coryphopterus nicholsi* (Teleostei: Gobiidae). *Gen. Comp. Endocrinol.* **118,** 273–283.

Laidley, C. W., and Thomas, P. (1994). Partial characterisation of a sex–steroid binding protein in the spotted seatrout (*Cynoscion nebulosus*). *Biol. Reprod.* **51,** 982–992.

Lank, D. B., Smith, C. M., Hanotte, O., Burke, T., and Cooke, F. (1995). Genetic polymorphism for alternative mating behaviour in lekking male ruff *Nature* **378,** 59–62.

Larsson, D. G., Sperry, T., and Thomas, P. (2002). Regulation of androgen receptors in Atlantic croaker brains by testosterone and estradiol. *Gen. Comp. Endocrinol.* **128,** 224–230.

Laumen, J., Pern, U., and Blüm, V. (1974). Investigations on the function and hormonal regulations of the anal appendices in *Blennius pavo. J. Exp. Zool.* **190,** 47–56.

Lee, J. S. F., and Bass, A. H. (2004). Effects of 11-ketotestosterone on brain, sonic muscle, and behaviour in type-II midshipman fish. *Horm. Behav.* **46,** 115–116.

Lee, Y. H., Du, J. L., Yueh, W. S., Lin, B. Y., Huang, J. D., Lee, C. Y., Lee, M. F., Lau, E. L., Lee, F. Y., Morrey, C., Nagahama, Y., and Chang, C. F. (2001). Sex change in the protandrous black porgy, *Acanthopagrus schlegeli*: A review in gonadal development, estradiol, estrogen receptor, aromatase activity and gonadotropin. *J. Exp. Zool.* **290,** 715–726.

Leitz, T. (1987). Social control of testicular steroidogenic capacities in the Siamese fighting fish *Betta splendens* Regan. *J. Exp. Zool.* **244,** 473–478.

Lejeune, P. (1985). Etude éco-étologique des comportements reproducteurs et sociaux des labridés Méditerranéens des genres *Symphodus* Rafinesque, 1810 et *Coris* Lacepede, 1802. *Cah. Ethol. Appl.* **5,** 1–208.

Liley, N. R., Breton, B., Fostier, A., and Tan, E. S. P. (1986). Endocrine changes associated with spawning behaviour and social stimuli in a wild population of rainbow trout (*Salmo gairdneri*) 1. Males. *Gen. Comp. Endocrinol.* **62,** 145–156.

Liley, N. R., Olsén, K. H., Foote, C. J., and van der Kraak, G. J. (1993). Endocrine changes associated with with spawning behaviour in male kokanee salmon (*Oncorhynchus nerka*) and the effects of anosmia. *Horm. Behav.* **27,** 470–487.

Liley, N. R., and Stacey, N. E. (1983). Hormones, pheromones, and reproductive behaviour in fish. *In* "Fish Physiology—Reproduction, Part B: Behaviour and Fertility Control" (Hoar, W. S., Randall, D. J., and Donaldson, E. M., Eds.), Vol. 9, pp. 1–63. Academic Press, New York.

Lin, X. W., and Peter, R. E. (1996). Expression of salmon gonadotropin–releasing hormone (GnRH) and chicken GnRH-II precursor messenger ribonucleic acids in the brain and ovary of goldfish. *Gen. Comp. Endocrinol.* **101,** 282–296.

Liu, S., Govoroun, M., D'Cotta, H., Ricordel, M. J., Lareyre, J.-J., McMeel, O. M., Smith, T., Nagahama, Y., and Guiguen, Y. (2000). Expression of cytochrome P45011β (11β-hydroxilase) gene during gondal sex differentiation and spermatogenesis in rainbow trout, *Oncorhynchus mykiss. J. Steroid Biochem. Mol. Biol.* **75,** 291–298.

Matsuda, M., Nagahama, Y., Shinomiya, A., Sato, T., Matsuda, C., Kobayashi, T., Morrey, C. E., Shibata, N., Asakawa, S., Shimizu, N., Hori, H., Hamaguchi, S., and Sakaizumi, M. (2002). DMY is a Y-specificDM-domain gene required for male development in the medaka fish. *Nature* **417,** 559–563.

Matsuyama, M., Adachi, S., Nagahama, Y., Kitajima, C., and Matsuura, S. (1991). Testicular development and serum levels of gonadal steroids during the annual reproductive cycle of captive Japanese sardine. *Jap. J. Ichthyol.* **37,** 381–390.

Mayer, I., Lundqvist, H., Berglund, I., Schmitz, M., Schulz, R., and Borg, B. (1990). Seasonal endocrine changes in Baltic salmon, *Salmo salar*, immature parr and mature male parr. I. Plasma levels of five androgens, 17α-hydroxy-20β-dihydroprogesterone, and 17β-estradiol. *Can. J. Zool.* **68,** 1360–1365.

Mayer, I., Rosenqvist, G., Borg, B., Ahnesjö, I., Berglund, A., and Schulz, R. W. (1993). Plasma levels of sex steroids in three species of pipefish (Syngnathidae). *Can. J. Zool.* **71,** 1903–1907.

Mayer, I., Liley, N. R., and Borg, B. (1994). Stimulation of spawning behaviour in castrated rainbow trout (*Oncorhynchus mykiss*) by 17α,20β-dihydroxy-4-pregnen-3-one, but not by 11-ketoandrostenedione. *Horm. Behav.* **28,** 181–190.

Mazzoldi, C., and Rasotto, M. B. (2002). Alternative male mating tactics in *Gobius niger*. *J Fish Biol.* **61,** 157–172.

Mazzoldi, C., Scaggiante, M., Ambrosin, E., and Rasotto, M. B. (2000). Mating system and alternative mating tactics in the grass goby *Zosterisessor ophiocephalus* (Teleostei: Gobiidae). *Mar. Biol.* **137,** 1041–1048.

Miguel-Queralt, S., Knowlton, M., Avvakumov, G. V., AL-Nouno, R., Kelly, G. M., and Hammond, G. L. (2004). Molecular and functional characterisation of sex hormone binding globulin in zebrafish. *Endocrinol.* **145,** 5221–5230.

Miranda, J. A., Oliveira, R. F., Carneiro, L. A., Santos, R. S., and Grober, M. S. (2003). Neurochemical correlates of male polymorphism and alternative reproductive tactics in the Azorean rock-pool blenny, *Parablennius parvicornis*. *Gen. Comp. Endocrinol.* **132,** 183–189.

Modesto, T., and Canário, A. V. M. (2003a). Morphometric changes and sex steroid levels during the annual reproductive cycle of the Lusitanean toadfish, *Halobatrachus didactylus*. *Gen. Comp. Endocrinol.* **131,** 220–231.

Modesto, T., and Canário, A. V. M. (2003b). Hormonal control of swimbladder sonic muscle dimorphism in the Lusitanian toadfish, *Halobatrachus didactylus*. *J. Exp. Biol.* **206,** 3467–3477.

Moore, M. C. (1991). Application of organisation-activation theory to alternative male reproductive strategies: A review. *Horm. Behav.* **25,** 154–179.

Moore, M. C., Hews, D. K., and Knapp, R. (1998). Hormonal control and evolution of alternative male phenotypes: Generalisations of models for sexual differentiation. *Am. Zool.* **38,** 133–151.

Morita, S., Matsuyama, M., and Kashiwagi, M. (1997). Seasonal changes of gonadal histology and serum steroid hormone levels in the bambooleaf wrasse *Pseudolabrus japonicus*. *Bull. Jpn. Soc. Sci. Fish./Nippon Suisan Gakkaishi* **63,** 694–700.

Morrey, C. E., and Nagahama, Y. (2000). 11β-hydroxilase and androgen receptor mRNA expression in the ovary, testis and brain of the protogynous hermaphrodite *Thalassoma duperrey*. *In* "Proceedings of the 6th International Symposium on the Reproductive Physiology of Fish" (Norgerg, B., Kjesbu, O. S., Taranger, G. L., Andersson, E., and Stefansson, S. O., Eds.), pp. 157–159. Bergen, Norway.

Munday, P., and Jones, G. (1998). Bi-directional sex change in a coral-dwelling goby. *Behav. Ecol. Sociobiol.* **43,** 371–377.

Nakamura, M., Mariko, T., and Nagahama, Y. (1994). Ultrastructure and *in vitro* steroidogenesis of the gonads in the protandrous anemonefish *Amphiprion frenatus*. *Jap. J. Ichthyol.* **41,** 47–56.

Nakamura, M., Kobayashi, T., Chang, X.-T., and Nagahama, Y. (1998). Gonadal sex differentiation in teleost fish. *J. Exp. Zool.* **281,** 362–372.

Nakamura, M., Bhandari, R. K., and Higa, M. (2003). The role estrogens play in sex differentiation and sex changes of fish. *Fish Physiol. Biochem.* **28,** 113–117.

Neat, F. C. (2001). Male parasitic spawning in two species of triplefin blenny (Tripterigiidae): Contrasts in demography, behaviour and gonadal characteristics. *Env. Biol. Fish* **61,** 57–64.

Neff, B. D. (2004). Increased performance of offspring sired by parasitic males in bluegill sunfish. *Behav. Ecol.* **15,** 327–331.

Neff, B. D., Fu, P., and Gross, M. R. (2003). Sperm investment and alternative mating tactics in bluegill sunfish (*Lepomis macrochirus*). *Behav. Ecol.* **14,** 634–641.

Nelson (1994). "Fishes of the World." John Wiley & Sons, New York.

Nelson, R. J. (2000). "An Introduction to Behavioural Endocrinology." Sinauer Associates, Sunderland, MA.

Norman, M. D., Finn, J., and Tregenza, T. (1999). Female impersonation as an alternative reproductive strategy in giant cuttlefish. *Proc. R. Soc. Lond. B* **266,** 1347–1349.

Oliveira, R. F. (2004). Social modulation of androgens in vertebrates: Mechanisms and function. *Adv. Study Behav.* **34,** 165–239.

Oliveira, R. F., and Almada, V. C. (1998a). Mating tactics and male-male courtship in the lek-breeding cichlid *Oreochromis mossambicus*. *J. Fish Biol.* **52,** 1115–1129.

Oliveira, R. F., and Almada, V. C. (1998b). Androgenisation of dominant males in a cichlid fish: Androgens mediate the social modulation of sexually dimorphic traits. *Ethology* **104,** 841–858.

Oliveira, R. F., and Almada, V. C. (1999). Male displaying characters, gonadal maturation and androgens in the cichlid fish *Oreochromis mossambicus*. *Acta Ethol.* **2,** 67–70.

Oliveira, R. F., and Canario, A. V. M. (2001). Hormones and social behaviour of cichlid fishes: A case study in the Mozambique tilapia. *J. Aquariculture and Aquat. Sci.* **9,** 187–207.

Oliveira, R. F., Almada, V. C., and Canario, A. V. M. (1996). Social modulation of sex steroid concentrations in the urine of male cichlid fish *Oreochromis mossambicus*. *Horm. Behav.* **30,** 2–12.

Oliveira, R. F., Almada, V. C., Gonçalves, E. J., Forsgren, E., and Canario, A. V. M. (2001a). Androgen levels and social interactions in breeding males of the peacock blenny. *J. Fish Biol.* **58,** 897–908.

Oliveira, R. F., Canario, A. V. M., and Grober, M. S. (2001b). Male sexual polymorphism, alternative reproductive tactics and androgens in combtooth blennies (Pisces: Blenniidae). *Horm. Behaviour* **40,** 266–275.

Oliveira, R. F., Canario, A. V. M., Grober, M. S., and Santos, R. S. (2001c). Endocrine correlates of alternative reproductive tactics and male polymorphism in the Azorean rock-pool blenny, *Parablennius sanguinolentus parvicornis*. *Gen. Comp. Endocrinol.* **121,** 278–288.

Oliveira, R. F., Carneiro, L. A., Gonçalves, D. M., Canario, A. V. M., and Grober, M. S. (2001d). 11-ketotestosterone inhibits the alternative mating tactic in sneaker males of the peacock blenny, *Salaria pavo*. *Brain Behav. Evol.* **58,** 28–37.

Oliveira, R. F., Carneiro, L. A., Canario, A. V. M., and Grober, M. S. (2001e). Effects of androgens on social behaviour and morphology of alternative reproductive males of the Azorean rock-pool blenny. *Horm. Behav.* **39,** 157–166.

Oliveira, R. F., Gonçalves, E. J., and Santos, R. S. (2001f). Gonadal investment of young males in two blenniid fishes with alternative mating tactics. *J. Fish Biol.* **59,** 459–462.

Oliveira, R. F., Lopes, M., Carneiro, L. A., and Canario, A. V. M. (2001g). Watching fights raises fish hormone levels. *Nature* **409,** 475.

Oliveira, R. F., Ros, A. F. H., Hirschenhauser, K., and Canario, A. V. M. (2001h). Androgens and mating systems in fish: intra- and interspecific analyses. *In* "Perspectives in Comparative Endocrinology: Unity and Diversity" (Goos, H. J., Rastogi, R. K., Vaudry, H., and Pierantoni, R., Eds.), pp. 203–215. Monduzzi Editore, Bologna.

Oliveira, R. F., Hirschenhauser, K., Carneiro, L. A., and Canario, A. V. M. (2002a). Social modulation of androgens in male teleost fish. *Comp. Biochem. Physiol.* **132B,** 203–215.

Oliveira, R. F., Miranda, J. A., Carvalho, N., Gonçalves, E. J., Grober, M. S., and Santos, R. S. (2002b). The relationship between the presence of satellite males and nest-holders' mating success in the Azorean rock-pool blenny, *Parablennius sanguinolentus parvicornis*. *Ethology* **108,** 223–235.

Oliveira, R. F., Hirschenhauser, K., Canario, A. V. M., and Taborsky, M. (2003). Androgen levels of reproductive competitors in a cooperatively breeding cichlid. *J. Fish Biol.* **63,** 1615–1620.

Oliveira, R. F., Ros, A. F. H., and Gonçalves, D. M. (2005). Intra-sexual variation in male reproduction in teleost fish: A comparative approach. *Horm. Behav.* in press.

Olsén, H. K., Järvi, J. T., Mayer, I., Petersson, E., and Kroon, F. (1998). Spawning behaviour and sex hormone levels in adult and precocious brown trout (*Salmo trutta* L.) males and the effect of anosmia. *Chemoecology* **8,** 9–17.

Parhar, I. (2002). Cell migration and evolutionary significance of GnRH subtypes. *Prog. Brain Res.* **141,** 3–17.

Parhar, I. (2003). Gonadotrophin-releasing hormone receptors: Neuroendocrine regulators and neuromodulators. *Fish Physiol. Biochem.* **28,** 13–18.

Pasmanik, M., and Callard, G. (1986). Characteristics of a testosterone-estradiol binding globulin (TEBG) in goldfish serum. *Biol. Reprod.* **35,** 838–845.

Pasmanik, M., and Callard, G. (1988). A high abundance androgen receptor in goldfish brain: Characteristics and seasonal changes. *Endocrinol.* **123,** 1162–1171.

Patzner, R. A., and Lahnsteiner, F. (1999). The accessory organs of the male reproductive system in Mediterranean Blennies (Blenniidae) in comparison with those of other blenniid fishes (tropical Blenniidae, Tripterygiidae, Labrisomidae, Clinidae, Chaenopsidae, Dactyloscopidae). *In* "Behaviour and Conservation of Littoral Fishes" (Almada, V. C., Oliveira, R. F., and Gonçalves, E. J., Eds.), pp. 179–228. ISPA, Lisboa.

Perril, S. A., Gerhardt, H. C., and Daniel, R. (1978). Sexual parasitism in the green tree frog, *Hyla cinerea*. *Science* **200,** 1179–1180.

Perry, A. N., and Grober, M. S. (2003). A model for social control of sex change: Interactions of behaviour, neuropetides, glucocorticoids, and sex steroids. *Horm. Behav.* **43,** 31–38.

Peter, R. E., Yu, K. L., Marchant, T. A., and Rosenblum, P. N. (1990). Direct neural regulation of the teleost adenohypophysis. *J. Exp. Zool.* **4,** 84–89.

Picciulin, M., Verginela, L., Spoto, M., and Ferrero, E. A. (2004). Colonial nesting and the importance of the brood size in male parasitic reproduction of the Mediterranean damselfish *Chromis chromis* (Pisces: Pomacentridae). *Env. Biol. Fish.* **70,** 23–30.

Pickford, G. E., and Strecker, E. L. (1977). The spawning reflex response of the killifish *Fundulus heteroclitus*; isotocin is relatively inactive in comparison with arginine vasotocin. *Gen. Comp. Endocrinol.* **32,** 132–137.

Reavis, R. H., and Grober, M. S. (1999). An integrative approach to sex change: Social, behavioural and neurochemical changes in *Lythrypnus dalli* (Pisces). *Acta Ethol.* **2,** 51–60.

Reinboth, R., and Becker, B. (1984). *In vitro* studies on steroid metabolism by gonadal tissues from ambisexual teleosts. I. Conversion of 14–C testosterone by males and females of the protogynous wrasse *Coris julis* L. *Gen. Comp. Endocrinol.* **55,** 245–250.

Remage-Healey, L., and Bass, A. H. (2004). Rapid, hierarchical modulation of vocal patterning by steroid hormones. *J. Neurosci.* **24,** 5892–5900.

Reynolds, J. D., Gross, M. R., and Coombs, M. J. (1993). Environmental conditions and male morphology determine alternative mating behaviour in Trinidadian guppies. *Anim. Behav.* **45,** 145–152.

Rhen, T., and Crews, D. (2002). Variation in reproductive behaviour within a sex: Neural systems and endocrine activation. *J. Neuroendocrinol.* **14,** 517–531.

Ros, A. F. H., Canario, A. V. M., Couto, E., Zeilstra, I., and Oliveira, R. F. (2003). Endocrine correlates of intra-specific variation in the mating system of the St. Peter's fish (*Sarotherodon galilaeus*). *Horm. Behav.* **44,** 365–373.

Ros, A. F. H., Bouton, N., Santos, R. S., and Oliveira, R. F. (2004a). Are aggressive fathers immunosupressed? Alternative reproductive tactics, androgens and immunocompetence in the Azorean rock-pool blenny. *Horm. Behav.* **46,** 120.

Ros, A. F. H., Bruintjes, R., Santos, R. S., Canario, A. V. M., and Oliveira, R. F. (2004b). The role of androgens in the trade–off between territorial and parental behaviour in the Azorean rock-pool blenny, *Parablennius parvicornis*. *Horm. Behav.* **46,** 491–497.

Rosen, D. E., and Tucker, A. (1961). Evolution of secondary sexual characters and sexual behaviour patterns in a family of viviparous fishes (Cyprinodontiformes: Poeciliidae). *Copeia* **1961,** 201–212.

Rouger, Y., and Liley, N. R. (1993). The effect of social environment on plasma hormones and availability of milt in spawning male rainbow trout (*Oncorhynchus mykiss* Walbaum). *Can. J. Zool.* **71,** 280–285.

Ruchon, F., Laugier, T., and Quignard, J. P. (1995). Alternative male reproductive strategies in the peacock blenny. *J. Fish Biol.* **47,** 826–840.

Ryan, M. J., Pease, C. M., and Morris, M. R. (1992). A genetic polymorphism in the swordtail *Xiphophorus nigrensis*: Testing the predictions of equal fitnesses. *Am. Nat.* **139,** 21–31.

Salek, S. J., Sullivan, C. V., and Godwin, J. (2002). Arginine vasotocin effects on courtship behaviour in male white perch (*Morone americana*). *Behav. Brain Res.* **133,** 177–183.

Santos, R. S. (1985). Parentais e satélites: Tácticas alternativas de acasalamento nos machos de *Blennius sanguinolentus* Pallas (Pisces: Blenniidae). *Arquipélago, Sér Ciênc. Nat.* **6,** 119–146.

Sargent, R. C., and Gross, M. R. (1993). William's principle: an explanation of parental care in teleost fishes. *In* "Behaviour of Teleost Fishes" (Pitcher, T. J., Ed.), pp. 333–361. Chapman and Hall, London.

Scaggiante, M., Grober, M. S., Lorenzi, V., and Rasotto, M. B. (2004). Changes in the male reproductive axis in response to social context in a gonochoristic gobiid, *Zosterisessor ophiocephalus* (Teleostei, Gobiidae), with alternative mating tactics. *Horm. Behav.* **46,** 607–617.

Schartl, M., Wilde, B., Schlupp, I., and Parzefall, J. (1995). Evolutionary origin of a parthenoform, the Amazon molly, *Poecilia formosa*, on the basis of a molecular genealogy. *Evolution* **49,** 827–835.

Schlinger, B. A., Greco, C., and Bass, A. H. (1999). Aromatase activity in the hindbrain and vocal control region of a teleost fish: Divergence among males with alternative reproductive tactics. *Proc. Roy. Soc. Lond. B* **266,** 131–136.

Schlupp, I., Parzefall, J., Epplen, J. T., Nanda, I., Schmid, M., and Schartl, M. (1992). Pseudomale behaviour and spontaneous masculinisation in the all-female teleost *Poecilia formosa* (Teleostei: Poeciliidae). *Behaviour* **122,** 88–104.

Schlupp, I., McKnab, R., and Ryan, M. J. (2001). Sexual harassment as a cost for molly females: Bigger males cost less. *Behaviour* **138,** 277–286.

Schulz, R. W., and Miura, T. (2002). Spermatogenesis and its endocrine regulation. *Fish Physiol. Biochem.* **26,** 43–56.

Scott, A. P., and Baynes, S. M. (1982). Plasma levels of sex steroids in relation to ovulation and spermiation in rainbow trout (*Salmo gairdneri*). *In* "Proceedings of the International Symposium on Reproductive Physiology of Fish" (Richter, C. J. J., and Goos, H. J. T., Eds.), pp. 103–106. PUDOC, Wageningen, Netherlands.

Semsar, K., and Godwin, J. (2003). Multiple mechanisms of phenotype development in the bluehead wrasse. *Horm. Behav.* **45,** 345–353.

Semsar, K. A., Kandel, F. L. M., and Godwin, J. (2001). Manipulations of the AVT system shift social status and related courtship and aggressive behaviour in the bluehead wrasse. *Horm. Behaviour* **40,** 21–31.

Shuster, S. M. (1992). The reproductive behaviour of α-, β- and γ-male morphs in *Paracerceis sculpta*, a marine isopod crustacean. *Behaviour* **121,** 231–258.
Shuster, S. M., and Wade, M. J. (2003). "Mating Systems and Strategies." Princeton University Press, Princeton.
Silverman, H. I. (1978). Changes in male courting frequency in pairs of the cichlid fish, *Sarotherodon (Tilapia) mossambicus*, with unlimited or with only visual contact. *Behav. Biol.* **23,** 189–196.
Simon, N. G. (2002). Hormonal processes in the development and expression of aggressive behaviour. *In* "Hormones, Brain and Behaviour, Vol. 1" (Pfaff, D. W., Arnold, A. P., Etgen, A. M., Farbach, S. E., and Rubin, R. T., Eds.), pp. 339–392. Academic Press, New York.
Sinervo, B., and Lively, C. M. (1996). The rock-paper-scissors game and the evolution of alternative male strategies. *Nature* **380,** 240–243.
Sinervo, B., and Svensson, E. (1998). Mechanistic and selective causes of life-history trade-offs and plasticity: Evolutionary physiology of the costs of reproduction. *Oikos* **83,** 432–442.
Sisneros, J. A., Forlano, P. M., Knapp, R., and Bass, A. H. (2004). Seasonal variation of steroid hormone levels in an intertidal-nesting fish, the vocal plainfin midshipman. *Gen. Comp. Endocrinol.* **136,** 101–116.
Sperry, T., and Thomas, P. (1999). Characterisation of two nuclear androgen receptors in Atlantic croaker: Comparison of their biochemical properties and binding specificities. *Endocrinol.* **140,** 1602–1611.
Sperry, T., and Thomas, P. (2000). Androgen binding profiles of two distinct nuclear androgen receptors in Atlantic croaker (*Micropogonias undulatus*). *Steroid Biochem. Mol. Biol.* **72,** 93–103.
Stacey, N. E. (1987). Roles of hormones and pheromones in fish reproductive behaviour. *In* "Psychobiology of Reproductive Behaviour: An Evolutionary Perspective" (Crews, D., Ed.), pp. 28–60. Prentice-Hall, Englewood Cliffs, NJ.
Stacey, N., and Kobayashi, M. (1996). Androgen induction of male sexual behaviours in female goldfish. *Horm. Behav.* **30,** 434–445.
Stacey, N. E., and Kyle, A. L. (1983). Effects of olfactory tract lesions on sexual and feeding behaviour in the goldfish. *Physiol. Behav.* **30,** 621–628.
Strüssman, C. A., and Nakamura, M. (2002). Morphology, endocrinology, and environmental modulation of gonadal sex differentiation in teleost fishes. *Fish Physiol. Biochem.* **26,** 13–29.
Sunobe, T., and Nakazono, A. (1993). Sex change in both directions by alteration of social dominance in *Trimma okinawae* (Pisces: Gobiidae). *Ethology* **94,** 339–345.
Taborsky, M. (1994). Sneakers, satellites, and helpers: parasitic and cooperative behaviour in fish reproduction. *Adv. Study Behav.* **23,** 1–100.
Taborsky, M. (1997). Bourgeois and parasitic tactics: Do we need collective, functional terms for alternative reproductive behaviours? *Behav. Ecol. Sociobiol.* **41,** 361–362.
Taborsky, M. (1998). Sperm competition in fish: 'Bourgeois' males and parasitic spawning. *TREE* **13,** 222–227.
Taborsky, M. (1999). Conflict or Cooperation: what determines optimal solutions to competition in fish reproduction? *In* "Behaviour and Conservation of Littoral Fishes" (Almada, V. C., Oliveira, R. F., and Gonçalves, E. J., Eds.), pp. 301–349. I.S.P.A., Lisboa.
Taborsky, M. (2001). The evolution of bourgeois, parasitic, and cooperative reproductive behaviours in fishes. *J. Hered.* **92,** 100–110.
Tanaka, H., Hirose, K., Nogami, K., Hattori, K., and Ishibashi, N. (1990). Sexual maturation and sex reversal in red spotted grouper, *Epinephelus akaara*. *Bull. Natl. Res. Inst. Aquacult.* **17,** 1–15.

Taru, M., Kanda, T., and Sunobe, T. (2002). Alternative mating tactics of the gobiid fish *Bathygobius fuscus*. *J. Ethol.* **20,** 9–12.

Thomas, P. (2000). Nuclear and membrane steroid receptors and their functions in teleost gonads. *In* "Proceedings of the 6th International Symposium on the Reproductive Physiology of Fish" (Norgerg, B., Kjesbu, O. S., Taranger, G. L., Andersson, E., and Stefansson, S. O., Eds.), pp. 149–156. Bergen, Norway.

Thomas, P. (2003). Rapid, nongenomic steroid actions initiated at the cell surface: Lessons from studies with fish. *Fish Physiol. Biochem.* **28,** 3–12.

Thresher, R. E. (1984). "Reproduction in Reef Fishes." T.F.H. Publications, Neptune City.

Tomkins, J. L., and Simmons, L. W. (2002). Measuring relative investment: A case study of testes investment in species with alternative male reproductive tactics. *Anim. Behav.* **63,** 1009–1016.

Trant, J. M., Gavasso, S., Ackers, J., Chung, B. C., and Place, A. R. (2001). Developmental expression of cytochrome P450 aromatase genes (CYP19A) and CYP19B) in zebrafish fry (*Danio rerio*). *J. Exp. Zool.* **290,** 475–483.

Tsai, C. L., Wang, L. H., Chang, C. F., and Kao, C. C. (2000). Effects of gonadal steroids on brain serotonergic and aromatase activity during the critical period of sexual differentiation in tilapia, *Oreochromis mossambicus*. *J. Neuroendocrinol.* **12,** 894–898.

Turner, G. (1993). Teleost mating behaviour. *In* "Behaviour of Teleost Fishes" (Pitcher, T. J., Ed.), pp. 307–331. Chapman & Hall, London.

Tyler, W. A. (1989). Optimal colony size in the Hawaiian Sergeant, *Abudefduf abdominalis* (Pisces: Pomacentridae). *Pacific Sci.* **43,** 204.

Uglem, I., Rosenqvist, G., and Schioler Wasslavik, H. (2000). Phenotypic variation between dimorphic males in corkwing wrasse (*Symphodus melops* L.). *J. Fish Biol.* **57,** 1–14.

Uglem, I., Mayer, I., and Rosenqvist, G. (2002). Variation in plasma steroids and reproductive traits in dimorphic males of corkwing wrasse (*Symphodus melops* L.). *Horm. Behav.* **41,** 396–404.

Yeung, W. S., and Chan, S. T. (1987). A radioimmunoassay study of the plasma levels of sex steroids in the protandrous, sex-reversing fish *Rhabdosargus sarba* (Sparidae). *Gen. Comp. Endocrinol.* **66,** 353–363.

Yu, K. L., Lin, X. W., Cunha Bastos, J., and Peter, R. E. (1997). Neural regulation of GnRH in teleost fishes. *In* "GnRH Neurons: Gene to Behaviour" (Parhar, I. S., and Sakuma, Y., Eds.), pp. 277–312. Brain Shuppan Publishing, Tokyo.

Warner, R. R. (1984). Mating behaviour and hermaphroditism in coral reef fishes. *Am. Sci.* **72,** 128–136.

Warner, R. R. (1985). Alternative mating behaviours in a coral reef fish: a life-history analysis. *In* "Proceedings of the 5th International Coral Reef Conference," pp. 145–150. Tahiti.

Warner, R. R. (1987). Female choice of sites versus mates in a coral reef fish, *Thalassoma bifasciatum*. *Anim. Behav.* **35,** 1470–1478.

Warner, R. R. (1990). Male versus female influences on mating-site determination in a coral reef fish. *Anim. Behav.* **39,** 540–548.

Warner, R. R., and Robertson, D. R. (1978). Sexual patterns in the Labroid fishes of the Western Caribbean, I: The wrasses (Labridae). *Smithsonian Contributions to Zoology* **254,** 1–27.

West-Eberhard, M. J. (1989). Phenotypic plasticity and the origins of diversity. *Ann. Rev. Ecol. Syst.* **20,** 249–278.

White, S. A., Kasten, T. L., Bond, C. T., Adelman, J. P., and Fernald, R. D. (1995). Three gonadotropin-releasing hormone genes in one organismsuggest novel roles for an ancient peptide. *Proc. Natl. Acad. Sci. USA* **92,** 8363–8367.

Wingfield, J. C., Hegner, R. E., Dufty, A. M., and Ball, G. F. (1990). The "challenge hypothesis": Theoretical implications for patterns of testosterone secretion, mating systems, and breeding strategies. *Am. Nat.* **136,** 829–846.

Wingfield, J. C., Jacobs, J. D., Soma, K., Maney, D. L., Hunt, K., Wisti-Peterson, D., Meddle, S., Ramenofsky, M., and Sullivan, K. (1999). Testosterone, aggression, and communication: Ecological bases of endocrine phenomena. *In* "The Design of Animal Communication" (Hauser, M. D., and Konishi, M., Eds.), pp. 257–283. MIT Press, Cambridge, MA.

Wingfield, J. C., Lynn, S. E., and Soma, K. K. (2001). Avoiding the 'costs' of testosterone: Ecological bases of hormone-behaviour interactions. *Brain Behav. Evol.* **57,** 239–251.

Zarkower, D. (2001). Establishing sexual dimorphism: Conservation amidst diversity? *Nature Rev.* **2,** 175–185.

Zhu, Y., Rice, C. D., Pang, Y., Pace, M., and Thomas, P. (2003). Cloning, expression, and characterisation of a membrane progestin receptor and evidence it is an intermediary in meiotic maturation in fish oocytes. *Proc. Natl. Acad. Sci. USA* **100,** 2231–2236.

Zimmerer, E. J., and Kallman, K. D. (1989). Genetic basis for alternative reproductive tactics in the pygmy swordtail, *Xiphophorus nigrensis*. *Evolution* **43,** 1298–1307.

Zupanc, G. K. H. (2001). Adult neurogenesis and neuronal regeneration in teleost fish. *Brain Behav. Evol.* **58,** 250–275.

Zupanc, G. K. H., and Lamprecht, J. (2000). Towards a cellular understanding of motivation: Structural reorganisation and biochemical switching as key mechanisms of behavioural plasticity. *Ethology* **106,** 467–477.

9

REPRODUCTIVE PHEROMONES

NORM STACEY
PETER SORENSEN

1. INTRODUCTION

Although it has long been known that fish commonly employ pheromones (chemical signals that pass between members of the same species; Sorensen and Wyatt, 2000) to coordinate many aspects of their reproductive biology, only in the past two decades has it become apparent that many of these cues are derived from, or are closely related to, hormones and other essential body metabolites (Liley, 1982; Stacey and Sorensen, 2002). This possibility was first raised by Døving (1976), who noted that hormonal products are preadapted to serve as pheromones because they are produced and released at relevant times. The first evidence for this possibility was the observation that bile acids (steroids produced by the liver and generally released in the bile) released by juvenile Arctic charr (*Salvelinus alpinus*) are potent odourants that appeared capable of attracting adults migrating

Behaviour and physiology of Fish: Volume 24
FISH PHYSIOLOGY

DOI: 10.1016/S1546-5098(05)24009-8

to spawning grounds (Døving and Selset, 1980). Soon thereafter, Colombo *et al.* (1980) showed that a 5β-reduced conjugated androgen (etiocholanolone glucuronide [Etio-g]) produced by male black gobies (*Gobius niger*) attracted ovulated conspecifics. Van Den Hurk and Lambert (1983) next proposed that ovarian steroid glucuronides function as sex pheromones in zebrafish (*Danio rerio*).

Clear proof that hormones function as pheromones came from studies showing that an oocyte maturation inducing steroid (MIS), 4-pregnen-17,20β-diol-3-one (17,20β-P) is released by female goldfish, *Carassius auratus*, and induces in males strong behavioural and endocrinological (i.e., pheromonal) effects (Stacey and Sorensen, 1986; Dulka *et al.*, 1987; Sorensen *et al.*, 1987, 1990). Since then, hormones and related compounds have been found to function as potent odourants which influence conspecific behaviour and/or physiology in a variety of fish species including African catfish, *Clarias gariepinus* (reviewed by Van Den Hurk and Resink, 1992), Atlantic salmon, *Salmo salar* (Moore and Scott, 1991), oriental weatherfish loach, *Misgurnus anguillicaudatus* (Kitamura *et al.*, 1994a), common carp, *Cyprinus carpio* (Irvine and Sorensen, 1993), and crucian carp, *Carassius carassius* (Bjerselius *et al.*, 1995). The sea lamprey has lately emerged as a model for understanding how bile acids may function as pheromones (e.g., Li *et al.*, 2003; Sorensen *et al.*, 2003).

Electrophysiological studies of olfactory function show that many bile acids, sex steroids, and prostaglandins (as well as their precursors and derivatives) are detected by fish, providing compelling evidence that fish commonly employ these chemicals as pheromones. Most of these studies have employed extracellular underwater electro-olfactogram (EOG) recording, which measures transepithelial voltage changes that reflect summed generator responses of olfactory receptor neurons (ORNs; Scott and Scott-Johnson, 2002). These studies have now and have shown that more than 100 fish species are acutely sensitive to steroids and prostaglandins (Kitamura *et al.*, 1994b; Cardwell and Stacey, 1995; Sorensen and Stacey, 1999; Stacey and Sorensen, 2002; Laberge and Hara, 2003a; Stacey *et al.*, 2003; Lower *et al.*, 2004). More recently, EOG studies have also discovered that sea lamprey (*Petromyzon marinus*) are extremely sensitive to bile acids (Li *et al.*, 1995; Li and Sorensen, 1997), which appear to act as pheromones mediating migration (Sorensen and Vrieze, 2003) and sexual attraction (Li *et al.*, 2002, 2003).

When considering pheromonal function in fish, it is important to consider that chemical signals propagate differently than other sensory cues. Whereas visual and acoustic signals are propagated with predictable speed and direction, chemicals are transmitted slowly and are subject to turbulence and variable flow, which can slow and distort the position, shape, size, and temporal nature of the cue (Weissburg, 2000). Moreover, chemical information can persist for days (Sorensen *et al.*, 2000) and thus potentially become separated from its source. Although mechanisms for orientating in pheromone plumes have been described in some invertebrates (e.g., crustaceans and moths), this issue is poorly understood in fish (Vickers, 2000). Pheromone function in water may also be complicated by other dissolved compounds including humics that can bind odourants (Hubbard *et al.*, 2002; Mesquita *et al.*, 2003), heavy metals that can disrupt neural function (Hansen *et al.*, 1999), and endocrine-disrupting chemicals that might mask natural signals (Kolodziej *et al.*, 2003).

Finally, it is important to consider that, by the nature of the information it processes, olfaction differs in important respects from other sensory modalities. In terms of basic design, the central nervous system encodes linear arrays of sight and sound information in intuitive spatial and temporal manners, whereas encoding of chemical information does not lend itself to such simple representation, greatly complicating its study. Moreover, the frequency and temporal information in an individual's acoustic and visual signals will often be broadly detectable by the sensory neurons of many species, but biologically relevant only to those (typically conspecifics) with appropriate central neural processing. In contrast, due to the acute specificity of fish olfactory receptor mechanisms, pheromones effectively activate discrete sensory channels that could be unique to conspecifics and evolve rapidly with only small change in receptor specificity.

2. TERMS AND CONCEPTS

Although it has been recognised since Darwin (1887) that animals respond to conspecific odours, Bethe (1932) first distinguished these special chemicals using the term *ectohormone*. Karlson and Lüscher (1959) later replaced this term with *pheromone* (literally "to transfer excitement"), which they defined as "substances that are excreted to the outside of an individual and received by a second individual of the same species in which they release a specific reaction, for example a definite behaviour or developmental process." However, this vague definition has caused much confusion (see Johnston, 2000). To clarify the pheromone concept for the study of fishes (see Sorensen and Stacey, 1999; Sorensen and Wyatt, 2000), the authors define pheromones as a substance, or mixture of substances, released by an individual, which evokes a specific and adaptive response in conspecifics, the expression of which does not require learning. This allows for the possibilities that pheromones need not always be species-specific, that the responses to them can be modified by experience, and that they can be mixtures, a situation commonly observed in insects (Sorensen *et al.*, 1998).

Further, the authors define a *reproductive pheromone* as a pheromone that induces any behavioural or physiological response associated with reproductive activity. Thus, cues that directly prepare maturing fish for spawning (e.g., evoke prespawning aggregation) or trigger spawning itself are included, whereas cues that are broadly effective across multiple life stages (aggregants, alarm cues, kin-recognition, etc.; Courtenay *et al.*, 1997; Olsén *et al.*, 2002a; Wisenden, 2003) are not. We also employ the terms *releaser* and *primer* (Wilson and Bossert, 1963) to distinguish between rapid behavioural effects and typically slower physiological effects induced by pheromones. We apply these terms only to pheromonal effects rather than to pheromones *per se* because, as noted by Wilson and Bossert (1963), "it is quite possible for the same pheromone to be both a releaser and a primer." For example, several reproductive pheromones exert both primer and releaser effects in goldfish (Dulka *et al.*, 1987; Sorensen *et al.*, 1989; Poling *et al.*, 2001).

Finally, the authors apply the term *hormonal pheromone* for any reproductive pheromone that contains at least one compound derived from a chemical pathway that produces hormones (internal chemical signals). This class of pheromones warrants special consideration both because most fish appear to employ hormonal cues as pheromones and because in so doing, special evolutionary links between internal and external chemical signals are implied to exist within this group. By our definition, hormonal pheromones can contain unmodified hormones, hormonal precursors, and/or their metabolites. Although almost all reproductive pheromones identified in fish are hormonal pheromones, important exceptions include the bile acid pheromones of the sea lamprey (Li *et al.*, 2002; Sorensen *et al.*, 2003), the putative L-amino acid pheromone of the rose bitterling, *Rhodeus ocellatus* (Kawabata, 1993), and tetrodotoxin pheromone of the puffer fish, *Fugu rubripes* (Matsumura, 1985). Others almost certainly exist.

The authors also propose (Stacey and Sorensen, 1991, 2002; Wisenden and Stacey, 2005) that the evolution of chemical communication involves three functionally distinct phases: *ancestral*, *spying*, and *communication* (Figure 9.1). The initial phase, in which individuals (*originators*) fortuitously release a metabolic product(s) that does not influence receivers, can evolve into the second, *spying*, if selection causes conspecifics (*receivers*) to detect and respond adaptively to the originator's released chemical(s), which we now term a pheromonal *cue*(s). In the spying phase, originators of cues remain unspecialised with respect to production and release of the pheromonal cue(s), regardless of whether they benefit from the receiver's response. In some instances, spying can develop into a third phase, *communication*, if originators also come to benefit from cue production, and develop specialisation(s) to produce it in more effective manner(s). Such specialised cues are termed *signals*. The authors believe that the use of seemingly unmodified

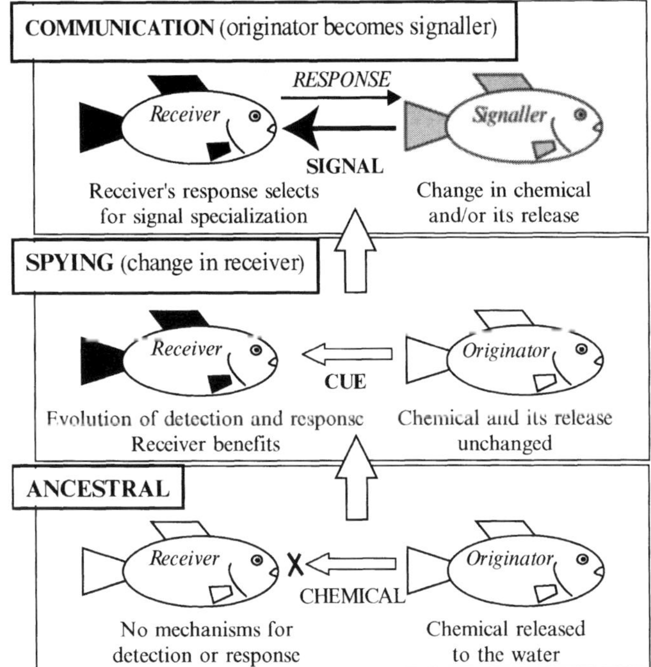

Fig. 9.1. Proposed stages in the evolution of hormonal pheromones. (Adapted from Stacey and Sorensen, 1991, 1999, 2002 with kind permission from Elsevier Ltd., and from Sorensen and Stacey, 1999 with kind permission from Springer-Verlag GmbH.)

hormones as pheromones by goldfish (Section 5.3.1) is an example of spying, whereas the use of Etio-g produced by the black goby mesorchial gland (Rasotto and Mazzoldi, 2002; Section 5.3.3) appears to be an example of communication.

For several reasons, this simple tripartite scheme appears to offer the most parsimonious explanation of pheromone evolution, at least for hormonal pheromones. It seems highly likely that pheromonal functions of sex steroids and prostaglandins were derived from their primitive hormonal functions for the simple reason that the chemical diversity of hormonal pheromones appears to be far greater than the remarkably conserved set of hormonal compounds retained not only by fish, but by all vertebrates. Further, it seems self-evident that evolutionary specialisations for pheromonal function in receivers must precede those in signalers, and that the establishment of spying creates opportunities for communication to evolve. A weakness in our terminology related to pheromone evolution is that the spying phase is defined by absence of specialisation in the pheromone originator, a situation that might be difficult to confirm. Nonetheless, we suggest that a pheromone released by male Pacific herring (*Clupea harengus pallasii*) functions in pheromonal spying, simply because the herring mating system appears to preclude the evolution of pheromonal communication.

Spawning of herring is triggered in both sexes by a milt (sperm and seminal fluid) pheromone that might contain steroids (Stacey and Hourston, 1982; Carolsfeld *et al.*, 1992), occurs in immense schools often kilometers in extent, and involves no evident coordination between the sexes, with males and females independently depositing gametes on benthic substrates. Because in this mating system males are highly unlikely to fertilise the eggs of the females they stimulate to spawn, there appears to be no mechanism whereby females can exert selective pressure for specialisation in male pheromone production.

The distinction between pheromonal communication and spying is central to such basic issues as the coevolution of signalers and receivers, and the mechanisms whereby species-specific pheromones might arise (Sorensen and Stacey, 1999). We suggest two concepts that might guide research in this important, poorly understood area. First, we postulate that the evolutionary origins of pheromones are stochastic, insofar as the particular conspecific chemicals to which receivers fortuitously evolve olfactory responsiveness will greatly influence further pheromone evolution. These new hormonal odourants could be incidental metabolites whose production and/or release might readily be modified as spying progresses to communication. However, they might also be key hormones, whose existing endogenous functions might provide physiological inertia preventing the transition to communication. Second, a species' mating system first defines the selection pressures that establish spying, and then confers the constraints and opportunities that determine if and how communication might arise. Given the diverse reproductive tactics and strategies of fish, the variety of ecological factors they encounter, and the stochastic factors that initiate the evolution of pheromone signals, we expect similar diversity in their reproductive pheromone systems.

3. OVERVIEW OF THE FISH OLFACTORY SYSTEM

Where studied in vertebrates, responsiveness to pheromones is mediated exclusively by the olfactory system (cranial nerve 1). As in other vertebrates, the fish olfactory system has three key components: *olfactory receptor neurons* (ORNs), which express olfactory receptor molecules and have their cell bodies in the olfactory epithelium; *olfactory bulbs*, to which most ORNs project and synapse with output neurons to form glomeruli; and *output neurons* (mitral cells) and their terminal fields in the telencephalon, hypothalamus, and other brain regions. This system has been reviewed elsewhere (Sorensen and Caprio, 1997; Laberge and Hara, 2001) and is only briefly summarised here. Olfaction is also considered in Chapter 2.

3.1. Olfactory Epithelium and Olfactory Receptor Neurons

The olfactory epithelium within the nasal capsules contains both non-sensory supporting cells and ORNs. Although there is evidence for sexual dimorphism in the olfactory epithelium of some fish, neither its cause nor functional significance is known. For example, males of some deep-sea fish have much larger olfactory epithelia than females (Baird *et al.*, 1990), and EOG responsiveness of some fish to prostaglandins is sexually dimorphic and androgen-dependent (Sorensen and Goetz, 1993; Cardwell *et al.*, 1995; Stacey *et al.*, 2003; Lower *et al.*, 2004; Section 5.4). Most fish possess at least three morphological classes of ORNs—ciliated, microvillous, and crypt—whereas sharks appear to have only microvillous and lamprey only ciliated (Zeiske *et al.*, 1992; Sorensen and Caprio, 1997; Morita and Finger, 1998; Hansen *et al.*, 2003). Although the function of these ORN types is unclear, neuroanatomical, pharmacological, and electrophysiological studies from zebrafish, channel catfish (*Ictalurus punctatus*), and crucian carp suggest some ciliated ORNs transduce responses to some bile acids (no established bile acid pheromones were tested), whereas both ciliated and microvillar ORNs may transduce responses to amino acids (feeding stimuli) (see Hansen *et al.*, 2003). Although no definitive data link ORN type to a known pheromonal compound, an ORN regeneration study in goldfish (Zippel *et al.*, 1997) suggests that some microvillar cells mediate some sex-pheromone responses.

Using probes based on G-protein linked receptors found in mammalian olfactory epithelia, three basic classes of olfactory receptors have been cloned in fish (Dryer and Berghard, 1999), a V2R (vomeronasal) type resembling a mGluR-type receptor in the mammalian vomeronasal system and suggested to mediate pheromone responses (Cao *et al.*, 1998; Naito *et al.*, 1998), a "VIR-like receptor" resembling another class of mammalian vomeronasal receptors (Pfister and Rodriguez, 2005), and a G-Olf type based on the 7-transmembrane receptor type first isolated from the rat (Buck and Axel, 1991; Ngai *et al.*, 1993a). The V2R receptor type appears to occur in microvillous cells (Cao *et al.*, 1998), and one that has been functionally cloned (in goldfish) binds the amino acid L-arginine with high specificity (Speca *et al.*, 1999). Channel catfish have been estimated to possess approximately 100 G-Olf receptor subtypes in six subfamilies, whereas mammals possess more than 1000, many of which may be pseudogenes (Rouquier *et al.*, 2000). *In situ* hybridisation using G-Olf receptor probes indicates fish ORNs (like those of mammals) express only one or a few subtypes, suggesting individual fish ORNs are likely specialised for detection of one or a few specific odours (Ngai *et al.*, 1993a,b; Hansen *et al.*, 2003, 2004).

Unfortunately, olfactory receptors for hormonal pheromones have yet to be cloned. However, using a membrane preparation of goldfish olfactory epithelium, Rosenblum *et al.* (1991) demonstrated that the goldfish steroid pheromone 17,20β-P binds not only with low capacity and high affinity, but also with a specificity that is expected of an olfactory receptor, and is similar to that expected from ovarian membranes on which this steroid functions as a maturation inducing hormone (Nagahama *et al.*, 1994). It is intriguing that a putative membrane-associated MIS cloned from spotted seatrout, *Cynoscion nebulosus* (Zhu *et al.*, 2003), appears to be a 7-transmembrane, G-protein–coupled receptor (Thomas, 2003).

3.2. Olfactory Bulbs and Tracts

ORNs project from the olfactory epithelium via the olfactory nerves to the olfactory bulbs, where the overwhelming majority of their axons terminate. In the bulbs, ORNs with common receptor types converge and synapse on mitral cells in the bulbar glomeruli, dense and complex aggregations of the neuropils of mitral and granule cells that receive inhibitory input from other glomeruli and the central nervous system (Sorensen and Caprio, 1997). From the bulbs, mitral cells project via the medial and lateral olfactory tracts to terminate in specific telencephalic and hypothalamic fields. Some nerves project from the periphery without synapsing in the bulb, such as the terminal nerve (cranial nerve 0), which Demski and Northcutt (1983) proposed to be specialised for pheromone detection. However, terminal nerve ablations do not cause evident reproductive deficits (Kobayashi *et al.*, 1994, 1997a; Yamamoto *et al.*, 1997), and single unit, extracellular recordings in goldfish show that pheromones and other odourants that affect olfactory bulbar units do not affect terminal nerve activity (Fujita *et al.*, 1991). A current hypothesis is that the terminal nerve modulates olfactory sensitivity (Eisthen *et al.*, 2000).

Evidence derived primarily from mammals indicates that glomeruli are functional units that encode odour cues and pass this information to the brain (Xu *et al.*, 2000). The situation in fish is complicated by the fact that mitral cells may project to many, relatively diffuse glomeruli. Food odours including amino acids appear to be encoded in a combinatorial fashion involving multiple glomeruli with overlapping sensitivities (Freidrich and Korsching, 1998; Hanson, 2001; Nikonov and Caprio, 2001) and incorporating a temporal component (Friedrich and Laurent, 2001). Processing of pheromonal responses in zebrafish (Friedrich and Korsching, 1998) and goldfish (Hanson *et al.*, 1998) appears to be restricted to certain bulbar areas. However, Laberge and Hara (2003b) report that, although lake whitefish (*Coregonus clupeaformis*) process amino acid stimuli in bulbar areas, putative pheromonal prostaglandins activate extrabulbar units in what is likely the medial olfactory tract. Nevertheless, all three studies indicate that fish process pheromonal information in some kind of specialised manner that differs from processing of feeding cues. Results of olfactory lesion and stimulation studies are consistent with this idea. For example, electrical stimulation of the cod (*Gadus morhua*) olfactory tract (Døving and Selset, 1980) and olfactory tract lesions in goldfish (Stacey and Kyle, 1983; Dulka and Stacey, 1990), crucian carp (Hamdani *et al.*, 2000; Weltzien *et al.*, 2003), and African catfish (Van Den Hurk and Resink, 1992) demonstrate that the lateral tracts mediate food responses, whereas the medial tracts mediate response to both reproductive and nonreproductive pheromones. Extracellular tract recordings in goldfish (Sorensen *et al.*, 1991) agree with these findings.

4. BILE ACID-DERIVED PHEROMONES OF SEA LAMPREY

The anadromous sea lamprey spawns in freshwater streams where its larvae spend 3–20 years filter-feeding before metamorphosing into a parasitic phase that enters the Atlantic ocean or large lakes to parasitise other fish. After approximately 1 year, parasites cease feeding, begin to mature and, aided by an olfactory system larger than their brain, migrate up suitable streams for spawning, in which females appear to use a sex pheromone to locate nesting spermiating males (Teeter, 1980). Lamprey pheromones have been extensively studied in the hope they can be employed to control lampreys in the Great Lakes where they threaten fisheries (Sorensen and Vrieze, 2003; Twohey *et al.*, 2003).

4.1. Migratory Pheromone

Maturing adult sea lamprey locate suitable spawning habitat using a pheromone released by stream-dwelling larvae. Behavioural studies (Vrieze and Sorensen, 2001) show that water from streams with larvae is more attractive to adults than water from streams without larvae, that adding larval odour to stream water increases its attractiveness, and that individual larvae (1 g body weight) create a large "active space" (the volume of water activated by pheromone) of at least 400 L in an hour. The pheromone is not species-specific, perhaps because all North American lamprey use similar spawning and nursery habitats and larvae do not benefit from attracting adults (Fine *et al.*, 2004).

The larval pheromone is a complex mixture that acts through specific olfactory receptors with detection thresholds ranging below 10^{-13} M (Sorensen *et al.*, 2003). Pheromone characterisation has relied strongly upon behavioural assays and alternating use of high-performance liquid chromatography (HPLC) fractionation/mass spectrometry to confirm purity, and EOG recording to screen for olfactory function. One compound isolated with this approach is petromyzonol sulfate (PS; 3α,12α,24-trihydroxy-5α-cholan-24-sulphate), a unique bile acid that, despite considerable olfactory activity, has only moderate behavioural activity on its own. A second lamprey bile acid, allocholic acid (ACA; 3α,7α,12α-trihydroxy-5α-cholan-24-oic acid), has good olfactory activity but little (if any) behavioural activity, whereas two unidentified and novel sulfated steroids of unknown origins have even greater behavioural and olfactory activity than PS and strongly compliment its actions (Sorensen *et al.*, 2003). Both PS and ACA are synthesised by the liver of larvae (but not of parasitic or adult phases) of various lamprey species (Polkinghorne *et al.*, 2001; Yun *et al.*, 2003a). Because larval gall bladders and bile ducts atrophy during metamorphosis, at which time PS and ACA synthesis ceases (Polkinghorne *et al.*, 2001), these compounds should serve as specific indicators of streams containing larvae and thus favorable spawning and nursery habitat. These bile acids likely evolved to facilitate digestion and to suppress disease, and are apparently released primarily in larval feces (Polkinghorne *et al.*, 2001). They also are detected with great sensitivity (olfactory detection thresholds <1 pM) by the olfactory organ of migratory adults (Li and Sorensen, 1997). A mixture of PS and ACA attracts maturing adults in laboratory mazes, but to a lesser extent than does whole larval odour, the activity of which is likely due to the unidentified odourants described above (Vrieze and Sorensen, 2001; Sorensen *et al.*, 2003). When fully mature, sea lampreys are not attracted to larval odour, suggesting sensitivity to it is regulated by endocrine state (Bjerselius *et al.*, 2000). Importantly, PS has been identified at pM concentrations in water from rivers with larval populations, providing direct evidence that this cue is a critical component of the prespawning migratory attractant of sea lamprey (Fine *et al.*, 2004). Together, these results demonstrate that a blend of conspecific (and likely congeneric) cues regulate stream selection and upstream migration of maturing adult sea lamprey, adaptive responses that increase the likelihood of locating suitable larval habitat. All ecological, behavioural, and physiological evidence strongly suggests this to be an example of spying.

4.2. Male Reproductive Pheromone

During upstream migration, lamprey undergo final maturation (spermiation and ovulation), lose behavioural responsiveness to the larval pheromone, and develop behavioural responsiveness to the odour of mature conspecifics of the opposite sex (Teeter, 1980; Bjerselius *et al.*, 2000). Behavioural tests using mature males placed in traps (see Teeter, 1980) indicate that male lamprey (which precede females to the spawning grounds) release a pheromone that attracts females. Further, Teeter (1980) reports strong behavioural responses to the urine of mature male lamprey.

3-Keto-petromyzonol-sulphate (3K-PS; 7α,12α,24-trihydroxy-3-one-5α-cholan-24-sulphate) appears to be a potent (10^{-12} M EOG threshold) component of the pheromone released by spermiated male sea lamprey, whereas 3-keto-allocholic acid (3K-ACA; 7α,12α-dihydroxy-5α-cholan-3-one-24-oic acid) may be a less potent (10^{-10} M threshold) component (Li *et al.*, 2002; Yun *et al.*, 2003b). Although both compounds are detected by the olfactory system (Figure 9.2), only 3K-PS has been investigated for behavioural activity. When added to a two-choice maze, 3K-PS attracts ovulated females and stimulates searching (Li *et al.*, 2002). However, the precise role of 3K-PS is unclear as there are no reports of its activity either being directly compared with that of whole male odour or being assessed over a range of natural concentrations. Nonetheless, immunoassay has shown that nonspermiated males do not release appreciable quantities of immunoreactive 3K-PS, whereas spermiated males release large quantities (~250 μg/h), estimated

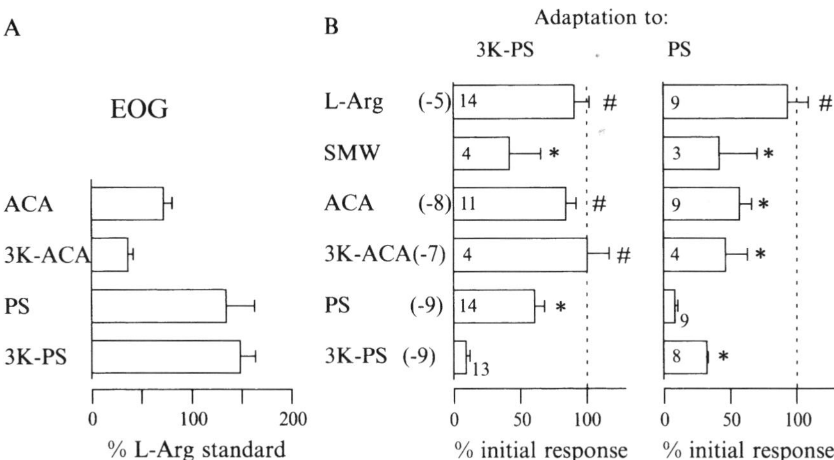

Fig. 9.2. Electro-olfactogram (EOG) responses induced in female sea lamprey by the lamprey bile acids allocholic acid (ACA), 3-keto-allocholic acid (3K-ACA), petromyzonol-sulfate (PS) and 3-keto-petromyzonol sulfate (3K-PS). (A) Magnitude of EOG response (mean ± SEM; as a percentage of 10 μM L-arginine standard) to 1 nM bile acids. (B) Results of EOG cross-adaptation experiments involving the four lamprey bile acids, the L-arginine standard, and holding water from spermiating male sea lamprey (SMW) * = partial adaptation, # = no adaptation. (Adapted from Siefkes and Li, 2004 with kind permission from the authors and Springer-Verlag GmbH.)

to produce a very large active space (>10^6 l/h; Li *et al.*, 2002; Yun *et al.*, 2002).

3K-PS appears to be produced in the liver of spermiated males (Li *et al.*, 2002) and released by the gills, which in mature males (but not females) develop glandular cells (Li *et al.*, 2003; Siefkes *et al.*, 2003) that appear specialised for pheromone release. Given the scant evidence for specialisation in fish pheromone systems, the clear possibility that the male sea lamprey pheromone functions as a communicative signal deserves consideration. EOG recording from female sea lamprey shows that concentrated water from spermiated males induces responses similar in size to those induced by 3K-PS (Siefkes and Li, 2004) and that PS (larval cue) and 3K-PS (mature male cue) stimulate different sets of olfactory receptor(s) with overlapping sensitivities. This lack of perfect specificity, coupled with the ability of lamprey to discriminate behaviourally between larval and adult odours (Bjerselius *et al.*, 2000), suggests that 3K-PS may be complimented by unknown pheromonal cues released by males.

5. HORMONAL PHEROMONES

Results of two experimental approaches indicate that fish have commonly evolved to use gonadal steroids and prostaglandins as pheromones (Sorensen and Stacey, 1999; Stacey and Sorensen, 2002). In the first approach (which has been taken with only a small number of studies) pheromone production, release, detection, and response are simultaneously examined in a model species, thereby addressing reasonable requirements to prove the existence of a pheromone (Sorensen and Stacey, 1999). In the second approach, EOG screening has been used to determine what hormonal compounds are detected, assuming that sensitive and specific olfactory responses are indicative of pheromonal function. Together, studies employing these approaches have described olfactory and behavioural responses to steroids and prostaglandins in fishes from four orders including Perciformes (black goby: Colombo *et al.*, 1980; round goby, *Neogobius melanostomus*: Murphy *et al.*, 2001; Zielinski *et al.*, 2003), Siluriformes (African catfish: Van Den Hurk and Resink, 1992), Cypriniformes (goldfish: Sorensen *et al.*, 1987, 1988, 1989, 1990; common carp: Irvine and Sorensen, 1993; Stacey *et al.*, 1994; crucian carp: Bjerselius *et al.*, 1995; oriental weatherfish loach: Kitamura *et al.*, 1994a; Ogata *et al.*, 1994; a cyprinid barb, *Puntius schwanenfeldi*: Cardwell *et al.*, 1995; roach, *Rutilus rutilus*: Lower *et al.*, 2004); Salmoniformes (Atlantic salmon: Moore, 1991; Moore and Waring, 1996: brown trout, *Salmo trutta*: Essington and Sorensen, 1996; Olsén *et al.*, 2000; Moore *et al.*, 2002; Laberge and Hara, 2003a; Arctic charr: Sveinsson and Hara, 1995, 2000; lake whitefish: Laberge and Hara, 2003a). Moreover, EOG screening studies, the more recent of which employed approximately 150 commercially available steroids and prostaglandins, show that most of the more than 100 species tested detect at least one sex steroid or prostaglandin at pM concentrations (Sorensen *et al.*, 1990; Cardwell and Stacey, 1995; Stacey and Cardwell, 1995; Stacey *et al.*, 1995; Murphy *et al.*, 2001; Stacey and Sorensen, 2002; Cole and Stacey, 2003; Narayanan and Stacey, 2003). These EOG studies also have found species that detect known hormonal products in another four orders (see Stacey and Sorensen, 2002) including Elopiformes (Indo-Pacific tarpon, *Megalops cyprinoides*), Characiformes (many species from family Characidae; Cardwell and Stacey, 1995), Gymnotiformes (brown ghost knifefish, *Apteronotus leptorhynchus*), and Osmeriformes (ayu, *Plecoglossus altivelis*; Kitamura *et al.*, 1994b). EOG screening likely underestimates the prevalence of steroids and prostaglandin detection, both because sensitivity to these compounds tends to be highly specific and because it is unlikely that naturally occurring odourants have been tested for all taxa. For example, a large number of steroids have failed to induce EOG response in the ruffe (*Gymnocephalus cernuus*), despite strong evidence that it uses an unknown 21-carbon steroid metabolite as a pheromone (Sorensen *et al.*, 2004; Section 5.3.3).

EOG studies reveal a remarkably close correlation between phylogeny and the patterns of compounds detected. In Cypriniformes, which contains two superfamilies, Cobitoidea and Cyprinoidea, patterns of olfactory sensitivity reflect phylogeny (Nelson, 1994). Although EOG shows that all of more than 80 tested cypriniform species detect prostaglandins, no species from the four cobitoid families detect conjugated or nonconjugated steroids, whereas only one of the seven cyprinid subfamilies tested (Acheilognathinae; bitterlings) failed to detect steroidal compounds. Moreover, although patterns of steroid detection vary considerably among cyprinid subfamilies tested, they are usually remarkably similar within subfamilies (Stacey and Cardwell, 1995; Stacey *et al.*, 1995; Stacey and Sorensen, 2002). Such patterns suggest that the evolution of hormonal pheromones has been constrained and that species specificity, if it has evolved, likely is achieved by altering ratios of a common odourant mixture and/or the use of other contextual cues (see Sorensen and Stacey, 1999).

5.1. Hormones and Related Compounds Serving as Hormonal Pheromones

The major reproductive events in the lives of many fish (maturation, ovulation, and spawning) are temporally correlated with changes in plasma levels of key reproductive hormones (gonadal steroids and prostaglandins) that should have predisposed them (and their precursors and metabolites) to serve as hormonal pheromones.

5.1.1. Gonadal Steroids

As in tetrapods, postpubertal activity of the brain-pituitary-gonad axis induces gonadal recrudescence and sexually dimorphic plasma concentrations of 17β-estradiol (estradiol) and 11-ketotestosterone (Borg, 1994; Weltzien *et al.*, 2004). Estradiol stimulates vitellogenesis (Patiño and Sullivan, 2002) and regulates luteinising hormone secretion (Trudeau, 1997). 11-Ketotestosterone regulates luteinising hormone secretion (Antonopoulou *et al.*, 1999), and induces male-typical morphology and behaviour (Borg, 1994), spermatogenesis (Schulz and Miura, 2002; Miura and Miura, 2003), and gonadal sex differentiation (Strüssmann and Nakamura, 2002).

Despite clear gender-typical changes in estradiol and 11-ketotestosterone in fish, plasma testosterone concentrations in females, which also increase during gonadal recrudescence, often are equal to or greater than those in males (Borg, 1994). Testosterone influences luteinising hormone secretion in male and female fish (Trudeau, 1997) and also induces male-typical phenotype, although where effects of exogenous testosterone and 11-ketotestosterone have been compared, 11-ketotestosterone typically is more potent (Borg, 1994; see also Chapter 8).

Released androgenic and estrogenic steroids are likely candidates for pheromonal function, because they should convey to conspecifics information on both gender and reproductive status. Indeed, 11-ketotestosterone is detected by the cyprinid *P. schwanenfeldi* (Cardwell *et al.*, 1995), testosterone sulphate is detected by the cichlid *H. burtoni* (Cole and Stacey, 2003), and estrone and free and conjugated estradiol are detected by *H. burtoni* (Cole and Stacey, 2003), the round goby (Murphy *et al.*, 2001), and numerous characiform species (Cardwell and Stacey, 1995; Stacey and Sorensen, 2002). However, only in goldfish (Sorensen *et al.*, 2005; Section 5.3.1) and Atlantic salmon (Moore, 1991; Section 5.3.2.b.i) is there information on both detection and biological functions of androgen-related pheromones.

Shortly before germinal vesicle migration and breakdown (GVBD) in females and spermiation in males, an abrupt surge release of luteinising hormone typically increases production of MISs. 17,20β-P is the female maturation inducing steroid (MIS) in a variety of species (Nagahama *et al.*, 1994), whereas 4-pregnen-17,20β,21-triol-3-one (17,20β,21-P; also termed 20β-S) is the MIS in others (Thomas, 1994). MIS acts through oocyte membrane receptors to induce oocyte maturation (resumption and completion of meiosis) (Thomas, 2003) and through nuclear receptors to induce ovulation (Patiño *et al.*, 2003). Although relatively little is known of MIS function in males, 17,20β-P binds to receptors on the sperm plasma membrane in spotted seatrout (Thomas *et al.*, 1997) and induces maturational acquisition of sperm motility in salmonids and Japanese eel (*Anguilla japonica*) (Miura and Miura, 2003).

EOG studies (Cardwell and Stacey, 1995; Stacey and Cardwell, 1995; Stacey *et al.*, 1995; Sorensen and Stacey, 1999; Stacey and Sorensen, 2002; Narayanan and Stacey, 2003) indicate that MISs and related 21-carbon steroids are detected by many cypriniforms, characiforms, and siluriforms as well as Atlantic salmon (Moore and Scott, 1992), and are more commonly used as pheromones than are 18- and 19-carbon steroids. If true, this might result not only from the fact that peak MIS production immediately precedes a critical life-history event, but also because the typically transient MIS production should provide unambiguous information about imminent spawning opportunity. However, even in the roach where MIS production appears to be high for some time prior to and following maturation in both sexes, glucuronidated 17,20β-P (17,20β-P-g) is a potent odourant (Lower *et al.*, 2004). The nature of detected MISs and MIS-related steroids varies widely: e.g., both conjugated and unconjugated 17,20β-P in goldfish (Kobayashi *et al.*, 2002), conjugated metabolites in African catfish (Van Den Hurk and Resink, 1992) and *Haplochromis* (Cole and Stacey, 2003) and unconjugated metabolites in *Synodontis* catfish (Narayanan and Stacey, 2003). Such diversity in putative MIS-derived pheromones raises the distinct possibility that some may have rather different functions than demonstrated in goldfish (Sorensen *et al.*, 2004).

5.1.2. Prostaglandins

Sensitive and specific EOG responses to prostaglandin $F_{2\alpha}$ and its metabolites provide strong though indirect evidence that hormonal pheromones derived from F-series prostaglandins are widespread, particularly amongst Cypriniformes, Characiformes, and Salmoniformes (Stacey and Cardwell, 1995; Stacey and Sorensen, 2002; Laberge and Hara, 2003b). With the exception of goldfish, however, identities and functions of pheromonal prostaglandin Fs are very poorly understood and have not been studied using definitive modern biochemical techniques such as mass spectrometry. In fish, as in a variety of vertebrates, prostaglandin $F_{2\alpha}$ plays a key role in ovulation, which is stimulated by prostaglandin $F_{2\alpha}$ injection, blocked by the prostaglandin synthetase inhibitor indomethacin, and consistently associated with increased levels of ovarian immunoreactive prostaglandin $F_{2\alpha}$ (Sorensen and Goetz, 1993; Kagawa *et al.*, 2003). The critical role of prostaglandins in ovulation is reflected in a study of circulating levels in the female goldfish, whose low preovulatory prostaglandin $F_{2\alpha}$ levels ($<$2 ng/ml) increase dramatically ($\sim$100 ng/ml) at ovulation (Sorensen *et al.*, 1995b).

This postovulatory prostaglandin $F_{2\alpha}$ increase is an apparent response to ovulated eggs in the reproductive tract (Section 5.2.2).

Immunoreactive prostaglandin F has been measured in the ovulatory urine and/or holding water of several fish species, but only for the goldfish has a specific metabolite (15-keto-prostaglandin $F_{2\alpha}$) been identified using definitive techniques (mass spectrometry; Sorensen *et al.*, 1995b). However, HPLC studies (Ogata *et al.*, 1994) indicate ovulated oriental weatherfish loach greatly increase release of putative 13,14-dihydro-15-keto-prostaglandin $F_{2\alpha}$, a proposed pheromone (Kitamura *et al.*, 1994a). Immunoreactive prostaglandin F released by male Arctic charr also is reported to function as a pheromone that stimulates nest construction by females (Sveinsson and Hara, 1995, 2000). Because prostaglandin pathways are complex and poorly characterised, it is important to realise that the small number of commercially available prostaglandin Fs that have been examined is unlikely to include all physiologically and pheromonally active prostaglandins used by fish. Clearly, much more information on the identities, sources, and metabolic fates of prostaglandins is required if this evidently significant aspect of fish hormonal pheromones is to be understood.

5.1.3. Patterns and Routes of Hormonal Pheromone Release

When released, steroid and prostaglandin hormones and related chemicals have the potential to transmit information to conspecifics because their synthesis is predictably linked with reproductive events. Indeed, changes in plasma concentrations of key ovarian and testicular steroids associated with seasonal gonadal recrudescence (Pacific herring; Koya *et al.*, 2002, 2003; Figure 9.3), final maturation (Stacey *et al.*, 1989; Figure 9.4; Inbaraj *et al.*, 1997), and the transition from sexual to parental activity (Oliveira *et al.*, 2002) are often sufficiently dramatic to serve as unambiguous indicators of altered reproductive condition. Unfortunately, with the exception of salmon and goldfish, little is known about how such plasma hormone changes are reflected in patterns of release.

Fish release steroids and prostaglandins (PGs) by four routes: gills, urine, gonadal fluids, and feces (bile) (Resink *et al.*, 1989; Scott *et al.*, 1991a,b; Scott and Liley, 1994; Oliveira *et al.*, 1996; Vermeirssen and Scott, 1996; Inbaraj *et al.*, 1997; Appelt and Sorensen, 1999; Sorensen *et al.*, 2000). Both the rates and routes of steroid release are associated with conjugation: free steroids are released rapidly by gills, and sulfated and glucuronidated steroids are released more slowly in urine and bile, respectively (Vermeirssen and Scott, 1996; Sorensen *et al.*, 2000). In goldfish, the best understood model, the result is that 17,20β-P synthesised during the periovulatory period generates at least two 17,20β-P-based pheromonal cues (Sorensen *et al.*, 1995a) that are released by different routes and over different time courses, and that

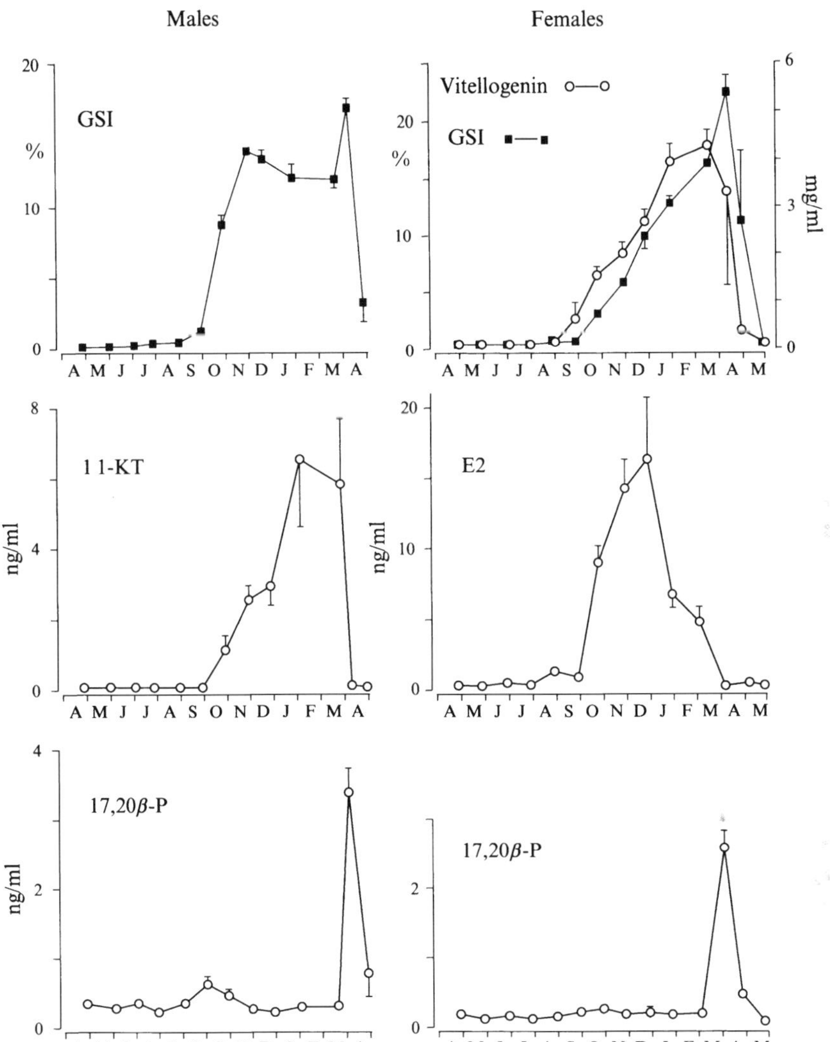

Fig. 9.3. Annual cycle of gonadosomatic index (GSI; gonad weight expressed as a percentage of body weight), and plasma concentrations of 17β-estradiol (E2), vitellogenin, 11-ketotestosterone (11-KT), and 4-pregnen-17α,20β-diol-3-one (17,20β-P) in captive Pacific herring (*Clupea pallassi*) undergoing their first reproductive cycle. (Adapted from Koya *et al.*, 2002, 2003 with kind permission from the authors and Blackwell Publishing.)

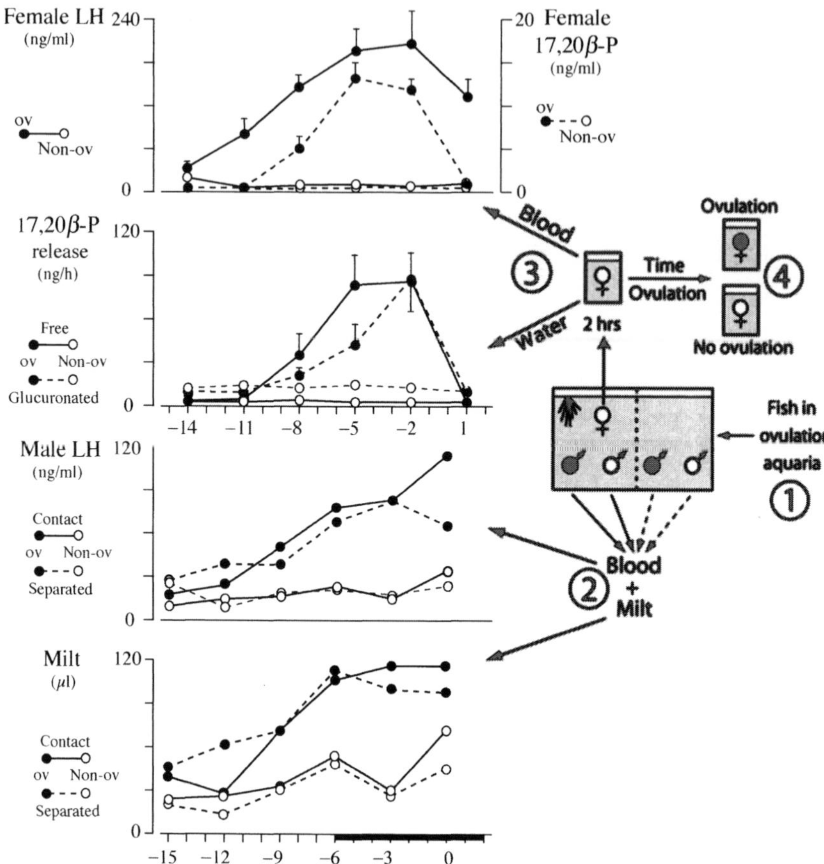

Fig. 9.4. Changes in serum luteinising hormone (LH) and milt volume in male goldfish held either with ovulatory females or separated from them by a perforated barrier. After being held overnight in conditions favourable to spontaneous ovulation (1), males were sampled for luteinising hormone and milt (2), and females were held for 2 hours to obtain holding water for 17,20β-P determination, bled for serum luteinising hormone and 17,20β-P determination (3), and checked frequently to determine the precise time of ovulation (4). Approximately half of the females ovulated. (Adapted from Stacey *et al.*, 1989 and Stacey and Sorensen, 1991 with kind permission from Elsevier Ltd.)

act through different olfactory receptor mechanisms to induce distinct biological responses (Scott and Sorensen, 1994; Sorensen and Scott, 1994; Section 5.3.1.b). In contrast, prostaglandins in female goldfish are primarily cleared in urinary pulses that coincide with entering spawning substrate (Appelt and Sorensen, 1999). It is unfortunate that so little is known about hormonal pheromone release, which obviously is a key aspect of pheromone function (e.g., Parks and Leblanc, 1998).

5.2. Hormonal Regulation of Reproductive Behaviours

As in other vertebrates, fish use reproductive tract hormones to synchronise reproductive behaviour with appropriate stages of gonadal development (Liley and Stacey, 1983; Borg, 1994). However, fish differ from many other vertebrates in that the hormones that induce behaviour-gonadal synchrony within the individual are frequently released to act as pheromones synchronising reproductive events between and among conspecifics (Kobayashi *et al.*, 2002; Stacey and Sorensen, 2002).

5.2.1. Hormonal Control of Male Behaviour

In male fish, male-typical behaviours are typically reduced or abolished by castration and restored in castrates (Borg, 1994), or induced in both juveniles (Cardwell *et al.*, 1995) and females (Stacey and Kobayashi, 1996; Kobayashi and Nakanishi, 1999) by exogenous androgen treatments. Studies that compare the behavioural potencies of testosterone and 11-ketotestosterone consistently find 11-ketotestosterone to be the more potent, such as in the three-spine stickleback, *Gasterosteus aculeatus* (Borg and Mayer, 1995) and goldfish (Stacey and Kobayashi, 1996). Although the temporal correlation between increased 11-ketotestosterone and male reproductive behaviours, and the low levels of 11-ketotestosterone levels in females, suggest this steroid and its metabolites should be good candidates for male pheromones, olfactory responses to 11-ketotestosterone have been observed only in the cyprinid *Puntius schwanenfeldi* (Cardwell *et al.*, 1995). However, testosterone has been proposed to function as a pheromone in Atlantic salmon (Moore, 1991; Section 5.3.2.b.i), whereas brook charr detect testosterone glucuronide (Sorensen and Essington, 1996), *H. burtoni* detect testosterone sulfate (Cole and Stacey, 2004), and male goldfish release and detect androstenedione (Sorensen *et al.*, 2005).

Plasma androgens increase not only during gonadal recrudescence but also when mature male fish are sexually active or involved with territorial activities (Oliveira *et al.*, 2002), raising the possibility of rapid modulations of androgen-based hormonal pheromones. Further, in species with paternal behaviour such as the stickleback, plasma androgens can decline during the transition from courtship to nesting (Oliveira *et al.*, 2002; Pall *et al.*, 2002a; Figure 9.5A). These androgen declines have been suggested to mediate the tradeoff between territorial aggression and paternal care (Oliveira *et al.*, 2002), although, in the stickleback, the lack of effect of androgen implants on spawning-induced decrease in courtship and increase in paternal behaviour

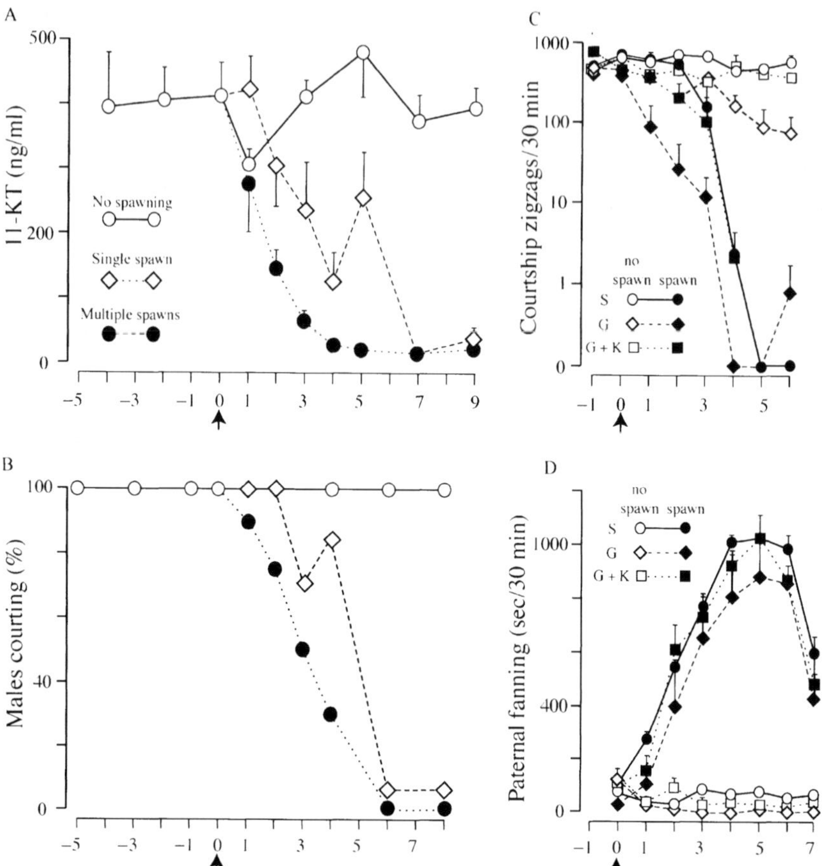

Fig. 9.5. Changes in male behaviour and 11-ketotestosterone (11-KT) with the transition from courtship to nesting in the male three-spine stickleback. (A) Serum 11-KT concentrations decrease dramatically and in a dose-dependent manner, in response to spawning on day 0. (B) Spawning on day 0 also decreases male courtship of additional females on following days. (C) Courtship behaviour remains high in nonspawned males following sham gonadectomy (S) or gonadectomy plus 11-ketoandrostenedione treatment (G + K), and declines gradually in gonadectomized males (G). In contrast, spawned males rapidly cease courting, regardless of surgical or steroid treatment. (D) Surgical and steroid treatments do not affect the onset of parental fanning in spawned males. (Adapted from Pall *et al.*, 2002a,b with kind permission from the authors and Elsevier Ltd.)

argue that androgens may be the result rather than the cause of the spawning to parental transition (Pall *et al.*, 2002b; Figure 9.5B).

In many fish, relationships among male hormones, behaviour, and pheromones are further complicated by their frequent use of alternate reproductive tactics (Taborsky, 2001; see also Chapter 8). For example, in diandric species (species exhibiting more than one mature male morph), divergent phenotypic adaptations include not only different morphology and behaviour but also different concentrations of plasma androgens, 11-ketotestosterone typically being higher in *bourgeois* males (males that monopolise females—sometimes known as *parental* males) than in *parasitic* males (typically smaller males that exploit investments of bourgeois males) (Borg, 1994; Taborsky, 2001; Uglem *et al.*, 2002; see also Chapter 8). These differences raise the possibility that male morphs employ different steroidal pheromones, a situation which appears to be the case in the black goby Section 5.3.3).

5.2.2. Hormonal Control of Female Behaviour

Although studied in only a few species of fishes, hormonal control of reproductive behaviour in females appears to be determined by their mode of reproduction. In the ovoviviparous guppy (*Poecilia reticulata*), the only internally fertilising fish studied, cyclic ovarian estrogen synthesis appears to synchronise cyclic sexual receptivity with sex pheromone production. Receptivity and release of an unidentified pheromone, which normally coincides with peak ovarian steroidogenesis for several days after parturition, are restored in ovariectomised or hypophysectomised females by exogenous estrogen (Liley, 1972; Meyer and Liley, 1982).

In contrast to the guppy, prostaglandin $F_{2\alpha}$ rather than estradiol induces female-spawning behaviour (behaviours associated with oviposition) in the few externally fertilising species that have been examined (Liley and Stacey, 1983; Sorensen and Goetz, 1993; Kobayashi *et al.*, 2002). Injection of prostaglandin $F_{2\alpha}$ rapidly induces female-spawning behaviour in several cypriniform (goldfish: Stacey, 1976; *Puntius* spp.: Liley and Tan, 1985; Cardwell *et al.*, 1995) and perciform species (*Cichlasoma bimaculatum*: Cole and Stacey, 1984; paradise fish, *Macropodus opercularis*: Villars *et al.*, 1985).

The effects of prostaglandin on female behaviour are best understood in the goldfish (Kobayashi *et al.*, 2002), in which studies were undertaken following the observation (Yamazaki, 1965) that females cease spawning if their ovulated eggs are stripped out. Subsequent experiments (Stacey and Liley, 1974) showed that spawning is extended by preventing egg release, and restored in stripped fish by injecting eggs into the ovarian cavity. The behavioural effect of eggs appears to be mediated by prostaglandin $F_{2\alpha}$ (Stacey, 1976), which increases in the plasma during the postovulatory period and in response to egg injection (Sorensen *et al.*, 1988, 1995b) and acts within the brain (Stacey and Peter, 1979). The effect of prostaglandin $F_{2\alpha}$ appears to be independent of sex steroids: prostaglandin $F_{2\alpha}$-induced female behaviour is equivalent in intact males and females (Stacey, 1981;

Stacey and Kyle, 1983) and in females is unaffected by ovariectomy or steroid treatment (Kobayashi and Stacey, 1993). Sorensen *et al.* (1988) first hypothesised that postovulatory production of prostaglandin $F_{2\alpha}$ has a pheromonal function in goldfish (Section 5.3.1.c) based on the observations that male sexual behaviour is dependent on olfaction (Stacey and Kyle, 1983), and that males appear to find prostaglandin $F_{2\alpha}$-injected, nonovulated females as attractive as ovulated females (Stacey, 1981).

Female reproductive hormones undoubtedly have been co-opted as sex pheromones because their synthesis and release is temporally correlated with key reproductive events. This is likely to account for the evidently widespread use of estradiol, MISs, and related steroids as hormonal pheromones (Cardwell and Stacey, 1995; Stacey and Cardwell, 1995; Murphy *et al.*, 2001; Stacey and Sorensen, 2002; Cole and Stacey, 2003; Sorensen *et al.*, 2004), given that production of these steroids is reliably associated with vitellogenesis and ovulation (Figure 9.3). Although the synthesis and behavioural functions of postovulatory prostaglandins have not been well studied, the fact that the olfactory systems of many dozens of cypriniform (Stacey *et al.*, 1995; Stacey and Cardwell, 1995), characiform (Cardwell and Stacey, 1995), and salmoniform fishes (e.g., Moore *et al.*, 2002; Laberge and Hara, 2003a) detect these compounds, suggests that many fish also use postovulatory prostaglandins as pheromones.

5.3. Current Models for Hormonal Pheromone Studies

Most recent hormonal pheromone research has involved three groups of fish: the goldfish and related carps within the subtribe Cyprini, the Salmonidae, and various perciform species. We refer the reader to earlier reviews of hormonal pheromones (Van Den Hurk and Resink, 1992; Sorensen and Stacey, 1999; Stacey and Sorensen, 2002) for information on species (e.g., zebrafish, African catfish, oriental weatherfish) that are not currently studied.

5.3.1. Goldfish and Related Carps

Goldfish, crucian, and common carp are temperate species that have little external sexual dimorphism, and live in mixed-sex, seemingly unstructured groups with no apparent territoriality, dominance hierarchies, or parental care. Gonadal recrudescence begins in winter and spawning occurs sporadically in spring and summer. At ovulation, which occurs near dawn, males compete for spawning access as females repeatedly enter aquatic vegetation to oviposit adhesive, undefended eggs over a period of several hours. We believe that the evolution of these species' hormonal pheromones has been strongly influenced by the fact they typically spawn in turbid water and employ a scramble competition mating system (Taborsky,

2001). Although much more is known about pheromone identity and function in goldfish than in related carps, all appear to employ very similar pheromone systems.

In goldfish, and likely in crucian and common carp, the timing of spawning is controlled by the female, in which increasing plasma testosterone in late vitellogenesis primes responsiveness to exogenous factors that induce ovulation (Kobayashi *et al.*, 1989) and thus initiates a cascade of events that synchronise behavioural and gonadal changes within and between the sexes. Postvitellogenic females exposed to warm temperatures, aquatic vegetation (Stacey *et al.* 1979a,b, 1989), or waterborne (pheromonal) 17,20β-P (Kobayashi *et al.*, 2002) exhibit a dramatic luteinising hormone surge that commences in midphotophase, induces synthesis of 17,20β-P (the goldfish MIS) and subsequent final oocyte maturation, and terminates at ovulation in the following night (Figure 9.4), at which point female spawning activity is triggered by prostaglandin $F_{2\alpha}$ synthesised in response to ovulated eggs in the oviduct (Sorensen *et al.*, 1988, 1995b). During the approximately 15-hour period between onset of the luteinising hormone surge and completion of spawning, females sequentially release a preovulatory steroid hormonal pheromone (Section 5.3.1.b) and a postovulatory prostaglandin-based hormonal pheromone (Section 5.3.1.c), both of which induce distinct and dramatic primer and releaser effects on males (Figure 9.6).

a. Female Recrudescence Pheromone. Female goldfish undergoing vitellogenesis release a "recrudescent" pheromone that attracts males (Kobayashi *et al.*, 2002; Figure 9.6). This pheromone (previously termed the *maturation* pheromone; Stacey and Sorensen, 2002) was first reported in urine of E2-treated female goldfish (Yamazaki and Watanabe, 1979; Yamazaki, 1990) and confirmed by our recent demonstration that contact and inspection behaviours among males is increased by the odour of estradiol-treated, ovariectomised females (Kobayashi *et al.*, 2002). The cue is unidentified and not likely to be estradiol or related 18-carbon steroids, which EOG studies (Sorensen *et al.*, 1987) show goldfish do not detect.

b. Female Preovulatory Steroid Pheromone. The preovulatory pheromone is a dynamic mixture whose multiple effects on males change with the shifting ratios of its three principal components (17,20β-P, its 20β-sulfated metabolite [17,20β-P-s], and androstenedione), each of which acts through separate olfactory receptor mechanisms (see Kobayashi *et al.*, 2002; Stacey and Sorensen, 2002). The preovulatory pheromone also contains less potent components (e.g., 20β-S and 17,20β-P-g) that amplify the effects of the mixture. During the luteinising hormone surge, the composition of the preovulatory pheromone changes systematically (Stacey *et al.*, 1989; Scott

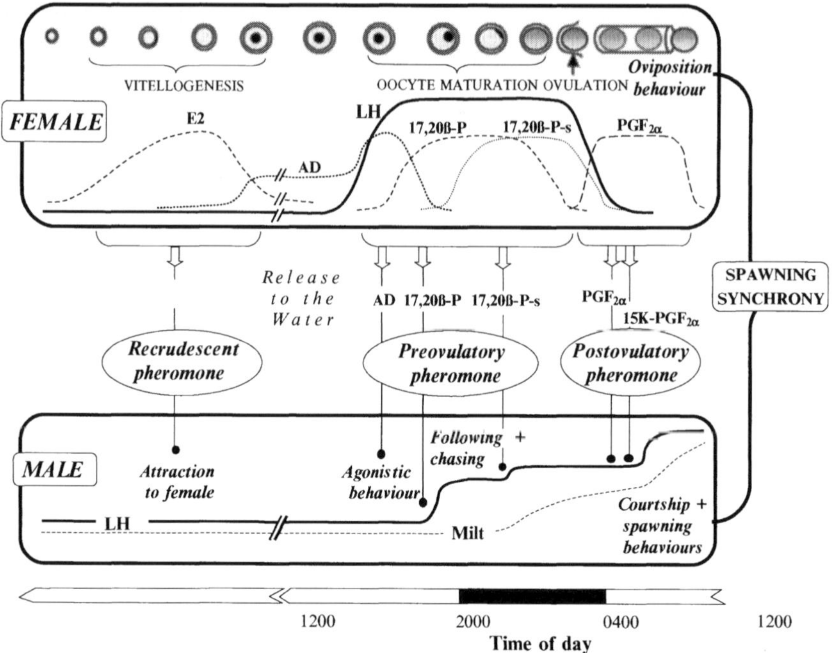

Fig. 9.6. Schematic model of female goldfish pheromones and their primer and releaser effects on males. 17β-Estradiol (E2) in vitellogenic females stimulates urinary release of an unidentified recrudescent pheromone that attracts males. In postvitellogenic females, exogenous cues induce a luteinising hormone surge stimulating release of a dynamic preovulatory pheromone containing androstenedione (AD), the maturation inducing steroid 17,20β-P, and its sulfated metabolite, 17,20β-P-s. Early in the luteinising hormone (LH) surge, androstenedione induces agonistic behaviours among males. As the 17,20β-P : androstenedione ratio increases, males increase luteinising hormone and begin to follow and chase conspecifics. Late in the luteinising hormone surge, 17,20β-P-s dominates the preovulatory pheromone mixture, enhancing its behavioural and endocrine effectiveness. Males exposed to the preovulatory pheromone increase both the quantity and quality of sperm in the sperm ducts prior to ovulation. At ovulation, eggs in the oviduct induce synthesis of prostaglandin $F_{2\alpha}$ ($PGF_{2\alpha}$), which acts in the brain to stimulate female sex behaviour and is released with its major metabolite (15K-$PGF_{2\alpha}$) as a postovulatory pheromone stimulating both male courtship and spawning behaviours and additional luteinising hormone increase. (Adapted from Kobayashi *et al.*, 2002 and Sorensen and Stacey, 2004 with kind permission of the Royal Society of New Zealand and Elsevier Ltd.)

and Sorensen, 1994; Sorensen and Scott, 1994). Soon after onset of the luteinising hormone surge, there is a rapid but temporary increase in androstenedione release, followed first by a dramatic increase in release of 21-carbon steroids (including 17,20β-P) during most of the surge, and then an equally dramatic decline in steroid release at ovulation, when spawning commences. The unconjugated steroids (androstenedione and 17,20β-P) are released across the gills, whereas sulfated (17,20β-P-s) and glucuronidated forms (17,20β-P-g) are released in urine (Sorensen *et al.*, 2000) and likely bile, respectively (Vermeirssen and Scott, 1996; Section 5.1.3).

The preovulatory pheromone 17,20β-P increases a male's reproductive performance by increasing both his behavioural competitiveness and the quantity and quality of his releasable sperm (Kobayashi *et al.*, 2002). Moreover, the pheromone's effects evidently change during the luteinising hormone surge in response to changes in release of its components. For example, androstenedione released early in the surge increases aggressive interactions between males while suppressing luteinising hormone release (Stacey, 1991; Poling *et al.*, 2001). In contrast, 17,20β-P released during midsurge rapidly increases plasma luteinising hormone (Dulka *et al.*, 1987; Zheng and Stacey, 1996, 1997) and induces subtle increases in locomotion and inspection behaviour among males that persist until spawing (DeFraipont and Sorensen, 1993; Poling *et al.*, 2001), whereas 17,20β-P-s released in urine pulses late in the surge also increases male luteinising hormone but stimulates intense, transient bouts of chasing and following (Sorensen *et al.*, 1995a; Poling *et al.*, 2001). At normal spawning temperatures, pheromone-induced luteinising hormone increase in males increases milt (sperm and seminal fluid) volume within 5 hours (Zheng and Stacey, 1996), likely by increasing synthesis of testicular 17,20β-P (Dulka *et al.*, 1987), which is known to promote testicular maturation in fish (Miura and Miura, 2003).

The culmination of exposure to the preovulatory pheromone is that males achieve dramatically greater reproductive success in competition with other males (Zheng *et al.*, 1997). Although pheromone-enhanced competitive behaviours (Zheng *et al.*, 1997) likely contribute to this effect, it also appears due to enhanced sperm function, because pheromonal 17,20β-P increases sperm motility and enhances paternity in *in vitro* fertilisation (DeFraipont and Sorensen, 1993; Zheng *et al.*, 1997).

c. Female Postovulatory Prostaglandin Pheromone. At ovulation, plasma luteinising hormone dramatically decreases, release of the preovulatory steroid pheromone dramatically decreases (Stacey *et al.*, 1989; Scott and Sorensen, 1994), and movement of ovulated eggs into the oviduct induces synthesis of prostaglandin $F_{2\alpha}$ that triggers spawning behaviour (Stacey, 1976; Sorensen *et al.*, 1995b). Coincidentally, females begin to release the postovulatory pheromone (prostaglandin $F_{2\alpha}$ and its more potent metabolite, 15-keto-prostaglandin $F_{2\alpha}$ [15K-prostaglandin $F_{2\alpha}$]) that attracts the male to the female, triggers male courtship, and activates mechanisms (different from those mediating responses to the preovulatory pheromone) that further increase releasable milt stores (Sorensen *et al.*, 1988, 1989; Zheng and Stacey, 1996, 1997).

EOG recording shows that the male olfactory epithelium detects prostaglandin $F_{2\alpha}$ and 15K-prostaglandin $F_{2\alpha}$ with great sensitivity (nM and pM thresholds, respectively), and through separate receptor mechanisms (Sorensen *et al.*, 1988). Unlike the situation with sex steroids, the olfactory epithelium of male goldfish (Sorensen and Goetz, 1993) and common carp (Irvine and Sorensen, 1993) is more sensitive to prostaglandin Fs than is that of females, a sexual dimorphism mediated by hormonal androgen in at least some other cyprinids (Cardwell *et al.*, 1995; Stacey *et al.*, 2003; Section 5.4). Immunoassay and mass spectrometry show that ovulated goldfish (20–30 g) release large quantities of pheromonal prostaglandin F (>200 ng/h; Sorensen *et al.*, 1988, 1995b). Moreover, the pheromone is released exclusively in urine pulses that coincide with female spawning activity (Appelt and Sorensen, 1999), raising the distinct possibility that selection has specialised the control of pheromone release to enhance female attractiveness.

Exposing grouped male goldfish to prostaglandin $F_{2\alpha}$ or 15K-prostaglandin $F_{2\alpha}$ induces a suite of responses (increased sociosexual interaction, luteinising hormone concentration, and milt volume) that is superficially comparable to that induced by pheromonal 17,20β-P, but differs significantly in underlying mechanisms. For example, although prostaglandin Fs induce more intense behavioural response than 17,20β-P, the response to prostaglandin Fs persists only for about 15 minutes, whereas that to 17,20β-P persists for hours (Sorensen *et al.*, 1989; DeFraipont and Sorensen, 1993; Poling *et al.*, 2001). Furthermore, pheromonal prostaglandin increases luteinising hormone and milt only amongst grouped males (Sorensen *et al.*, 1989), whereas 17,20β-P also affects isolated fish (Fraser and Stacey, 2002). Because responses to pheromonal prostaglandin Fs thus appear to depend on, and may even be driven by, the social interactions they induce, they typically have been studied indirectly by observing responses of males spawning with nonovulated, prostaglandin $F_{2\alpha}$-injected females that are known to release prostaglandin $F_{2\alpha}$ and 15K-prostaglandin $F_{2\alpha}$ (Sorensen *et al.*, 1995b). These studies reveal that although 17,20β-P increases milt volume only by increasing luteinising hormone, pheromonal prostaglandins also increase milt through an extrapituitary pathway and with a shorter latency (<1 hour) than the mechanism activated by 17,20β-P (Zheng and Stacey, 1996). Pheromonal prostaglandin Fs and 17,20β-P also differ in the neuroendocrine mechanisms by which they induce luteinising hormone release because dopamine receptor agonists block 17,20β-P-induced, but not pheromonal prostaglandin-induced, luteinising hormone and milt responses (Zheng and Stacey, 1997). Although prostaglandin $F_{2\alpha}$ and 15K-prostaglandin $F_{2\alpha}$ are more potent when tested as a mixture, whether these compounds act synergistically and/or have different pheromonal functions in the goldfish has yet to be determined.

In summary, male goldfish respond to the changing odour of periovulatory females with a suite of behavioural and physiological responses that enhance reproductive success. Although elements of the goldfish pheromone system may have been modified through domestication, it appears remarkably similar to those of the closely related (and nondomesticated) crucian and common carp (Bjerselius and Olsén, 1993; Irvine and Sorensen, 1993; Stacey *et al.*, 1994; Bjerselius *et al.*, 1995). Recent studies in *Barilius bendelisis*, a more distantly related cyprinid, suggest that preovulatory 17,20β-P-s release also rapidly increases the quantity and quality of releasable sperm (Bhatt and Sajwan, 2001). The roach also releases and detects both unconjugated and glucuronidated 17,20β-P (Lower *et al.*, 2004), although the functional significance is unknown. Given that olfactory detection of MISs and related 21-carbon steroids is widespread among the cyprinoid cyprinids (Stacey and Sorensen, 2002), this diverse group could provide valuable insight into the evolution of MIS-based hormonal pheromones.

d. Intrasexual Pheromones. In addition to responding to preovulatory and postovulatory female pheromones, male goldfish also respond strongly to the presence and reproductive condition of male conspecifics (Stacey *et al.*, 2001; Fraser and Stacey, 2002; Sorensen *et al.*, 2005). For example, if one male in a group is stimulated to increase sperm stores, by either gonadotropin injection or exposure to 17,20β-P, untreated males in the group also increase their stores (Stacey *et al.*, 2001; Fraser and Stacey, 2002). These results indicate that male goldfish synchronise their milt production with impending ovulation by monitoring female condition both directly (by detecting the preovulatory steroid pheromone) and indirectly (by detecting cues the female has stimulated other males to release). Further, males isolated from a group of unstimulated males dramatically increase their milt volume within 24 hours, indicating they normally suppress milt production in response to an inhibitory male cue(s) (Fraser and Stacey, 2002). These milt responses to males appear functionally distinct from those to females. For example, although both the female's preovulatory and postovulatory hormonal pheromones increase male luteinising hormone and rapidly increase milt (within 5 hours and 30 minutes, respectively; Zheng and Stacey, 1996), the milt increases in response to stimulated males and isolation have much greater latencies (12 and 24 hours, respectively) and are not preceded by increased luteinising hormone (Stacey *et al.*, 2001; Fraser and Stacey, 2002).

Recent analyses of male holding water (Sorensen *et al.*, 2005) indicate that steroidal mixtures dominated by androstenedione are responsible both for inhibition of milt production and behaviour in unstimulated male groups

(Stacey, 1991; Fraser and Stacey, 2002; Poling *et al.*, 2001). Sexually active males release more androstenedione than sexually maturing males or even females (Sorensen *et al.*, 2005). Moreover, in dramatic contrast to the female preovulatory pheromone (which has a C19-to-C21 ratio of 1:7), males release androstenedione in the virtual absence of C21 steroids (C19-to-C21 ratio of 10:1), suggesting that, if a pheromone from stimulated males induces the milt response of grouped males (Stacey *et al.*, 2001), it is unlikely to be comprised of conventional 21-carbon steroids (Sorensen *et al.*, 2005). Waterborne androstenedione also stimulates male pushing, evidently a component of male–male competition (Poling *et al.*, 2001); this male cue, which is likely perceived as a mixture rather than a specific compound (androstenedione), thus has both priming and releasing effects.

Although little is understood about female responses to hormonal pheromones, anosmic females undergo apparently normal vitellogenesis, ovulation, and spawning behaviour (Stacey and Kyle, 1983; Kobayashi *et al.*, 1994, 1997a), suggesting that reproductive functions of females are less dependent on conspecific odours than those of males. Nevertheless, the olfactory systems of females and males are sensitive to steroids, and waterborne 17,20β-P increases the occurrence of ovulation (Kobayashi *et al.*, 2002), an effect likely reflecting female response to the odour of preovulatory females (Kobayashi *et al.*, 2002), given that females release far more 17,20β-P than do males (Sorensen and Scott, 1994; Sorensen *et al.*, 2005) and that occurrence of ovulation appears unaffected by the presence of males (Stacey *et al.*, 1979b). Although the adaptive value of the ovulatory response to pheromonal 17,20β-P is unclear, it nonetheless is consistent with observations that females ovulate synchronously both in the field and laboratory (Kobayashi *et al.*, 1988).

e. Goldfish Pheromones As a Chemical Network. Although our understanding of goldfish pheromonal function derives from dyadic and small group studies in laboratory aquaria, it is important to consider that wild goldfish and their relatives normally live in shoals in relatively large bodies of water. Fish in these conditions are expected to routinely encounter each other's pheromone plumes (e.g., Sorensen *et al.*, 2000), effectively creating dynamic, pheromonally-mediated information networks that link the reproductive functions of the many individuals within local populations (Wisenden and Stacey, 2005). We speculate that these networks routinely result in populations of fish that share an endocrine state characterised by extended periods of hormonal stasis interrupted at spawning by synchronous hormonal fluctuations. Specifically, we propose that for the majority of the reproductive season (when stimulatory cues from preovulatory or ovulated females are absent) male goldfish function as both originators and receivers of inhibitory cues (e.g., androstenedione) that suppress sperm production (and perhaps egg production) by maintaining basal luteinising hormone (Stacey *et al.*, 2001; Fraser and Stacey, 2002). However, when exogenous stimuli trigger female preovulatory luteinising hormone surge(s) and subsequent pheromone release, this stable, negative-feedback situation is rapidly and transiently perturbed, as exposed males and nonovulatory females in turn increase their luteinising hormone, amplifying the original stimulus, and promoting synchronous final maturation (ovulation and increased sperm stores) among many individuals. Upon completion of spawning, the postovulatory pheromone stimulus for luteinising hormone increase is removed, and basal hormone levels are restored as males resume release of inhibitory cues.

Although knowledge of the precise interactions among individuals in such chemically mediated networks awaits further research, it already is clear that the very nature of hormonal pheromones forces us to broaden our perspective of classical endocrine function in two significant ways. The first is to recognise that actions of hormones need not be strictly endogenous, but can extend outside the animal via pheromonal pathways to affect conspecifics. The second is to recognise that situations where hormonal pheromones alter hormone synthesis (and thus hormone release) provide a mechanism for the evolution of pheromonally-mediated feedback directly linking conspecific reproductive endocrine systems. From this perspective, it is ironic that some hormonal pheromones might well be considered to be *ectohormones*, the term originally proposed for pheromones by Bethe (1932). Understanding hormonal pheromonal networks could be of practical significance because they could conceivably be manipulated in aquaculture situations and in the wild to control invasive species (Sorensen and Stacey, 2004).

5.3.2. Salmonid Fishes

The Salmonidae contains three subfamilies of northern hemisphere species (Coregoninae, whitefish; Salmoninae, salmon and trout; Thymallinae, grayling) that have anadromous and/or lacustrine or riverine lifestyles, and often exhibit diandry involving large bourgeois males and precociously mature males. Reproductive pheromones have not been studied in Thymallinae but have received considerable attention in the other salmonid subfamilies.

a. Subfamily Coregoninae. In lake whitefish, 15K-prostaglandin $F_{2\alpha}$ and 13,14-dihydro-prostaglandin $F_{2\alpha}$, induce large EOG and behavioural responses in males indicative of pheromonal function (Hara and Zhang, 1997; LaBerge and Hara, 2003a; Figure 9.7). It has been suggested that these responses to prostaglandins may be mediated by an extrabulbar pathway

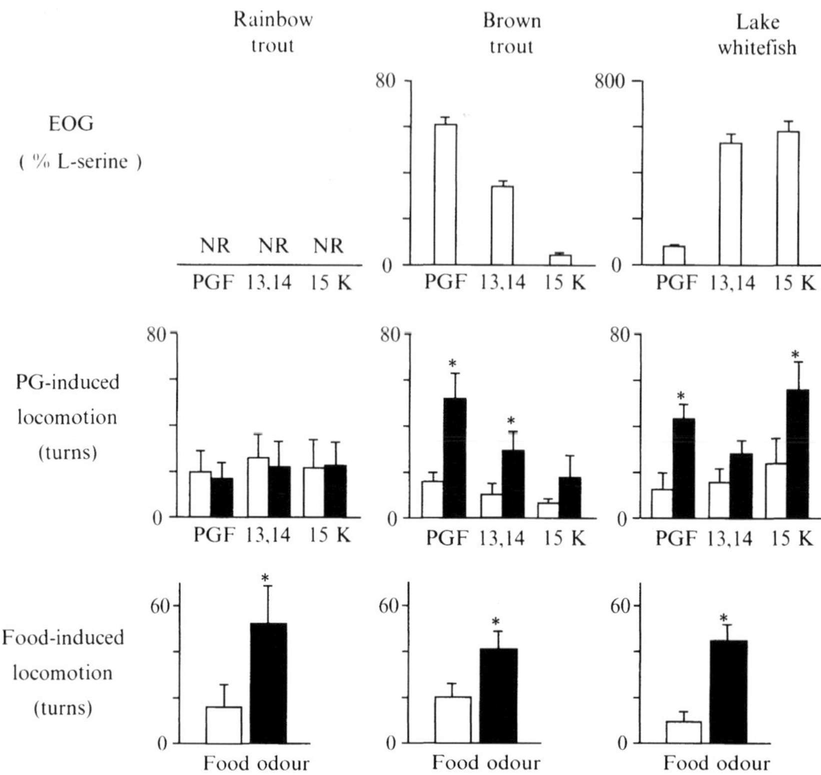

Fig. 9.7. Correlation between olfactory and behavioural responsiveness to prostaglandin (PG) in three salmonids. Top panels: At 10-nM concentration, rainbow trout exhibit no EOG response (NR) to prostaglandin $F_{2\alpha}$ (PGF), 13,14-dihydro-prostaglandin $F_{2\alpha}$ (13,14) or 15-keto-prostaglandin $F_{2\alpha}$ (15 K), whereas brown trout and lake whitefish respond with markedly different EOG magnitudes (mean ± SEM; as a percentage of 10-μM L-serine standard). Middle panels: With the exception of 13,14-dihydro-prostaglandin $F_{2\alpha}$ in lake whitefish (which induced large EOG but no behavioural response), swimming turns following exposure to detectable prostaglandin Fs (filled bars) were significantly greater than prior to prostaglandin F exposure (open bars). Bottom panels: In the same apparatus used to test prostaglandin F responses, all species exhibited more swimming turns following food odour exposure (filled bars) than prior to food odour exposure (open bars). (Adapted from Laberge and Hara, 2003a with kind permission of the authors and Blackwell Publishing.)

because F prostaglandins do not elicit measurable field potentials from the olfactory bulb, at least using standard integration (Section 3.2). Lake whitefish also detect the 5β-reduced androgen, etiocholanolone glucuronide (Etio-g), but its function is unknown (Hara and Zhang, 1997).

b. Subfamily Salmoninae. Hormonal pheromones have been studied in three salmonid genera, *Oncorhynchus*, *Salvelinus*, and particularly *Salmo*.

i. Genus Salmo. Consistent with their close taxonomic relationship and ability to hybridise (Youngson *et al.*, 1992), the Atlantic salmon and the brown trout appear to use very similar hormonal pheromone systems (Olsén *et al.*, 2000), the specific nature of which is unresolved (see also Stacey and Sorensen, 2002).

ATLANTIC SALMON. Studies using the convenient precociously mature male parr model indicate these fish use sex steroids (testosterone, 17,20β-s) and prostaglandin Fs as sex pheromones. Important aspects of this complex system unfortunately remain unclear, in part because some studies have not sought, or been able, to support each other. Nonetheless, clear evidence for priming effects of a prostaglandin $F_{2\alpha}$ pheromone comes from studies showing that waterborne prostaglandin $F_{2\alpha}$ increases plasma hormones in mature male parr (Moore and Waring, 1995, 1996; Olsén *et al.*, 2001), and that prostaglandin Fs are detected by the olfactory system with high sensitivity, perhaps in a seasonally dependent manner (Moore and Waring, 1996). Somewhat less clear are the origin(s) and function(s) of prostaglandin $F_{2\alpha}$. Thus, whereas Moore and Waring (1996) report that ovulated salmon urine contains significant immunoreactive prostaglandin $F_{2\alpha}$, to which they attribute priming activity, Olsén *et al.* (2001, 2002b) find much higher prostaglandin $F_{2\alpha}$ concentrations in ovarian fluid, to which they attribute the majority of the priming activity. Interestingly, Olsén *et al.* (2002b) also report that although ovulated salmon urine has only weak priming activity, it has strong behavioural activity, which they attribute to unknown compounds. Notably, definitive biochemical techniques have yet to be applied to the study of salmonid prostaglandin synthesis to determine the actual identities of compounds released by any route. In any case, together with the finding that a carbamate pesticide both reduces EOG responses to prostaglandin $F_{2\alpha}$ (Chapter 10) and blocks prostaglandin-induced endocrine responses (Waring and Moore, 1997), these studies all support the possibility that a prostaglandin F pheromone in urine and/or ovulatory fluid induces important priming effects in the precocious male Atlantic salmon.

Although it has also been suggested that sex steroids function as pheromones in Atlantic salmon, the evidence is from a number of isolated studies and is less clear. For example, EOG recordings (Moore and Scott, 1991) show that precocious males are extremely and specifically sensitive to testosterone, but only for a brief period prior to spawning, a situation not observed in any other species, even trout. Testosterone also attracts male parr (Moore, 1991), but does not affect milt volume or hormone concentrations (Waring *et al.*, 1996). EOG recordings also indicate that mature male parr do not normally detect 17,20β-P-s but become extremely sensitive to this steroid immediately following even brief (5-second) exposure to ovulated female urine (Moore and Scott, 1992), a rapid sensitisation of the olfactory epithelium for which there appears to be no precedent. Other studies have failed to identify a function for this compound or explain the etiology of this unique phenomenon (Waring and Moore, 1995; Waring *et al.*, 1996). It will be interesting to determine whether responses to steroids occur only in parasitic parr, or in bourgeois males as well.

Brown Trout. Recent studies using parasitic and bourgeois males reveal both similarities and differences between putative hormonal pheromones of brown trout and Atlantic salmon (Essington and Sorensen, 1996; Olsén *et al.*, 2000; Moore *et al.*, 2002; Laberge and Hara, 2003a). In particular, pheromonal activity in both the trout and salmon has been associated with F prostaglandins, whereas no function has yet been found for any sex steroid in trout, except perhaps for Etio-g for which only EOG activity has been described (Hara and Zhang, 1997; Laberge and Hara, 2003a). Indeed, even when brown trout are treated with androgenic hormones, their olfactory epithelium is insensitive to sex steroids (Essington and Sorensen, 1996).

Bodily fluids from ovulated female brown trout do, however, induce pheromonal responses in male trout that appear similar to those in Atlantic salmon and can be at least partially attributed to F prostaglandins. Thus, both precocious male brown trout and Atlantic salmon increase circulating hormone levels when exposed to a mixture of urine and ovulatory fluids from conspecific and heterospecific females (Olsén *et al.*, 2000). Detailed studies of male brown trout suggest that most of the priming activity is attributable to ovulatory fluid which, also like salmon, contains much higher levels of immunoreactive prostaglandin F (>175 ng/ml) than urine (<5 ng/ml) (Moore *et al.*, 2002). Further, as in Atlantic salmon, high EOG sensitivity (1 pM threshold) to prostaglandin Fs provides strong evidence that brown trout employ prostaglandin Fs as sex pheromones (Essington and Sorensen, 1996; Hara and Zhang, 1997; Moore *et al.*, 2002; Laberge and Hara, 2003a; Figure 9.7). However, although four EOG studies agree on the relative olfactory potencies of prostaglandin Fs in male brown trout (e.g., prostaglandin $F_{2\alpha} \geq$ prostaglandin $F_{1\alpha} >$ 15K-prostaglandin $F_{2\alpha} >$ 13,14-dihydro-15K-prostaglandin $F_{2\alpha}$), and that a single receptor mechanism may be involved in both trout and salmon (Laberge and Hara, 2003a; Moore and Waring, 1996), they disagree on whether there might be a relationship between maturational state and EOG response. In particular, although Moore *et al.* (2002) find that only precociously mature parr detect prostaglandin Fs (immature parr had no EOG response even at 10 μM), Laberge and Hara (2003a) report that prostaglandin Fs induce similar EOG responses in undifferentiated juveniles, adult females, and bourgeois males (Section 5.4.), findings consistent with those of Essington and Sorensen

(1996) in brown trout and of Sveinsson and Hara (2000) in Arctic charr (see Section 5.3.2.c.ii). In any case, a preponderance of evidence shows that prostaglandin Fs have pheromonal activity in brown trout. Thus, Moore *et al.* (2002) report that both prostaglandin $F_{1\alpha}$ and prostaglandin $F_{2\alpha}$ increase male parr hormone and milt levels, similar to effects they describe in the Atlantic salmon. However, Laberge and Hara (2003a) report that prostaglandin Fs also have behavioural activity in trout; for example, prostaglandin Fs increased swimming and turning activities in mature bourgeois male trout (Figure 9.7) and induced digging and nest probing in females. Although it will be important to identify the specific prostaglandin Fs released by brown trout, and to determine their precise origins and activities in all life history stages/genders, it is clear that, as in salmon, this class of compounds exerts important pheromonal effects that warrant further study.

ii. Genus *Salvelinus* (Charrs). Evidence for prostaglandin F pheromones in charrs comes from EOG studies showing that prostaglandin $F_{2\alpha}$ is detected by Arctic, lake (*S. namaycush*), and brook charr (*S. fontinalis*) (Essington and Sorensen, 1996; Hara and Zhang, 1997; Sveinsson and Hara, 2000). With the exception of Etio-g detected by all tested salmonids (Essington and Sorensen, 1996; Hara and Zhang, 1997; Laberge and Hara, 2003a) and testosterone glucuronide detected in brook charr (Essington and Sorensen, 1996), charrs have not been found to detect any sex steroids or their metabolites. EOG responsiveness to prostaglandin $F_{2\alpha}$ appears unaffected by gender or maturity in bourgeois Arctic (Sveinsson and Hara, 2000) and brook charr (Essington and Sorensen, 1996), consistent with the study of brown trout by Laberge and Hara (2003a), but in marked contrast to other studies of brown trout (Moore *et al.*, 2002) and precocial male Atlantic salmon (e.g., Moore and Waring, 1995).

Ovulated Arctic charr are attracted to prostaglandin $F_{2\alpha}$ and bourgeois male Arctic charr are reported to release immunoreactive prostaglandin Fs (Sveinsson and Hara, 2000), a scenario seemingly different than that proposed for *Salmo,* whose males are suggested to respond to prostaglandin Fs released by females (e.g., Olsén *et al.*, 2002b; Moore *et al.*, 2002). Such differences are surprising, given that brook charr and brown trout hybridise (Sorensen *et al.*, 1995c).

Bile acids released by stream-resident fish have been suggested to serve as pheromonal attractants for maturing adult charr that frequently have an anadromous life history. A variety of studies (Døving *et al.*, 1980; Zhang *et al.*, 2001) show that Arctic charr and lake charr release and detect various bile acids, which also induce behavioural responses (Selset and Døving, 1980; Jones and Hara, 1985). However, it remains to be determined which specific bile acids released by charrs exert these effects or what their precise behavioural significance might be.

iii. Genus *Oncorhynchus* (Pacific Salmon). Sex pheromones of *Oncorhynchus* appear to be rather different than those of *Salmo* and *Salvelinus*. Ovulatory fluid from rainbow trout has long been known to attract males (Emanuel and Dodson, 1979), which also exhibit rapid endocrine responses both to bile fluid (containing both bile and sex steroid conjugates; Vermeirssen and Scott, 2001) and urine from ovulated females (Scott *et al.*, 1994; Vermeirssen *et al.*, 1997). Most recently, Yambe and Yamazaki (2001) find that responses to urinary pheromone(s) are androgen-dependent. Urine-induced, androgen-dependent behaviours also occur in masu salmon (*O. masou*; Yambe *et al.*, 1999, 2003; Yambe and Yamazaki, 2000; Section 5.4.). *Oncorhynchus* species thus appear to release a variety of sex pheromones of undetermined identity by several routes.

Although Dittman and Quinn (1994) report that 17,20β-P repels chinook salmon (*O. tshawytscha*), and one EOG study (Pottinger and Moore, 1997) indicates that immature rainbow trout detect testosterone, all other studies of *Oncorhynchus* spp. show that (with the exception of Etio-g) rainbow trout, chinook, and amago salmon (*O. rhodurus*) do not detect a range of prostaglandins and steroids including testosterone and 17,20β-P (Kitamura *et al.*, 1994b; Hara and Zhang, 1997; Laberge and Hara, 2003a; Dittman and Sorensen, unpublished results). The urinary pheromone employed by masu salmon does not have chemical characteristics of known sex steroids or prostaglandins, although ovulated masu urine contains a modest concentration of immunoreactive prostaglandin $F_{2\alpha}$ (Yambe *et al.*, 1999). In summary, the present evidence indicates either that, as suggested by Laberge and Hara (2003a), *Oncorhynchus* have lost the use of prostaglandin Fs (Figure 9.7) or that, as suggested by Yambe *et al.* (1999), this group employs novel prostaglandins or other compounds.

5.3.3. Perciform Fishes

Order Perciformes (>9,000 species) contains 40% of extant teleost fishes (Nelson, 1994) and is the dominant vertebrate group in oceans and tropical freshwaters. However, despite their ecological and economic importance and the fact that hormonal pheromones were first described in a perciform (black goby; Colombo *et al.*, 1980), little is known about the reproductive pheromones of this taxon. Most information comes from studies of two gobies (family Gobiidae) that provide evidence for olfactory sensitivity to sex steroids and suggest a biological function for the putative steroidal pheromones. As well, EOG studies show that an African cichlid (*Haplochromis burtoni*; family Cichlidae) detects and discriminates between a number of conjugated sex steroids (Robison *et al.*, 1998; Cole and Stacey, 2003), whereas behavioural studies of the ruffe (family Percidae: Sorensen *et al.*, 2004) suggest it employs novel and unidentified steroids as male pheromones.

a. The Black and Round Gobies. Goby mating systems typically involve male territoriality, nest defense, and paternal behaviours, and can also involve alternative male tactics in which parasitic males fertilise eggs in the nest of bourgeois males (e.g., Rasotto and Mazzoldi, 2002; Immler *et al.*, 2004; Chapter 8). Male gobies possess a mesorchial gland, a specialised, nonspermatogenic portion of the testis containing a high concentration of interstitial (Leydig) cells producing primarily 5β-reduced androgen conjugates, in particular Etio-g and Etio-sulphate (see Locatello *et al.*, 2002). EOG and behavioural studies provide strong evidence that in the round and black gobies Etio-g and other steroids function as hormonal pheromones (Colombo *et al.*, 1980; Murphy *et al.*, 2001), possibly a result of female mate choice. Indeed, specialisation of pheromone production is expected in species where females respond to odours of nest-guarding males (e.g., Laumen *et al.*, 1974; Gonçalves *et al.*, 2002; Rasotto and Mazzoldi, 2002).

EOG recordings using a large number of commercially available sex steroids and prostaglandins show that male and female round gobies do not respond to prostaglandin Fs, but exhibit sexually isomorphic responses to more than a dozen free and conjugated 18-, 19-, and 21-carbon steroids (including Etio-g) that appear to be discriminated by four olfactory receptor mechanisms for which the most potent known odourants are estrone, 17β-estradiol-3β-glucuronide, etiocholanolone (Etio), and dehydroepiandrosterone-3-sulphate (Murphy *et al.*, 2001; Figure 9.8A and B). Unfortunately, it is not yet known if round gobies release any of these putative pheromonal steroids.

Although in laboratory tests none of the steroids detected by round gobies induced overt reproductive behaviours, males increased the frequency of ventilation (opercular and buccal pumping) when exposed to Etio, estrone, or 17β-estradiol-3β-glucuronide, and females exhibited a similar response to Etio (Murphy *et al.*, 2001; Figure 9.8C), a dimorphism that is androgen-dependent (Murphy and Stacey, 2002; Section 5.4). The ventilation increase induced by steroid odourants may facilitate odour detection by increasing water flow through the olfactory organ, as has been proposed in other benthic fish (Nevitt, 1991). Importantly, because the ventilation response adapts within 15 minutes, it provides a simple behavioural assay that demonstrates the steroids discriminated at the sensory level also are discriminated behaviourally (Murphy *et al.*, 2001; Figure 9.8D). Although Etio (or Etio-g) likely attracts female round gobies to male nests, as in the black goby (Colombo *et al.*, 1980), the functions of other steroids detected by this species are not known.

In the black goby, where males employ alternate reproductive tactics (Rasotto and Mazzoldi, 2002), bourgeois and parasitic males appear to release distinct odours. Thus, bourgeois male odour induces aggression in nest-holding males, whereas parasitic male odour does not (Locatello *et al.*,

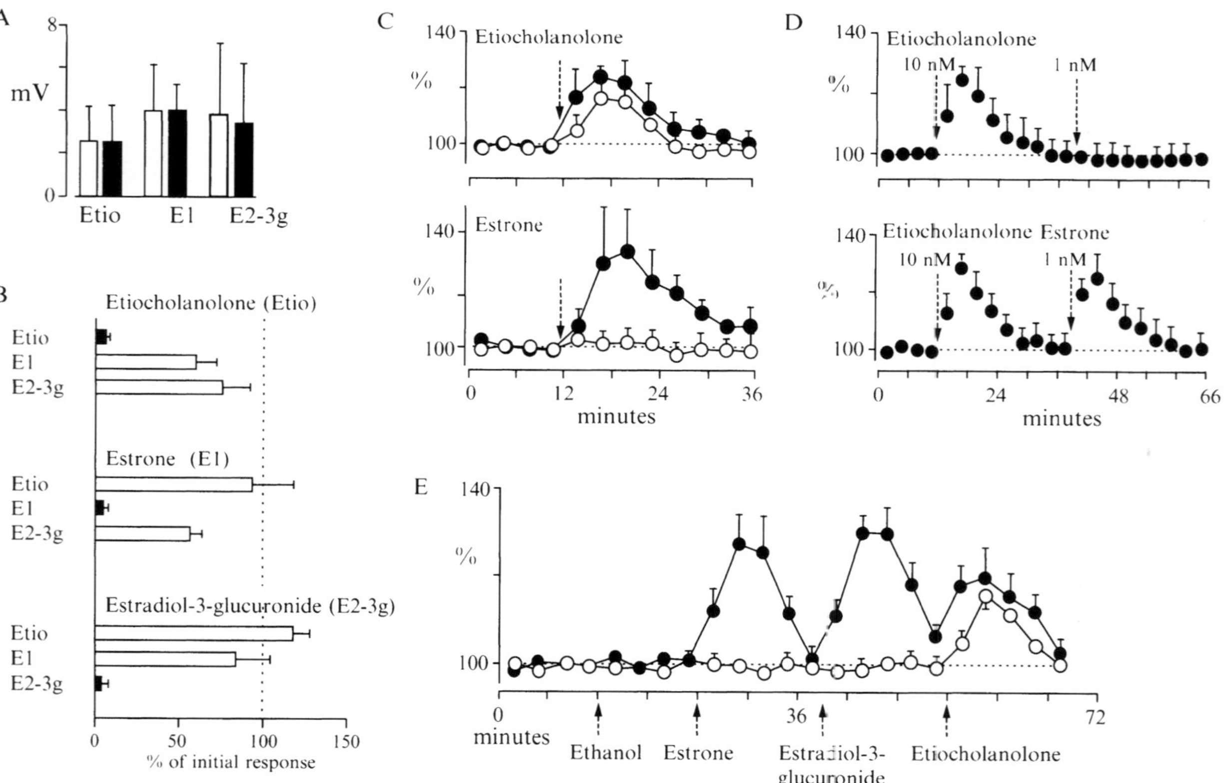

A
mV
8
0
Etio
E1
E2-3g
B
Etiocholanolone (Etio)
Etio
E1
E2-3g
Estrone (E1)
Etio
E1
E2-3g
Estradiol-3-glucuronide (E2-3g)
Etio
E1
E2-3g
0
50
100
150
% of initial response
C
140
%
100
Etiocholanolone
Estrone
0
12
24
36
minutes
D
140
%
100
Etiocholanolone
10 nM
1 nM
Etiocholanolone Estrone
10 nM
1 nM
0
24
48
66
minutes
E
140
%
100
0
36
72
minutes
Ethanol
Estrone
Estradiol-3-glucuronide
Etiocholanolone

2002; Figure 9.9). Although the basis for these differential effects is not known, the much larger mesorchial glands of bourgeois males might suggest differential release of Etio-g or other steroids (Rasotto and Mazzoldi, 2002; Figure 9.9). The fact that gobies are a speciose and widely available group that are well-suited to laboratory studies suggests they will become increasingly valuable model species for hormonal pheromone studies.

b. The Eurasian Ruffe. The ruffe (family Percidae) was accidentally introduced into the Great Lakes with detrimental effects on native fauna. Ruffe alarm and female reproductive pheromones are the best understood of any percid because they have been examined with the objective of population control (Maniak *et al.*, 2000; Sorensen *et al.*, 2004). Typical of percids, ruffe are nonterritorial, spawn only once in the spring (at night and often in turbid waters), do not employ specific spawning substrate, are sexually monomorphic and, as with the goldfish, appear to employ a promiscuous mating system in which males engage in scramble competition.

Bioassays using female ruffe holding water show that, during final oocyte maturation, females release a behaviourally active pheromone, but cease doing so at ovulation (Sorensen *et al.*, 2004). Pheromone release is closely correlated with circulating levels of 20β-S (17,20β,21-P), the proposed MIS for this species and other perciforms (Thomas, 1994). Further, the pheromone appears to be derived from 20β-S because injecting this steroid into females (but not males) rapidly induces pheromone release via the urine whereas 17,20βP injection has no such effect (Sorensen *et al.*, 2004). Because EOG studies indicate ruffe do not detect any commercially available precursors or metabolites of either 17,20βP or 20β-S (or other steroids and

---

Fig. 9.8. EOG and behavioural responses (mean $\pm$ SEM) of round gobies to waterborne steroids. (A) Male (filled bars) and female (open bars) exhibit equivalent EOG responses to etiocholanolone (Etio), estrone (E1), and 17β-estradiol-3-glucuronide (E2–3g). (B) EOG cross-adaptation studies indicate Etio, E1, and E2-3g are detected by separate olfactory receptor mechanisms. Top: During adaptation to 100 nM Etio, EOG response to a 10 nM Etio pulse is drastically reduced (filled bar), whereas responses to E1 and E2–3g (open bars) are unaffected. Middle and bottom: Similar results during adapation to E1 and E2–3g. (C) Both males (filled circles) and females (empty circles) exhibit a transient increase in ventilation frequency (presented as a percentage of mean pre-exposure frequency) in response to 10 nM waterborne Etio, whereas only males respond to 10 nM E1. (D) Males that are behaviourally adapted to 10 nM Etio do not increase ventilation in response to a 10% increase in Etio concentration, but do increase ventilation in response to an equivalent amount of E1. (E) Females implanted with methyltestosterone capsules for 18 days (filled circles) do not increase ventilation in response to ethanol vehicle (EtOH) but do increase ventilation in response to 1 nM E1, E2-3g and Etio; control females implanted with empty capsules (open circles) respond only to Etio. (Adapted from Murphy *et al.*, 2001; Murphy and Stacey, 2002; Stacey *et al.*, 2003 with kind permission of Springer-Verlag GmbH, NRC Research Press, and Elsevier Ltd.)

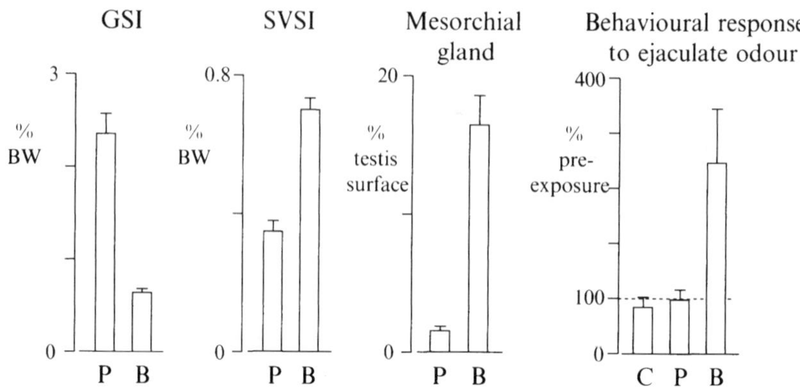

Fig. 9.9. In bourgeois (territorial and paternal) male black gobies (B), gonadosomatic indices (GSI; gonad weight as a percentage of body weight [%BW]) are smaller than in parasitic males (P), but seminal vesicle somatic indices (SVSI; seminal vesicle weight as a percentage of BW) are larger, and the mesorchial gland covers a greater proportion of the gonadal surface. Bourgeois males exhibit defensive (tail beating) behaviour (presented as a percentage of pre-exposure level; dashed line) in response to the odour of bourgeois, but not of parasitic males. C, blank water control. (Adapted from Locatello *et al.*, 2002; Rasotto and Mazzoldi, 2002 with permission of the authors, Blackwell Publishers, and John Wiley and Sons.)

prostaglandin Fs), it seems reasonable to hypothesise that the ruffe pheromone is an unknown metabolite of 20β-S that stimulates prespawning aggregation (Sorensen *et al.*, 2004).

If so, an important lesson from the ruffe is that the steroid, prostaglandin, and bile acid pheromones of many fish are likely novel metabolites that cannot be elucidated by EOG and immunoassay using known, commercially available chemicals as test stimuli (see Section 5 introduction). Another potential example of unknown steroid metabolites functioning as pheromones is the Mozambique tilapia (*Oreochromis mossambicus*), where EOG responses to fractionated body odours suggest steroid sulfates have pheromonal function, but no tested steroid has olfactory activity (Frade *et al.*, 2002). In these, as in many other aspects of fish pheromone research, our poor understanding of hormone and bile acid metabolism is clearly a major impediment to progress.

5.4. Hormonal Modulation of Pheromone Responses

It is logical to expect that physiological mechanisms have evolved to synchronise pheromonal responsiveness with those life-history stages in which reproductive responses to conspecifics are adaptive. In the case of bile acid-derived pheromones of sea lamprey, for example, only maturing adults respond to the migratory pheromone and, at the time of ovulation, females respond only to the male sex pheromone (Bjerselius *et al.*, 2000; Section 4). Although the physiological basis for these changes in sea lamprey is unknown, in some teleost fishes it is clear that the reproductive endocrine system serves both as the source of pheromones and as the regulator of responses to them. The physiological basis of endocrine regulation of pheromonal responsiveness is not well understood, but evidently can involve actions at both peripheral and central levels.

The best evidence for central regulation comes from species in which a pheromone induces equivalent peripheral (EOG) olfactory responses in both sexes, but a biological response in only one. These studies have exploited the adult sexual bipotentiality of the fish brain (Kobayashi *et al.*, 2002) to determine if pheromone-induced responses that are normally seen only in males also can be induced in androgen-treated females. In the round goby, for example, where male and female exhibit equivalent EOG responses to Etio, estrone, and 17β-estradiol-3β-glucuronide (Figure 9.8A), males increase ventilation (sniffing) in response to all three steroids, whereas females respond only to Etio (Murphy *et al.*, 2001). However, within 2 weeks of methyl-testosterone implant, female round gobies not only exhibit male-typical ventilation responses but also behaviourally discriminate (assessed by sniffing) between Etio, estrone, and 17β-estradiol-3β-glucuronide (Murphy and Stacey, 2002; Figure 9.8E). Also in the goldfish, where pheromonal 17,20β-P induces equivalent EOG responses in males and females (Figure 9.10A) but gender-typical luteinising hormone responses, females treated with 11-ketotestosterone exhibit male-typical luteinising hormone response to pheromone exposure (Kobayashi *et al.*, 1997b; Figure 9.10G).

EOG studies also provide clear evidence that gender-typical pheromone response in some species might be attributable to androgenic actions on the olfactory epithelium. Mature male goldfish (Sorensen and Goetz, 1993), red fin sharks (*Epalzeorhynchus frenatus*; Stacey *et al.*, 2003) and tinfoil barbs (Cardwell *et al.*, 1995; Figure 9.10B) have lower EOG thresholds and larger EOG responses to prostaglandin Fs than do mature conspecific females, whereas both genders exhibit similar EOG responses to sex steroids (Stacey *et al.*, 2003; Figure 9.10C and D). Circulating androgens likely mediate these adult gender differences in olfactory response, because treating juveniles with androgen markedly enhances response to prostaglandin Fs without affecting EOG responses to steroids (Figure 9.10H–J). In the cyprinid *Barilius bendelisis*, androgen treatment also enhances behavioural response to 15K-prostaglandin $F_{2\alpha}$ and an ovarian extract believed to contain steroid sulfates (Bhatt *et al.*, 2002), although it is not clear where or how androgen treatment has this effect. EOG studies in both lake whitefish and brown

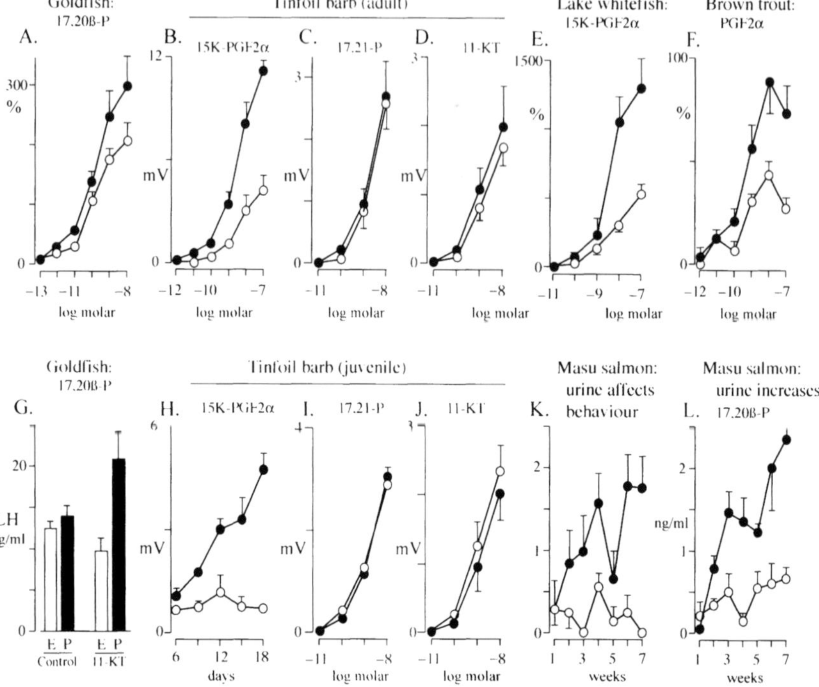

Fig. 9.10. (A–F) Gender difference in EOG response (as mV or percentage of a 10 μM L-amino acid standard) to putative hormonal pheromones (males and females represented by filled and empty circles, respectively). (G–L) Effects of androgens on EOG and behavioural responses (androgen-treated and control fish represented by filled and empty circles, respectively, in H–J). (A) Mature male and female goldfish exhibit equivalent EOG response to 17,20β-P. (B–D) Mature male and female tinfoil barbs also exhibit equivalent EOG response to steroids, 17,21-P and 11-ketotestosterone (11-KT), but males are more responsive to 15-keto-prostaglandin $F_{2\alpha}$ (15K-$PGF_{2\alpha}$). (E–F) In mature lake whitefish and brown trout, EOG threshold to 15K-$PGF_{2\alpha}$ and prostaglandin $PGF_{2\alpha}$ ($PGF_{2\alpha}$) respectively, is similar in males and females, but males have larger EOG amplitudes. (G) 11-KT implant induces male-typical luteinising hormone (LH) response to water-borne 17,20β-P (P) in female goldfish; E, ethanol control odour. (H–J) In juvenile tinfoil barbs, androgen implants increase magnitude of EOG response (mV) to 15K-$PGF_{2\alpha}$ but do not affect response to 17,21-P or 11-ketotestosterone. (K–L) Immature masu salmon parr normally do not respond to ovulated female urine. However, in parr treated with androgen in the diet, ovulated urine (filled circles) induces greater behavioural (K) and endocrine (L) response than does urine from nonovulated females (empty circles). (Adapted from Sorensen *et al.*, 1987; Cardwell *et al.*, 1995; Kobayashi *et al.*, 1997; Stacey and Sorensen, 2002; Laberge and Hara, 2003a; Stacey *et al.*, 2003; Yambe *et al.*, 2003 with kind permission of the authors, Blackwell Publishers, Elsevier Ltd., NRC Research Press, and Springer-Verlag GmbH.)

trout indicate olfactory threshold to prostaglandin Fs is similar in males and females, although males exhibit larger responses than females at suprathreshold concentrations (Laberge and Hara, 2003a; Figure 9.10E and F), the significance of which is also not understood. In the masu salmon, precocious maturity of male parr induces attraction to ovulated urine (Yambe *et al.*, 1999); androgen appears to mediate not only this behavioural change (Yambe and Yamazaki, 2001; Figure 9.10K) but also a steroidal primer response to urine (Yambe *et al.*, 2003; Figure 9.10L).

The physiological mechanisms mediating these peripheral and central androgenic effects on pheromonal responsiveness are completely unknown. However, it already is clear that these effects play a major part in pheromone function and are widespread in current model species.

6. SUMMARY

Although it has been recognised for many years that pheromones play important roles in fish reproduction, only recently has progress in identifying three chemical classes of reproductive pheromones (bile acids, sex steroids, and F prostaglandins) in a variety of fish provided clear insight into both the complexity and diversity of their functions. Although bile acids are produced, released, and detected by many fishes, only in the sea lamprey is their function understood. Here, different sets of bile acids function as attractants for migratory adults and as a male sex pheromone, raising the possibility that this class of compounds might also have multiple, endocrine-associated functions in many other species. Better understood are the sex steroid and prostaglandin pheromones (hormonal pheromones), which are widespread if not ubiquitous among the bony fishes, where they are detected with great sensitivity and specificity, and induce potent and specific behavioural and physiological effects in such major taxa as carps and salmonids. Because these hormonal pheromone systems chemically link an individual's endocrine system with the nervous systems of conspecifics, they challenge classical concepts that sex hormone actions are restricted to the individual. Indeed, the interrelationships among hormones and pheromones appear so pervasive that it now seems prudent to assume that many aspects of fish reproductive endocrine function can be understood only in the context of the social group. Despite these exciting advances, major aspects of reproductive pheromone function remain almost unexplored. Remarkably little is known about the metabolism and release of bile acids, sex steroids, and, particularly, prostaglandins. Even less is known about how these cues might function as mixtures that could contribute to species specific function; progress in these latter areas is essential if we are to determine the true nature of fish reproductive pheromones.

ACKNOWLEDGEMENTS

Norm Stacey gratefully acknowledges many years of support from the Natural Sciences and Engineering Research Council of Canada (NSERC). Peter Sorensen thanks the Alberta Heritage Foundation for Medical Research, Minnesota Agricultural Experiment Station, Minnesota Sea Grant, Great Lakes Fishery Commission, The National Institutes of Health (NIH/DC03792), and the National Science Foundation (NSF/IBN9723798), all of which have generously supported research on fish hormonal pheromones over the past 2 decades. We also are grateful for the conscientious efforts of two anonymous reviewers whose comments greatly improved an earlier version of this paper.

REFERENCES

Antonopoulou, E., Swanson, P., Mayer, I., and Borg, B. (1999). Feedback control of gonadotropins in Atlantic salmon, *Salmo salar*, male parr. *Gen. Comp. Endocrinol.* **114,** 142–150.

Appelt, C. W., and Sorensen, P. W. (1999). Freshwater fish release urinary pheromones in a pulsatile manner. *In* "Advances in Chemical Signals in Vertebrates" (Johnston, R. E., Muller-Schwarze, D., and Sorensen, P. W., Eds.), pp. 247–256. Kluwer Academic/Plenum Publishers, New York.

Baird, R. C., Jumper, G. Y., and Gallaher, E. E. (1990). Sexual dimorphism and demography in two species of oceanic midwater fishes (Stomiiformes: Sternoptychidae) from the eastern Gulf of Mexico. *Bull. Mar. Sci.* **47,** 561–566.

Bethe, V. A. (1932). Vernachlässigte hormone. *Naturwissenschaften* **20,** 177–181.

Bhatt, J. P., and Sajwan, M. S. (2001). Ovarian steroid sulphate functions as priming pheromone in male *Barilius bendelisis* (Ham.). *J. Biosci.* **26,** 253–263.

Bhatt, J. P., Kandwal, J. S., and Nautiyal, R. (2002). Water temperature and pH influence olfactory sensitivity to pre-ovulatory and post-ovulatory ovarian pheromones in male *Barilius bendelisis. J. Biosci.* **27,** 273–281.

Bjerselius, R., and Olsén, K. H. (1993). A study of the olfactory sensitivity of crucian carp (*Carassius carassius*) and goldfish (*Carassius auratus*) to 17α,20β-dihydroxyprogesterone and prostaglandin F2α *Chem. Senses* **18,** 427–436.

Bjerselius, R., Olsén, K. H., and Zheng, W. (1995). Endocrine, gonadal and behavioural responses of male crucian carp (*Carassius carassius*) to the hormonal pheromone 17α,20β-dihydroxy-4-pregnen-3-one. *Chem. Senses* **20,** 221–230.

Bjerselius, R., Li, W., Teeter, J. H., Selye, J. G., Johnsen, P. B., Maniak, P. J., Grant, G. C., Polkinghorne, C. N., and Sorensen, P. W. (2000). Direct behavioural evidence that unique bile acids released by larval sea lamprey (*Petromyzon marinus*) function as a migratory pheromone. *Can. J. Fish. Aquatic Sci.* **57,** 557–569.

Borg, B. (1994). Androgens in teleost fishes. *Comp. Biochem. Physiol. C* **109,** 219–245.

Borg, B., and Mayer, I. (1995). Androgens and behaviour in the three-spined stickleback. *Behaviour* **132,** 1025–1035.

Buck, L., and Axel, R. (1991). A novel, multigene family may encode odourant receptors: A molecular basis for odour recognition. *Cell* **65,** 171–187.

Cao, Y., Oh, B. C., and Stryer, L. (1998). Cloning and localisation of two multigene receptor families in goldfish olfactory epithelium. *Proc. Nat. Acad. Sci.* **95,** 11987–11992.

Cardwell, J. R., and Stacey, N. E. (1995). Hormonal sex pheromones in characiform fishes: An evolutionary case study. *In* "Fish Pheromones: Origins and Modes of Action" (Canário, M., Power, A. V., and D. M., Eds.), pp. 47–55. University of Algarve Press, Faro, Portugal.

Cardwell, J. R., Stacey, N. E., Tan, E. S. P., McAdam, D. S. O., and Lang, S. L. C. (1995). Androgen increases olfactory receptor response to a vertebrate sex pheromone. *J. Comp. Physiol. A* **176,** 55–61.

Carolsfeld, J., Sherwood, N. M., Kyle, A. L., Magnus, T. H., Pleasance, S., and Kreiberg, H. (1992). Characterisation of a spawning pheromone from Pacific herring. *In* "Chemical Signals in Vertebrates" (Doty, R. L., and Muller-Schwarze, D., Eds.), Vol. 6, p. 271. Plenum Press, New York.

Cole, K. S., and Stacey, N. E. (1984). Prostaglandin induction of female spawning behaviour in *Cichlosoma bimaculatum* (Pisces, Cichlidae). *Horm. Behav.* **18,** 235–248.

Cole, T. B., and Stacey, N. E. (2003). Olfactory and endocrine response to steroids in an African cichlid fish, *Haplochromis burtoni. Fish Physiol. Biochem.* **28,** 265–266.

Colombo, L., Marconato, A., Belvedere, P. C., and Frisco, C. (1980). Endocrinology of teleost reproduction: A testicular steroid pheromone in the black goby, *Gobius jozo* L. *Boll. Zool.* **47,** 355–364.

Courtenay, S. C., Quinn, T. P., Dupuis, H. M. C., Groot, C., and Larkin, P. A. (1997). Factors affecting the recognition of population-specific odours by juvenile coho salmon. *J. Fish Biol.* **50,** 1042–1060.

Darwin, C. (1887). "The descent of man and selection in relation to sex, 2nd edition." John Murray, London.

DeFraipont, M., and Sorensen, P. W. (1993). Exposure to the pheromone 17α,20β-dihydroxy-4-pregnen-3-one enhances the behavioural spawning success, sperm production, and sperm motility of male goldfish. *Anim. Behav.* **46,** 245–256.

Demski, L. S., and Northcutt, R. G. (1983). The terminal nerve: A new chemosensory system in the vertebrates. *Science* **202,** 435–437.

Dittman, A. H., and Quinn, T. P. (1994). Avoidance of a putative pheromone, 17α,20β-dihydroxy-4-pregnen-3-one, by precociously mature chinook salmon (*Oncorhynchus tshawytscha*). *Can. J. Zool.* **72,** 215–219.

Døving, K. (1976). Evolutionary trends in olfaction. *In* "The Structure-Activity Relationships in Olfaction" (Benz, G., Ed.), pp. 149–159. IRL Press, London.

Døving, K. B., and Selset, R. (1980). Behaviour patterns in cod released by electrical stimulation of olfactory tract bundlets. *Science* **207,** 559–560.

Døving, K. J. B., Selset, R., and Thommesen, G. (1980). Olfactory sensitivity to bile acids in salmonid fishes. *Acta Scand. Physiol.* **108,** 123–131.

Dryer, L., and Berghard, A. (1999). Odourant receptors: A plethora of G-protein-coupled receptors. *Trends Pharm. Sci.* **20,** 413–417.

Dulka, J. G., and Stacey, N. E. (1990). Effects of olfactory tract lesions on gonadotropin and milt responses to the female sex pheromone, 17α,20β-dihydroxy-4-pregnen-3-one, in male goldfish. *J. Exp. Zool.* **257,** 223–229.

Dulka, J. G., Stacey, N. E., Sorensen, P. W., and Van Der Kraak, G. J. (1987). A sex steroid pheromone synchronises male-female spawning readiness in goldfish. *Nature* **325,** 251–253.

Eisthen, H. L., Delay, R. J., Wirsig-Wiechmann, C., and Dionne, V. E. (2000). Neuromodulatory effects of gonadotropin releasing hormone on olfactory receptor neurons. *J. Neurosci.* **20,** 3947–9955.

Emanuel, M. E., and Dodson, J. J. (1979). Modification of rheotropic behaviour of male rainbow trout (*Salmo gairdneri*) by ovarian fluid. *J. Fish. Res. Bd. Can.* **36,** 63–68.

Essington, T. E., and Sorensen, P. W. (1996). Overlapping sensitivities of brook trout and brown trout to putative hormonal pheromones. *J. Fish Biol.* **48,** 1027–1029.

Fine, J. M., Vrieze, L. A., and Sorensen, P. W. (2004). Evidence that petromyzontid lampreys employ a common migratory pheromone that is partially comprised of bile acids. *J. Chem. Ecol.* **30,** 2091–2110.

Frade, P., Hubbard, P. C., Barata, E. N., and Canário, A. V. M. (2002). Olfactory sensitivity of the Mozambique tilapia to conspecific odours. *J. Fish Biol.* **61,** 1239–1254.

Fraser, E. J., and Stacey, N. (2002). Isolation increases milt production in goldfish. *J. Exp.Zool.* **293,** 511–524.

Friedrich, R. W., and Korsching, S. I. (1998). Chemotopic, combinatorial, and noncombinatorial odourant representations in the olfactory bulb revealed using a voltage-sensitive axon tracer. *J. Neurosci.* **18,** 9977–9988.

Friedrich, R. W., and Laurent, G. (2001). Dynamic optimisation of odour representations by slow temporal patterning of mitral cell activity. *Science* **291,** 889–891.

Fujita, I., Sorensen, P. W., Stacey, N. E., and Hara, T. J. (1991). The olfactory system, not the terminal nerve, functions as the primary chemosensory pathway mediating responses to sex pheromones in goldfish. *Brain Behav. Evol.* **38,** 313–321.

Gonçalves, D. M., Barala, E. N., Oliveira, R. F., and Canário, A. V. M. (2002). The role of male visual and chemical cues on the activation of female courtship behaviour in the sex-role reversed peacock blenny. *J. Fish Biol.* **61,** 90–105.

Hamdani, E-H., Stabell, O. B., Alexander, G., and Døving, K. B. (2000). Alarm reaction in the crucian carp is mediated by the medial bundle of the medial olfactory tract. *Chem. Senses* **25,** 103–109.

Hansen, J. A., Rose, J. D., Jenkins, R. A., Gerow, K. G., and Bergman, H. L. (1999). Chinook salmon (*Oncorhynchus tshawytscha*) and rainbow trout (*Oncorhynchus mykiss*) exposed to copper: Neurophysiological and histological effects on the olfactory system. *Environ. Toxicol. Chem.* **18,** 1979–1991.

Hansen, A., Rolen, S. H., Anderson, K., Morita, Y., Caprio, J., and Finger, T. E. (2003). Correlation between olfactory receptor cell type and function in the channel catfish. *J. Neurosci.* **23,** 9328–9339.

Hansen, A., Anderson, K. T., and Finger, T. E. (2004). Differential distribution of olfactory receptor neurons in goldfish: Structural and molecular correlates. *J. Comp. Neurol.* **477,** 347–359.

Hanson, L. R. (2001). Neural coding of identified sex pheromones in the goldfish. *Carassius auratus.* Ph.D. Thesis, University of Minnesota.

Hanson, L. R., Sorensen, P. W., and Cohen, Y. (1998). Sex pheromones and amino acids evoke distinctly different spatial patterns of electrical activity in the goldfish olfactory bulb. *Ann. N. Y. Acad. Sci.* **855,** 521–524.

Hara, T. J., and Zhang, C. (1997). Topographic bulbar projections and dual neural pathways of the primary olfactory neurons in salmonid fishes. *Neuroscience* **82,** 301–313.

Hubbard, P. C., Barata, E. N., and Canário, A. V. M. (2002). Possible disruption of pheromonal communication by humic acid in the goldfish, *Carassius auratus. Aquat. Toxicol.* **60,** 169–183.

Immler, S., Mazzoldi, C., and Rasotto, M. B. (2004). From sneaker to parental male: Change of reproductive traits in the black goby, *Gobius niger* (Teleostei, Gobiidae). *J. Exp. Zool. A* **301,** 177–185.

Inbaraj, R. M., Scott, A. P., and Vermeirssen, E. L. M. (1997). Use of a radioimmunoassay which detects C21 steroids with a 5β-reduced, 3α-hydroxylated configuration to identify and measure steroids involved in final oocyte maturation in female plaice (*Pleuronectes platessa*). *Gen. Comp. Endocrinol.* **105,** 50–61.

Irvine, I. A. S., and Sorensen, P. W. (1993). Acute olfactory sensitivity of wild common carp, *Cyprinus carpio*, to goldfish sex pheromones is influenced by gonadal maturity. *Can. J. Zool.* **71,** 2199–2210.

Johnston, R. E. (2000). Chemical communication and pheromones: The types of signals and the role of the vomeronasal system. *In* "The Neurobiology of Taste and Smell" (Finger, T. E., Silver, W. L., and Restrepo, D., Eds.), 2nd ed, pp. 101–127. Wiley-Liss, New York.

Jones, K. A., and Hara, T. J. (1985). Behavioural responses of fishes to chemical cues: Results from a new bioassay. *J. Fish Biol.* **27,** 495–504.

Kagawa, H., Tanaka, H., Unuma, T., Ohta, H., Gen, K., and Okuzawa, K. (2003). Role of prostaglandin in the control of ovulation in the Japanese eel *Anguilla japonica*. *Fish. Science* **69,** 234–241.

Karlson, P., and Lüscher, M. (1959). 'Pheromones': A new term for a class of biologically active substances. *Nature* **183,** 55 56.

Kawabata, K. (1993). Induction of sexual behaviour in male fish (*Rhodeus ocellatus ocellatus*) by amino acids. *Amino Acids* **5,** 323–327.

Kitamura, S., Ogata, H., and Takashima, F. (1994a). Activities of F-type prostaglandins as releaser sex pheromones in cobitide loach, *Misgurnus anguillicaudatus*. *Comp. Bioch. Physiol. A* **107,** 161–169.

Kitamura, S., Ogata, H., and Takashima, F. (1994b). Olfactory responses of several species of teleost to F-prostaglandins. *Comp. Biochem. Physiol. A* **107,** 463–467.

Kobayashi, M., and Nakanishi, T. (1999). 11-Ketotestosterone induces male-type sexual behaviour and gonadotropin secretion in gynogenetic crucian carp, *Carassius auratus langsdorfii*. *Gen. Comp. Endocrinol.* **115,** 178–187.

Kobayashi, M., and Stacey, N. E. (1993). Prostaglandin-induced female spawning behaviour in goldfish (*Carassius auratus*) appears independent of ovarian influence. *Horm. Behav.* **27,** 38–55.

Kobayashi, M., Aida, K., and Hanyu, I. (1988). Hormone changes during the ovulatory cycle in goldfish. *Gen. Comp. Endocrinol.* **69,** 301–307.

Kobayashi, M., Aida, K., and Hanyu, I. (1989). Induction of gonadotropin surge by steroid hormone implantation in ovariectomised and sexually regressed female goldfish. *Gen. Comp. Endocrinol.* **73,** 469–476.

Kobayashi, M., Amano, M., Kim., M.-H., Furukawa, K., Hasegawa, Y., and Aida, K. (1994). Gonadotropin-releasing hormones of terminal nerve origin are not essential to ovarian development and ovulation in goldfish. *Gen. Comp. Endocrinol.* **95,** 192–200.

Kobayashi, M., Amano, M., Kim, M. H., Yoshiura, Y., Sohn, Y. C., Suetake, H., and Aida, K. (1997a). Gonadotropin-releasing hormone and gonadotropin in goldfish and masu salmon. *Fish Physiol. Biochem.* **17,** 1–8.

Kobayashi, M., Furukawa, K., Kim, M-H., and Aida, K. (1997b). Induction of male-type gonadotropin secretion by implantation of 11-ketotestosterone in female goldfish. *Gen. Comp. Endocrinol.* **108,** 434–445.

Kobayashi, M., Sorensen, P. W., and Stacey, N. E. (2002). Hormonal and pheromonal control of spawning behaviour in the goldfish. *Fish Physiol. Biochem.* **26,** 71–84.

Kolodziej, E. P., Gray, J. L., and Sedlak, D. L. (2003). Quantification of steroid hormones with pheromonal properties in municipal wastewater effluent. *Environ. Tox. Anal. Chem.* **22,** 2622–2629.

Koya, Y., Soyano, K., Yamamoto, K., Obana, H., and Matsubara, T. (2002). Testicular development and serum profiles of steroid hormone levels in captive male Pacific herring *Clupea pallasii* during their first maturational cycle. *Fish. Sci.* **68,** 1099–1105.

Koya, Y., Soyano, K., Yamamoto, K., Obana, H., and Matsubara, T. (2003). Oocyte development and serum profiles of vitellogenin and steroid hormone levels in captive female Pacific herring *Clupea pallasii* during their first maturational cycle. *Fish. Sci.* **69,** 137–145.

Laberge, F., and Hara, T. J. (2001). Neurobiology of fish olfaction: A review. *Brain Res. Rev.* **36,** 46–59.

Laberge, F., and Hara, T. J. (2003a). Behavioural and electrophysiological responses to F-prostaglandins, putative spawning pheromones, in three salmonid fishes. *J. Fish Biol.* **62,** 206–221.

Laberge, F., and Hara, T. J. (2003b). Non-oscillatory discharges of an F-prostaglandin responsive neuron population in the olfactory bulb-telencephalon transition area in lake whitefish. *Neuroscience* **116,** 1089–1095.

Laumen, T. J., Pern, U., and Blum, V. (1974). Investigations on the function and hormonal regulation of the anal appendices in *Blennius pavo* (Risso). *J. Exp. Zool.* **190,** 47–56.

Li, W., and Sorensen, P. W. (1997). Highly independent olfactory receptor sites for conspecific bile acids in the sea lamprey, *Petromyzon marinus. J. Comp. Physiol. A* **180,** 429–438.

Li, W., Sorensen, P. W., and Gallaher, D. D. (1995). The olfactory system of migratory adult sea lamprey (*Petromyzon marinus*) is specifically and acutely sensitive to unique bile acids released by conspecific larvae. *J. Gen. Physiol.* **105,** 569–587.

Li, W., Scott, A. P., Siefkes, M. J., Yan, H., Liu, Q., Yun, S.-S., and Gage, D. A. (2002). Bile acid secreted by male sea lamprey that acts as a sex pheromone. *Science* **296,** 138–141.

Li, W., Scott, A. P., Siefkes, M. J., Yun, S.-S., and Zielinski, B. (2003). A male pheromone in the sea lamprey: An overview. *Fish Physiol. Biochem.* **28,** 259–262.

Liley, N. R. (1972). The effects of estrogens and other steroids on the sexual behaviour of the female guppy, *Poecilia reticulata. Gen. Comp. Endocrinol. Suppl.* **3,** 542–552.

Liley, N. R. (1982). Chemical communication in fish. *Can. J. Fish. Aquatic Sci.* **39,** 22–35.

Liley, N. R., and Stacey, N. E. (1983). Hormones, pheromones, and reproductive behaviour in fish. *In* "Fish Physiology" (Hoar, W. S., Randall, D. J., and Donaldson, E. M., Eds.), Vol. 9B, pp. 1–63. Academic Press, New York.

Liley, N. R., and Tan, E. S. P. (1985). The induction of spawning behaviour in *Puntius gonionotus* (Bleeker) by treatment with prostaglandin F2α *J. Fish Biol.* **26,** 491–502.

Locatello, L., Mazzoldi, C., and Rasotto, M. B. (2002). Ejaculate of sneaker males is pheromonally inconspicuous in the black goby, *Gobius niger* (Teleostei, Gobiidae). *J. Exp. Zool.* **293,** 601–605.

Lower, N., Scott, A. P., and Moore, A. (2004). Release of sex steroids into the water by roach. *J. Fish Biol.* **64,** 16–33.

Maniak, P. J., Lossing, R., and Sorensen, P. W. (2000). Injured Eurasian ruffe, *Gymnocephalus cernuus*, release an alarm pheromone which may prove useful in their control. *J. Great Lakes Res.* **26,** 183–195.

Matsumura, K. (1985). Tetrodotoxin as a pheromone. *Nature* **378,** 563–564.

Mesquita, R. M. R. S., Canario, A. V. M., and Melo, E. (2003). Partition of fish pheromones between water and aggregates of humic acids. Consequences for sexual signaling. *Environ. Sci. Technol.* **37,** 742–746.

Meyer, J. H., and Liley, N. R. (1982). The control of production of a sexual pheromone in the guppy, *Poecilia reticulata. Can. J. Zool.* **60,** 1505–1510.

Miura, T., and Miura, C. I. (2003). Molecular control mechanisms of fish spermatogenesis. *Fish Physiol. Biochem.* **28,** 181–186.

Moore, A. (1991). Behavioural and physiological responses of precocious male Atlantic salmon (*Salmo salar* L.) parr to testosterone. *In* "Proceedings of the Fourth International Symposium on the Reproductive Physiology of Fish" (Scott, A. P., Sumpter, J. P., Kime, D. E., and Rolfe, M. S., Eds.), pp. 194–196. FishSymp 91, Sheffield.

Moore, A., and Scott, A. P. (1991). Testosterone is a potent odourant in precocious male Atlantic salmon (*Salmo salar* L.) parr. *Phil. Trans. Roy. Soc. Lond. B* **332,** 241–244.

Moore, A., and Scott, A. P. (1992). 17α,20β-dihydroxy-4-pregnen-3-one-20-sulphate is a potent odourant in precocious male Atlantic salmon parr which have been pre-exposed to the urine of ovulated females. *Proc. Roy. Soc. Lond. Ser. B Biol. Sci.* **249,** 205–209.

Moore, A., and Waring, C. P. (1995). Seasonal changes in olfactory sensitivity of mature male Atlantic salmon (*Salmo salar* L.) parr to prostaglandins. *In* "Proceedings of the Fifth International Symposium on the Reproductive Physiology of Fish" (Goetz, F. W., and Thomas, P., Eds.), p. 273. Fish Symposium 95, Austin, Texas.

Moore, A., and Waring, C. P. (1996). Electrophysiological and endocrinological evidence that F-series prostaglandins function as priming pheromones in mature male Atlantic salmon (*Salmo salar*) parr. *J. Exp. Biol.* **199,** 2307–2316.

Moore, A., Olsén, K. H., Lower, N., and Kindahl, H. (2002). The role of F-series prostaglandins as reproductive priming pheromones in the brown trout. *J. Fish Biol.* **60,** 613–624.

Morita, Y., and Finger, T. E. (1998). Differential projections of ciliated and microvillous olfactory receptor cells in the catfish, *Ictalurus punctatus*. *J. Comp. Neurol.* **398,** 539–550.

Murphy, C. A., and Stacey, N. E. (2002). Methyl-testosterone induces male-typical behavioural responses to putative steroidal pheromones in female round gobies (*Neogobius melanostomus*). *Horm. Behav.* **42,** 109–115.

Murphy, C. A., Stacey, N., and Corkum, L. D. (2001). Putative steroidal pheromones in the round goby, *Neogobius melanostomus*: Olfactory and behavioural responses. *J. Chem. Ecol.* **27,** 443–470.

Nagahama, Y., Yoshikuni, M., Yamashita, M., and Tanaka, M. (1994). Regulation of oocyte maturation in fish. *In* "Fish Physiology" (Sherwood, N. M., and Hew, C. L., Eds.), Vol. 13, pp. 393–439. Academic Press, New York.

Naito, T., Saito, Y., Yumamoto, J., Nozaki, Y., Tomura, K., Hazama, M., Nakanishi, S., and Breener, S. (1998). Putative pheromone receptors related to the Ca^{2+} sensing receptor in *Fugu*. *Proc. Natl. Acad. Sci. USA* **95,** 5178–5181.

Narayanan, A., and Stacey, N. E. (2003). Olfactory responses to putative steroidal pheromones in allopatric and sympatric species of mochokid catfish. *Fish Physiol. Biochem.* **28,** 275–276.

Nelson, J. S. (1994). "Fishes of the World, 4th ed." Wiley-Interscience, New York.

Nevitt, G. A. (1991). Do fish sniff? A new mechanism of olfactory sampling in pleuronectid flounders. *J. Exp. Biol.* **157,** 1–18.

Ngai, J., Dowling, M. M., Buck, L., Axel, R., and Chess, A. (1993a). The family of genes encoding odourant receptors in the channel catfish. *Cell* **72,** 657–666.

Ngai, J., Chess, A., Dowling, M. M., Necles, N., Macagno, E. R., and Axel, R. (1993b). Coding of olfactory information: Topography of odourant receptor expression in the catfish olfactory epithelium. *Cell* **72,** 667–680.

Nikonov, A. A., and Caprio, J. (2001). Electrophysiological evidence for a chemotopy of biologically relevant odours in the olfactory bulb of the channel catfish. *J. Neurophysiol.* **86,** 1869–1876.

Ogata, H., Kitamura, S., and Takashima, F. (1994). Release of 13,14-dihydro-15-keto-prostaglandin F2α, a sex pheromone, to water by cobitid loach following ovulatory stimulation. *Fish. Sci.* **60,** 143–148.

Oliveira, R. F., Almada, V. C., and Canário, A. V. (1996). Social modulation of sex steroid concentrations in the urine of male cichlid fish *Oreochromis mossambicus*. *Horm. Behav.* **30,** 2–12.

Oliveira, R., Hirschenhauser, K., Carneiro, L. A., and Canário, A. V. M. (2002). Social modulation of androgen levels in male teleost fish. *Comp. Biochem. Physiol. B* **132,** 203–215.

Olsén, K. H., Bjerselius, R., Petersson, E., Jarva, E., Mayer, I., and Hedenskog, M. (2000). Lack of species-specific primer effects of odours from female Atlantic salmon, *Salmo salar*, and brown trout, *Salmo trutta*. *Oikos* **88,** 213–220.

Olsén, K. H., Bjerselius, R., Mayer, I., and Kindahl, H. (2001). Both ovarian fluid and female urine increase sex steroid levels in mature Atlantic salmon (*Salmo salar* L.) male parr. *J. Chem. Ecol.* **27,** 2337–2349.

Olsén, K. H., Grahn, M., and Lohm, J. (2002a). Influence of MHC on sibling discrimination in Arctic charr, *Salvelinus alpinus* (L.). *J. Chem. Ecol.* **28,** 783–796.

Olsén, K. H., Johanssen, A.-K., Bjerselius, R., Mayer, I., and Kindahl, H. (2002b). Mature Atlantic salmon (*Salmo salar* L.) male parr are attracted to ovulated female urine but not ovarian fluid. *J. Chem. Ecol.* **28,** 29–40.

Pall, M. K., Mayer, I., and Borg, B. (2002a). Androgen and behaviour in the three-spined stickleback, *Gasterosteus aculeatus*. I. Changes in 11-ketotestosterone levels during the nesting cycle. *Horm. Behav.* **41,** 377–383.

Pall, M. K., Mayer, I., and Borg, B. (2002b). Androgen and behaviour in the three-spined stickleback, *Gasterosteus aculeatus*. II. Castration and 11-ketoandrostenedione effects on courtship and parental care during the nesting cycle. *Horm. Behav.* **42,** 337–344.

Parks, L. G., and Leblanc, G. A. (1998). Involvement of multiple biotransformation processes in the metabolic elimination of testosterone by juvenile and adult fathead minnows (*Pimephales promelas*). *Gen. Comp. Endocrinol.* **112,** 69–79.

Patiño, R., and Sullivan, C. G. (2002). Ovarian follicle growth, maturation, and ovulation in teleost fish. *Fish Physiol. Biochem.* **26,** 57–70.

Patiño, R., Thomas, P., and Yoshizaki, G. (2003). Ovarian follicle maturation and ovulation: An integrated perspective. *Fish Physiol. Biochem.* **28,** 305–308.

Pfister, P., and Rodriguez, I. (2005). Olfactory expression of a single and highly variable V1r pheromone receptor-like gene in fish *species*. *Proc. Natl. Acad. Sci.* **102,** 5489–5494.

Poling, K. R., Fraser, E. J., and Sorensen, P. W. (2001). The three steroidal components of the goldfish preovulatory pheromone signal evoke different behaviours in males. *Comp. Biochem. Physiol. B* **129,** 645–651.

Polkinghorne, C. N., Olson, J. N., Gallaher, D. G., and Sorensen, P. W. (2001). Larval sea lamprey release two unique bile acids to the water at a rate sufficient to produce detectable riverine pheromone plumes. *Fish Physiol. Biochem.* **24,** 15–30.

Pottinger, T. G., and Moore, A. (1997). Characterisation of putative steroid receptors in the membrane, cytosol and nuclear fractions from the olfactory tissue of brown and rainbow trout. *Fish Physiol. Biochem.* **16,** 45–63.

Rasotto, M. B., and Mazzoldi, C. (2002). Male traits associated with alternative reproductive tactics in *Gobius niger*. *J. Fish Biol.* **61,** 173–184.

Resink, J. W., Schoonen, W. G. E. J., Alpers, P. C. H., File, D. M., Notenbloom, C. D., Van Den Hurk, R., and Van Oordt, P. G. W. J. (1989). The chemical nature of sex attracting pheromones from the seminal vesicle of the African catfish, *Clarias gariepinus*. *Aquaculture* **83,** 137–151.

Robison, R. R., Fernald, R. D., and Stacey, N. E. (1998). The olfactory system of a cichlid fish responds to steroidal compounds. *J. Fish Biol.* **53,** 226–229.

Rouquier, S., Blancher, A., and Giorgi, D. (2000). The olfactory receptor gene repertoire in primates and mouse: Evidence for reduction of the functional fraction in primates. *PNAS* **97,** 2870–2874.

Rosenblum, P. M., Sorensen, P. W., Stacey, N. E., and Peter, R. E. (1991). Binding of the steroidal pheromone 17α,20β-dihydroxy-4-pregnen-3-one to goldfish (*Carassius auratus*) olfactory epithelium membrane preparations. *Chem. Senses* **16,** 143–154.

Schulz, R. W., and Miura, T. (2002). Spermatogenesis and its endocrine regulation. *Fish Physiol. Biochem.* **26,** 43–56.

Scott, A. P., and Liley, N. R. (1994). Dynamics of excretion of 17α,20β-dihydroxy-4-pregnen-3-one-20-sulphate, and of the glucuronides of testosterone and 17β–oestradiol, by urine of reproductively mature male and female rainbow trout (*Oncorhynchus mykiss*). *J. Fish Biol.* **44,** 117–129.

Scott, A. P., and Sorensen, P. W. (1994). Time course of release of pheromonally active steroids and their conjugates by ovulatory goldfish. *Gen. Comp. Endocrinol.* **96,** 309–323.

Scott, A. P., Sherwood, N. M., Canário, A. V. M., and Warby, C. M. (1991a). Identification of free and conjugated steroids, including cortisol and 17α-20β-dihydroxy-4-pregnen-3-one, in the milt of Pacific herring, *Clupea harengus pallasi. Can. J. Zool.* **69,** 104–110.

Scott, A. P., Canário, A. V. M., Sherwood, N. M., and Warby, C. M. (1991b). Levels of steroids, including cortisol and 17α,20β-dihydroxy-4-pregnen-3-one, in plasma, seminal fluid and urine of Pacific herring (*Clupea harengus pallasi*) and North Sea plaice (*Pleuronectes platessa* L.). *Can. J. Zool.* **69,** 111–116.

Scott, A. P., Liley, N. R., and Vermeirssen, E. L. M. (1994). Urine of reproductively mature female rainbow trout, *Oncorhynchus mykiss* (Walbaum), contains a priming pheromone which enhances plasma levels of sex steroids and gonadotrophin II in males. *J. Fish Biol.* **44,** 131–147.

Scott, J. W., and Scott-Johnson, P. E. (2002). The electroolfactogram: A review of its history and uses. *Microscop. Res. Tech.* **58,** 152–160.

Selset, R., and Døving, K. B. (1980). Behaviour of mature anadromous char (*Salvelinus alpinus* L) towards odourants produced by smolts of their own populations. *Acta Physiol. Scand.* **108,** 113–122.

Siefkes, M. J., and Li, W. (2004). Electrophysiological evidence for detection and discrimination of pheromonal bile acids by the olfactory epithelium of female sea lampreys (*Petromyzon marinus*). *J. Comp. Physiol. A* **190,** 193–199.

Siefkes, M. J., Scott, A. P., Zielinski, B., Yun, S.-S., and Li., W. (2003). Male sea lampreys, *Petromyzon marinus*, excrete a sex pheromone from gill epithelia. *Biol. Reprod.* **69,** 125–132.

Sorensen, P. W., and Caprio, J. (1997). Chemoreception. *In* "The Physiology of Fishes, Second Edition" (Evans, D. H., Ed.), pp. 375–405. CRC Press, Boca Raton.

Sorensen, P. W., and Goetz, F. W. (1993). Pheromonal and reproductive function of F-prostaglandins and their metabolites in teleost fish. *J. Lipid Mediators* **6,** 385–393.

Sorensen, P. W., and Scott, A. P. (1994). The evolution of hormonal sex pheromones in teleost fish: Poor correlation between the pattern of steroid release by goldfish and olfactory sensitivity suggests that these cues evolved as a result of chemical spying rather than signal specialisation. *Acta Scand. Physiol.* **152,** 191–205.

Sorensen, P. W., and Stacey, N. E. (1999). Evolution and specialisation of fish hormonal pheromones. *In* "Advances in Chemical Signals in Vertebrates" (Johnston, R. E., Muller-Schwarze, D., and Sorensen, P. W., Eds.), pp. 15–47. Kluwer Academic/Plenum Publishers, New York.

Sorensen, P. W., and Stacey, N. E. (2004). Brief review of fish pheromones and discussion of their possible uses in the control of non-indigenous teleost fishes. *New Zeal. J. Mar. Freshw. Res.* **38,** 399–417.

Sorensen, P. W., and Vrieze, L. A. (2003). The chemical ecology and potential application of the sea lamprey migratory pheromone. *J. Great Lakes Res.* **29** (Suppl. 1), 66–84.

Sorensen, P. W., and Wyatt, J. (2000). Pheromones. *In* "The Corsini Encyclopedia of Psychology and Behavioural Science, Third edition" (Craighead, W. E., and Nemeroff, C. B., Eds.), pp. 1193–1195. John Wiley & Sons, New York.

Sorensen, P. W., Hara, T. J., and Stacey, N. E. (1987). Extreme olfactory sensitivity of mature and gonadally-regressed goldfish to a potent steroidal pheromone, 17α,20β-dihydroxy-4-pregnen-3-one. *J. Comp. Physiol. A* **160,** 305–313.

Sorensen, P. W., Hara, T. J., Stacey, N. E., and Goetz, F. W. (1988). F prostaglandins function as potent olfactory stimulants that comprise the postovulatory female sex pheromone in goldfish. *Biol. Reprod.* **39,** 1039–1050.

Sorensen, P. W., Chamberlain, K. J., and Stacey, N. E. (1989). Differing behavioural and endocrinological effects of two female sex pheromones on male goldfish. *Horm. Behav.* **23,** 317–332.

Sorensen, P. W., Hara, T. J., Stacey, N. E., and Dulka, J. G. (1990). Extreme olfactory specificity of the male goldfish to the preovulatory steroidal pheromone 17α,20β-dihydroxy-4-pregnen-3-one. *J. Comp. Physiol. A* **166,** 373–383.

Sorensen, P. W., Hara, T. J., and Stacey, N. E. (1991). Sex pheromones selectively stimulate the medial olfactory tracts of male goldfish. *Brain Res.* **558,** 343–347.

Sorensen, P. W., Scott, A. P., Stacey, N. E., and Bowdin, L. (1995a). Sulfated 17α,20β-dihydroxy-4-pregnen-3-one functions as a potent and specific olfactory stimulant with pheromonal actions in the goldfish. *Gen. Comp. Endocrinol.* **100,** 128–142.

Sorensen, P. W., Brash, A. R., Goetz, F. W., Kellner, R. G., Bowdin, L., and Vrieze, L. A. (1995b). Origins and functions of F prostaglandins as hormones and pheromones in the goldfish. *In* "Proceedings of the Fifth International Symposium on the Reproductive Physiology of Fish" (Goetz, F. W., and Thomas, P., Eds.), pp. 244–248. Fish Symposium 95, Austin Texas.

Sorensen, P. W., Cardwell, J. R., Essington, T., and Weigel, D. E. (1995c). Reproductive interactions between brook and brown trout in a small Minnesota stream. *Can. J. Fish. Aquatic Sci.* **52,** 1958–1965.

Sorensen, P. W., Christensen, T. A., and Stacey, N. E. (1998). Discrimination of pheromonal cues in fish: Emerging parallels with insects. *Curr. Opin. Neurobiol.* **8,** 458–467.

Sorensen, P. W., Scott, A. P., and Kihslinger, R. L. (2000). How common hormonal metabolites function as relatively specific pheromonal signals in goldfish. *In* "Proceedings of the Sixth International Symposium on the Reproductive Physiology of Fish" (Norberg, B., Ksebu, O. S., Taranger, G. L., Andersson, E., and Stefanson, S. O., Eds.), pp. 125–128. John Grieg AS, Bergen.

Sorensen, P. W., Vrieze, L. A., and Fine, J. M. (2003). A multi-component migratory pheromone in the sea lamprey. *Fish Physiol. Biochem.* **28,** 253–257.

Sorensen, P. W., Murphy, C. A., Loomis, K., Maniak, P., and Thomas, P. (2004). Evidence that 4-pregnen-17,20β,21-triol-3-one functions as a maturation-inducing hormone and pheromone precursor in the percid fish, *Gymnocephalus cernuus. Gen. Comp. Endocrinol.* **139,** 1–11.

Sorensen, P. W., Pinillos, M., and Scott, A. P. (2005). Sexually mature male goldfish release large quantities of androstenedione into the water where it functions as a pheromone. *Gen. Comp. Endocrinol.* **140,** 164–175.

Speca, D. J., Lin, D. M., Sorensen, P. W., Isacoff, E. Y., Ngai, J., and Dittman, H. (1999). Functional identification of a goldfish odourant receptor. *Neuron* **23,** 487–498.

Stacey, N. E. (1976). Effects of indomethacin and prostaglandins on spawning behaviour of female goldfish. *Prostaglandins* **12,** 113–128.

Stacey, N. E. (1981). Hormonal regulation of female reproductive behaviour in fish. *Amer. Zool.* **21,** 305–316.

Stacey, N. E. (1991). Hormonal pheromones in fish: Status and prospects. *In* "Proceedings of the Fourth International Symposium on the Reproductive Physiology of Fish" (Scott, A. P., Sumpter, J. P., Kime, D. S., and Rolfe, M. S., Eds.), pp. 177–181. FishSymp 91, Sheffield.

Stacey, N. E. (2003). Hormones, pheromones, and reproductive behaviour. *Fish Physiol.Biochem.* **28,** 229–235.

Stacey, N. E., and Cardwell, J. R. (1995). Hormones as sex pheromones in fish: Widespread distribution among freshwater species. *In* "Proceedings of the Fifth International

Symposium on the Reproductive Physiology of Fish" (Goetz, F. W., and Thomas, P., Eds.), pp. 244–248. Symposium 95, Austin, Texas.

Stacey, N. E., and Hourston, A. H. (1982). Spawning and feeding behaviour of captive Pacific herring, *Clupea harengus pallasi*. *Can. J. Fish. Aquatic Sci.* **39,** 489–498.

Stacey, N. E., and Kobayashi, M. (1996). Androgen induction of male sexual behaviours in female goldfish. *Horm. Behav.* **30,** 434–445.

Stacey, N. E., and Kyle, A. L. (1983). Effects of olfactory tract lesions on sexual and feeding behaviour of goldfish. *Physiol. Behav.* **30,** 621–628.

Stacey, N. E., and Liley, N. R. (1974). Regulation of spawning behaviour in the female goldfish. *Nature* **247,** 71–72.

Stacey, N. E., and Peter, R. E. (1979). Central action of prostaglandins in spawning behaviour of female goldfish. *Physiol. Behav.* **22,** 1191–1196.

Stacey, N. E., and Sorensen, P. W. (1986). 17α,20β-dihydroxy-4-pregnen-3-one: A steroidal primer pheromone which increases milt volume in the goldfish. *Can. J. Zool.* **64,** 2412–2417.

Stacey, N. E., and Sorensen, P. W. (1991). Function and evolution of fish hormonal pheromones. *In* "Biochemistry and Molecular Biology of Fishes: Phylogenetic and Biochemical Perspectives" (Hochachka, P. W., and Mommsen, T. P., Eds.), Vol. 1, pp. 109–135. Elsevier, New York.

Stacey, N. E., and Sorensen, P. W. (1999). Pheromones, fish. *In* "Encyclopedia of Reproduction" (Knobil, E., and Neill, J. D., Eds.), Vol. 3, pp. 748–755. Academic Press, New York.

Stacey, N. E., and Sorensen, P. W. (2002). Fish hormonal pheromones. *In* "Hormones, Brain, and Behaviour" (Pfaff, D. W., Arnold, A. P., Etgen, A., Fahrbach, S., and Rubin, R., Eds.), Vol. 2, pp. 375–435. Academic Press, New York.

Stacey, N. E., Cook, A. F., and Peter, R. E. (1979a). Ovulatory surge of gonadotropin in the goldfish, *Carassius auratus*. *Gen. Comp. Endocrinol.* **37,** 246–249.

Stacey, N. E., Cook, A. F., and Peter, R. E. (1979b). Spontaneous and gonadotropin-induced ovulation in the goldfish, *Carassius auratus* L.: Effects of external factors. *J. Fish Biol.* **15,** 349–361.

Stacey, N. E., Sorensen, P. W., Van Der Kraak, G. J., and Dulka, J. G. (1989). Direct evidence that 17α,20β-dihydroxy-4-pregnen-3-one functions as a goldfish primer pheromone: Preovulatory release is closely associated with male endocrine responses. *Gen. Comp. Endocrinol.* **75,** 62–70.

Stacey, N. E., Zheng, W. B., and Cardwell, J. R. (1994). Milt production in common carp (*Cyprinus carpio*): Stimulation by a goldfish steroid pheromone. *Aquaculture* **127,** 265–276.

Stacey, N. E., Cardwell, J. R., and Murphy, C. (1995). Hormonal pheromones in freshwater fishes: Preliminary results of an electro-olfactogram survey. *In* "Fish Pheromones: Origins and Modes of Action" (Canario, A. V. M., and Power, D. M., Eds.), pp. 47–55. University of Algarve Press, Faro, Portugal.

Stacey, N. E., Fraser, E. J., Sorensen, P. W., and Van Der Kraak, G. J. (2001). Milt production in goldfish: Regulation by multiple social stimuli. *Comp. Biochem. Physiol. C* **130,** 467–476.

Stacey, N. E., Chojnacki, A., Narayanan, A., Cole, T. B., and Murphy, C. A. (2003). Hormonally-derived sex pheromones in fish: Exogenous cues and signals from gonad to brain. *Can. J. Physiol. Pharmacol.* **81,** 329–341.

Strüssmann, C. A., and Nakamura, M. (2002). Morphology, endocrinology, and environmental modulation of gonadal sex differentiation in teleost fishes. *Fish Physiol. Biochem.* **26,** 13–29.

Sveinsson, T., and Hara, T. J. (1995). Mature males of Arctic char, *Salvelinus alpinus*, release F-type prostaglandins to attract conspecific mature females and stimulate their spawning behaviour. *Environ. Biol. Fishes* **42,** 253–266.

Sveinsson, T., and Hara, T. J. (2000). Olfactory sensitivity and specificity of arctic char, *Salvelinus alpinus*, to a putative male pheromone, prostaglandin F2α *Physiol. Behav.* **69,** 301–307.

Taborsky, M. (2001). The evolution of bourgeois, parasitic, and cooperative reproductive behaviours in fishes. *J. Hered.* **92,** 100–110.

Teeter, J. (1980). Pheromone communication in sea lampreys (*Petromyzon marinus*): Implications for population management. *Can J. Fish. Aquat. Sci.* **37,** 2123–2132.

Thomas, P. (1994). Hormonal control of final oocyte maturation in sciaenid Fishes. *In* "Perspectives in Comparative Endocrinology" (Davey, K. G., Peter, R. E., and Tobe, S., Eds.), pp. 619–662. National Research Council of Canada, Ottawa.

Thomas, P. (2003). Rapid, nongenomic steroid actions initiated at the cell surface: Lessons from studies in fish. *Fish Physiol. Biochem.* **28,** 3–12.

Thomas, P., Breckenridge-Miller, D., and Detweiler, C. (1997). Binding characteristics and regulation of the 17,20β,21-trihydroxy-4-pregnen-3-one (20β-S) receptor on testicular and sperm plasma membranes of spotted seatrout (*Cynoscion nebulosus*). *Fish Physiol. Biochem.* **17,** 109–116.

Trudeau, V. L. (1997). Neuroendocrine regulation of gonadotrophin II release and gonadal growth in goldfish. *Rev. Reprod.* **2,** 55–68.

Twohey, M. B., Sorensen, P. W., and Li, W. (2003). Possible applications of pheromones in an integrated sea lamprey control program. *J. Great Lakes Res.* **29** (Suppl. 1), 794–800.

Uglem, I., Mayer, I., and Rosenqvist, G. (2002). Variation in plasma steroids and reproductive traits in dimorphic males of corkwing wrasse (*Symphodus melops* L.). *Horm. Behav.* **41,** 396–404.

Van Den Hurk, R., and Lambert, J. G. D. (1983). Ovarian steroid glucuronides function as sex pheromones for male zebrafish, *Brachydanio rerio*. *Can. J. Zool.* **61,** 2381–2387.

Van Den Hurk, R., and Resink, J. W. (1992). Male reproductive system as sex pheromone producer in teleost fish. *J. Exp. Zool.* **261,** 204–213.

Vermeirssen, E. L. M., and Scott, A. P. (1996). Excretion of free and conjugated steroids in rainbow trout (*Oncorhynchus mykiss*): Evidence for branchial excretion of the maturation–inducing steroid 17α,20β-dihydroxy-4-pregnen-3-one. *Gen. Comp. Endocr.* **101,** 180–194.

Vermeirssen, E. L. M., and Scott, A. P. (2001). Male priming pheromone is present in bile, as well as urine, of female rainbow trout. *J. Fish Biol.* **58,** 1039–1045.

Vermeirssen, E. L. M., Scott, A. P., and Liley, N. R. (1997). Female rainbow trout urine contains a pheromone which causes a rapid rise in plasma 17,20β–dihydroxy-4-pregnen-3-one levels and milt amounts in males. *J. Fish Biol.* **50,** 107–119.

Vickers, N. J. (2000). Mechanisms of animal navigation in odour plumes. *Biol. Bull.* **198,** 203–212.

Villars, T. A., Hale, N., and Chapnick, D. (1985). Prostaglandin F2α stimulates reproductive behaviour of female paradise fish (*Macropodus opercularis*). *Horm. Behav.* **19,** 21–35.

Vrieze, L. A., and Sorensen, P. W. (2001). Laboratory assessment of the role of a larval pheromone and natural stream odour in spawning stream localisation by migratory sea lamprey (*Petromyzon marinus*). *Can. J. Fish. Aquatic Sci.* **58,** 2374–2385.

Waring, C., and Moore, A. (1995). F-series prostaglandins have a pheromonal priming effect on mature male Atlantic salmon (*Salmo salar*) parr. *In* "Proceedings of the Fifth International Symposium on the Reproductive Physiology of Fish" (Goetz, F. W., and Thomas, P., Eds.), pp. 255–257. Fish Symposium 95, Austin, Texas.

Waring, C. P., and Moore, A. (1997). Sublethal effects of a carbamate pesticide on pheromonal mediated endocrine function in mature male Atlantic salmon (*Salmo salar* L.) parr. *Fish Physiol. Biochem.* **17,** 203–211.

Waring, C. P., Moore, A., and Scott, A. P. (1996). Milt and endocrine responses of mature male Atlantic salmon (*Salmo salar* L.) parr to water-borne testosterone, 17α,20β-dihydroxy-4-pregnen-3-one 20-sulfate, and the urines from adult female and male salmon. *Gen. Comp. Endocrinol.* **103,** 142–149.

Weissburg, M. J. (2000). The fluid dynamical context of chemosensory behaviour. *Biol. Bull.* **198,** 188–202.

Weltzien, F-A., Hoglund, E., Hamdani, E. H., and Døving, K. B. (2003). Does the lateral bundle of the medial olfactory tract mediate reproductive behaviour in male crucian carp? *Chem. Senses* **28,** 293–300.

Weltzien, F.-A., Andersson, E., Andersen, Ø., Shalchian-Tabrizi, K., and Norberg, B. (2004). The brain-pituitary-gonad axis in male teleosts, with special emphasis on flatfish (Pleuronectiformes). *Comp. Biochem. Physiol. A* **137,** 447–477.

Wilson, E. O., and Bossert, W. H. (1963). Chemical communication among animals. *In* "Recent Progress in Hormone Research" (Pincus, G., Ed.), Vol. 19, pp. 673–716. Academic Press, New York.

Wisenden, B. D. (2003). Chemically mediated strategies to counter predation. *In* "Sensory Assessment of the Aquatic Environment" (Collin, S. P., and Marshall, N. J., Eds.), pp. 236–251. Springer-Verlag, New York.

Wisenden, B. D., and Stacey, N. E. (2005). Fish semiochemicals and the evolution of communication networks. *In* "Animal Communication Networks" (McGregor, P., Ed.), pp 540–567. Cambridge University Press, Cambridge.

Xu, F., Greer, C. A., and Shepherd, G. M. (2000). Odour maps in the olfactory bulb. *J. Comp. Neurol.* **422,** 489–495.

Yamamoto, N., Oka, Y., and Kawashima, S. (1997). Lesions of gonadotropin-releasing hormone-immunoreactive terminal nerve cells: Effects on the reproductive behaviour of male dwarf gouramis. *Neuroendocrinology* **65,** 403–412.

Yamazaki, F. (1965). Endocrinological studies on the reproduction of the female goldfish, *Carassius auratus* L., with special reference to the function of the pituitary gland. *Mem. Fac. Fisher. Hokkaido Univ.* **13,** 1–64.

Yamazaki, F. (1990). The role of urine in sex discrimination in the goldfish *Carassius auratus*. *Bull. Fac. Fish. Hokkaido Univ.* **41,** 155–161.

Yamazaki, F., and Watanabe, K. (1979). The role of sex hormones in sex recognition during spawning behaviour of the goldfish, *Carassius auratus* L. *Proc. Indian Natl. Sci. Acad., B* **45,** 505–511.

Yambe, H., and Yamazaki, F. (2000). Urine of ovulated female masu salmon attracts immature male parr treated with methyltestosterone. *J. Fish Biol.* **57,** 1058–1064.

Yambe, H., and Yamazaki, F. (2001). A releaser pheromone that attracts methyltestosterone-treated immature fish in the urine of ovulated female trout. *Fish. Sci.* **67,** 214–220.

Yambe, H., Shindo, M., and Yamazaki, F. (1999). A releaser pheromone that attracts males in the urine of mature female masu salmon. *J. Fish Biol.* **55,** 158–171.

Yambe, H., Munakata, A., Kitamura, S., Aida, K., and Fusetani, N. (2003). Methyltestosterone induces male sensitivity to both primer and releaser pheromones in the urine of ovulated female masu salmon. *Fish Physiol. Biochem.* **28,** 279–280.

Youngson, A. F., Knox, D., and Johnstone, R. (1992). Wild adult hybrids of *Salmo salar* L. and *Salmo trutta* L. *J. Fish Biol.* **40,** 817–820.

Yun, S.-S., Scott, A. P., Siefkes, M., and Li, W. (2002). Development and application of an ELISA for a sex pheromone released by the male sea lamprey (*Petromyzon marinus* L.). *Gen. Comp. Endocrinol* **129,** 163–170.

Yun, S.-S., Scott, A. P., Bayer, J. M., Seelye, J. G., Close, D. A., and Li, W. (2003a). HPLC and ELISA analyses of larval bile acids from Pacific and western brook lampreys. *Steroids* **68,** 515–523.

Yun, S.-S., Scott, A. P., and Li, W. (2003b). Pheromones of the male sea lamprey, *Petromyzon marinus* L: Structural studies on a new compound, 3-keto allocholic acid, and 3-keto petromyzonol sulfate. *Steroids* **68,** 297–304.

Zeiske, E., Theisen, B., and Breuker, H. (1992). Structure development, and evolutionary aspects of the peripheral olfactory system. *In* "Fish Chemoreception" (Hara, T. J., Ed.), pp. 13–39. Chapman and Hall, London.

Zhang, C., Brown, S. B., and Hara, T. J. (2001). Biochemical and physiological evidence that bile acids produced and released by lake char (*Salvelinus namaycush*) function as chemical signals. *J. Comp. Physiol. B* **171,** 161–171.

Zheng, W., and Stacey, N. E. (1996). Two mechanisms for increasing milt volume in male goldfish. *J. Exp. Zool.* **276,** 287–295.

Zheng, W., and Stacey, N. E. (1997). A steroidal pheromone and spawning stimuli act via different neuroendocrine mechanisms to increase gonadotropin and milt volume in male goldfish (*Carassius auratus*). *Gen. Comp. Endocrinol.* **105,** 228–235.

Zheng, W., Strobeck, C., and Stacey, N. E. (1997). The steroid pheromone 17α,20β-dihydroxy-4-pregnen-3-one increases fertility and paternity in goldfish. *J. Exp. Biol.* **200,** 2833–2840.

Zhu, Y., Rice, C. D., Pang, Y., Pace, M., and Thomas, P. (2003). Cloning, expression, and characterisation of a membrane progestin receptor and evidence it is an intermediary in meiotic maturation of fish oocytes. *Proc. Natl. Acad. Sci, USA* **100,** 2231–2236.

Zielinski, B., Arbuckle, W., Belanger, A., Corkum, L. D., Li, W., and Scott, A. P. (2003). Evidence for the release of sex pheromones by male round gobies (*Neogobius melanostomus*). *Fish Physiol. Biochem.* **28,** 237–239.

Zippel, H. P., Sorensen, P. W., and Hansen, A. (1997). High correlation between microvillous olfactory receptor cell abundance and sensitivity to pheromones in olfactory nerve-sectioned goldfish. *J. Comp. Physiol. A* **180,** 39–52.

10

ANTHROPOGENIC IMPACTS UPON BEHAVIOUR AND PHYSIOLOGY

KATHERINE A. SLOMAN
ROD W. WILSON

1. INTRODUCTION

As has been described in great detail in the preceding chapters, the physiology and behaviour of fish are closely associated, but this relationship presents a fragile link open to interference from outside disturbances. The

Behaviour and Physiology of Fish: Volume 24
FISH PHYSIOLOGY

DOI: 10.1016/S1546-5098(05)24010-4

aquatic ecosystem has been altered by human activities in many ways, and so a volume on the interactions between fish physiology and behaviour would not be complete if it did not address how human activities can, and do, influence these interactions. As one of the most prevalent forms of anthropogenic disruption, chemical pollution will be the main focus of the current Chapter. Chemical contaminants are sourced from many places including households, industry, and agriculture and can be roughly divided into several groups. Inorganic chemicals (including metals), organic pollutants, radioisotopes, and, more recently, pharmaceuticals all make their way to the aquatic environments that fish inhabit and are thus a major concern for the health of aquatic ecosystems. Table 10.1 contains a list of chemicals that will be addressed in this Chapter with abbreviations and the major uses of these compounds. Less researched are acoustic disturbances (e.g., sonar systems, seismic air guns, underwater explosions, boat noise, and sonic booms), which are considered briefly in Chapter 2, Section 6. Physical disturbances such as habitat destruction from activities such as trawling, agriculture, and water abstraction, as well as the release of transgenic fish are other potential problems initiated by man. However, these are beyond the scope of the current review.

Many chemical contaminants target specific physiological systems and exert their effects on behaviour via physiological pathways. In some cases, behavioural changes may be detectable before physiological changes are manifested, and the use of fish behaviour as a sensitive and ecologically relevant tool for monitoring thresholds of effect has long been suggested (Little *et al.*, 1985; Rand, 1985; Beitinger, 1990). This Chapter is not intended to be an extensive overview of the volumes of literature available on anthropogenic effects on fish biology. Instead, it will look at the interactions between fish behaviour and physiology discussed in the preceding Chapters and consider some examples of how these interactions can be disrupted. Understudied and overlooked areas of behavioural toxicology are highlighted as potential directions for future research.

2. EFFECTS OF CONTAMINANTS ON COGNITION

2.1. Interference with Learning and Awareness

The effects of anthropogenic contaminants on cognition have been investigated mainly through studying conditioned responses in the laboratory. For example, using a conditional avoidance technique, Weir and Hine (1970) looked at retention of a conditioned response in goldfish, *Carassius auratus*, after exposure to sodium arsenate (Na_2HAsO_4), lead nitrate ($Pb(NO_3)_2$),

Table 10.1
Compounds Used in Fish Toxicity Studies With Their Abbreviations, Chemical Structures and Major Uses

Common name	Abbreviation/chemical structure	Major uses
Aroclor 1254	$C_{12}H_9Cl$	Aroclor is the registered trademark for the Monsanto Chemical Company for their PCB mixtures. Hundreds of industrial and commercial applications
Atrazine	2-chloro-4-ethylamino-6-isopropoylamino-*s*-triazine	A pre- and postemergence herbicide for the control of annual and perennial weeds
Bis (tributyltin) oxide	TBTO/ $C_{24}H_{54}OSn_2$	A tributyltin (TBT) compound used as a biocide, particularly in antifouling paints
Cadmium	Cd	A component of Ni-Cd batteries, TV sets, semiconductors, electroplating and control rods and shields within nuclear reactors
Carbaryl	1-napthyl-N-methyl carbamate	A broad spectrum carbamate insecticide
Carbofuran	2,3-dihydro-2,2-dimethyl-7-benzofuranyl methyl carbamate	A water-soluble carbamate insecticide used as an alternative to organophosphate chemicals
Chlorpyrifos	O,O-diethyl O-3,5,6-trichloro-2-pyridyl phosphorothioate	A broad spectrum organophosphate insecticide
Copper	Cu	A component of electrical generators, wiring, radio and TV sets, home heating systems, water pipes and fungicides
Crude oil	Mixture of hydrocarbons	A naturally occurring substance that is then refined into different usable products
Cypermethrin	cis, trans-3-(2,2-dichlorovinyl)-2,2-dimethylcyclopropanecarboxylate	A type II pyrethroid pesticide used in sheep dips and also to treat sea lice in aquaculture
Diazinon	O,O-diethyl 0-2-isopropyl-6-methyl (pyrimidine-4-yl)phosphorothioate	Organophosphate pesticide used to control household and public health insects
Dichlorodiphenyl trichloroethane	DDT/$C_{14}H_9Cl_5$	An organochlorine insecticide

(continued)

Table 10.1 *(continued)*

Common name	Abbreviation/chemical structure	Major uses
Dichlorvos	DDVP/ 2,2-dichlorovinyl dimethylphosphate	An organophosphorus insecticide
2,4-dinitrophenol	$C_6H_4N_2O_5$	A wood preservative and insecticide, also used in photochemicals and explosives and as an acid-base indicator
Diuron	N-(3,4-dichlorophenyl)-N,N-dimethyl urea	A substituted urea herbicide
Endosulfan (Thiodan)	6,7,8,9,10,10-hexachloro-1,5,5a,6,9,9a-hexahydro-6,9-methano-2,4,3-benzodioxathiopin-3-oxide	An organochlorine insecticide-miticide
17β-estradiol	E_2	An endogenous female hormone and a component of municipal sewage
Ethambutol	2,2′-(Ethylenediimine)-di-1-butanol	A tuberculostatic drug
Ethinyl oestradiol	EE_2	A synthetic estrogen used in the contraceptive pill.
Fenitrothion (Sumithion)	O,O-dimethyl O-(3-methyl-4-nitrophenyl phosphorothioate	An organophosphate insecticide and miticide
Fenvalerate (Pydrin)	Cyano(3-phenoxyphenyl) methyl 4-chloroalpha-(1-methylethyl) benzeneacetate	A synthetic pyrethroid insecticide
Flutamide	2-methyl-N-[4-nitro-3 (trifluoromethyl) phenyl] propanamide	An acetanilid, non-steroidal anti-androgen used in treatment of prostate cancer
Gentamicin sulphate	$C_{21}H_{43}N_5O_7$; $C_{20}H_4N_5O_7$; $C_{19}H_{39}N_5O_7$; $C_{20}H_{41}N_5O_7$	A broad spectrum bactericidal aminoglycoside antibiotic

Lead	Pb	A component of lead-acid batteries, piping, weights, roofing and lead cames in stain glass windows
Lindane	γ-HCH/ Hexachlorocyclohexane; benzene hexachloride	An organochlorine pesticide
Malathion	Diethyl (dimethoxy thiophosphorylthio) succinate	A nonsystemic, wide-spectrum organophosphate insecticide
Manganese	Mn	A component of dry cell batteries, used in iron and steel making and aluminum alloys
Mercury	Hg	A component of barometers, manometers, thermometers, mercury-vapor lamps and electric switches
Methylmercury	MeHg	A fungicide MeHg mainly occurs as a result of environmental transformation from inorganic Hg
Nickel	Ni	A metal used in the production of alloy steels, particularly stainless steel, gas turbines, coins, paints and rocket engines
Nicosulfuron	2-[(4,6-dimethoxpyrimidin-2ylcar bomoyl)sulfamoyl]-*N*-*N*-dimethylnicotinamide	A general use pesticice
No. 2 Fuel oil	Mixture of hydrocarbons	An oil with various energy uses including home heating.
1-octanol	$C_8H_{18}O$	A component of perfumes, pesticides, flavourings, essential oils, resistant coatings and linings.
ortho, para-dichlorodiphenyl trichloroethane	o,p′-DDT/$C_{14}H_9Cl_5$	An isomer of DDT
Paraquat	1,1′-dimethyl-4,4′bipyridinium	A quaternary nitrogen herbicide
Pentachlorophenol	PCP	A chlorinated hydrocarbon insecticide and fungicide.
Phenol	C_6H_6O	A component of the production of phenolic resins, and in plywood, construction, automotive and appliance industries

(*continued*)

Table 10.1 (*continued*)

Common name	Abbreviation/chemical structure	Major uses
Dichlorodiphenyl dichloroethane	p,p′-DDE / $C_{14}H_8Cl_4$	A DDT metabolite
Selenium dioxide	SeO_2	A component of electronics, ceramics, rubber vulcanising and photocopying
Sodium arsenate	Na_2HAsO_4	An insecticide (e.g., ant poison)
Sodium-dodecylbenzene sulphonate	$C_{18}H_{29}NaO_3S$	A component of many detergents
Sodium lauryl sulphate	SLS/ $C_{12}H_{25}NaO_4S$	An anionic detergent used in shampoos, hair conditioners and bubble baths
4-*tert*-octylphenol	$C_{14}H_{22}O$/ 4-(1,1,3,3-tetramethylbutyl) phenol	An intermediate in the production of surfactants and formaldehyde resins and formed from the breakdown of octylphenol ethoxylates
Thiobencarb	5-4-chlorobenzyldiethyl (thiocarbamate)	A selective herbicide.
Vinclozolin	3-(3,5-dichlorophenyl)-5-methyl-vinyl-1,3-oxazolidine-2,4-dione	A dicarboximide fungicide
Zinc	Zn	A metal used for galvanising, construction, brass and pharmaceuticals

mercuric chloride ($HgCl_2$), or selenium dioxide (SeO_2; see Table 10.1). Goldfish were placed in a conditioned avoidance apparatus. To obtain a correct score, when swimming in the part of the apparatus where light appeared, the fish were required to move past the midline of the tank to the dark, unelectrified side. If this did not occur within 3 seconds, then the fish received an electric shock. All contaminants impaired performance at levels below the concentration at which 1% mortality of the population occurred following 48-hour exposure and 7 subsequent days in clean water. Of the four contaminants investigated, mercury (Hg) had the greatest effect, causing disruption of conditioning at 3 μg/l. In similar experiments, Hg and aromatic hydrocarbons have been found to impair conditioned learning (Salzinger *et al.*, 1973; Purdy, 1989), whereas experiments with dichlorodiphenyltrichloroethane (DDT) have yielded conflicting results (Anderson and Peterson, 1969; McNicholl and Mackay, 1975a,b).

Low and high doses of chlorpyrifos (10 and 100 ng/ml on days 1–5 postfertilisation) had significant persisting effects on spatial discrimination learning in zebrafish, *Danio rerio* (Levin *et al.*, 2003). Zebrafish were given the choice of swimming into either the left or right compartment of a maze tank and the choice accuracy calculated. Swimming into the "wrong" compartment resulted in a confinement period. Average choice accuracy decreased with increasing concentration of chloropyrifos exposure, a behavioural impairment that lasted through to adulthood.

The physiological basis for behavioural learning in fish is complex (see Chapter 1) and thus understanding how these behaviours are affected by contaminants at the physiological level is not a simple matter. Disruption of memory, perception (i.e., sense organs), and locomotion could all play roles in the observed effects of toxicants on learning. A potential candidate for the inhibition of learning is the interference with cholinesterase activity (see Section 2.2.1; Hatfield and Johansen, 1972). Other malfunctions in the brain and the central nervous system (CNS) could also be involved.

2.2. Physiological Disruption of Brain Function

2.2.1. Neurotransmitters

The blood–brain barrier may provide a certain amount of protection against pollutants, but many contaminants can either cross this barrier or bypass this route by entering the brain directly via the olfactory system. Commonly measured indicators of disrupted brain function include compounds associated with nerve transmission. Cholinesterases (ChEs) can be divided into two types of enzymes: acetylcholinesterase (AChE) and butyrylcholinesterase (BChE). AChE degrades the neurotransmitter acetylcholine

(ACh) to end cholinergic neural transmission. Acetylcholine synapses are particularly associated with the spinal cord and the brain and participate in a number of central circuits, including those involved in motor coordination. BChE can also hydrolyze acetylcholine, although its preferred substrate is butyrylcholine. The majority of toxicological studies have measured either total brain cholinesterase (ChE) activity or AChE activity because less is known about the occurrence and distribution of BChE in fish (Chuiko, 2000). Compounds that result in the inhibition of AChE cause an accumulation of acetylcholine, eliciting cholinergic toxicity due to continuous stimulation of cholinergic receptors throughout the nervous system. AChE inhibition is commonly associated with exposure to organophosphate pesticides (Sandhal and Jenkins, 2002). There are many examples of compounds that affect AChE activity including carbofuran, diuron, nicosulfuron, paraquat, chlorpyrifos, and copper sulphate ($CuSO_4$) (Table 10.1; Nemcsók *et al.*, 1984; Sturm *et al.*, 1999; Bretaud *et al.*, 2000; Chuiko *et al.*, 2002; Flammarion *et al.*, 2002). Dose-dependent inhibition of AChE activity can vary among tissues. For example, high concentrations of organophosphates inhibit levels of AChE activity in the brain and muscle to similar levels but lower concentrations produce a higher inhibition in the brain than in the muscle (Straus and Chambers, 1995; Sancho *et al.*, 1997).

European eels, *Anguilla anguilla*, exposed to sublethal concentrations of thiobencarb for 4 days displayed inhibition of AChE activity in the brain, muscle, and gill tissues (Fernánadez-Vega *et al.*, 2002). Depuration in clean water resulted in a recovery of activity to about 60–80% of control values after 4 days. Behavioural effects were also observed, including reduced motility, loss of equilibrium, uncoordinated movements, and an increase in the levels of respiratory frequency. Additionally, European eels exposed to sublethal fenitrothion for 4 days suffered a 44% reduction in brain AChE activity at 0.02 mg/l and 64% at 0.04 mg/l (Sancho *et al.*, 1997). One week of recovery in clean water was not sufficient for AChE activity to return to control levels. The resultant behavioural changes in this study included erratic swimming and convulsions.

Exposure to malathion and diazinon, two well-known cholinesterase-inhibiting chemicals (Table 10.1), affected the swimming behaviour of rainbow trout, *Oncorhynchus mykiss* (Brewer *et al.*, 2001). Following 24 hours of exposure to malathion, trout showed decreases in distance swum and swimming speed and tended to swim in a more linear path than control fish. These behavioural changes were more pronounced after 96 hours of exposure but were eliminated after 48 hours in clean water. Diazinon caused similar symptoms to malathion but without subsequent recovery following 48 hours of depuration (the period of time spent in clean water with the potential for contaminant release). Behaviour correlated with brain ChE activity. In fish exposed to malathion, greater inhibition of brain ChE activity resulted in lower swimming speed and distance swum ($r^2 > 0.9$). A negative correlation existed between ChE activity and the rate of turning following malathion exposure ($r^2 = 0.7$). In contrast, behavioural parameters were less correlated to ChE activity in diazinon-exposed fish, although there was still a significant relationship between swimming behaviour and brain ChE activity ($r^2 = 0.4$; $p < 0.01$).

Beauvais *et al.* (2001) also examined the relationship between swimming behaviour and brain ChE activity in rainbow trout exposed to either carbaryl or cadmium (Cd). In addition, they measured the number and affinity of muscarinic cholinergic receptors (MChR). Brain ChE activity and swimming speed both decreased with increasing concentrations of carbaryl (188–750 μg/l) but there was no correlation between ChE activity and behaviour in trout exposed to 2.5 or 5.0 μg/l Cd. Neither chemical affected MChR binding affinity, and MChR number did not appear to downregulate.

These studies present correlative evidence that changes in swimming behaviour are linked with changes in brain cholinesterase activity. In addition, bream (*Abramis brama*) exposed to sublethal concentrations of the organophosphorus insecticide dichlorvos (DDVP) showed both decreased food consumption and inhibited AChE activity in the brain (Pavlov *et al.*, 1992), with behavioural effects apparent after 96 hours exposure to only one-fifteenth of the 48-hour LC50 (1.87 mg/l) DDVP. Intraperitoneal injection of drugs that block cholinergic receptors (e.g., atropine and TMB-4 [dipiroxim, 1,1′-3-methylene-bis (4-formilpiridine bromide) dioxim] recovered the feeding activity to that of control fish. In addition, TMB-4 resulted in recovery of AChE activity, suggesting a direct link between impaired feeding behaviour and lowered brain AChE activities.

2.2.2. Brain Monoamines

Links between brain monoaminergic systems and stressful situations have been established (Winberg and Nilsson, 1993; see Chapter 5 for further details) and monoaminergic neuronal systems, particularly the serotonergic neurons, are well conserved within the vertebrate phylum (Parent, 1984). Examples of biogenic amines include serotonin (5-HT; 5-hydroxytryptamine), 5-hydroxyindoleacetic acid (5-HIAA; the major metabolite of 5-HT), dopamine (DA), homovanillic acid (HVA; a DA metabolite), 3,4-dihydroxyphenylacetic acid (DOPAC; a DA metabolite), norepinephrine (NE), and 3-methoxy, 4-hydroxyphenyl glycol (MHPG; an NE metabolite).

Waterborne exposure of rainbow trout to 0.05 and 0.1 mg/l of lindane (γ-HCH; an organochlorine pesticide; see Table 10.1) for 2 hours caused a dose-dependent decrease in the hepatic concentration of tryptophan (the amino-acid precursor of 5-HT), whereas blood serum tryptophan levels increased (Rozados *et al.*, 1991). Brain tryptophan concentrations were not affected by lindane exposure but 5-HT increased significantly in the hypothalamus and decreased in the optic tectum (see Chapter 1 for details on brain structure). Variations in effects of lindane among different brain regions were attributed to a specific mode of action and could relate to specific alterations in behaviour. In contrast, rainbow trout given intraperitoneal coconut oil implants containing 0.05 mg of lindane showed decreased concentrations of tryptophan in the hypothalamus (Aldegunde *et al.*, 1999). Decreased concentrations of both 5-HIAA and 5-HT were also found in the hypothalamus indicating a lowering of hypothalamic serotonergic metabolism. In the telencephalon of lindane-implanted fish, levels of tryptophan were unaltered but concentrations of 5-HT were elevated, resulting in lower 5-HIAA/5-HT ratios. It appears, therefore, that the decrease in serotonergic activity in the telencephalon is not related to tryptophan availability but rather a decrease in 5-HT synthesis or release. Levels of brain lindane were higher in implanted fish than control fish but no discrimination between different regions of the brain was made when measuring the brain accumulation of this pesticide.

Khan and Thomas (1996) considered the effects of a dietary PCB (polychlorinated biphenyl) mixture (Aroclor 1254) on monoamine concentrations in regions of the hypothalamus in male Atlantic croaker (*Micropogonias undulatus*). Similar to the effects of lindane exposure seen by Aldegunde *et al.* (1999) but in contrast to the work of Rozados *et al.* (1991), a significant decline in 5-HT was seen in the preoptic-anterior hypothalamus and medial and posterior hypothalamus. Concentrations of other monoamines were also measured in response to this 30-day dietary exposure to Aroclor 1254 (0.1 mg per 100 g body weight) and DA levels were also seen to decline. Both the 5-HIAA/5-HT and DOPAC/DA ratios increased in the hypothalamus following exposure to Aroclor.

Methyl mercury (MeHg), like many organic chemicals, is capable of crossing the blood-brain barrier and has been suggested as a potential explanation for differences in brain monoamine concentrations seen among mummichog (*Fundulus heteroclitus*) populations from different habitats. Heads of untreated mummichog larvae from a site polluted by a mixture of chemicals including MeHg had higher levels of DA and its metabolites (DOPAC and HVA) than larvae from a "clean" site (Zhou *et al.*, 1999). However, no differences in concentrations of 5-HT or 5-HIAA/5-HT ratios were seen between populations. Dopaminergic and serotonergic activity were suppressed in mummichog larvae from the clean site when they were subsequently exposed to 10 μg/l MeHg after hatching (Zhou *et al.*, 1999).

Concentrations of 5-HT decreased in a dose-dependent fashion in the hypothalamus of male tilapia, *Oreochromis mossambicus*, following a

6-month exposure to 0.015 and 0.05 mg/l $HgCl_2$ (Tsai *et al.*, 1995) beginning 7 days posthatching. Lead (Pb) caused an increase in whole brain levels of 5-HT and NE in fathead minnows, *Pimephales promelas* (Weber *et al.*, 1991), which was associated with changes in feeding behaviour, but Munkittrick *et al.* (1990) found little evidence of brain monoamine disruption following exposure of white sucker (*Catostomus commersoni*) to waterborne copper (Cu) and zinc (Zn) associated with mining wastes. De Boeck *et al.* (1995) exposed juvenile common carp, *Cyprinus carpio*, to Cu concentrations of 14, 22, and 54 μg/l for 1 week and observed dose-dependent falls in 5-HT and DA concentrations in the telencephalon, with approximately 50% of losses occurring following exposure to the highest Cu concentrations. Less dramatic decreases in these monoamines were also seen in the hypothalamus and brainstem. No effects of Cu on 5-HIAA concentrations were seen.

2.2.3. Brain Pathologies

Other toxicant-induced changes in the brain and central nervous system can occur, as well as alterations in brain neurotransmitters and monoamines. At a cellular level within neurons, toxicants can interfere with aerobic metabolism, protein synthesis, intermediate metabolism, and permeability of cell membranes. Damage to axonal transport, non-neuronal cells, and blood capillaries will also affect nervous system function (Buckley, 1998). Handy (2003) reported edema (swelling) and the formation of vacuoles in the cell body layers of the tectum close to the hypothalamic region of rainbow trout exposed for 6 weeks to dietary copper (1000 mg Cu/kg). Berntssen *et al.* (2003) found that 5 mg/kg of MeHg (dietary) increased by twofold the antioxidant enzyme superoxide dismutase (SOD) in the brains of Atlantic salmon, and 10 mg/kg caused a sevenfold increase in lipid peroxidative products and a subsequent 1.5-fold decrease in antioxidant enzymes. Brains of fish fed 10 mg/kg MeHg also showed pathological damage and reduced monoamine oxidase activity; these fish also showed reduced postfeeding activity.

2.3. Linking Alterations in Brain Function with Behavioural Changes

In general, contaminants have the ability to interfere with cholinesterase activity, brain monoamine concentrations, and/or to cause physical damage to brain tissue. Many compounds decrease brain ChE activity; in some cases, this has been correlated to changes in swimming (Beauvais *et al.*, 2000, 2001; Brewer *et al.*, 2001) and feeding behaviour (Weber *et al.*, 1991; Pavlov *et al.*, 1992). The effects of aquatic contaminants on brain monoamines are less clear and in some cases conflicting. Some of the apparent conflict in results may be explained by the fact that many substances in the central nervous system show endogenous rhythms (Handy, 2003), so it is possible that different sampling times are affecting the results. Changes in brain monoamine concentrations are also dependent upon brain region. Although there is evidence that monoamines play an important role in behavioural processes and there is increasing evidence for physiological disruption of these systems by toxicants, there appears to be a surprising lack of data relating toxicological changes in brain monoamines to changes in behaviour. We have presented evidence for contaminants affecting learning behaviours and also examples of toxicants interfering with brain function. There is, however, a large gap in our knowledge of how these studies interrelate and further research is now essential to understand how physiological changes in brain function affect cognition.

3. EFFECTS OF CONTAMINANTS ON THE SENSES

Olfaction, vision, sound, mechanoreception, and electroreception all provide fish with valuable information about their surrounding environment and allow effective communication between conspecifics (see also Chapter 2). Little research has been conducted into the effects of toxicants on communication and subsequent behavioural effects but physiological disruption of sensory systems obviously has the potential to affect communication pathways. The important role of communication in predator/prey relations, social behaviour, and reproduction means that disruption of the sensory systems has far-reaching implications. The small number of papers that have considered both toxicant-induced changes in sensory systems and subsequent behavioural effects are summarised in Table 10.2.

3.1. Olfaction

The olfactory apparatus is particularly vulnerable to waterborne contaminants because it is in direct contact with the external medium and toxic damage to this sense organ occurs easily. A method regularly used to monitor the damaging effects of waterborne pollutants on the olfactory apparatus is to consider how the normal electrical response of the olfactory epithelium (measured as an electro-olfactogram [EOG]) in response to odourants varies following exposure to contaminants. Many of these odourants (e.g., L-serine) are known to elicit neurological and avoidance behaviour responses in fish (Rehnberg and Schreck, 1986). Histological examination of the olfactory apparatus can yield evidence of toxic effects, in particular by considering changes in olfactory receptor number and affinity.

Table 10.2

Studies That Have Considered Both the Physiological/Morphological Injury Caused to Sensory Systems Through Toxicant Exposure and Subsequent Behavioural Changes

Sense	Pollutant	Species	Exposure concentration	Duration of exposure	Route of exposure	Injury	Behavioural effect	Reference
Olfaction	Cu	Colorado pikeminnow	66 μg/l	24 hours	Waterborne	Decrease in ciliated olfactory receptor cells	Decreased response to alarm pheromone	Beyers and Farmer (2001)
Olfaction	Cd	Rainbow trout	2 μg/l	7 days	Waterborne	Accumulation of Cd in olfactory system	Decreased response to alarm pheromone	Scott *et al.* (2003)
Olfaction	Detergent	Yellow bullheads	4–5 mg/l	24 hours to 4 weeks	Waterborne	Thickening of sensory cell borders in olfactory rosette folds	Reduced chemical detection of food	Bardach *et al.* (1965)
Taste	Detergent	Yellow bullheads	0.5–1.0 mg/l	24 hours to 4 weeks	Waterborne	50% of taste buds eroded	Reduced chemical detection of food	Bardach *et al.* (1965)

The electrical responses of the fish olfactory epithelia in response to olfactory stimulants such as L-serine or L-alanine can be studied by immobilising the fish by light anesthesia and removing the skin flaps around the nares to allow direct dispensing of solutions onto the epithelium (Hara *et al.*, 1976; Baatrup *et al.*, 1990; Winberg *et al.*, 1992). Using this technique, it has been demonstrated that Hg depresses the olfactory response to stimulants (Hara *et al.*, 1976) with inorganic $HgCl_2$ and organic CH_3HgCl (methylmercuric chloride) exerting very different reversibility, the latter having a greater element of irreversibility (Baatrup *et al.*, 1990). Cu can also decrease olfactory bulb responses in a dose and time dependent manner (Hara *et al.*, 1976; Winberg *et al.*, 1992; Hansen *et al.*, 1999). Organic contaminants such as the anionic detergent sodium lauryl sulphate (Hara and Thompson, 1978) and diazinon (Moore and Waring, 1996) have similar effects. Interference with the response to a variety of olfactory stimuli ranging from amino acids to sex pheromones suggests that disruption of olfaction can have implications for many behaviours, ranging from feeding to reproduction.

Histological methods show that the number of olfactory receptors decreases in Chinook salmon, *Oncorhynchus tshawytscha*, exposed to >50 μg/l Cu and in rainbow trout exposed to ≥ 200 μg/l for 1 hour (Hansen *et al.*, 1999). Three classes of receptors were detected in the olfactory tissue of both species: microvillar receptors with 1- to 2-μm diameter apical knobs, ciliated receptors with dendrites having 1- to 2-μm diameter apical knobs, and a type II ciliated receptor having a 3.5- to 5-μm flat dendrite apex containing both cilia and microvilli. Exposure to 25 μg/l of Cu for 4 hours reduced the number of olfactory receptors in both species. Histological examination showed that the loss of receptors was due to cellular necrosis. Copper-induced damage was only evident in ciliated type I and microvillar receptors. Ciliated type II receptors and other cells in the epithelium were not visibly damaged. Metals may also disrupt olfaction by decreasing the affinity of odourants for their receptors or by forming metal-odour complexes. For example, Hg can inhibit serine-receptor binding, which has been linked to changes in behavioural avoidance of serine by coho salmon, *Oncorhynchus kisutch*, (Rehnberg and Schreck, 1986). When considering the effects of contaminants on the olfactory system, it should be remembered that many of the sensory cells are replaced on a regular basis (e.g., see Zeni *et al.*, 1995) and so recovery from contaminant injury may be possible, especially if a depuration period is allowed.

Not only is the olfactory sense vulnerable to damage from direct contact with a contaminated external medium, but the olfactory apparatus can also transport contaminants along nerve connections towards the brain. Using autoradiography, Tjälve *et al.* (1986) considered the whole body distribution of Cd during waterborne exposure of brown trout (*Salmo trutta*) and found that the tissues accumulating the highest levels of Cd were the olfactory apparatus, the gills, and the trunk kidney. Following waterborne exposure, Cd was present in the epithelium of the olfactory rosette, the olfactory nerve, and the anterior part of the olfactory bulb of the brain. A similar pattern of accumulation occurred in the olfactory systems of pike (*Esox lucius*) exposed to waterborne Hg (Borg-Neczak and Tjälve, 1996). Transport of Cd in the olfactory nerves is believed to be as a Cd-metallothionein (CdMT) complex (Tallkvist *et al.*, 2002) that moves at a maximal velocity of 2.38 ± 0.1 mm/h at 10 °C by active axoplasmic transportation, being similar to the transport of proteins and amino acids rather than simple diffusion (Gottofrey and Tjälve, 1991).

Similar studies in rats have found that manganese (Mn) can migrate via secondary and tertiary olfactory pathways and via further connections into most parts of the brain and spinal cord (Tjälve *et al.*, 1996). This is in contrast to Cd, which cannot pass the synaptic junctions between the primary and secondary neurons. Nickel (Ni), again in contrast to Cd, is taken up in the olfactory epithelium and can pass to secondary and tertiary olfactory neurons (Henriksson *et al.*, 1997; Tallkvist *et al.*, 1998). Ni is transported through the olfactory system at a maximal rate of 0.13 mm/h at 10 °C, indicating that it travels by a class of slow axonal transport bound to both particulate and soluble cytosolic constituents (Tallkvist *et al.*, 1998). Organometals can also undergo axonal transport. Rouleau *et al.* (2003) demonstrated that ^{113}Sn-labelled tributyltin (TBT) could reach the brain via sensory systems and also across the blood–brain barrier, with "hot spots" of TBT accumulation occurring in parts of the brain receiving sensory nerves from water-exposed sensory organs.

Therefore, direct contact between waterborne contaminants and the olfactory system has the potential to transport contaminants to the brain. Although Cd and Hg are transported along the primary olfactory pathways to the olfactory bulbs but can not go any further, Ni can pass as far as secondary and tertiary neurons and Mn and TBT have the capacity to be passed transneuronally to other parts of the brain.

3.2. Vision

Vision is an important sense for fish, particularly diurnal species, with many behavioural communication displays utilising visual signals; examples such as luminescence (Sasaki *et al.*, 2003) and reflective communication (Rowe and Denton, 1997) are deeply reliant upon this sense. Changes in turbidity can occur naturally but can also be accentuated by activities such as dredging, sewage disposal, and soil erosion through deforestation. Increases in turbidity can cause serious problems for visual communication and disrupt essential behaviours (Sweka and Hartman, 2003). Disturbances in mate choice, sexual selection, and reproductive isolation in cichlid species of Lake Victoria have been attributed to constraints in colour vision resulting from increased turbidity (Seehausen *et al.*, 1997).

Toxicants may specifically target the visual system. Neurotransmitters within the fish retina play a role in retinal signal processing. In particular, GABA has been indicated as an important retinal neurotransmitter (Douglas and Djamgoz, 1990). Although little is understood regarding the effects of toxicants on eye function and visual processing, there is some evidence to suggest toxicant disruption of visual neurotransmitters. The high nerve activity associated with the visual system may also increase sensitivity to toxicant exposure. Thiobencarb exposure (0.22 mg/l for 96 hours) had significant inhibitory effects on total and specific acetylcholinesterase (AChE) activity in the eyes of yellow eel (*Anguilla anguilla*; Sancho *et al.*, 2000). AChE activity was reduced by up to 70%, and was still depressed after 1 week of recovery in clean water. Parental exposure to *ortho, para*-DDT (*o,p'*-DDT; 2 or 10 μg/l/day in the diet for 1 month) affected the responses of larval Atlantic croaker to a visual stimulus (Faulk *et al.*, 1999). Mean and maximum burst swimming speeds were decreased in *o,p'*-DDT-exposed fish. Other chemicals may also have the potential to interfere with optic nerve function (e.g., the tuberculostatic drug ethambutol; Parkyn and Hawryshyn, 1999).

3.3. Octavolateral System

The auditory, vestibular, and lateral line systems make up the octavolateral system of fish and are often considered together as they share a common sensory cell type (Schellart and Wubbels, 1998; see also Chapter 4). Thus toxicants have the potential to interfere with these systems by similar mechanisms. Acoustic communication is particularly important for nocturnal species where visual communication is reduced, and the lateral line is used for communication in shoaling fish where information about the location of other fish or objects is perceived by this pressure-sensitive organ.

Ototoxins are chemicals that can damage the hair cells in the ear and lateral line and have been utilised in particular by researchers to investigate the physiological and behavioural role of the lateral line. Examples of ototoxins include the antibiotics gentamicin sulphate and streptomycin and the diuretic drug amiloride, all of which are known to block lateral-line function (Wiersinga-Post and van Netten, 1998).

A concentration of 0.002% (~20 mg/l) of the aminoglycoside antibiotic gentamicin sulphate was sufficient to cause damage to the sensory hair cells in the lateral line receptors of oscars (*Astronotus ocellatus*; Song *et al.*, 1995). However, scanning electron microscopy (SEM) revealed that only one class of the lateral line receptors, the canal neuromasts, were affected but there was no evidence of damage to the superficial neuromasts. The pattern of cell destruction seen by Song *et al.* (1995) is similar to that previously seen in the fish ear. Although Song *et al.* (1995) did not consider behavioural effects of these pharmaceuticals, interference with lateral-line function has clear behavioural implications. Rheotaxis trials were carried out by Baker and Montgomery (2001) on banded kokopu (*Galaxias fasciatus*) by looking at the orientation of fish in a current. These experiments were conducted in the dark using infrared light so that the fish had no visual cues. Fish were exposed to 1 or 2 μg/l of Cd and interference with lateral-line function occurred after exposure to the higher concentration. A positive control using streptomycin sulphate was also used, as this compound is known to entirely block lateral line function. Cadmium-induced interference with the lateral line was reversible after 14 days in clean water.

There is obvious potential for contaminants to interfere with lateral line and auditory function. In general, the use of ototoxic pharmaceuticals in fish research has helped our fundamental understanding of lateral-line function. However, with increasing concern regarding pharmaceuticals in the environment (Hirsch *et al.*, 1999), the occurrence of unidentified ototoxic compounds in the environment should perhaps be considered. For example, the antibiotic gentamicin, which has clear behavioural effects in fish, has been detected in hospital wastewater at concentrations ranging from 0.4–7.6 μg/l (Löffler and Ternes, 2003).

As well as studies investigating the effects of toxicants on the octavolateral system, there has been considerable research into the effects of noise pollution on fish behaviour (Knudsen *et al.*, 1997; Wardle *et al.*, 2001; Maes *et al.*, 2004). The effects of anthropogenic acoustic disturbances on fish behaviour are also discussed in Chapter 2.

3.4. Taste

Another sense important in determining fish behaviour is that of taste. Disruption of taste can have severe implications for feeding behaviours. Both "hard" and "soft" detergents caused disintegration of the central portion of the taste buds in yellow bullheads, *Ictalurus natali* (Bardach *et al.*, 1965). Action potentials from affected taste buds diminished after exposure to detergents and chemical detection of food was reduced. Catfish (*Ictalurus sp.*) exposed to 3 mg/l of an anionic detergent (Na-dodecylbenzene sulphonate) for 15 days showed histological changes in the mandibular barbels (Zeni *et al.*, 1995). The taste buds were detached from the surrounding epidermis and protruded outwards. No interruption of this degenerative process had occurred after 3 days transfer to clean water, but after 6, 9, and 12 days in clean water the authors noted a large number of taste buds with a lesser degree of damage.

4. EFFECTS OF CONTAMINANTS ON PREDATOR-PREY RELATIONSHIPS AND PARASITE INTERACTIONS

4.1. Predator Avoidance and Foraging Behaviours

Interactions between predator and prey are complex and although predation is usually density-dependent (Sutherland, 1997), it is very difficult to predict the effects of toxicological interference on these relationships using lab-based studies. Many fish species are piscivorous but will also be preyed upon by other fish species. Any contaminants entering the aquatic environment therefore have the potential to affect more than one trophic level. Although top predators external to the aquatic environment, such as birds and mammals, may not be directly exposed to waterborne contaminants, dietary exposure through eating contaminated fish may have effects further up the food chain.

Numerous studies have considered either the effects of chemical pollutants on the ability of fish to avoid predation (predator avoidance) or foraging behaviour (predatory ability; Tables 10.3 and 10.4, respectively). All of the studies given in Table 10.3 report a decreased ability to avoid predation caused by a suite of different chemicals. For example, guppies (*Poecilia reticulata*) exposed to pentachlorophenol (PCP; Table 10.1) have a reduced ability to avoid largemouth bass (*Micropterus salmoides*) predation (Brown *et al.*, 1985). Additionally, largemouth bass exposed to a lower concentration of PCP have a reduced ability to feed on live brine shrimp (*Artemia* spp.; Brown *et al.*, 1987). Many other examples of reduced foraging ability in response to toxicants exist (Table 10.4).

These studies in combination allow the conclusion that predator–prey relationships can be influenced by aquatic toxicants. But the simplicity of the studies makes it very difficult to predict how these relationships might be affected and any subsequent population effects. When both predator and prey are exposed, the situation could become far more complex, depending on the relative sensitivity of predator and prey species and the local and migratory movements of both. Studies that have exposed both predator and prey simultaneously have found that the feeding behaviour of exposed predators may still be disadvantaged when feeding on exposed prey (Sandheinrich and Atchison, 1989). Thus, in some cases either the predator

Table 10.3

Studies That Have Considered the Effects of Chemical Contaminants on the Ability of Fish to Behaviourally Respond to a Predation Threat

Pollutant	Species	Exposure concentration	Duration of exposure	Exposure route	Lowest concentration/shortest time effect seen	End-point measured	Reference
Sumithion (Fenitrothion)	Atlantic salmon	0.1 and 1 mg/l	24 hours	Waterborne	1 mg/l	Increase in vulnerability to predation by brook trout (*Salvelinus fontinalis*)	Hatfield and Anderson (1972)
Endosulfan	Medaka	0.5, 1, 1.5 μg/l	24 hours	Waterborne	1 μg/l	Consumed quicker by bluegill	Carlson *et al.* (1988)
Fenvalerate	Medaka	0.4–1.6 μg/l	24 hours	Waterborne	0.9 μg/l	Consumed quicker by bluegill	Carlson *et al.* (1988)
1-octanol	Medaka	8.9–17.8 mg/l	24 hours	Waterborne	7.8 mg/l	Consumed quicker by bluegill	Carlson *et al.* (1988)
2,4-dinitrophenol	Medaka	7.5–14.5 mg/l	24 hours	Waterborne	10 mg/l	Consumed quicker by bluegill	Carlson *et al.* (1988)
PCP	Guppy	100, 500, 700 μg/l	4 weeks	Waterborne	100 μg/l	Reduced ability to avoid predation by largemouth bass	Brown *et al.* (1985)

(*continued*)

Table 10.3 (*continued*)

Pollutant	Species	Exposure concentration	Duration of exposure	Exposure route	Lowest concentration/ shortest time effect seen	End-point measured	Reference
EE_2	Stickleback	100 ng/l	~1 month	Waterborne	NA	Exposed fish were more active under predation risk	Bell (2004)
Bis(tributyltin)-oxide	Stickleback	3, 9, 27 μg/l	24 hours	Waterborne	3 μg/l	Decreasing ability to immediately respond to visually stimulated predator attack	Wibe *et al.* (2001)
MeHg	Mummichog	5 and 10 μg/l	Fertilisation to hatching or 1 week as larvae	Waterborne	10 μg/l	More susceptible to predation by grass shrimp or 1-year old mummichogs	Weis and Weis (1995a)
Hg	Mosquito fish (*Gambusia affinis*)	0.005–0.1 mg/l	24 hours	Waterborne	0.01 mg/l	Increased predation rate by largemouth bass	Kania and O'Hara (1974)

Hg	Golden shiner (*Notemigonus crysoleucas*)	455 and 959 ng/g	90 days	Dietary	455 ng/g	Exposed fish formed less compact schools after predator disturbance	Webber and Haines (2003)
Cd	Fathead minnow	0.013–0.5 mg/l	48 hours to 21 days	Waterborne	0.375 mg/l for 48 hours 0.025 mg/l for 21 days	Proportion of fish eaten by largemouth bass increased	Sullivan *et al.* (1978)
Pb	Mummichog	0.1–1.0 mg/l	2–4 weeks	Waterborne	1.0 mg/l at 4 weeks	Increased vulnerability to predation by grass shrimp	Weis and Weis (1998)

Table 10.4
Studies That Have Considered the Effects of Chemical Contaminants on the Ability of Fish Foraging or Predatory Behaviour

Pollutant	Species	Exposure concentration	Duration of exposure	Exposure route	Lowest concentration/ shortest time effect seen	Endpoint measured	Reference
PCP	Largemouth bass	1.6–88 μg/l	8 weeks	Waterborne	67 μg/l	Significantly fewer feeding acts and lower rate of capture of brine shimp (*Artemia*) and *Daphnia pulex*	Brown *et al.* (1987)
PCP	Largemouth bass	50 μg/l	14 days	Waterborne	NA	Less efficient at feeding on live Poecilid fish, *Xiphorous maculatas* and *X. helleri*	Mathers *et al.* (1985)
Sea-water-soluble fraction of Cook Inlet Crude Oil	Coho salmon	230–530 μg/l total hydrocarbons	17 days	Waterborne	NA	Reduced feeding on live rainbow trout fry	Folmar *et al.* (1981)
No. 2 fuel oil	Coho salmon	800 μg/l total hydrocarbons	17 days	Waterborne	NA	Reduced feeding on live rainbow trout fry	Folmar and Hodgins (1982)
Aroclor 1254	Coho salmon	150 g/kg	Single injection	Intraperitoneal injection	NA	Reduced feeding on live rainbow trout fry	Folmar and Hodgins (1982)

MeHg	Mummichog	2, 5, 10 μg/l	Fertilisation to hatching	Waterborne	10 μg/l	Slower ability to capture *Artemia salina* nauplii	Weis and Weis (1995b)
MeHg	Zebrafish	5–15 μg/l	Fertilisation to hatching	Waterborne	10 μg/l	Ability to capture *Paramecium* was reduced	Samson *et al.* (2001)
Pb	Mummichog	0.1, 0.3, 1 mg/l	4 weeks	Waterborne	1 mg/l	Ability to capture *Artemia salina* nauplii decreased	Weis and Weis (1998)
Cd	Lake charr (*Salvelinus namaycush*)	0.5 and 5 μg/l	106–112 days	Waterborne	0.5 μg/l	Reduced consumption of fathead minnow	Kislalioglu *et al.* (1996)
Cd	Bluegill	30–240 μg/l	21–22 days	Waterborne	37.3 μg/l	Attack rates on *Daphnia* were reduced	Bryan *et al.* (1995)
Cd	Brook trout	0.5 and 5 μg/l	30 days	Waterborne	0.5 μg/l	Decrease in ability to capture mayfly (*Baetis tricaudatus*) nymphs.	Riddell *et al.* (2005).

or prey may be more severely impacted by a pollutant (Hedtke and Norris, 1980). The potential also exists for toxicants to make fish less susceptible to predation, for example, by reducing activity and therefore the chances of being detected (although this in turn would reduce foraging ability). More studies are clearly necessary to better understand and predict the effects of contaminants on the complexity of predator–prey relations.

The majority of toxicological studies consider the effects of a single chemical and yet this is clearly not representative of the cocktail of chemicals present in the aquatic environment. In a suite of studies, interlinking laboratory and field-based approaches, Weis and colleagues considered the impacts of combined aquatic pollutants on the susceptibility of mummichogs to predation. These studies focused on comparisons between a clean site (Tuckerton) and a polluted site (Piles Creek), in New Jersey, located in a salt marsh near many industrial works. Piles Creek has elevated levels of many pollutants in its sediments including Hg, Cu, Cd, Zn, and polyaromatic hydrocarbons (PAHs) (Weis and Khan, 1990).

Mummichogs from Piles Creek had a reduced ability to capture prey and to avoid being eaten compared to Tuckerton mummichogs (Smith and Weis, 1997). Mummichogs taken from the clean site at Tuckerton and exposed for 3 weeks to water, sediment, and food from Piles Creek experienced a substantial decrease in prey capture ability. However, Piles Creek fish allowed to depurate for 10 weeks in clean water did not increase their ability to catch grass shrimp, *Palamonetes pugio*. Brain accumulation of Hg was high in Piles Creek fish and did not decrease following depuration in clean water. Tuckerton fish had lower brain Hg concentrations but when exposed to Piles Creek conditions showed a corresponding increase in brain Hg concentration (Figure 10.1). Thus the foraging and predator avoidance behaviours of mummichog appeared to correlate with brain Hg concentrations.

Pollutants may not only affect the predator–prey relationships of the exposed generation but can have transgenerational effects. One-month-old mummichog larvae bred from adults taken from a polluted site were more susceptible to predation than larvae bred from adults from a clean site (Zhou and Weis, 1999). When embryos and larvae (bred from adults of both clean and polluted sites) were subsequently exposed to MeHg in the lab, both groups of larvae had a decreased ability to avoid predation by yearling mummichogs. However, larvae bred from adults from the clean site were more resistant to MeHg exposure and least vulnerable to predation (Zhou and Weis, 1998). Thus parental exposure altered not only the ability of offspring to avoid predation but also the effects of subsequent toxicant exposure.

Other studies on the behaviour of offspring from adults that have been exposed to contaminants have yielded conflicting results. No effect on predator avoidance or feeding behaviour of fry was seen in lake trout,

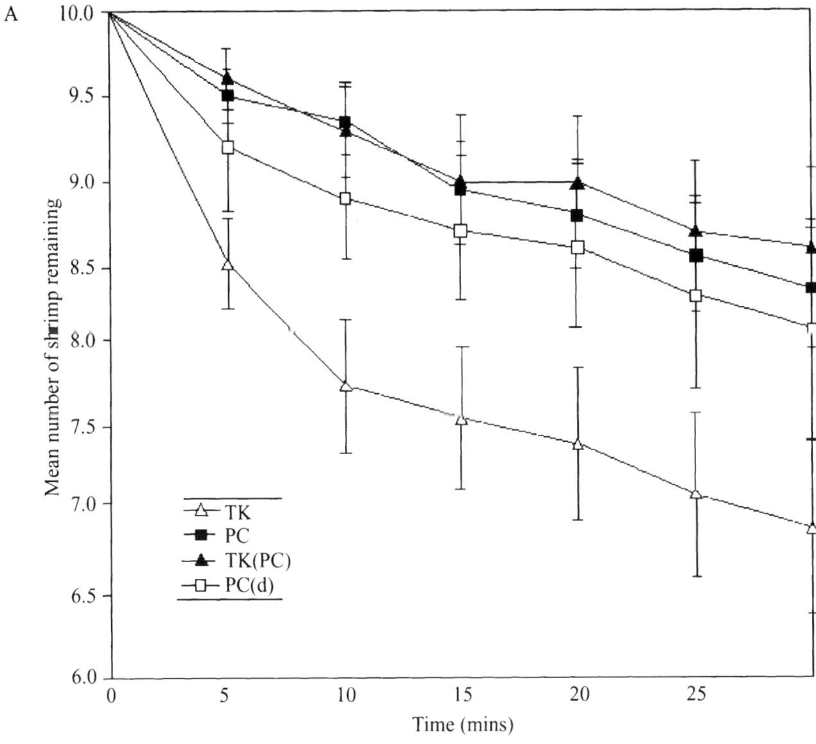
A
10.0
9.5
9.0
8.5
8.0
7.5
7.0
6.5
6.0
Mean number of shrimp remaining
TK
PC
TK(PC)
PC(d)
0
5
10
15
20
25
Time (mins)

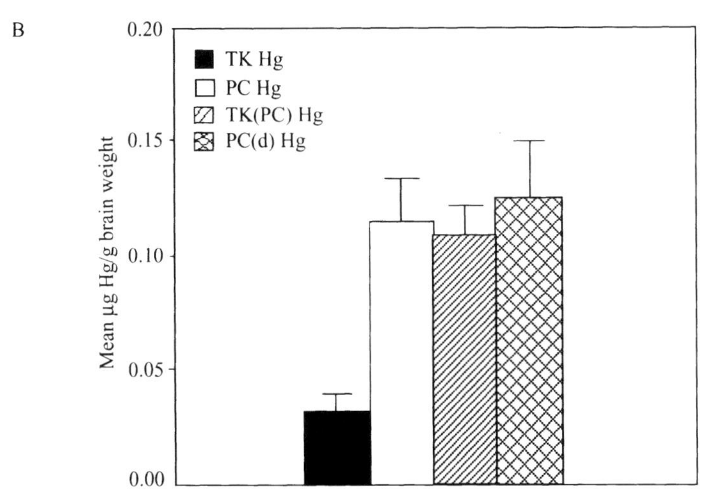
B
0.20
0.15
0.10
0.05
0.00
Mean µg Hg/g brain weight
TK Hg
PC Hg
TK(PC) Hg
PC(d) Hg

Salvelinus namaycush, following parental exposure to PCBs and DDT through field exposure in Lake Michigan and contaminant-soaked food pellets (Savino *et al.*, 1993). Larvae from adult Atlantic croaker fed Aroclor 1254 (0.4 mg/kg fish/day for 2 weeks) during the final stages of gonadal recrudescence exhibited decreased startle responses, possibly making them more susceptible to predation (McCarthy *et al.*, 2003).

In addition to the overall complexity associated with predator/prey relations, there are inherent problems with using predator/prey behaviours as end-points in toxicological studies. The ability to avoid being eaten, or to catch food to eat, consists of a series of behaviours rather than one fixed act. For example, foraging involves locating, identifying, chasing, capturing, handling, and consuming behaviours and toxicants can interfere with one or many of these stages. Additionally, such parameters can be influenced by interpopulation differences in behaviour (Weis *et al.*, 2001), different food rations (Mathers *et al.*, 1985), different-sized prey (Bryan *et al.*, 1995), acclimation (Weis and Khan, 1990), and genetic differences between populations (Zhou *et al.*, 2001). Chemical contamination may alter prey distributions and can result in decreased food conversion efficiencies (Mathers *et al.*, 1985). Nevertheless, it is clear that many chemical contaminants do have the potential to alter both predator avoidance and foraging ability both through direct exposure and through transgenerational effects. Any attempt to use these behavioural changes as indicators of exposure therefore necessitates an understanding of the physiological processes underlying these behaviours.

4.2. Linking Physiological Mechanisms with Altered Predator–Prey Behaviours

Olfaction plays a crucial role in predator–prey interactions and several studies have associated the physiological effects of contaminants on the olfactory system with changes in predator/prey behaviours. Perhaps one of

---

Fig. 10.1. In this experiment, the foraging ability of mummichogs from a polluted environment (Piles Creek [PC]), a clean environment (Tuckerton [TK]), the polluted environment but given a 10-week depuration period in clean water (PC[d]), and from the clean environment but exposed to PC conditions for 3 weeks (TK[PC]) was considered. (A) Mean number of shrimp remaining ($\pm$ SEM) after predation experiments with Piles Creek (PC) and Tuckerton (TK) fish, TK(PC) and PC(d). PC and TK were significantly different after 5 minutes ($p < 0.02$) and remained so after 30 minutes ($p < 0.05$); PC(d) were not significantly different in their feeding behaviour from TK ($p > 0.1$) or PC ($p > 0.75$) fish. There was a significant reduction in feeding ability in TK(PC). (B) Levels of brain Hg (μg Hg/g brain $\pm$ SEM) in each group of fish tested. PC had significantly higher concentrations of Hg compared to TK fish ($p < 0.001$); PC(d) were not significantly different from PC fish, neither were TK(PC). (Adapted from Smith and Weis, 1997 with permission from Elsevier and the authors.)

the easiest methods of linking impaired predator avoidance behaviour specifically to olfactory damage is to consider the response of exposed fish to alarm pheromone. Alarm pheromone is released from damaged skin of a prey fish and acts as a warning signal to other conspecifics. Addition of alarm pheromone to tank water characteristically results in antipredator behaviours (see Chapter 2, Section 5 and Chapter 3, Section 3.4). Olfactory ability, defined as the behavioural response to alarm pheromone, decreased with increasing Cu concentration in Colorado pikeminnow (*Ptychocheilus lucius*) (Beyers and Farmer, 2001). By calculating the probability of response to skin homogenate as a function of exposure concentration, it was demonstrated that exposure to Cu concentrations of less than 66 μg/l resulted in a greater effect on the olfactory system following 24 hours of exposure than seen following 96 hours of exposure. This would suggest that the fish were able to both acclimate and recover from the toxic effects of Cu even during continued exposure. Scanning electron microscopy (SEM) of the olfactory apparatus showed that the ciliated receptor cells seen in control fish were absent in fish immediately after Cu exposure, providing a potential disrupting mechanism for behavioural responses to alarm pheromone. These cells were regenerated after 96 hours of exposure, paralleling the reduced effect at this time.

Diazinon exposure (1.0 μg/l) affected the olfactory-mediated behavioural response of chinook salmon to alarm pheromone (Scholz *et al.*, 2000) but swimming behaviour and visually guided food capture were not affected at this concentration. Waterborne Cd exposure (2 μg/l) for 7 days eliminated normal antipredator responses to alarm pheromone in rainbow trout (Scott *et al.*, 2003). Exposures of shorter duration or lower concentrations had no effect. These Cd-induced changes in behaviour were clearly mediated via the olfactory system as 7-day exposure to Cd via the diet (resulting in similar tissue burdens of the metal as found during waterborne exposure) had no effect on antipredator behaviour. Scott *et al.* (2003) also used autoradiography to illustrate Cd accumulation in the olfactory system during waterborne exposure (Figure 10.2).

Another way in which contaminants may disrupt behavioural responses to alarm pheromone was demonstrated by Leduc *et al.* (2003) who found that weakly acidic conditions (pH 6.0) reduced the normal antipredator response seen in juvenile pumpkinseed (*Lepomis gibbosus*) in response to chemical alarm cues. Under weakly acidic conditions the cyprinid alarm cue (hypoxanthine-3-*N*-oxide; Brown *et al.*, 2001) is chemically altered with an associated loss of function. Although pumpkinseed are centrarchids, it is possible that here the alarm signal is being eliminated before detection occurs. This should be an important consideration for additional studies on response to alarm pheromone. Hubbard *et al.* (2002) demonstrated that pheromonal cues (the pheromones 17α, 20β-dihydroxy-4-pregnen-3-one and prostaglandin

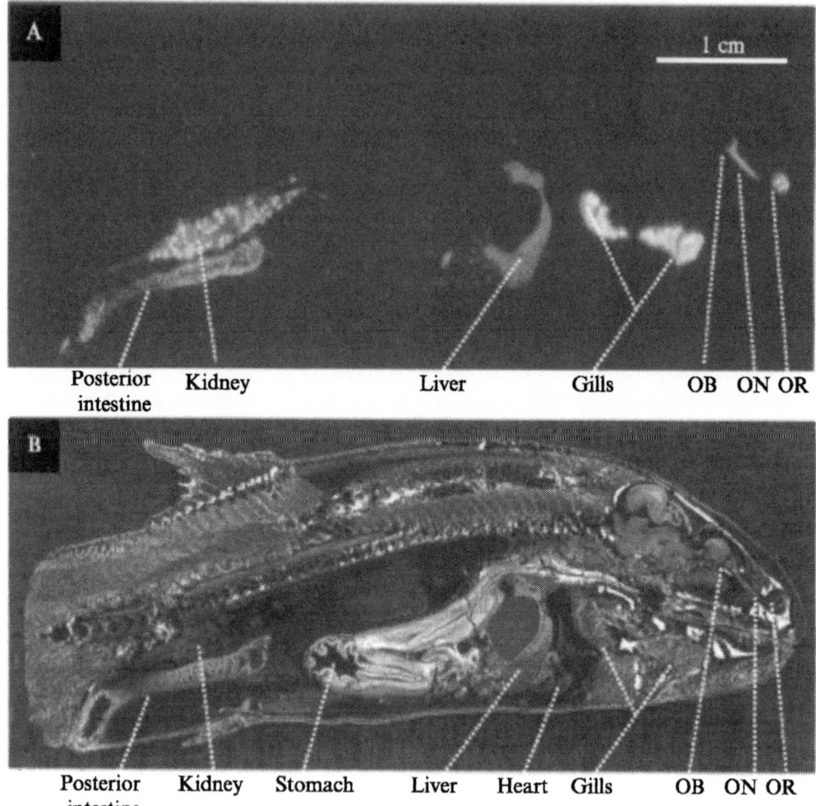

Fig. 10.2. (A) Sagittal whole-body autoradiogram showing ^{109}Cd accumulation in rainbow trout after 7-day Cd exposure followed by a 2-day depuration in control water. (B) Corresponding whole-body tissue section for comparison of location of organs within the body. It can be seen that cadmium accumulated in the posterior intestine, kidney, liver, gills, olfactory rosette (OR), olfactory nerve (ON) and olfactory bulb (OB). (Adapted from Scott *et al.*, 2003.)

$F_{2\alpha}$) can be altered by the presence of humic acid and become unavailable for olfactory detection by goldfish (see also Chapter 3, Section 3.4 and Chapter 9, Section 1).

If pheromonal cues warning of predator presence are detected, contaminants can still interfere with the ability of a fish to avoid being eaten. Fish generally possess a pair of Mauthner cells located either side of the brainstem. The cells are neurons with large dendrites that receive information from the sensory systems and are intrinsically involved in the "escape" response (see also Chapter 3). Exposure of 21- to 32-day-old medaka, *Oryzias latipes*, to carbaryl and phenol affected transmission from Mauthner cell to motoneuron while exposure to chloropyrifos, carbaryl, phenol and 2,4-dinitrophenol had neuromuscular effects (Carlson *et al.*, 1998). Physiological signs of poisoning by chlorpyrifos and carbaryl (e.g., muscle weakness and fatigue) were consistent with elevation of acetylcholine (ACh) in neuromuscular synapses and these chemicals may increase susceptibility to predation due to reduced efficiency in the escape response. Both behavioural and electrophysiological effects of endosulfan were consistent with disruption of glycine and GABA-mediated synapses. In particular, endosulfan-exposed medaka were hyperactive when disturbed.

Physiological disturbances have also been related to changes in foraging behaviour. Yellow bullheads exposed for 4 weeks to detergent showed a decreased foraging ability (Bardach *et al.*, 1965). Detergent-treated bullheads only found food when it was in their swimming path, unlike control fish that could actively search for food, suggesting that the chemical detection of food was reduced in exposed fish. However, detergent-exposed fish were hyperactive and thus increased the number of times food was encountered by chance. Histological preparations of olfactory rosettes, barbels, and taste buds in fish demonstrated that the olfactory epithelium seemed more resistant to detergent exposure than the taste buds. Concentrations between 4 and 5 mg/l caused a thickening of the borders of sensory cells in the folds of the olfactory rosette. Histological damage lagged behind functional impairment of chemoreception and the change in feeding behaviour happened at concentrations lower than those where histological damage could be detected. Disruption of brain monoamine concentrations in fish with toxicant-induced reductions in feeding ability have also been noted (Weber *et al.*, 1991; Smith *et al.*, 1995), although more data is needed to support these correlative studies.

In comparison to the large number of studies that have tested the effect of aquatic contaminants on predatory and foraging behaviour, it is clear that much is still unknown regarding the physiological disruptions underlying these behavioural changes. Future research must strive not only to address the complexity of disruption to predator–prey systems but also to understand the physiological processes underlying these complex behaviours.

4.3. Parasites

Predators must be avoided to increase the chances of survival, but fish are also susceptible to parasites which can have adverse physiological and behavioural consequences for the fish (see Chapter 4). Research into the effects of contaminants on the interactions between fish and parasites is lacking. However, it is important to acknowledge that anthropogenic disturbances can influence the occurrence and prevalence of parasites in fish populations (Khan and Thulin, 1991) and have the potential to interact with behavioural changes associated with host–parasite interactions.

Synergism of parasitic infections with environmental contaminants has been demonstrated (Pascoe and Cram, 1977; Khan *et al.*, 1986; Khan, 1987) and is perhaps not surprising as both toxicants and parasites can reduce immunocompetence and increase stress levels, thus rendering fish more susceptible to further stressors. However, some parasites may be decreased by environmental pollutants. For example, Narasimhamurti and Kalavati (1984) found that the myxozoan gill parasite *Henneguya waltairensis* was absent among tanks of the freshwater catfish *Channa punctatus* exposed to sewage but present in tanks of control fish. High salinity, low oxygen levels, and increased turbidity were suggested as the explaining factors of decreased infections of this ectoparasite. The use of compounds such as $CuSO_4$, which is toxic to fish in large doses, as a prophylactic treatment in fish culture against a variety of fish parasites also shows how differential sensitivity to toxicants interferes with the normal host–parasite relationship. Parasitic infections thus appear to be an important consideration in behavioural toxicological studies and vice versa (i.e., toxicants in studies on the effects of parasites on the behaviour and physiology of fish).

5. EFFECTS OF CONTAMINANTS ON SOCIAL BEHAVIOUR AND HIERARCHIES

Social hierarchies based upon dominant–subordinate behaviours are an expression of individual differences in competitive ability and are responsible in part for reducing severe aggression among conspecifics and enhancing exploitation of food resources and may lead to population stability (Gurney and Nisbet, 1979). For salmonids at least, there is a clear potential for contaminants (especially those targeting the neuroendocrine systems) to disrupt behaviour and interfere with the establishment and/or maintenance of social hierarchies. If the motivation or ability of fish to socially compete is disrupted, then dominance hierarchies may be altered or their formation prevented. Not only does this have potential implications for population stability, but within mixed-species hierarchies the competitive ability of one species may be more susceptible to toxicant exposure, leading to competitive exclusion of that species from the habitat. Despite this potential, the effects of environmental contaminants on social hierarchies are currently poorly understood, even though such investigations should give ecologically relevant information for predictions about population effects.

5.1. How Social Rank May Affect Susceptibility to Pollution

Social interactions and aggression among immature juvenile salmonid fish for feeding territories results in the formation of social hierarchies. The "social stress" associated with social status leads to distinct physiological differences between dominant and subordinate fish, at least when held in pairs (dyads). Subordinate fish characteristically display increased blood plasma cortisol concentrations (Øverli *et al.*, 1999; Sloman *et al.*, 2001), higher standard metabolic rates (Sloman *et al.*, 2000), together with decreased growth rates, condition (Pottinger and Pickering, 1992), and resistance to disease (Peters *et al.*, 1980). Thus the position or rank of a fish within a social hierarchy can influence its behaviour and physiology. Brain monoamine concentrations can also vary with social rank, in particular subordinate fish demonstrate elevated 5HIAA/5-HT ratios (Winberg and Nilsson, 1993; see also Chapter 5). Both Cd and Pb are known to disrupt this relationship between physiology and social rank in dominant–subordinate pairs of rainbow trout (Sloman *et al.*, 2005).

In view of the clear correlation of physiological variables and social rank in nonexposed salmonids, it has further been hypothesised that there may be rank-related differences in the sensitivity to environmental toxicants. For example, between pairs of bluegills, *Lepomis macrochirus*, Sparks *et al.* (1972) found that dominant fish survived exposure to Zn longer than subordinate fish. This effect of social rank could be alleviated by the addition of a shelter suggesting a link between intensity of aggression and differential response to toxicants.

As a dramatic example to demonstrate how social rank can affect sublethal susceptibility to toxicants, the rate of uptake of Cu in rainbow trout was shown to be ~20-fold higher in subordinate fish than in dominant fish when held in pairs (Figure 10.3A). When fish were held in social groups of 10 individuals exposed to 30 μg/l Cu for 1 week, the liver copper concentration was inversely correlated with specific growth rate (i.e., highest in the slower growing, more subordinate fish; Sloman *et al.*, 2002). A similar relationship, though less pronounced, has been found for pairs of trout exposed to silver (Ag), with subordinates having twofold higher Ag uptake rates compared to dominant fish (Figure 10.3B; Sloman *et al.*, 2003a). Enhanced uptake of a toxicant in subordinate fish could simply reflect the greater stress (and plasma cortisol/catecholamine concentrations) or elevated metabolic rate of subordinates (Sloman *et al.*, 2000) leading to greater "functional" exposure of the uptake surface to the toxicant. This would be similar to faster metal uptake rates found in smaller fish with higher mass-specific metabolic rates (Carpenter, 1930;

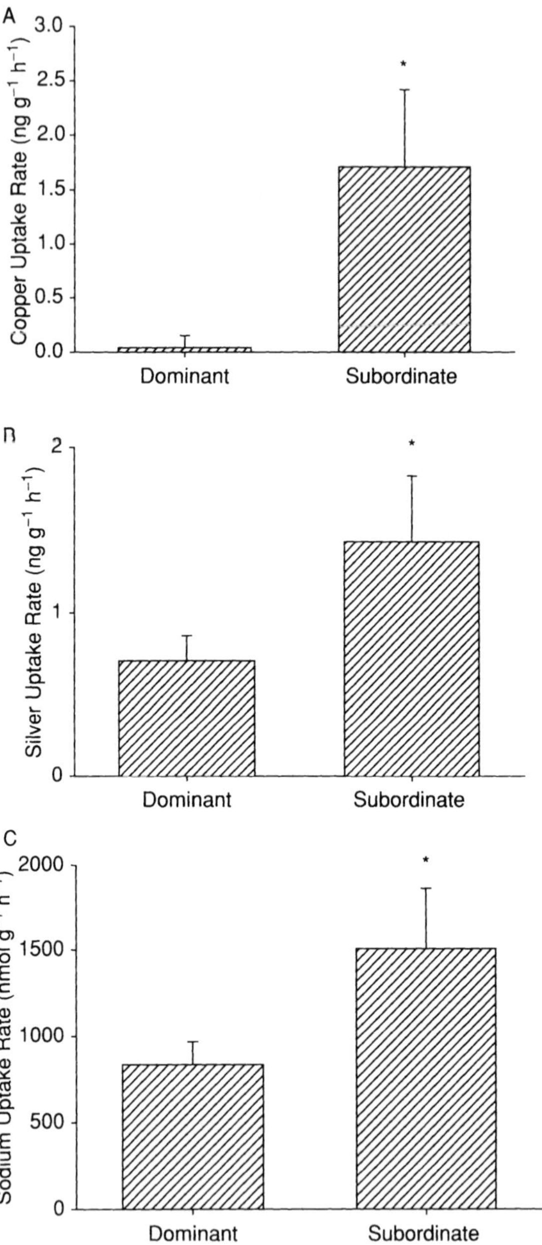
A
Copper Uptake Rate (ng g^{-1} h^{-1})
3.0
2.5
2.0
1.5
1.0
0.5
0.0
*
Dominant
Subordinate
B
Silver Uptake Rate (ng g^{-1} h^{-1})
2
1
0
*
Dominant
Subordinate
C
Sodium Uptake Rate (nmol g^{-1} h^{-1})
2000
1500
1000
500
0
*
Dominant
Subordinate

Sorensen, 1991). If this explanation were correct then it could apply to many, if not all, contaminants that are absorbed across the gills.

However, further study has shown that the rates of uptake of both Cu and Ag in trout are not strongly correlated with metabolic rate or whole-body cortisol, but instead they are tightly coupled to the branchial uptake rates of sodium (Na^+), which are substantially elevated in subordinate fish (Sloman *et al.*, 2003a; Figure 10.3C). Both Cu and Ag appear to share a common entry pathway to Na^+ at the apical membrane of fish gills (i.e., a Na^+ channel; Bury and Wood, 1999; Grosell and Wood, 2002). The enhanced uptake of these toxic metals in subordinate fish therefore seems to be a byproduct of a several-fold higher Na^+ uptake rate (Sloman *et al.*, 2003a). The reason for enhanced Na^+ uptake rate in subordinates may indirectly involve higher metabolic rate, as this would also induce a greater diffusive loss of ions across the gills (Gonzalez and McDonald, 1992) and/or excretion via the urine (e.g., exercise-induced diuresis; Wood and Randall, 1973). Indeed, increased Na^+ loss across both the gills and kidney of subordinate trout has been documented (Sloman *et al.*, 2004). However, the monopolisation of food by dominant fish would also reduce the dietary intake of sodium in subordinates, requiring greater active branchial uptake to compensate.

Such a generalised phenomenon, of subordinates being more prone to the uptake of toxicants, could have far-reaching consequences for the structure of hierarchies and population stability in the wild. However, the influence of social rank on the uptake of toxicants does not appear to be this simple once other contaminants are considered. In contrast to the studies on Cu and Ag (Sloman *et al.*, 2002, 2003a), Cd has been found to accumulate more rapidly in the gills of dominant, rather than subordinate, trout when exposed to 2 μg/l for 1 week (Sloman *et al.*, 2003b). Whereas Cu and Ag are believed to enter the gills in parallel with Na^+ uptake, at least some Cd is accumulated via the calcium uptake pathway through chloride cells of freshwater fish (Verbost *et al.*, 1989; Wickland Glynn *et al.*, 1994). Branchial uptake of calcium seems to correlate well with the need for growth, which is

Fig. 10.3. The mean uptake rates (± SEM) for copper (A), silver (B), and sodium (C) by dominant and subordinate rainbow trout. Size-matched fish (means ranging from 1.64 ± 0.09 g to 3.69 ± 0.5 g) were held in pairs (N = 8 pairs per treatment) for 48 hours under control conditions to establish dominance hierarchies. After a further 48-hour exposure to either 20 μg/l copper or 0.2 μg/l silver, the uptake rates of each metal were measured simultaneously with the uptake rate of sodium by the addition of either ^{64}Cu and ^{22}Na, or ^{110m}Ag and ^{24}Na. *Significant difference from the uptake rate in the dominant fish ($p < 0.05$). (Adapted from Sloman *et al.*, 2002, with permission from Alliance Communications Group and from Sloman *et al.*, 2003, with permission from Elsevier.)

higher in dominant fish (associated with higher food intake) and hence they accumulate more Cd.

The relationship between social rank and the sensitivity to toxicants can therefore be opposite for different toxicants. Routes of exposure can also influence the sensitivity of different social ranks. If a toxicant is present in the diet, then a dominant fish that eats a greater proportion of the diet is likely to accumulate toxicants faster. However, these examples demonstrate an important principle that has so far been ignored in toxicology: the potential for toxicants to selectively target either end of the social hierarchy spectrum. The effects this might have on the structure of hierarchies and population stability in the wild are unknown and present an exciting line of future research.

5.2. Social Behaviours

Not only does the sensitivity to toxicants have the potential to vary with position in the social hierarchy, but toxicants also have the ability to interfere with the behavioural processes involved in social behaviour, specifically the formation and stability of social hierarchies. Several studies have used paired contests, where one fish will become dominant and the other subordinate (largely as a result of greater success of the former in aggression contests) to test the potential for contaminants to interfere with social behaviour. Sloman *et al.* (2003b,c) showed that even acute (24-hour) exposure to 2–3 μg/l of the neurotoxic metal Cd can significantly reduce aggression between pairs of exposed fish, and speed up the formation of social hierarchies among groups of exposed fish. Furthermore, 24–72 hours of exposure to this level of Cd rendered fish much more likely to become subordinate when paired with nonexposed fish, with clear implications for the competitive ability of trout following acute exposure to Cd in the wild. The mechanism behind these acute effects of Cd on social behaviour is thought to involve its ability to accumulate in, and disrupt the function of, the olfactory system (Scott *et al.*, 2003), and possibly the inability to detect conspecific odours that enhance aggression (Sloman *et al.*, 2003b). However, although the olfactory neurons are capable of regeneration and normal social behaviours can recover from acute Cd exposure (e.g., after 5 days of depuration in clean water), the extent of its interference with the maintenance of social hierarchy during chronic exposures remains to be determined.

Sloman *et al.* (2003c) also compared the effects of acute (24-hour) exposure to Cd with that of four other toxic metals (Cu, Ni, Zn, and Pb) on the ability of rainbow trout to form social relationships. Although Pb had a tendency to increase successful aggressive attacks in pairs (though not significant), it is notable that only the example given above (i.e., Cd) was found to significantly alter social behaviour in these pairs. This could relate to the fact that of these five metals only Cd (and Pb) are known neurotoxic agents (Sorensen, 1991), and hence fits with the initial hypothesis stated above that contaminants known to target the neuroendocrine systems have great potential to disrupt the establishment of social hierarchies. However, again the maintenance or stability of social hierarchies during chronic exposure to any of these metals remains unknown. Dietary contaminants also have the potential to affect hierarchy formation. Dyads of rainbow trout fed a copper-contaminated diet (730 mg Cu/kg dry weight of feed) formed dominant–subordinate relations faster with less aggression than seen in interactions between control fish (Campbell *et al.*, 2005).

5.3. Energetics and Social Behaviour

Other chemicals not considered as neurotoxins have been associated with changes in social behaviour. Johnsson *et al.* (2003) found that exposure to paper mill effluent reduced territorial aggression in rainbow trout fry, and Henry and Atchison (1986) found that among groups of bluegills the behaviour of the most subordinate and the most dominant fish were most affected by exposure to Cu. Subordinate fish had higher cough, yawn, and fin-flick frequencies during the treatment and the dominant fish had the next highest frequencies. The aggression levels of the dominant fish also decreased after Cu exposure. It may be that these effects on aggression and social behaviour can be explained by energetic constraints caused by the toxicants, rather than specific targeting of the nervous system per se.

Some evidence for this explanation is provided by studies on the chronic exposure of fish to different toxic metals. Several studies have demonstrated that toxic metals that incur a metabolic cost (e.g., reduced aerobic capacity or elevated basal/routine metabolism) can cause changes in spontaneous activity in fish, possibly influencing social interactions as a result. For example, both waterborne aluminium (Cleveland *et al.*, 1986; Wilson *et al.*, 1994; Smith and Haines, 1995; Al; Allin and Wilson, 1999, 2000) and dietary Cu (Handy *et al.*, 1999; Campbell *et al.*, 2002) have been found to reduce metabolically expensive burst (anaerobic) activity and increase the time spent either stationary or at slower swimming speeds when fish are held in social groups (Figure 10.4; Allin and Wilson, 1999, 2000; Campbell *et al.*, 2002). In the case of Al, this reduced spontaneous activity has been attributed to its respiratory toxicity, specifically the changes in gill morphometrics and impaired respiratory function (Wilson *et al.*, 1994; Allin and Wilson, 1999, 2000). It is also known that during sublethal exposure to Al, the reduction in spontaneous burst activity within the whole social group is largely associated with a reduced rate of aggressive attacks initiated by the

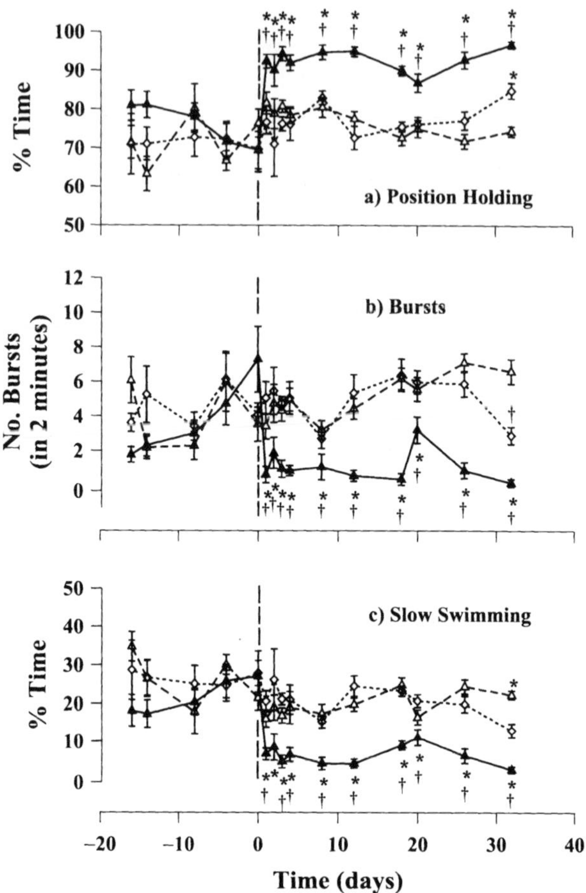

Fig. 10.4. Mean values (± SEM) for different swimming behaviours measured in groups of rainbow trout (15 fish per tank and three replicate tanks per treatment) exposed to one of three treatments. Prior to Day 0 all fish were maintained in circumneutral, artificial soft water (pH 6.4). Open diamonds joined by dotted line represent control fish maintained in circumneutral water (pH 6.4) without any added Al. Open triangles joined by a dashed line represent fish exposed to sub-lethal low pH (5.2) in the absence of Al. Filled triangles joined by a solid line represent fish exposed to sublethal low pH (5.2) in the presence of 30 μg/l of Al. The chronic sublethal exposure to Al caused a sustained reduction in metabolically expensive bursts (sudden movements of at least two body lengths per second, most likely to be anaerobic muscle use) and the time spent cruising at aerobic speeds (slow swimming). *Significant difference ($p < 0.05$) from the neutral control group. †Significant difference ($p < 0.05$) from the acid only group. (Adapted from Allin and Wilson, 1999, with permission from NRC.)

most dominant fish in each tank (Figure 10.5; Owen and Wilson, unpublished data). In other words, the most dominant fish causes the majority of the burst activity within the group (e.g., causing at least one, and often two, subordinate fish to use burst activity to escape an attack), so when aggressive challenges are reduced in the dominant fish due to the toxicant, this expensive activity declines accordingly within the whole social group.

Interestingly, exposure to hypoxia (with the obvious metabolic constraint this imposes) has a similar effect to the above respiratory toxicants, such as to reduce both aggressive behaviour and spontaneous activity (Kramer, 1987; Sneddon and Yerbury, 2004). Although social behaviours have not often been the primary focus of toxicology studies, in hindsight there are many other toxicants that are known to alter spontaneous activity (carbaryl and cadmium: Beauvais *et al.*, 2001; malathion and diazinon: Brewer *et al.*, 2001; and also reviewed by Beitinger, 1990 and Little and Finger, 1990) that could well exert this effect through changes in social interactions in groups of fish.

Campbell *et al.* (2002) also suggested that the effects of excess dietary Cu on spontaneous activity could be explained by the suppression of normal social behaviour, and found that this was further expressed in changes (or loss) in the circadian periodicity of activity. There is clear potential for toxicants that interfere with endocrine systems involved with such behavioural rhythms (e.g., melatonin, see Chapter 6) to cause similar disruption to social behaviour. However, to the authors' knowledge there is little data published on toxicants known to specifically target such endocrine/behaviour systems (but see Handy, 2003). Indeed there is scant information available on the relationship between social interactions and circadian rhythms even in the absence of toxicants (e.g., Chen *et al.*, 2002).

5.4. Energetics and Migratory Behaviour

The potential of contaminants to upset metabolic processes has implications not only for social behaviour but also other energetically demanding processes such as migration. The homing ability of fish to return their natal streams is disrupted by contaminants that interfere with olfaction (Scholz *et al.*, 2000) but the potential for the migration and dispersal of fish to be affected by contaminants that alter metabolism and energy allocation has yet to be considered. Brett (1958) classified environmental stressors in terms of their effect on aerobic metabolism as limiting (reducing maximum oxygen uptake) and/or loading (increasing the metabolic cost of routine maintenance; i.e., higher oxygen uptake at rest or at different swimming speeds). Both limiting and loading stressors can obviously impact upon swimming performance by either reducing the oxygen delivery to muscles

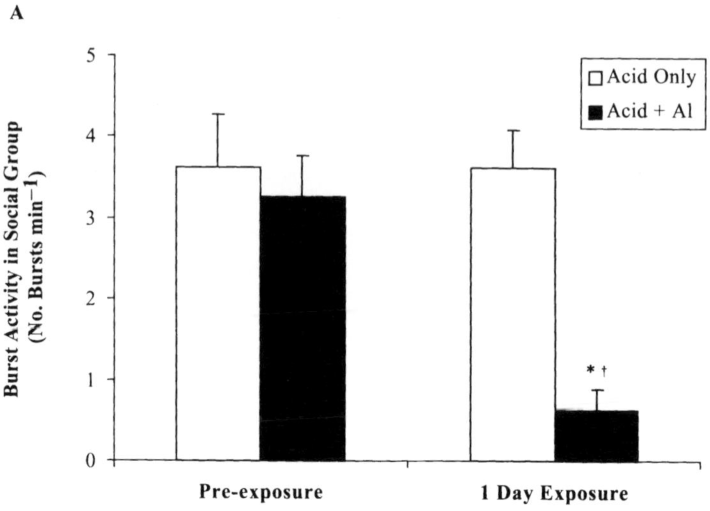

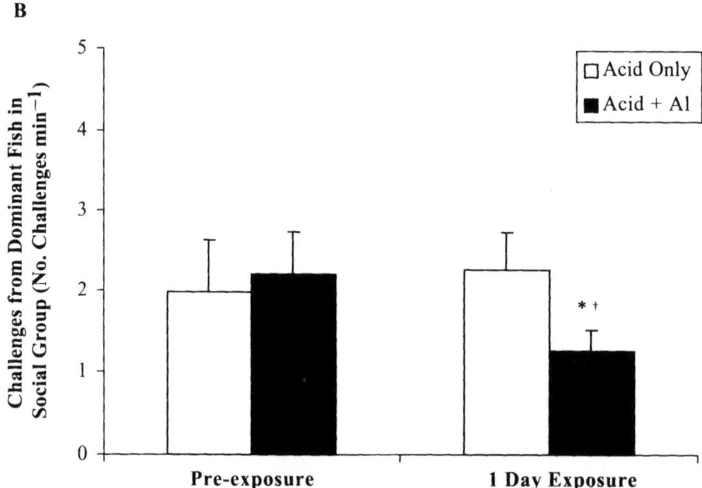

Fig. 10.5. Mean values (± SEM) for (A) burst swimming activity (sudden movements of at least two body lengths per second, most likely to be anaerobic muscle use) within the whole social group, and (B) aggressive challenges initiated by the most dominant fish in the social group. Data are for brown trout both prior to exposure, and after 1 day exposure to either sublethal low pH in the absence of Al (open bars) or in the presence of sublethal Al at 13 μg/l (solid bars). Low pH alone had no effect, but exposure to Al dramatically reduces both the burst activity within the whole social group, and the aggressive challenges initiated by the most dominant fish. *Significant difference ($p < 0.05$) from the relevant preexposure data. †Significant difference ($p < 0.05$) from the acid only group at the same time point. (Owen and Wilson, unpublished data.)

at the maximum speed (limiting stressors) or diverting oxygen away from working muscles at all speeds (loading stressors). There are a number of examples of toxicants that have been shown to have either or both these effects on swimming performance in fish (e.g., Al: Wilson *et al.*, 1994; Ni: Pane *et al.*, 2004; Cu: Waiwood and Beamish, 1978; Beaumont *et al.*, 1995) and therefore have the potential to influence migration and dispersion. Although there are many studies that have used radiotelemetry to study the migration of smolting salmonids (e.g., Dieperink *et al.*, 2002; see also Chapter 7), we are not aware of any that have examined the potential for toxicants to influence this particularly energetic behaviour.

In some species, migration to spawning grounds involves movement from hyposaline (i.e., freshwater) to hypersaline waters (e.g., estuaries or marine environments). In these cases, migration will also be vulnerable to toxicants that can interfere with ion and osmotic regulation. There are numerous toxicants known to impair ion regulation in freshwater fish (e.g., metals such as Al, Cu, Zn, Cd, Ag; see reviews by Wood, 1992; Wood *et al.*, 1999) and exposure of salmon and tilapia to organic chemicals such as 4-nonylphenol, atrazine, and estradiol in freshwater has been shown to impair the hypo-osmoregulatory ability of these fish when artificially transferred to seawater (Moore *et al.*, 2003; Madsen *et al.*, 2004; Waring and Moore, 2004; Vijayan *et al.*, 2001). However, once again very little is known about the influence of exposure to such toxicants on migratory behaviour, as opposed to just the osmoregulatory potential of euryhaline migratory fish. One exception is the study of Madsen *et al.* (2004) in which smolting Atlantic salmon were exposed to either 17-β estradiol or 4-nonylphenol by repeated injection with 2 μg/g or 120 μg/g, respectively, over a 20-day period in freshwater, before being released into a river. Compared to an appropriate control group, both the exposure groups suffered a significant delay and reduced success (presumably due to greater predation by herons) during a subsequent 3.2-km downstream migration to a trap site. Thus, a toxicant can have negative impacts on both the physiological capacity and the appropriate behaviour required for successful migrations between freshwater and seawater.

There is clearly a need to investigate the potential for other toxicants that target the neuro- and endocrine systems, as well as less specific toxicants that can influence energetics and activity, to interfere with social behaviour and initiation of social hierarchies. Furthermore, there is currently very little information available concerning the effects of chronic exposures on the maintenance of already established hierarchies, and hence the potential for interference with the long term stability of populations.

5.5. Schooling Behaviour

As dominance hierarchies make up the majority of research into the social behaviour of fish, there is a tendency to overlook other social behaviours when considering the effects of toxicants. Less is known about the effects of contaminants on schooling, a social behaviour vital to the survival of many fish species. Koltes (1985) recorded schools of Atlantic silverside (*Menidia menidia*) on video and compared activity before and after exposure to concentrations of Cu ranging from 0–100 μg/l. Exposure to Cu increased swimming speed, decreased the rate of change of direction, and also decreased the distances between nearest neighbour within the schools. In addition, schools swam more in parallel orientation following contamination. The physiological mechanisms involved in the disruption of these behaviours warrant further research. In particular, the increasing concern regarding the release of pharmaceuticals into the environment, many of which are ototoxic, may raise some interesting questions for behaviours that are particularly dependent upon orientation and lateral-line function.

6. EFFECTS OF CONTAMINANTS ON REPRODUCTION

Many fish species show morphological sexual characteristics that can be used in sexual displays to attract mates and may produce sexual pheromones to both attract and prime potential partners (Liley and Stacey, 1983; Chapter 9). The prominent fat pads on the heads of male fathead minnows and associated face tubercles are one such example of female attractants (Harries *et al*., 2000) The ability to build and defend nests is a compulsory component of sexual reproduction for some species such as the stickleback, *Gasterosteus aculeatus*, requiring the physiological ability to produce the nest glue spiggin (Katsiadaki *et al*., 2002). Successful reproduction will also depend on spawning and courtship behaviours and, for migratory species such as Atlantic salmon, the ability to return to natal streams (as discussed in Chapter 7). The sexual plasticity of some fish species resulting in alternative reproductive tactics and sex change of some individuals within a breeding group (Hourigan *et al*., 1991; Chapter 8) is yet another factor involved in successful reproduction.

Arguably some of the most important questions concerning the toxicity of a compound focus around reproductive capacity. Do toxicological properties of aquatic contaminants impair the ability of fish to reproduce and produce viable offspring that can also successfully reproduce? Repeatable tests for the presence of contaminants have been designed around the breeding characteristics of certain fish species (e.g., Harries *et al*., 2000).

Many studies have illustrated the potential of aquatic toxicants to alter either physiological or behavioural endpoints (see also reviews by Jones and Reynolds, 1997; Scott and Sloman, 2004). Table 10.5 presents a suite of toxicants that in particular cause decreases in courtship behaviours, spawning success, and gonadosomatic indices. For example, vinclozolin, *p, p′*-dichlorodiphenyldichloroethane (*p,p′*-DDE), and flutamide disrupt the courtship behaviour of guppies and interfere with male reproductive physiology (Bayley *et al.*, 2002). Toxicant-induced changes in parental care have also been extensively reviewed by Jones and Reynolds (1997).

Transgenerational studies have considered not only survival of offspring from exposed parents but also the reproductive success of the next generation. Offspring of MeHg-exposed mummichogs displayed reduced reproductive ability and altered sex ratios (Matta *et al.*, 2001). Exposure of adults to moderate concentrations of MeHg (equivalent to 0.44–1.0 μg/g measured in adult carcasses) produced a skew towards more male offspring and more females at higher concentrations (equivalent to <1.0 μg/g measured in adult carcasses). Although exposure of fathead minnows to MeHg did affect spawning success, Hammerschmidt *et al.* (2002) found no relationship between concentrations of Hg and developmental and hatching success of embryos of 7-day survival and growth of larvae. In contrast, eggs laid by fathead minnows exposed to 0.5 mg/l Pb for 4 weeks were less viable than those laid by control fish (Weber, 1993). Eggs of Japanese medaka exposed to 25 and 30 μg/l of octylphenol from 1 day to 6 months posthatch showed various developmental defects including circulatory problems, incomplete eye development, and failure to inflate swim bladders on hatching (Gray *et al.*, 1999). Transgenerational effects on physiological reproductive parameters have thus been demonstrated, but there is a now a need to consider transgenerational effects on reproductive behaviours in more detail, as has been conducted for predator–prey interactions.

6.1. Potential Physiological Causes of Altered Reproductive Behaviour

Many of the toxicological disruptions to physiological mechanisms discussed already have the potential to interfere with reproductive behaviour. Sensory systems are important for detecting the attractive nature of partners and environmental cues used to synchronise mating behaviours. Olfaction is vital not only for return migrations to natal streams but also to ensure detection of priming hormones. As covered in Section 3.1, many aquatic contaminants interfere with the electrical responses of the olfactory epithelia in response to stimulants. Female Atlantic salmon urine and prostaglandin $F_{2\alpha}$ ($PGF_{2\alpha}$) act as olfactory cues, both exerting a "priming" influence on males resulting in elevations of sex steroids

Table 10.5
Studies That Have Considered the Effects of Chemical Contaminants on the Physiology and Behaviour of Fish Reproduction

Pollutant	Species	Exposure concentration	Duration of exposure	Exposure route	Lowest concentration/ shortest time effect seen	Endpoint measured	Reference
Vinclozolin	Guppy	0.1, 10 μg/mg	Birth to adulthood	Dietary	0.1 μg/mg	Reduction in orange colouration, inhibited gonopodium development, reduced sperm count and suppressed courtship behaviour	Bayley *et al.* (2002)
p,p′-DDE	Guppy	0.01, 0.1 μg/mg	Birth to adulthood	Dietary	0.01 μg/mg	Reduction in orange colouration, inhibited gonopodium development, reduced sperm count and suppressed courtship behaviour	Bayley *et al.* (2002)
Flutamide	Guppy	0.01, 1.0 μg/mg	Birth to adulthood	Dietary	0.01 μg/mg	Reduction in orange colouration, inhibited gonopodium development, reduced sperm count and suppressed courtship behaviour	Bayley *et al.* (2002)
4-*tert*-octylphenol	Guppy	150 μg/l	4 weeks	Waterborne	NA	Decrease in rate and intensity of sexual displays	Bayley *et al.* (1999)
Octylphenol	Japanese medaka	10, 25, 50, 100 μg/l	1 day posthatch to 6 months posthatch	Waterborne	10 μg/l	Reduction in courtship behaviour and overall reproductive success	Gray *et al.* (1999)

Vinclozolin	Guppy	1, 10, 100 μg/mg	30 days	Dietary	1 μg/mg	Disruption in courtship behaviour, reduction in ejaculated sperm and reduction in orange-yellow colouration	Baatrup and Junge (2001)
p,p'-DDE	Guppy	0.1, 1, 10 μg/mg	30 days	Dietary	0.1 μg/mg	Disruption in courtship behaviour, reduction in ejaculated sperm, and reduction in orange-yellow colouration	Baatrup and Junge (2001)
Flutamide	Guppy	1, 10, 100 μg/mg	30 days	Dietary	1 μg/mg	Disruption in courtship behaviour, reduction in ejaculated sperm and reduction in orange-yellow colouration	Baatrup and Junge (2001)
Phenol	Guppy	10 mg/l	Up to 30 days	Waterborne	NA	Reduction in courtship behaviour	Colgan *et al.* (1982)
Lindane	Guppy	1 μg/l	1 week	Waterborne	NA	Decrease in courtship activities	Schröder and Peters (1988)
Pb	Fathead minnow	0.5 mg/l	4 weeks	Waterborne	NA	Suppressed spermatocyte production, retarded ovarian development, decreased number of eggs oviposited and reduced nest preparation and maintenance	Weber (1993)

(17,20β-dihydroxy-4-pregnen-3-one, testosterone and 11-ketotestosterone) and expressible milt (Chapter 9). Exposure of male salmon to carbofuran, atrazine, cypermethrin, or diazinon decreases the olfactory response to $PGF_{2\alpha}$ (Moore and Waring, 1996, 1998, 2001; Waring and Moore, 1997) and the physiological priming effects of female pheromones. To see whether there were direct effects of these compounds on the testes or whether the reduction in priming effects was mediated by olfactory disruption, testes were removed from control and exposed fish and incubated *in vitro* in the presence of salmon pituitary extract. In control animals, this resulted in the release of sex steroids. Cypermethrin and carbofuran did not influence *in vitro* steroid production and therefore are likely to reduce priming of male salmon solely via olfactory disruption. However, exposure to atrazine had a direct effect on the testes, as well as on olfactory function, altering both 17,20β-dihydroxy-4-pregnen-3-one and androgen secretion *in vitro*.

Reproductive behaviour is also influenced by brain neurotransmitters and, as discussed previously, many aquatic contaminants are neurotoxic. With particular reference to reproductive physiology, Khan and Thomas (2000) considered the effects of Aroclor 1254 and Pb on the 5-HT pathway. In goldfish and Atlantic croaker, 5-HT stimulates maturational gonadotropin secretion (GTH II), possibly through its actions on the gonadotropin releasing hormone (GnRH) system in the preoptic-anterior hypothalamic area and pituitary gland (Khan and Thomas, 1992). The rate-limiting enzyme in 5-HT synthesis, tryptophan hydroxylase, and the catabolic enzyme monoamine oxidase were measured in the hypothalamus of Atlantic croaker following dietary exposure to these compounds. Aroclor 1254 significantly inhibited tryptophan hydroxylase resulting in a decline in 5-HT, but monoamine oxidase activity was not affected. Only slight decreases in tryptophan hydroxylase activity were associated with Pb exposure and were accompanied by minor increases in monoamine oxidase activity. Khan and Thomas (2000) also found a significant inhibition of the GTH II response when Pb and Aroclor 1254 exposed fish were intraperitoneally administered with a luteinising hormone-releasing hormone analog (LHRHa; 5 ng/g body weight). The authors deduced that reduced 5-HT synthesis may result in a disruption of the 5-HT-GnRH pathway that controls GTH II secretion and gonadal growth.

Interference with endogenous reproductive hormonal pathways is a major way in which aquatic toxicants could interfere with reproductive behaviours, whether via neurotransmitters or acting directly upon hormone synthesis or receptor function. The current concern over the abundance of endocrine disrupting chemicals (EDCs) that can mimic endogenous hormones or interfere with their production has focused primarily on reproductive hormones. Certainly, synthetic estrogens such as ethinyl estradiol (EE_2)

can alter gonad development (Jobling *et al.*, 1998), vitellogenin synthesis (Harries *et al.*, 1997), and secondary sexual characteristics in male fish (Harries *et al.*, 2000), the latter perhaps being more relevant to behavioural changes. The lability of the sexual differentiation process in the early life stages of some fish species (e.g., zebrafish) can also make them susceptible to EDCs (Fenske and Segner, 2004). The hormonal control of the development of secondary sexual characteristics and reproductive behaviour is also vulnerable to disruption from these types of chemicals, but surprisingly few studies have considered disruption of hormone levels and/or brain neurotransmitters in conjunction with changes in reproductive behaviour.

Bell (2001) correlated plasma sex hormone concentrations and behaviours in male sticklebacks and found a disruption in male reproductive behaviour following exposure to EE_2. However, the results from this study were insufficient to relate changes in behaviour associated with EE_2 exposure to hormonal changes. Given the important implications of fish reproductive behaviours in ensuring the next generation and ultimate survival of the population (Jones and Reynolds, 1997), there is an obvious gap in our understanding of how the behavioural changes are influenced by physiology. Although there are many potential ways in which altered physiology may affect reproductive behaviour, little work conducted to date has aimed to link these two areas.

7. CONCLUSION

The interdependence of behaviour and physiology throughout the entire lifecycle of an individual fish presents many ways in which anthropogenic effects upon its environment can wield detrimental effects. It is clear from reviewing the literature that there are many areas that remain unresearched. What effects do contaminants have on host–parasite interactions? What do individual differences in physiology induced by social status mean for population stability and ultimately survival? What impact do contaminants have on an ecosystem if predator and prey are affected either similarly or differentially? How do contaminant-induced changes in endocrine systems and energy expenditure affect important behaviours such as circadian rhythms and migration? How are changes in reproductive behaviours related to changes in physiology? The preceding nine chapters illustrate the strong link between behaviour and physiology in fish and the challenge now is to apply this knowledge effectively to ecotoxicological studies. There is a need to continue research that furthers our understanding of how behaviour and physiology interact during toxicant exposure and to measure both in conjunction rather than separately to provide a more holistic understanding and better predictive capabilities. It is also imperative that we start to develop toxicity and chronic effect tests that take into account the natural environment of the fish as much as possible, incorporating the existence of social hierarchies, parasites, and an understanding that the whole ecosystem will likely be affected, not just one species at a time.

ACKNOWLEDGEMENTS

We would like to thank two anonymous referees and Sigal Balshine for their valuable input to this manuscript.

REFERENCES

Aldegunde, M., Soengas, J. L., Ruibal, C., and Andrés, M. D. (1999). Effects of chronic exposure to γ-HCH (Lindane) on brain serotonergic and gabaergic systems, and serum cortisol and thyroxine levels of rainbow trout, *Oncorhynchus mykiss*. *Fish Physiol. Biochem.* **20,** 325–330.

Allin, C. J., and Wilson, R. W. (1999). Behavioural and metabolic effects of chronic exposure to sublethal aluminum in acidic soft water in juvenile rainbow trout (*Oncorhynchus mykiss*). *Can. J. Fish. Aquat. Sci.* **56,** 670–678.

Allin, C. J., and Wilson, R. W. (2000). Effects of pre-acclimation to aluminium on the physiology and swimming behaviour of juvenile rainbow trout (*Oncorhynchus mykiss*) during a pulsed exposure. *Aquat. Toxicol.* **51,** 213–224.

Anderson, J. M., and Peterson, M. R. (1969). DDT: Sublethal effects on brook trout nervous system. *Science* **164,** 440–441.

Baatrup, E., and Junge, M. (2001). Antiandrogenic pesticides disrupt sexual characteristics in the adult male guppy (*Poecilia reticulata*). *Environ. Health Persp.* **109,** 1063–1070.

Baatrup, E., Døving, K. B., and Winberg, S. (1990). Differential effects of mercurial compounds on the electroolfactogram (EOG) of salmon (*Salmo salar* L.). *Ecotoxicol. Environ. Safety* **20,** 269–276.

Baker, C. F., and Montgomery, J. C. (2001). Sensory deficits induced by cadmium in banded kokopu, *Galaxias fasciatus*, juveniles. *Environ. Biol. Fish.* **62,** 455–464.

Bardach, J. E., Fujiya, M., and Holl, A. (1965). Detergents: Effects on the chemical senses of the fish *Ictalurus natalis* (le Sueur). *Science* **148,** 1605–1607.

Bayley, M., Nielsen, J. R., and Baatrup, E. (1999). Guppy sexual behaviour as an effect biomarker of estrogen mimics. *Ecotoxicol. Environ. Safety* **43,** 68–73.

Bayley, M., Junge, M., and Baatrup, E. (2002). Exposure of juvenile guppies to three antiandrogens causes demasculinisation and a reduced sperm count in adult males. *Aquat. Toxicol.* **56,** 227–239.

Beaumont, M. W., Butler, P. J., and Taylor, E. W. (1995). Exposure of brown trout, *Salmo trutta*, to sub-lethal copper concentrations in soft acidic water and its effect upon sustained swimming performance. *Aquat. Toxicol.* **33,** 45–63.

Beauvais, S. L., Jones, S. B., Brewer, S. K., and Little, E. E. (2000). Physiological measures of neurotoxicity of diazinon and malathion to larval rainbow trout (*Oncorhynchus mykiss*) and their correlation with behavioural measures. *Environ. Toxicol. Chem.* **19,** 1875–1880.

Beauvais, S. L., Jones, S. B., Parris, J. T., Brewer, S. K., and Little, E. E. (2001). Cholinergic and behavioural neurotoxicity of carbaryl and cadmium to larval rainbow trout (*Oncorhynchus mykiss*). *Ecotoxicol. Environ. Safety* **49,** 84–90.

Beitinger, T. L. (1990). Behavioural reactions for the assessment of stress in fishes. *J. Great Lakes Res.* **16,** 495–528.

Bell, A. M. (2001). Effects of an endocrine disrupter on courtship and aggressive behaviour of male three-spined stickleback, *Gasterosteus aculeatus*. *Anim. Behav.* **62,** 775–780.

Bell, A. M. (2004). An endocrine disruptor increases growth and risky behaviour in threespined stickleback (*Gasterosteus aculeatus*). *Horm. Behav.* **45,** 108–114.

Berntssen, M. H. G., Aatland, A., and Handy, R. D. (2003). Chronic dietary mercury exposure causes oxidative stress, brain lesions, and altered behaviour in Atlantic salmon (*Salmo salar*) parr. *Aquat. Toxicol.* **65,** 55–72.

Beyers, D. W., and Farmer, M. S. (2001). Effects of copper on olfaction of Colorado pikeminnow. *Environ. Toxicol. Chem.* **20,** 907–912.

Borg-Neczak, K., and Tjälve, H. (1996). Uptake of $^{203}Hg^{2+}$ in the olfactory system in pike. *Toxicol. Letters* **84,** 107–112.

Bretaud, S., Toutant, J. P., and Saglio, P. (2000). Effects of carbofuran, diuron, and nicosulfuron on acetylcholinesterase activity in goldfish (*Carassius auratus*). *Ecotoxicol. Environ. Safety* **47,** 117–124.

Brett, J. R. (1958). Implications and assessment of environmental stress. *In* "The Investigation of Fish-Power Problems" (Larkin, P. A., Ed.), pp. 69–93. Institute of Fisheries, University of British Columbia.

Brewer, S. K., Little, E. E., De Lonay, A. J., Beauvais, S. L., Jones, S. B., and Ellersieck, M. R. (2001). Behavioural dysfunctions correlate to altered physiology in rainbow trout (*Oncorhynchus mykiss*) exposed to cholinesterase-inhibiting chemicals. *Arch. Environ. Contam. Toxicol.* **40,** 70–76.

Brown, G. E., Adrian, J. C., and Shih, M. L. (2001). Behavioural responses of fathead minnows to hypoxanthine-3-*N*-oxide at varying concentrations. *J. Fish Biol.* **58,** 1465–1470.

Brown, J. A., Johansen, P. H., Colgan, P. W., and Mathers, R. A. (1985). Changes in the predator-avoidance behaviour of juvenile guppies (*Poecilia reticulata*) exposed to pentachlorophenol. *Can. J. Zool.* **63,** 2001–2005.

Brown, J. A., Johansen, P. H., Colgan, P. W., and Mathers, R. A. (1987). Impairment of early feeding behaviour of largemouth bass by pentachlorophenol exposure: A preliminary assessment. *Trans. Am. Fish. Soc.* **116,** 71–78.

Bryan, M. D., Atchison, G. J., and Sandheinrich, M. B. (1995). Effects of cadmium on the foraging behaviour and growth of juvenile bluegill, *Lepomis macrochirus*. *Can. J. Fish. Aquat. Sci.* **52,** 1630–1638.

Buckley, P. (1998). The nervous system. *In* "Target Organ Pathology" (Turton, J., and Hooson, J., Eds.), pp. 273–310. Taylor and Francis, London.

Bury, N. R., and Wood, C. M. (1999). Mechanism of branchial apical silver uptake by rainbow trout is via the proton-coupled Na^+ channel. *Am. J. Physiol.* **277,** R1385–R1391.

Campbell, H. A., Handy, R. D., and Sims, D. W. (2002). Increased metabolic cost of swimming and consequent alterations to circadian activity in rainbow trout (*Oncorhynchus mykiss*) exposed to dietary copper. *Can. J. Fish. Aquat. Sci.* **59,** 768–777.

Campbell, H.A., Handy, R.D., and Sims, D.W. (2005). Shifts in a fish's resource holding power during a contact paired interaction: The influence of a copper contaminated diet in rainbow trout. *Physiol. Biochem. Zool.* **78,** 706–714.

Carlson, R. W., Bradbury, S. P., Drummond, R. A., and Hammermeister, D. E. (1998). Neurological effects on startle response and escape from predation by medaka exposed to organic chemicals. *Aquat. Toxicol.* **43,** 51–68.

Carpenter, K. E. (1930). Further research on the action of metallic salts on fishes. *J. Exp. Zool.* **56,** 407.

Chen, W-M., Naruse, M., and Tabata, M. (2002). The effect of social interactions on circadian self-feeding rhythms in rainbow trout *Oncorhynchus mykiss* Walbaum. *Physiol. Behav.* **76,** 281–287.

Chuiko, G. M. (2000). Comparative study of acetylcholinesterase and butyrylcholinesterase in brain and serum of several freshwater fish: Specific activities and *in vitro* inhibition by DDVP, an organophosphorus pesticide. *Comp. Biochem. Physiol.* **127C,** 233–242.

Chuiko, G. M., Podgornaya, V. A., and Lavrikova, I. V. (2002). Organophosphorus compound O,O-dimethyl-O-(2,2-dichlorovynyl)-phosphate as selective inhibitor for separate determination of acetylcholinesterase and butyrylcholinesterase activities in the roach *Rutilus rutilus*. *J. Evol. Biochem. Physiol.* **38,** 264–269.

Colgan, P. W., Cross, J. A., and Johansen, P. H. (1982). Guppy behaviour during exposure to a sub-lethal concentration of phenol. *Bull. Environ. Contam. Toxicol.* **28,** 20–27.

Cleveland, L., Little, E. E., Hamilton, S. J., Buckler, D. R., and Hunn, J. B. (1986). Interactive toxicity of aluminium and acidity to early life stages of brook trout. *Trans. Am. Fish. Soc.* **115,** 610–620.

De Boeck, G., Nilsson, G. E., Elofsson, U., Vlaeminck, A., and Blust, R. (1995). Brain monoamine levels and energy status in common carp (*Cyprinus carpio*) after exposure to sublethal levels of copper. *Aquat. Toxicol.* **33,** 265–277.

Dieperink, C., Bak, B. D., Pedersen, L.-F., Pedersen, M. I., and Pedersen, S. (2002). Predation on Atlantic salmon and sea trout during their first days as postsmolts. *J. Fish Biol.* **61,** 848–852.

Douglas, R. H., and Djamgoz, M. B. A. (1990). "The Visual System of Fish." Chapman and Hall, Cambridge.

Faulk, C. K., Fuiman, L. A., and Thomas, P. (1999). Parental exposure to *ortho,para*-dichlorodiphenyltrichloroethane impairs survival skills of Atlantic croaker (*Micropogonias undulatus*) larvae. *Environ. Toxicol. Chem.* **18,** 254–262.

Fenske, M., and Segner, H. (2004). Aromatase modulation alters gonadal differentiation in developing zebrafish (*Danio rerio*). *Aquat. Toxicol.* **67,** 105–126.

Fernández-Vega, C., Sancho, E., Ferrando, M. D., and Andreu, E. (2002). Thiobencarb-induced changes in acetylcholinesterase activity of the fish *Anguilla anguilla*. *Pest. Biochem. Physiol.* **72,** 55–63.

Flammarion, P., Noury, P., and Garric, J. (2002). The measurement of cholinesterase activities as a biomarker in chub (*Leuciscus cephalus*): The fish length should not be ignored. *Env. Pollut.* **120,** 325–330.

Folmar, L. C., and Hodgins, H. O. (1982). Effects of aroclor 1254 and No. 2 fuel oil, singly and in combination, on predator-prey interactions in coho salmon (*Oncorhynchus kisutch*). *Bull. Environ. Contam. Toxicol.* **29,** 24–28.

Folmar, L. C., Craddock, D. R., Blackwell, J. W., Joyce, G., and Hodgins, H. O. (1981). Effects of petroleum exposure on predatory behaviour of coho salmon (*Oncorhynchus kisutch*). *Bull. Environ. Contam. Toxicol.* **27,** 458–462.

Gonzalez, R. J., and McDonald, D. G. (1992). The relationship between oxygen consumption and ion loss in a fresh-water fish. *J. Exp. Biol.* **163,** 317–332.

Gottofrey, J., and Tjälve, H. (1991). Axonal transport of cadmium in the olfactory nerve of the pike. *Pharm. Toxicol.* **69,** 242–252.

Gray, M. A., Teather, K. L., and Metcalfe, C. D. (1999). Reproductive success and behaviour of Japanese medaka (*Oryzias latipes*) exposed to 4-*tert*-octylphenol. *Environ. Toxicol. Chem.* **18,** 2587–2594.

Grosell, M., and Wood, C. M. (2002). Copper uptake across rainbow trout gills: Mechanisms of apical entry. *J. Exp. Biol.* **205,** 1179–1188.

Gurney, W. S. C., and Nisbet, R. M. (1979). Ecological stability and social hierarchy. *Theoret. Pop. Biol.* **16,** 48–80.

Hammerschmidt, C. R., Sandheinrich, M. B., Wiener, J. G., and Rada, R. G. (2002). Effects of dietary methylmercury on reproduction of fathead minnows. *Environ. Sci. Technol.* **36,** 877–883.

Handy, R. D. (2003). Chronic effects of copper exposure versus endocrine toxicity: Two sides of the same toxicological process? *Comp. Biochem. Physiol.* **135A,** 25–38.

Handy, R. D., Sims, D. W., Giles, A., Campbell, H. A., and Musonda, M. M. (1999). Metabolic trade-off between locomotion and detoxification for maintenance of blood chemistry and growth parameters by rainbow trout (*Oncorhynchus mykiss*) during chronic dietary exposure to copper. *Aquat. Toxicol.* **47,** 23–41.

Hansen, J. A., Rose, J. D., Jenkins, R. A., Gerow, K. G., and Bergman, H. L. (1999). Chinook salmon (*Oncorhynchus tshawytscha*) and rainbow trout (*Oncoryhnchus mykiss*) exposed to copper: Neurophysiological and histological effects on the olfactory system. *Environ. Toxicol. Chem.* **18,** 1979–1991.

Hara, T. J., and Thompson, B. E. (1978). The reaction of whitefish, *Coregonus clupeaformis*, to the anionic detergent sodium lauryl sulphate and its effects on their olfactory responses. *Water Res.* **12,** 893–897.

Hara, T. J., Law, Y. M. C., and Macdonald, S. (1976). Effects of mercury and copper on the olfactory response in rainbow trout, *Salmo gairdneri*. *J. Fish. Res. Bd Can.* **33,** 1568–1573.

Harries, J. E., Sheahan, D. A., Jobling, S., Matthiessen, P., Neall, P., Sumpter, J. P., Tylor, T., and Zaman, N. (1997). Estrogenic activity in five United Kingdom rivers detected by measurement of vitellogenesis in caged male trout. *Environ. Toxicol. Chem.* **16,** 534–542.

Harries, J. E., Runnalls, T., Hill, E., Harris, C. A., Maddix, S., Sumpter, J. P., and Tyler, C. R. (2000). Development of a reproductive performance test for endocrine disrupting chemicals using pair-breeding fathead minnows (*Pimephales promales*). *Environ. Sci. Technol.* **34,** 3003–3011.

Hatfield, C. T., and Anderson, J. M. (1972). Effects of two insecticides on the vulnerability of Atlantic salmon (*Salmo salar*) parr to brook trout (*Salvelinus fontinalis*) predation. *J. Fish. Res. Bd Can.* **29,** 27–29.

Hatfield, C. T., and Johansen, P. H. (1972). Effects of four insecticides on the ability of Atlantic salmon parr (*Salmo salar*) to learn and retain a simple conditioned response. *J. Fish. Res. Bd Can.* **29,** 315–321.

Hedke, J. L., and Norris, L. A. (1980). Effect of ammonium chloride on predatory consumption rates of brook trout (*Salvelinus fontinalis*) on juvenile chinook salmon (*Oncorhynchus tshawytscha*) in laboratory streams. *Bull. Environ. Contamin. Toxicol.* **24,** 81–89.

Henriksson, J., Tallkvist, J., and Tjälve, H. (1997). Uptake of nickel into the brain via olfactory neurons in rats. *Toxicol. Lett.* **91,** 153–162.

Henry, M. G., and Atchison, G. J. (1986). Behavioural changes in social groups of bluegills exposed to copper. *Trans. Am. Fish. Soc.* **115,** 590–595.

Hirsch, R., Ternes, T. A., Haberer, K., and Kratz, K. L. (1999). Occurrence of antibiotics in the aquatic environment. *Sci. Total Environ.* **225,** 109–118.

Hourigan, T. F., Nakamura, N., Nagahama, Y., Yamauchi, K., and Grau, E. G. (1991). Histology, ultrastructure, and *in vitro* steroidogenesis of the testes of two male phenotypes of the protogynous fish, *Thalassoma duperrey* (Labridae). *Gen. Comp. Endo.* **83,** 193–217.

Hubbard, P. C., Barata, E. N., and Canario, A. V. M. (2002). Possible disruption of pheromonal communication by humic acid in the goldfish, *Carassius auratus*. *Aquat. Toxicol.* **60,** 169–183.

Jobling, S., Nolan, M., Tyler, C. R., Brighty, G., and Sumpter, J. P. (1998). Widespread sexual disruption in wild fish. *Environ. Sci. Technol.* **32,** 2498–2506.

Johnsson, J. I., Parkkonen, J., and Förlin, L. (2003). Reduced territorial defence in rainbow trout fry exposed to a paper mill effluent: Using the mirror image stimulation test as a behavioural bioassay. *J. Fish Biol.* **62,** 959–964.

Jones, J. C., and Reynolds, J. D. (1997). Effects of pollution on reproductive behaviour of fishes. *Rev. Fish Biol. Fish.* **7,** 463–491.

Kania, H. J., and O'Hara, J. (1974). Behavioural alterations in a simple predator-prey system due to sublethal exposure to mercury. *Trans. Am. Fish. Soc.* **1,** 134–136.

Katsiadaki, I., Scott, A. P., Hurst, M. R., Matthiessen, P., and Mayer, I. (2002). Detection of environmental androgens: A novel method based on enzyme-linked immunosorbent assay of spiggin, the stickleback (*Gasterosteus aculeatus*) glue protein. *Environ. Toxicol. Chem.* **21,** 1946–1954.

Khan, I. A., and Thomas, P. (1992). Stimulatory effects of serotonin on maturational gonadotropin release in the Atlantic croaker, *Micropogonias undulates. Gen. Comp. Endo.* **88,** 388–396.

Khan, I. A., and Thomas, P. (1996). Disruption of neuroendocrine function in Atlantic croaker exposed to Aroclor 1254. *Mar. Environ. Res.* **42,** 145–149.

Khan, I. A., and Thomas, P. (2000). Lead and Aroclor 1254 disrupt reproductive neuroendocrine function in Atlantic croaker. *Mar. Environ. Res.* **50,** 119–123.

Khan, R. A. (1987). Effects of chronic exposure to petroleum hydrocarbons on two species of marine fish infected with a hemoprotozoan, *Trypanosoma murmanensis. Can. J. Zool.* **65,** 2703–2709.

Khan, R. A., and Thulin, J. (1991). Influence of pollution on parasites of aquatic animals. *Ad. Parasitol.* **30,** 201–238.

Khan, R. A., Bowring, W. R., Burgeois, C., Lear, H., and Pippy, J. H. (1986). Myxosporean parasites of marine fish from the continental shelf off Newfoundland and Labrador. *Can. J. Zool.* **64,** 2218–2226.

Kislalioglu, M., Scherer, E., and McNicol, R. E. (1996). Effects of cadmium on foraging behaviour of lake charr, *Salvelinus namaycush. Environ. Biol. Fish.* **46,** 75–82.

Koltes, K. H. (1985). Effects of sublethal copper concentrations on the structure and activity of Atlantic silverside schools. *Trans. Am. Fish. Soc.* **114,** 413–422.

Knudsen, F. R., Schreck, C. B., Knapp, S. M., Enger, P. S., and Sand, O. (1997). Infrasound produces flight and avoidance responses in Pacific juvenile salmonids. *J. Fish Biol.* **51,** 824–829.

Kramer, D. L. (1987). Dissolved oxygen and fish behaviour. *Env. Biol. Fishes* **18,** 81–92.

Leduc, A. O. H. C., Noseworthy, M. K., Adrian, J. C., Jr., and Brown, G. E. (2003). Detection of conspecific and heterospecific alarm signals by juvenile pumpkinseed under weak acidic conditions. *J. Fish Biol.* **63,** 1331–1336.

Levin, E. D., Chrysanthis, E., Yacisin, K., and Linney, E. (2003). Chlorpyrifos exposure of developing zebrafish: Effects on survival and long-term effects on response latency and spatial discrimination. *Neurotoxicol. Teratol.* **25,** 51–57.

Liley, N. R., and Stacey, N. E. (1983). Hormones, Pheromones, and Reproductive Behaviour in Fish. *In* "Fish Physiology Volume IXB: Reproduction: Behaviour and Fertility Control" (Hoar, W. S., Randall, D. J., and Donaldson, E. M., Eds.), pp. 1–63. Academic Press, New York.

Little, E. E., and Finger, S. E. (1990). Swimming behaviour as an indicator of sublethal toxicity in fish. *Environ. Toxicol. Chem.* **9,** 13–19.

Little, E. E., Archeski, R. D., Flerov, B. A., and Kozlovskaya, V. I. (1985). Behavioural indicators of sublethal toxicity in rainbow trout. *Arch. Environ. Contam. Toxicol.* **19,** 380–385.

Löffler, D., and Ternes, T. A. (2003). Analytical method for the determination of the aminoglycoside gentamicin in hospital wastewater via liquid chromatography-electrospray-tandem mass spectrometry. *J. Chromatography A* **1000,** 583–588.

Madsen, S. S., Skovbølling, S., Nielsen, C., and Korsgaard, B. (2004). 17-β estradiol and 4-nonylphenol delay smolt development and downstream migration in Atlantic salmon, *Salmo salar*. *Aquat. Toxicol.* **68,** 109–120.

Maes, J., Turnpenny, A. W. H., Lambert, D. R., Nedwell, J. R., Parmentier, A., and Ollevier, F. (2004). Field evaluation of a sound system to reduce estuarine fish intake rates at a power plant cooling water inlet. *J. Fish Biol.* **64,** 938–946.

Mathers, R. A., Brown, J. A., and Johansen, P. H. (1985). The growth and feeding behaviour responses of largemouth bass (*Micropterus salmoides*) exposed to PCP. *Aquat. Toxicol.* **6,** 157–164.

Matta, M. B., Linse, J., Cairncross, C., Francendese, L., and Kocan, R. M. (2001) Reproductive and transgenerational effects of methylmercury or Aroclor 1268 on *Fundulus heteroclitus*. *Environ. Toxicol. Chem.* **20,** 327–335.

McCarthy, I. D., Fuiman, L. A., and Alvarez, M. C. (2003). Aroclor 1254 affects growth and survival skills of Atlantic croaker *Micropogonias undulatus* larvae. *Mar. Ecol. Prog. Ser.* **252,** 295–301.

McNicholl, P. G., and Mackay, W. C. (1975a). Effect of DDT and M.S. 222 on learning in a simple conditioned response in rainbow trout (*Salmo gairdneri*). *J. Fish. Res. Bd Can.* **32,** 661–665.

McNicholl, P. G., and Mac Kay, W. C. (1975b). Effect of DDT on discriminating ability of rainbow trout (*Salmo gairdneri*). *J. Fish. Res. Bd Can.* **32,** 785–788.

Moore, A., and Waring, C. P. (1996). Sublethal effects of the pesticide Diazinon on olfactory function in mature male Atlantic salmon parr. *J. Fish Biol.* **48,** 758–775.

Moore, A., and Waring, C. P. (1998). Mechanistic effects of a triazine pesticide on reproductive endocrine function in mature male Atlantic salmon (*Salmo salar* L.) parr. *Pest. Biochem. Physiol.* **62,** 41–50.

Moore, A., and Waring, C. P. (2001). The effects of a synthetic pyrethroid pesticide on some aspects of reproduction in Atlantic salmon (*Salmo salar* L.). *Aquat. Toxicol.* **52,** 1–12.

Moore, A., Scott, A. P., Lower, N., Katsiadaki, I., and Greenwood, L. (2003). The effects of 4-nonylphenol and atrazine on Atlantic salmon (*Salmo salar*) smolts. *Aquacult.* **222,** 253–263.

Munkittrick, K. R., Martin, R. J., and Dixon, D. G. (1990). Seasonal changes in whole brain amine levels of white sucker exposed to elevated levels of copper and zinc. *Can. J. Zool.* **68,** 869–873.

Narasimhamurti, C. C., and Kalavati, C. (1984). Seasonal variation of the myxosporidian, *Henneuya waltairensis* parasitic in the gills of the freshwater fish, *Channa punctatus* Bl. *Archiv für Protistenkunde* **128,** 351–356.

Nemcsók, J., Németh, Á., Buzás, Z. S., and Boross, L. (1984). Effects of copper, zinc and paraquat on acetylcholinersterase activity in carp (*Cyprinus carpio* L.). *Aquat. Toxicol.* **5,** 23–31.

Øverli, Ø., Harris, C. A., and Winberg, S. (1999). Short-term effects of fights for social dominance and the establishment of dominant-subordinate relationships on brain monoamines and cortisol in rainbow trout. *Brain, Behav. Evolution* **54,** 263–275.

Pane, E. F., Haque, A., Goss, G. G., and Wood, C. M. (2004). The physiological consequences of exposure to chronic, sublethal waterborne nickel in rainbow trout (*Oncorhynchus mykiss*): Exercise vs. resting physiology. *J. Exp. Biol.* **207,** 1249–1261.

Parent, A. (1984). Functional anatomy and evolution of monoaminergic systems. *Am. Zool.* **24,** 783–790.

Parkyn, D. C., and Hawryshyn, C. W. (1999). Ethambutol affects the spectral and polarisation sensitivity of on-responses in the optic nerve of rainbow trout. *Vision Res.* **39,** 4145–4151.

Pascoe, D., and Cram, P. (1977). The effect of parasitism on the toxicity of cadmium to the three-spined stickleback, *Gasterosteus aculeatus* L. *J. Fish Biol.* **10,** 467–472.

Pavlov, D. D., Chuiko, G. M., Gerassimov, Y. V., and Tonkopiy, V. D. (1992). Feeding behaviour and brain acetylcholinesterase activity in bream (*Abramis brama* L.) as affected by DDVP, an organophosphorus insecticide. *Comp. Biochem. Physiol.* **103C,** 563–568.

Peters, G., Delventhal, H., and Klinger, H. (1980). Physiological and morphological effects of social stress in the eel, (*Anguilla anguilla* L.). *Arch. Fisch. Wiss.* **30,** 157–180.

Pottinger, T. G., and Pickering, A. D. (1992). The influence of social interaction on the acclimation of rainbow trout, *Oncorhynchus mykiss* (Walbaum) to chronic stress. *J. Exp. Biol.* **41,** 435–447.

Purdy, J. E. (1989). The effects of brief exposure to aromatic hydrocarbons on feeding and avoidance behaviour in coho salmon, *Oncorhynchus kisutch. J. Fish Biol.* **34,** 621–629.

Rand, G. M. (1985). Behaviour. *In* "Fundamentals of Aquatic Toxicology: Methods and Applications" (Rand, G. M., and Petrocelli, S. R., Eds.), pp. 221–263. Hemisphere Publishing Corporation, London.

Rehnberg, B. C., and Schreck, C. B. (1986). Acute metal toxicity of olfaction in coho salmon: Behaviour, receptors, and odor-metal complexation. *Bull. Environ. Contam. Toxicol.* **36,** 579–586.

Riddell, D. J., Culp, J. M., and Baird, D. J. (2005). Behavioural responses of sublethal cadmium exposure within an experimental aquatic food web. *Environ. Toxicol. Chem.* **24,** 431–441.

Rouleau, C., Xiong, Z. H., Pacepavicius, G., and Huang, G. L. (2003). Uptake of waterborne tributyltin in the brain of fish: Axonal transport as a proposed mechanism. *Environ. Sci. Technol.* **37,** 3298–3302.

Rozados, M. V., Andrés, M. D., and Aldegunde, M. A. (1991). Preliminary studies on the acute effect of lindane (γ-HCH) on brain serotoninergic system in rainbow trout *Oncorhynchus mykiss. Aquat. Toxicol.* **19,** 33–40.

Rowe, D. M., and Denton, E. J. (1997). The physical basis of reflective communication between fish, with special reference to the horse mackeral, *Trachurus trachurus. Phil. Trans. Biol. Sci.* **352,** 531–549.

Salzinger, K., Fairhurst, S. P., Freimark, S. J., and Wolkoff, F. D. (1973). Behaviour of the goldfish as an early warning system for the presence of pollutants in water. *J. Environ. Syst.* **3,** 27–40.

Samson, J. C., Goodridge, R., Olobatuyi, F., and Weis, J. S. (2001). Delayed effects of embryonic exposure of zebrafish (*Danio rerio*) to methylmercury (MeHg). *Aquat. Toxicol.* **51,** 369–376.

Sancho, E., Ferrando, M. D., and Andreu, E. (1997). Response and recovery of brain acetylcholinesterase activity in the European Eel, *Anguilla anguilla*, exposed to fenitrothion. *Ecotoxicol. Environ. Safety* **38,** 205–209.

Sancho, E., Fernández-Vega, C., Sanchez, M., Ferrando, M. D., and Andreu-Moliner, E. (2000). Alterations on AChE activity of the fish *Anguilla anguilla* as response to herbicide-contaminated water. *Ecotoxicol. Environ. Safety* **46,** 57–63.

Sandhal, J. F., and Jenkins, J. J. (2002). Pacific steelhead (*Oncorhynchus mykiss*) exposed to chlorpyrifos: Benchmark concentration estimates for acetylcholinesterase inhibition. *Environ. Toxicol. Chem.* **21,** 2452–2458.

Sandheinrich, M. B., and Atchison, G. J. (1989). Sublethal copper effects on bluegill, *Lepomis macrochirus*, foraging behaviour. *Can. J. Fish. Aquat. Sci.* **46,** 1977–1985.

Sasaki, A., Ikejima, K., Aoki, S., Azuma, N., Kashimura, N., and Wada, M. (2003). Field evidence for bioluminescent signaling in the pony fish, *Leiognathus elongates. Environ. Biol. Fish.* **66,** 307–311.

Savino, J. F., Henry, M. G., and Kincaid, H. L. (1993). Factors affecting feeding behaviour and survival of juvenile lake trout in the Great Lakes. *Trans. Am. Fish. Soc.* **122,** 366–377.

Schellart, N. A. M., and Wubbels, R. J. (1998). The auditory and mechanosensory lateral line system. *In* "The Physiology of Fishes" (Evans, D. H., Ed.), 2nd ed., pp. 283–312. CRC Press, Boca Raton.

Scholz, N. L., Truelove, N. K., French, B. L., Berejikian, B. A., Quinn, T. P., Casillas, E., and Collier, T. K. (2000). Diazinon disrupts antipredator and homing behaviours in chinook salmon (*Oncorhynchus tshawytscha*). *Can. J.Fish. Aquat. Sci.* **57,** 1911–1918.

Schröder, J. H., and Peters, K. (1988). Differential courtship activity of competeing guppy males (*Poecilia reticulata* Peters; Pisces: Poeciliidae) as an indicator for low concentrations of aquatic pollutants. *Bull. Environ. Contam. Toxicol.* **40,** 396–404.

Scott, G. R., and Sloman, K. A. (2004). The effects of environmental pollutants on complex fish behaviour: Integrating behavioural and physiological indicators of toxicity. *Aquat. Toxicol.* **68,** 369–392.

Scott, G. R., Sloman, K. A., Rouleau, C., and Wood, C. M. (2003). Cadmium disrupts behavioural and physiological responses to alarm substance in juvenile rainbow trout (*Oncorhynchus mykiss*). *J. Exp. Biol.* **206,** 1779–1790.

Seehausen, O., van Alphen, J. J. M., and Witte, F. (1997). Cichlid fish diversity threatened by eutrophication that curbs sexual selection. *Science* **277,** 1808–1810.

Sloman, K. A., Motherwell, G., O'Connor, K. I., and Taylor, A. C. (2000). The effect of social stress on the standard metabolic rate (SMR) of brown trout, *Salmo trutta. Fish Physiol. Biochem.* **23,** 49–53.

Sloman, K. A., Metcalfe, N. B., Taylor, A. C., and Gilmour, K. M. (2001). Plasma cortisol concentrations before and after social stress in rainbow trout and brown trout. *Physiol. Biochem. Zool.* **74,** 383–389.

Sloman, K. A., Baker, D. W., Wood, C. M., and McDonald, D. G. (2002). Social interactions affect physiological consequences of sublethal copper exposure in rainbow trout, *Oncorhynchus mykiss. Environ. Toxicol. Chem.* **21,** 1255–1263.

Sloman, K. A., Morgan, T. P., McDonald, D. G., and Wood, C. M. (2003a). Socially-induced changes in sodium regulation affect the uptake of water-borne copper and silver in the rainbow trout, *Oncorhynchus mykiss. Comp. Biochem. Physiol. C* **135,** 393–403.

Sloman, K. A., Scott, G. R., Diao, Z, Rouleau, C., Wood, C. M., and McDonald, D. G. (2003b). Cadmium affects the social behaviour of rainbow trout, *Oncorhynchus mykiss. Aquat. Toxicol.* **65,** 171–185.

Sloman, K. A., Baker, D. W., Ho, C. G., McDonald, D. G., and Wood, C. M. (2003c). The effects of trace metal exposure on agonistic encounters in juvenile rainbow trout, *Oncorhynchus mykiss. Aquat. Toxicol.* **63,** 187–196.

Sloman, K. A., Scott, G. R., McDonald, D. G., and Wood, C. M. (2004). Diminished social status affects ionoregulation at the gills and kidney in rainbow trout (*Oncorhynchus mykiss*). *Can. J. Fish. Aquat. Sci.* **61,** 618–626.

Sloman, K. A., Lepage, O., Rogers, J. T., Wood, C. M., and Winberg, S. (2005). Socially-mediated differences in brain monoamines: Effects of trace metal contaminants. *Aquat. Toxicol.* **71,** 237–247.

Smith, G. M., and Weis, J. S. (1997). Predator-prey relationships in mummichogs (*Fundulus heteroclitus* (L.)): Effects of living in a polluted environment. *J. Exp. Mar. Biol. Ecol.* **209,** 75–87.

Smith, G. M., Khan, A. T., Weis, J. S., and Weis, P. (1995). Behaviour and brain chemistry correlates in mummichogs (*Fundulus heteroclitus*) from polluted and unpolluted environments. *Mar. Environ. Res.* **39,** 329–333.

Smith, T. R., and Haines, T. A. (1995). Mortality, growth, swimming activity and gill morphology of brook trout (*Salmo fontinalis*) and Atlantic salmon (*Salmo salar*) exposed to low pH with and without aluminium. *Environ. Pollut.* **90,** 33–40.

Sneddon, L. U., and Yerbury, J. (2004). Differences in response to hypoxia in the three-spined stickleback from lotic and lentic localities: Dominance and an anaerobic metabolite. *J. Fish Biol.* **64**(3), 799–804.

Song, J., Yan, H. Y., and Popper, A. N. (1995). Damage and recovery of hair cells in fish canal (but not superficial) neuromasts after gentamicin exposure. *Hearing Res.* **91,** 63–71.

Sorensen, E. M. B. (1991). "Metal poisoning in fish." CRC Press, Boston.

Sparks, R. E., Waller, W. T., and Cairns, J. (1972). Effect of shelters on the resistance of dominant and submissive bluegills (*Lepomis macrochirus*) to a lethal concentration of zinc. *J. Fish. Res. Bd Can.* **29,** 1356–1358.

Straus, D. L., and Chambers, J. E. (1995). Inhibition of acetylcholinesterase and aliesterases of fingerling channel catfish by chlorpyrifos, parathion, and *S,S,S*-tributyl phosphorotrithioate (DEF). *Aquat. Toxicol.* **33,** 311–324.

Sturm, A., Wogram, J., Hansen, P. D., and Liess, M. (1999). Potential use of cholinesterase in monitoring low levels of organophosphates in small streams: Natural variability in three-spined stickleback (*Gasterosteus aculeatus*) and relation to pollution. *Environ. Toxicol. Chem.* **18,** 194–200.

Sullivan, J. F., Atchison, G. J., Kolar, D. J., and McIntosh, A. W. (1978). Changes in the predator-prey behaviour of fathead minnows (*Pimephales promelas*) and largemouth bass (*Micropterus salmoides*) caused by cadmium. *J. Fish. Res. Bd Can.* **35,** 446–451.

Sutherland, W. J. (1997). "From Individual Behaviour to Population Ecology." Oxford University Press, Oxford.

Sweka, J. A., and Hartman, K. J. (2003). Reduction of reactive distance and foraging success in smallmouth bass, *Micropterus dolomieu*, exposed to elevated turbidity levels. *Environ. Biol. Fish.* **67,** 341–347.

Tallkvist, J., Henriksson, J., d'Argy, R., and Tjälve, H. (1998). Transport and subcellular distribution of nickel in the olfactory system of pikes and rats. *Toxicol. Sci.* **43,** 196–203.

Tallkvist, J., Persson, E., Henriksson, J., and Tjälve, H. (2002). Cadmium-metallothionein interactions in the olfactory pathways of rats and pikes. *Toxicol. Sci.* **67,** 108–113.

Tjälve, H., Gottofrey, J., and Björklund, I. (1986). Tissue disposition of $^{109}Cd^{2+}$ in the brown trout (*Salmo trutta*) studied by autoradiography and impulse counting. *Toxicol. Environ. Chem.* **12,** 31–45.

Tjälve, H., Henriksson, J., Tallkvist, J., Larsson, B. S., and Lindquist, G. N. (1996). Uptake of manganese and cadmium from the nasal mucosa into the central nervous system via olfactory pathways in rats. *Pharm. Toxicol.* **79,** 347–356.

Tsai, C. L., Jang, T. H., and Wang, L. H. (1995). Effects of mercury on serotonin concentration in the brain of tilapia, *Oreochromis mossambicus*. *Neurosci. Letters* **184,** 208–211.

Verbost, P. M., Van Rooji, J., Flik, G., Lock, R. A. C., and Wendelaar Bonga, S. E. (1989). The movement of cadmium through freshwater trout branchial epithelium and its interference with calcium transport. *J. Exp. Biol.* **145,** 185–197.

Vijayan, M. M., Takemura, A., and Mommsen, T. P. (2001). Estradiol impairs hyposmoregulatory capacity in the euryhaline tilapia, *Oreochromis mossambicus*. *Am. J. Physiol. -Reg. Int. Comp. Physiol.* **281,** R1161–R1168.

Waiwood, K. G., and Beamish, F. W. H. (1978). Effects of copper, pH and hardness on the critical swimming performance of rainbow trout (*Salmo gairdneri* Richardson). *Water Res.* **12,** 611–619.

Wardle, C. S., Carter, T. J., Urquhart, G. G., Johnstone, A. D. F., Ziolkowski, A. M., Hampson, G., and Mackie, D. (2001). Effects of seismic air guns on marine fish. *Continental Shelf Res.* **21,** 1005–1027.

Waring, C. P., and Moore, A. (1997). Sublethal effects of a carbamate pesticide on pheromonal mediated endocrine function in mature male Atlantic salmon (*Salmo salar* L.) parr. *Fish Physiol. Biochem.* **17,** 203–211.

Waring, C. P., and Moore, A. (2004). The effect of atrazine on Atlantic salmon (*Salmo salar*) smolts in fresh water and after sea water transfer. *Aquat. Toxicol.* **66,** 93–104.

Webber, H. M., and Haines, T. A. (2003). Mercury effects on predator avoidance behaviour of a forage fish, golden shiner (*Notemigonus crysoleucas*). *Environ. Toxicol. Chem.* **22,** 1556–1561.

Weber, D. N., Russo, A., Seale, D. B., and Spieler, R. E. (1991). Waterborne lead affects feeding abilities and neurotransmitter levels of juvenile fathead minnows (*Pimephales promelas*). *Aquat. Toxicol.* **21,** 71–80.

Weber, D. N. (1993). Exposure to sublethal levels of waterborne lead alters reproductive behaviour patterns in fathead minnows (*Pimephales promelas*). *Neurotoxicol.* **14,** 347–358.

Weir, P. A., and Hine, C. H. (1970). Effects of various metals on behaviour of conditioned goldfish. *Arch. Environ. Health* **20,** 45–51.

Weis, J. S., and Khan, A. A. (1990). Effects of mercury on the feeding behaviour of the mummichog, *Fundulus heteroclitus* from a polluted habitat. *Mar. Environ. Res.* **30,** 243–249.

Weis, J. S., and Weis, P. (1995a). Swimming performance and predator avoidance by mummichog (*Fundulus heteroclitus*) larvae after embryonic or larval exposure to methylmercury. *Can. J. Fish. Aquat. Sci.* **52,** 2168–2173.

Weis, J. S., and Weis, P. (1995b). Effects of embryonic exposure to methylmercury on larval prey–capture ability in the mummichog, *Fundulus heteroclitus*. *Environ. Toxicol. Chem.* **14,** 153–156.

Weis, J. S., and Weis, P. (1998). Effects of exposure to lead on behaviour of mummichog (*Fundulus heteroclitus* L.) larvae. *J. Exp. Mar. Biol. Ecol.* **222,** 1–10.

Weis, J. S., Samson, J., Zhou, T., Skurnick, J., and Weis, P. (2001). Prey capture ability of mummichogs (*Fundulus heteroclitus*) as a behavioural biomarker for contaminants in estuarine systems. *Can. J. Fish. Aquat. Sci.* **58,** 1442–1452.

Wibe, A. E., Nordtug, T., and Jenssen, B. M. (2001). Effects of bis(tributyltin)oxide on antipredator behaviour in threespine stickleback *Gasterosteus aculeatus* L. *Chemosphere* **44,** 475–481.

Wickland Glynn, A., Norrgren, L., and Müssener, Å. (1994). Differences in uptake of inorganic mercury and cadmium in the gills of the zebrafish *Brachydanio rerio*. *Aquat. Toxicol.* **30,** 13–26.

Wiersinga-Post, J. E. C., and van Netten, S. M. (1998). Amiloride causes changes in the mechanical properties of hair cell bundles in the fish lateral line similar to those induced by dihydrostreptomycin. *Proc. R. Soc. Lond. B* **265,** 615–623.

Wilson, R. W., Bergman, H. L., and Wood, C. M. (1994). Metabolic costs and physiological consequences of acclimation to aluminum in juvenile rainbow trout (*Oncorhynchus mykiss*). 2: Gill morphology, swimming performance, and aerobic scope. *Can. J. Fish. Aquat. Sci.* **51,** 527–535.

Winberg, S., and Nilsson, G. E. (1993). Roles of brain monoamine neurotransmitters in agonistic behaviour and stress reactions, with particular reference to fish. *Comp. Biochem. Physiol.* **106C,** 597–614.

Winberg, S., Bjerselius, R., Baatrup, E., and Døving, K. B. (1992). The effect of Cu (II) on the electro-olfactogram (EOG) of the Atlantic salmon (*Salmo salar* L) in artificial freshwater of varying inorganic carbon concentrations. *Ecotoxicol. Environ. Safety* **24,** 167–178.

Wood, C. M. (1992). Flux measurements as indices of H^+ and metal effects on freshwater fish. *Aquat. Toxicol.* **22,** 239–264.

Wood, C. M., and Randall, D. J. (1973). The influence of swimming activity on sodium balance in the rainbow trout (*Salmo gairdneri*). *J. Comp. Physiol.* **82,** 207–233.

Wood, C. M., Playle, R. C., and Hogstrand, C. (1999). Physiology and modeling of mechanisms of silver uptake and toxicity in fish. *Environ. Toxicol. Chem.* **18,** 71–83.

Zeni, C., Caligiuri, A. S., and Bovolenta, M. R. (1995). Damage and recovery of *Ictalurus* sp. barbel taste buds exposed to sublethal concentrations of an anionic detergent. *Aquat. Toxicol.* **31,** 113–123.

Zhou, T., and Weis, J. S. (1998). Swimming behaviour and predator avoidance in three populations of *Fundulus heteroclitus* larvae after embryonic and/or larval exposure to methylmercury. *Aquat. Toxicol.* **43,** 131–148.

Zhou, T., and Weis, J. S. (1999). Predator avoidance in mummichog larvae from a polluted habitat. *J. Fish Biol.* **54,** 44–57.

Zhou, T., Rademacher, D. J., Steinpreis, R. E., and Weis, J. S. (1999). Neurotransmitter levels in two populations of larval *Fundulus heteroclitus* after methylmercury exposure. *Comp. Biochem. Physiol. C* **124,** 287–294.

Zhou, T., Scali, R., and Weis, J. S. (2001). Effects of methlymercuy on ontogeny of prey capture ability and growth in three populations of larval *Fundulus heteroclitus*. *Arch. Environ. Contamin. Toxicol.* **41,** 47–54.

INDEX

A

D

E

M

R

T

V

W

Y

Z

OTHER VOLUMES IN THE FISH PHYSIOLOGY SERIES

Printed and bound by CPI Group (UK) Ltd, Croydon, CR0 4YY

10/05/2025

01866500-0001